NUCLEAR REACTOR KINETICS

McGRAW-HILL INTERNATIONAL BOOK COMPANY

New York
St. Louis
San Francisco
Auckland
Bogotá
Guatemala
Hamburg
Johannesburg
Lisbon
London
Madrid
Mexico
Montreal
New Delhi
Panama
Paris
San Juan
São Paulo
Singapore
Sydney
Tokyo
Toronto

MILTON ASH

Manager, Nuclear Survivability/Vulnerability
Litton Guidance and Control Systems Division
Woodland Hills
California USA

Nuclear Reactor Kinetics

Second Edition

This book was set in Times Roman, Series 327

British Library Cataloguing in Publication Data

Ash, Milton
Nuclear reactor kinetics.—2nd ed.
1. Nuclear reactor kinetics
I. Title
621.48′31 QC786.6.K5 78-40791

ISBN 0-07-002380-8

NUCLEAR REACTOR KINETICS

1 2 3 4 5 MPMP 8 1 0 7 9

Printed and bound in the United States of America

To my wife, Shulamite, and children,
Avner, Miriam, Reuel

CONTENTS

PREFACE

After a variegated history of more than twenty years, nuclear reactor physics has come of age. Aside from the wealth of technological experience and theoretical knowledge amassed in this field, there have been more than just a few books published on its various aspects. Also, some of these have been sufficiently comprehensive to include the principal features of the entire field. However, this book is of the former variety. It attempts to treat reactor kinetics from the mathematical physics point of view and in what A. Weinberg calls "the scholarly tradition of nuclear reactor theory."

Academically, this book is written at the advanced undergraduate, or immediate graduate, level in physics or mathematics. It presupposes that the reader has some familiarity with reactor fundamentals whether through job experience or course work. Also, an acquaintance with the methods and techniques of classical and quantum mechanics is helpful. However, only the methodologies involved therein are meant to be emphasized, because they provide useful approaches to the exposition of many of the concepts in reactor statics and kinetics as well. The great body of reactor theory is closest to being an affair of classical statistical mechanics and, for example, touches on the substance of quantum mechanics only tangentially.

The problems at the end of each chapter should be considered as an intimate part of the text material. They include additional points, as well as an enhancement of developments made in the text proper. Not working the problems as one reads the text will result in a loss of an important facet of the over-all portrayal.

Effort has been made not only to include the more conventional approaches to topics in reactor kinetics but at the same time to stress some of the new and powerful mathematical tools that can be used to cope with certain difficult kinetics problems. For example, the conventional transfer function analysis and synthesis of control as it applies to reactor systems is treated only in an introductory manner. For more details there are texts available, and references to them are given.

Chapter 1 begins with the introduction of delayed neutrons, which characterizes most initial investigations into the area of reactor kinetics. Then the important reactor kinetics parameters are defined and discussed. This is followed by a derivation of the kinetics equations from the monoenergetic (thermal) transport equation. This approach is taken to acquaint the reader with the uncomfortable number of assumptions that must be invoked to derive the diffusion equation, and in turn the kinetics equations, from the transport equation. A more general derivation of the kinetics equation from transport theory is given in Chap. 7.

Chapter 2 discusses the transformation of the kinetics equations, a system of seven first-order ordinary differential equations, into two agglomerate forms. One is an integrodifferential equation, and the other an integral equation of the Volterra type. Solutions are investigated for the general cases of constant (step function), linear, and periodic reactivity variations. The inhour equation in its various forms is obtained. Simple approximate solutions to fit particular considerations are included. The transfer function relationships are then obtained by linearizing the above "zero-power" integral equation form of the kinetics equations.

Transfer functions and their role in reactor stability determination are considered in Chap. 3. They are examined from both the zero-power viewpoint, implying no reactivity feedback mechanisms such as those that occur at rated power, and operation that includes the introduction of power reactivity feedback. The various means of experimentally determining transfer functions, including a discussion of noise methods, are considered. The nonlinear reactor kinetics aspects are introduced in Chap. 3 by using the temperature reactivity effect as the nonlinear element. Describing functions, the next step from transfer functions toward nonlinearity, are also discussed.

Chapter 4 continues with advanced stability concepts stemming from mechanical analogues in classical mechanics as derived from variational principles. Various reactor stability arguments including those that are a function of reactor power, and stability in the large in the sense of Liapunov, are also examined.

Chapter 5, on the monte carlo method and its application to reactor theory, may be extending the purview of reactor kinetics. However, with an eye to the future, and with the more powerful large-scale digital computing machinery becoming available, monte carlo methods will probably be refined to the point where they will mirror, with high accuracy, neutron populations in complicated reactor configurations in a matter of minutes. It is with these thoughts in mind, as well as attempting to secure a place in the nuclear reactor theory curriculum for this imaginative technique, that the chapter was included.

Chapter 6, on dynamic programming, and especially its application to realistic reactor control synthesis, is included to make the reader familiar with one of the newer disciplines for treating, among other things, this type of reactor kinetics problem. These are the kinds of problems that one must face when seriously attempting to control and extract power from modern (and novel) reactors. Satellite reactors, nuclear rocket reactors, and profusely boiling liquid reactors are just three examples.

With the inherently nonlinear temporal nature of the reactor realized at this point, spatially dependent kinetics are discussed in Chap. 7. These include neutron waves, spatial flux and xenon oscillations, and reactor space-time accident analysis. The idea of the importance function is introduced, and the reactor kinetics equations—formally identical with those obtained in Chaps. 1 and 2—are rederived in full generality by using the elegant conceptualization implied by the importance or adjoint flux.

Chapters 8 and 9 consist of a survey of various reactor archetypes which are deemed interesting from the reactor kinetics standpoint. Included in Chap. 8 are discussions of the fast and fast-thermal coupled reactor, nuclear rocket reactor, and circulating fuel reactor. Chapter 9 continues with the boiling-liquid-moderated reactor and the various types of pulsed reactors.

Upwards of two-thirds of this text has been used as material for a one-semester concurrent course in reactor kinetics at University of California at Los Angeles (UCLA) taught by the author during 1962–1963. The students were mainly seniors, graduate students, and their counterparts from industry in the metropolitan Los Angeles area. The suggested route, and the one that has been used, for such a one-semester course is Chaps. 1, 2, 3, 7, and appropriate portions of the remaining chapters depending on the class situation and the propensities of the instructor. For a two-semester (one-year) course, the complete text could be used, especially if supplemented with pertinent reactor experiments available.

I would like to thank Dr. W. C. Biel, Vice President of System Development Corporation, who graciously appropriated the wherewithal for the typing of the manuscript. It is a pleasure to acknowledge the continued assistance of Dr. T. E. Hicks, chairman of the Nuclear Engineering Department of UCLA, who made available his various facilities. Also, without the cooperation of Mrs. Mary Tebbetts and Miss Suzy Wakeman of the UCLA Production Department, Mrs. Ilana Maik and Miss Clara Mizrahi of the Soreq Research Establishment, Yavne, Israel, and my son Avner, this book would have been impossible. Finally, to my former supervisor, Harry Harman, Manager of the Systems Simulation Research Laboratories of the System Development Corporation, who encouraged me at the outset to write this book, my fondest thanks.

Milton Ash

PREFACE TO THE SECOND EDITION

This revision, or second edition, of *Nuclear Reactor Kinetics* constitutes a major overhaul of the first edition. Material has been added, deleted, revised, reformatted, and reinterpreted in an attempt to make this edition as timely as possible, consistent with normal and natural time lags in the authoring, editing, and publishing milieu. A chapter by chapter detailing of the text follows.

In Chap. 1 (Delayed Neutrons and Derivation of Kinetics Equations), the derivation of the reactor kinetics equations in their various point model comprehensive forms, such as the kinetics integral equation, from the monoenergetic transport equation via the diffusion equation route has been amplified to explain more clearly the various steps and approximations involved.

Chapter 2 (The Point Model of Reactor Kinetics) has been expanded to include extra topics, such as inhour period measurement error considerations, and an amplification of the topic of point model neutron bursts.

Chapter 3 (Reactor Transfer Functions) now includes a discussion of the period transfer function as developed from the same perturbational basis that yields the usual transfer function.

Chapter 4 (Stability) is a radical rewrite of that in the first edition. It emphasizes more of the linear stability theory, with the nonlinear reactor aspects deferred to later discussions. Included are such items as the root locus method for transfer functions containing pure lag factors. Further topics include the stability of Liapounov and others of the Soviet school.

Chapter 5 (State Space Kinetics) is a new chapter discussing approaches to pertinant kinetics analyses alternate to the transfer function and its ramifications.

This chapter also contains a comparison between the Bellman principle of optimality and the Pontryagin maximum principle. Applications of the corresponding methods of analysis are given, as well as a basic theoretical comparison, culminating in the time dependent xenon shutdown problem.

Chapter 6 (Importance and Kinetics Revisited) is also a new chapter which discusses the elegant concept of Importance and its applications to reactor kinetics. Certain philosophic interpretations are presented. A rederivation of the point model kinetics equations from the fundamental transport equations, both the forward and adjoint, a la Henry is also presented. This includes basic interpretations of the principal kinetics parameters such as the neutron lifetime. Further delineations are provided using various functionals in the variational calculus context.

Chapter 7 (Space-Dependent Kinetics) is partially a reformulation of the material of the same chapter in the first edition. However, more-recent computational kinetics methods and results are also presented.

Chapter 8 (Kinetics of Reactor Prototypes—I) is similar in format to that in the first edition, but with further material, plus the addition of the kinetics of the Spectral Shift Control Reactor.

Chapter 9 (Kinetics of Reactor Prototypes—II) is also similar in format to that in the first edition, but also including further material. Examples are the kinetics of four additional reactor types. They are the Gaseous/Cavity reactor, KEWB pulsed reactor, Microreactors, and the SNAP series of space borne reactors.

Chapter 10 (Statistical Kinetics) is the third new chapter. It is a discussion of emerging aspects of the statistical nature of reactor kinetics with appropriate applications. These include startup statistics, weak source systematics, and applications of the Langevin technique.

More problem exercises have been added to all of the chapters, so that in many cases discussion of certain kinetics details are relegated to problems at the end of the chapters, in keeping with the pedagogical directions of the first edition. That is, the serious reader can achieve an increased level of understanding if he or she works most of the exercises.

Unfortunately, in endeavors such as this, errors of all types occur. Ultimately, all that can be said is they are all mine.

M. S. Ash
Santa Monica
1979

CHAPTER

ONE

DELAYED NEUTRONS AND DERIVATION OF KINETICS EQUATIONS

1-1 DELAYED NEUTRONS

The genesis of nuclear reactor kinetics—or reactor dynamics, as it is referred to in certain circles—lies in the observation that when uranium or plutonium fissions, certain fission fragment nuclei yield neutrons as a result of their radioactive decay subsequent to fission. These delayed neutrons are in addition to the neutrons that are immediately born of a fission, which are termed prompt neutrons. Such fission fragments are unstable and decay by emitting a high-energy β particle (electron) of several MeV of energy with a characteristic half-life. The half-lives range from milliseconds to minutes.

Upon emission of a fast β particle from the excited fission fragment nucleus, the resultant nucleus decays by emitting the "excess" neutron in $\sim 10^{-14}$ sec. Since this time is infinitesimally small compared to the β-particle half-life, the half-life for emission of the delayed neutron as reckoned from the instant of fission is essentially that of the β particle.

During the days of the Manhattan Project, the available experimental data for ^{235}U, for example, could delineate roughly five distinct delayed neutron half-lives corresponding to five fission fragment isotopes. Since then another half-life has been established, so that the number of delayed neutron half-life groups is usually taken as six, even though claims are made that more exist.

A parameter β is defined to be the fractional yield of delayed neutrons per fission neutron. It is a function of the particular fissionable isotope. It also depends on such quantities as the energy spectrum of the delayed neutrons, as emitted from their precursors, and the importance function of the particular reactor. These dependencies, discussed in detail in Sec. 6-6, result in an effective fractional yield $\bar{\beta} = \gamma\beta$, where γ is a constant.

For a ^{235}U-fueled reactor, $\beta \cong 0.0064$.[1] β_i is the yield of the ith group of delayed neutrons, so that

$$\sum_{i=1}^{6} \beta_i = \beta$$

Analogously

$$\bar{\beta}_i = \gamma_i \beta_i \qquad \text{and} \qquad \sum_{i=1}^{6} \gamma_i \beta_i = \bar{\beta}$$

The relative yield

$$a_i = \frac{\beta_i}{\beta} \qquad \text{with} \qquad \sum_{i=1}^{6} a_i = 1$$

The measurement of β_i and β has become more accurate owing to the advent of very high power reactors giving better counting statistics and less contamination from the delayed neutrons of fast neutron fissionable isotopes such as ^{238}U. Table 1-1 shows this trend for ^{235}U. Table 1-2 depicts a set of 1958 delayed neutron parameters.

For a nuclear reactor to remain critical at a given steady-state power, its neutron population must be essentially constant. If the phenomenon of delayed neutrons were nonexistent, it would be difficult to conceive of a stable reactor. This is because a slight increase in the neutron population would cause the population to multiply exponentially at an exceedingly rapid rate with disastrous results. Fortunately, delayed neutrons do exist, so the fact that their characteristic half-lives are of reasonable length can be exploited with respect to reactor control.

Table 1-1 Delayed neutron parameter trends for ^{235}U†

Redman and Saxon, 1947		Hughes et al., 1948 ($^{235}U_3O_8$)		Keepin and Wimett, 1955 (99.9% ^{235}U)	
$\tau_{1/2}$, sec	a_i	$\tau_{1/2}$, sec	a_i	$\tau_{1/2}$, sec	a_i
55.0	0.035	55.6 ± 0.2	0.034 ± 0.009	54.3 ± 0.9	0.036 ± 0.006
22.47	0.240	22.0 ± 0.2	0.220 ± 0.023	21.7 ± 0.8	0.210 ± 0.019
4.43	0.327	4.51 ± 0.1	0.282 ± 0.017	5.87 ± 0.17	0.192 ± 0.027
1.13	0.398	1.52 ± 0.05	0.319 ± 0.017	2.18 ± 0.07	0.409 ± 0.022
		0.43 ± 0.05	0.112 ± 0.011	0.46 ± 0.03	0.135 ± 0.008
		0.05 ± 0.02	0.033	0.13 ± 0.02	0.018 ± 0.004
$\beta = 0.0078$		$\beta = 0.00755$		$\beta = 0.00701 \pm 0.00028$	

† After Keepin.[2]

Table 1-2 Delayed neutron yields for ^{233}U, ^{235}U, ^{238}U, ^{239}Pu, and ^{240}Pu†

Mean life τ_i, sec	Decay constant λ_i, sec^{-1}	Yield fraction β_i	Relative yield $a_i = \beta_i/\beta$
		^{233}U; $\beta = 0.0026$	
0.321	3.12	0.000061	0.023
0.788	1.27	0.000194	0.075
3.30	0.303	0.000845	0.325
7.65	0.131	0.000604	0.232
29.9	0.0335	0.00073	0.281
79.5	0.0126	0.000024	0.009
		^{235}U; $\beta = 0.0064$	
0.258	3.88	0.000168	0.026
0.715	1.40	0.000824	0.129
3.22	0.311	0.00263	0.411
8.65	0.116	0.00121	0.189
31.5	0.0317	0.00137	0.214
78.7	0.0127	0.000246	0.038
		^{238}U; $\beta = 0.0148$	
0.249	4.02	0.00111	0.075
0.709	1.41	0.00333	0.225
2.79	0.358	0.00574	0.388
7.19	0.139	0.00240	0.162
31.15	0.0321	0.00203	0.137
75.76	0.0132	0.00019	0.013
		^{239}Pu; $\beta = 0.0020$	
0.312	3.21	0.000073	0.037
0.793	1.26	0.000216	0.108
3.02	0.331	0.000687	0.344
7.50	0.133	0.000452	0.226
32.2	0.0311	0.000584	0.292
77.5	0.0129	0.000080	0.040
		^{240}Pu; $\beta = 0.0026$	
0.248	4.04	0.00008	0.024
0.735	1.36	0.00033	0.119
3.003	0.333	0.00091	0.315
7.41	0.135	0.00050	0.162
32.26	0.031	0.00071	0.238
76.92	0.013	0.00007	0.028

† After Keepin and Wimett.[3]

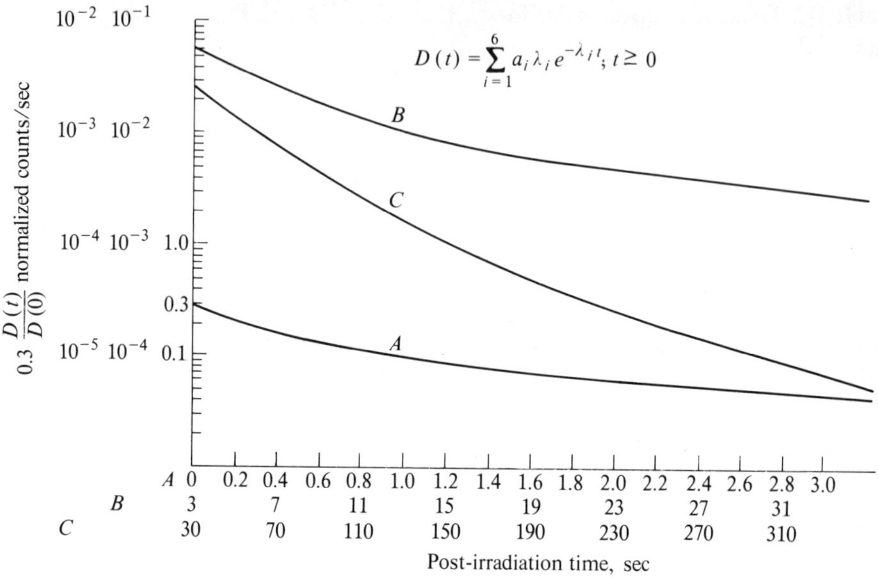

Figure 1-1 Delayed neutron count rate for ^{235}U.

That is, the reactor is kept slightly subcritical with respect to prompt neutrons. The additional population required to bring the reactor to criticality is supplied by the delayed neutrons, which make up approximately 0.64 percent of the total neutron population in ^{235}U. By virtue of their half-lives, however, the delayed neutrons provide a time bottleneck corresponding to an average reactor response time of the order of 10 sec, which is usually ample for the incorporation of the desired control in the conventional reactor control system.

1-2 DELAYED NEUTRON EXPERIMENTS

It is of interest to investigate in detail the experimental and analytical methods used to determine the decay constants and yields of the delayed neutron precursors. The typical method is to neutron-irradiate a sample of ^{235}U, for example, in a reactor. When the sample is in the reactor, it is fissioning at an equilibrium rate corresponding to the reactor operating flux level. Therefore, the delayed neutron density is assumed to attain an equilibrium rate corresponding to the reactor operating flux level. When the sample is removed from the reactor, essentially no further fissions occur within it,† and therefore it no longer begets prompt neutrons. The neutrons that are emitted from the sample subsequent to its removal from the reactor are then delayed neutrons only. The count rate $D(t)$ of these neutrons as a function of time after irradiation is plotted in Fig. 1-1.

† Except, of course, the relatively few due to the delayed neutrons.

To facilitate the accurate determination of count rate for short times, the sample must be quickly transferred from the reactor to the neutron counter, which is exterior to the reactor. A popular scheme is to shoot the sample through the reactor by means of a pneumatic tube threading it and to have the counter stationed near the orifice of the tube.

Since $D(t)$ represents the composite rate of delayed neutrons from the totality of β-emitting fission fragments, it can be expressed analytically as a linear combination of exponential terms, each corresponding to the contribution from an individual precursor. By using high-speed electronic data-processing machinery (EDPM), $D(t)$ can be synthesized by an elaborate least-square fit to the experimental data.[2] The early calculational method of constructing the linear combination of negative exponential terms with their coefficients, which add up to $D(t)$, is called "peeling." That is, since the ordinate of Fig. 1-1 is logarithmic, the appropriate sum of tangents of slope $-\lambda_i$ (decay constant of ith delayed neutron group $i = 1, \ldots, 6$) to $D(t)$ corresponds to the sum of exponential terms in its synthesis. The intercepts of the tangents with the $D(t)$ ordinate give the respective coefficients. If $D(t)$ is normalized to one delayed neutron, the coefficients are given by $a_i\lambda_i$.

$$D(t) = \sum_{i=1}^{6} a_i\lambda_i \exp(-\lambda_i t)$$

is the probability density function. That is, $D(t)\,dt$ can be interpreted as the probability that a delayed neutron will be emitted from some precursor between t and $t + dt$ sec after the sample is removed from the reactor.

The cumulative density function is

$$\int_0^t D(\tau)\,d\tau = \sum_{i=1}^{6} a_i[1 - \exp(-\lambda_i t)]$$

For later use, we define the delayed neutron decay function

$$f(t) = \int_t^\infty D(\tau)\,d\tau = 1 - \int_0^t D(\tau)\,d\tau = \sum_{i=1}^{6} a_i \exp(-\lambda_i t) \qquad t \geq 0$$

which is the (complementary) probability that a delayed neutron has not been emitted prior to time t measured from the sample removal time. $f(t)$ is plotted in Fig. 1-2.

The peeling procedure, beginning with the algebraically smallest decay constant for best accuracy, consists of subtracting the ordinates of the particular tangent from the total curve as sketched in Fig. 1-3. Continuing in this manner, including judicious replotting of the residue curve after each peeling, yields the decomposition of $D(t) = -\dot{f}(t)$. Then the delayed neutron decay function $f(t)$ is at hand. For ^{235}U, Table 1-2 gives, for $t \geq 0$

$$f(t) = 0.026\,e^{-3.88t} + 0.129\,e^{-1.40t} + 0.411\,e^{-0.311t} + 0.189\,e^{-0.116t} + 0.214\,e^{-0.0317t} + 0.038\,e^{-0.0127t} \qquad (1\text{-}1)$$

$f(t)$ is assumed to vanish identically for $t < 0$.

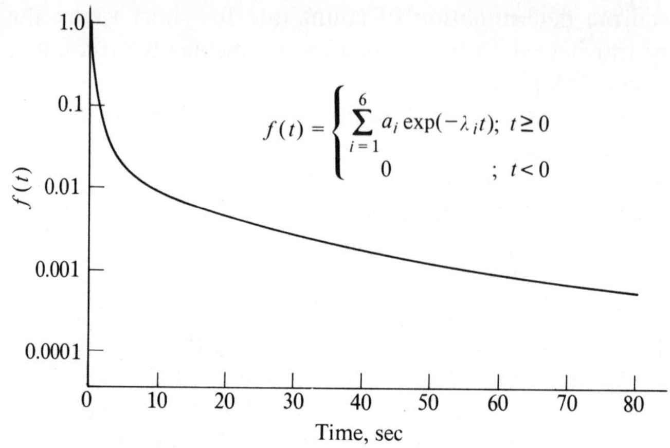

Figure 1-2 Delayed neutron decay function for ^{235}U.

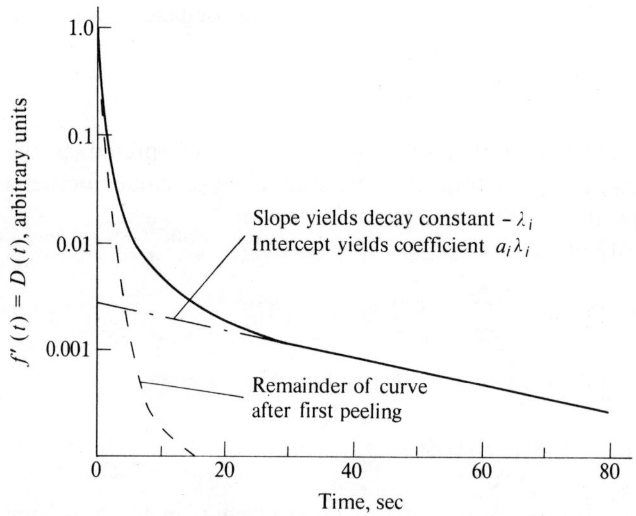

Figure 1-3 Peeling the algebraically smallest decay constant from the experimental decay function.

1-3 DELAYED PHOTONEUTRONS

Analogously to the delayed neutrons that are available as a result of β-emitting fission products, delayed photoneutrons are produced as a result of the following γ-ray photon reactions in the reactor moderator:

$$\gamma + X_{\text{mod}}^{A} \rightarrow n + X_{\text{mod}}^{A-1} \tag{1-2}$$

These photoneutrons are produced in a time very short compared to the half-lives of the γ-emitting fission product fragments whose γ's initiate the above reactions.

In the same sense as that of the β-emitter delayed neutrons, the delayed photoneutrons appear to have half-lives corresponding to those of the fission product γ emitters.

For most nuclei, except those of very low mass number, the binding energy per nucleon is about 8 MeV. However, for deuterium and beryllium, which are common reactor moderators, the binding energy is 2.23 and 1.67 MeV, respectively. These threshold energies fall well within the spectrum of fission product γ's, which lies essentially in the 0.1 to 6-MeV range. Then in Be- and D_2O-moderated reactors, photoneutrons can be produced in quantity. The half-lives and percent photoneutron yields per prompt neutron are given in Table 1-3.

As seen from Table 1-3, the total delayed photoneutron yield from deuterium is 0.0010 part of a fission neutron compared with 0.0064 delayed neutrons per fission neutron for ^{235}U. Further, the weighted mean life of the ordinary delayed neutrons for ^{235}U is

$$\bar{\tau} = \sum_{i=1}^{6} \frac{a_i}{\lambda_i} \cong 12.8 \text{ sec} \tag{1-3}$$

while the weighted mean life for delayed photoneutrons for a D_2O-moderated reactor is $\sim$1,445 sec, and that for a Be moderated reactor is $\sim$665.77 sec. In such reactors, the response time is appreciably longer than in a reactor which generates no photoneutrons. If fueled with ^{235}U, the corresponding aggregate delayed neutron mean life would be approximately 250 sec when moderated with heavy water.

Neutrons are also available from spontaneous fissioning of uranium and plutonium. For example, there are about 20 spontaneous fissions per gramme per hour in ^{238}U. This can be calculated by using the well-known barrier penetration considerations. That is, the probability of leakage per unit time

Table 1-3 Delayed photoneutron parameters for deuterium and beryllium

Deuterium†		Beryllium‡	
Half-life, sec	Percent yield per fission neutron	Half-life, sec	Percent yield per fission neutron
2.5	0.06320	2.18	0.0143
41.0	0.01980	4.55	0.233
144.0	0.00680	22.2	0.166
462.0	0.00326	56.5	0.0357
1,620.0	0.00200	188.3	0.0015
5,950.0	0.00226	396.5	0.0026
1.58×10^4	0.00031	1,214.0	0.0017
1.9×10^5	0.00010	3,906.0	0.0012
1.1×10^6	0.00005	6,636.0	0.0009

† After Bernstein et al.[5]

‡ After Krasin et al.[6]

through a volcano-shaped nuclear potential barrier can be computed by using simple quantum theory in the same way that α-particle decay times are calculated (G. Gamow, 1928). In general, neutrons from spontaneous fissions are so few that they can be neglected, except under certain startup considerations.

1-4 PROMPT NEUTRON LIFETIME

A second important reactor kinetics parameter (β being the first) is the prompt neutron lifetime. For illustrative purposes, consider a thermal reactor fueled by ^{235}U. Then, a mean of about 2.5 prompt neutrons are released per fission. The prompt neutrons are emitted with a fission spectrum[7] which ranges from a few electron volts to upwards of 14 MeV, with a most probable energy of 0.72 MeV and a mean of about 2 MeV.

Essentially all neutrons born of fission undergo collisions with, and lose kinetic energy to, the reactor moderator. In a thermal reactor most neutrons are moderated to thermal energies where they endure further collisions until they are absorbed or leak from the system. Some of the absorptions occur in fuel, where they may cause fissions and result in a next generation of prompt neutrons. The cycle time per generation is termed the *neutron generation time* or *prompt neutron lifetime*. In a thermal reactor, only about 1 to 10 percent of this time is spent in moderating the neutron from its fission energy to thermal energies. Most ($\sim$95 percent) of this time is spent as a thermal neutron undergoing collisions with the moderator, leaking, etc. The prompt neutron lifetime ranges from about 10 to 1,000 μsec for thermal reactors. For fast reactors, which possess no moderator, the prompt neutron lifetime ranges from approximately 10 nsec to 10 μsec. With regard to reactor kinetics calculations, the prompt neutrons can be sometimes construed as an additional group of "delayed" neutrons but with a mean lifetime corresponding to the prompt neutron lifetime. Experimental and calculational methods for obtaining the prompt neutron lifetime are discussed in Chap. 7.

1-5 REACTOR INVARIANTS

When investigating the theoretical aspects of a discipline, as in reactor theory, it is desirable to search out the important invariants. These help provide the cornerstones upon which the theory can be advanced. As a consequence of the preceding section, an invariant important in reactor kinetics is the ratio of prompt neutron lifetime to the effective reproduction factor. For a thermal reactor, one can define such a ratio as the prompt generation time Λ, or prompt lifetime, as†

$$\Lambda = \frac{l}{\beta k_{\text{eff}}} \tag{1-4}$$

† For calculational purposes, an excellent approximation to the prompt lifetime is simply $\Lambda \cong l/\beta$, since $k_{\text{eff}} \cong 1$ for most reactivity excursions of interest.

where $l = [v\Sigma_a(1 + L^2B_0^2)]^{-1}$, the prompt neutron lifetime for a thermal reactor.[8] β is inserted for mathematical convenience, as will be realized later. $k_{\text{eff}} = k_\infty[\exp(-\tau B_0^2)]/(1 + L^2B_0^2)$ is the effective reproduction factor for a finite thermal reactor assuming a Fermi slowing-down model. Σ_a is the total macroscopic absorption cross section, including fuel, for thermal neutrons. B_0^2 is the value of buckling, corresponding to the persistent flux mode (lowest eigenvalue). $k_\infty = \eta f \varepsilon p$, the usual four-factor formula for the reproduction factor of an infinite reactor. With $\eta f = \nu\Sigma_f/\Sigma_a$, where Σ_f is the macroscopic fission cross section, Eq. (1-4) becomes

$$\Lambda = [\beta\nu\Sigma_f v\varepsilon p \exp(-\tau B_0^2)]^{-1} \tag{1-5}$$

where v is the velocity of thermal neutrons. Σ_a changes as the control rods are moved in and out or when other short-time configuration changes occur. However, Λ is independent of Σ_a, while l is not. Actually Λ is not a true invariant, since Σ_f changes slowly as the reactor consumes fuel† (burnup). The remaining factors are essentially constant with time. For fast reactors, $\varepsilon = p = \exp(-\tau B_0^2) \cong 1$ so that, analogously, Λ (fast) $= (\beta\nu\Sigma_f v_f)^{-1}$. v_f is the mean velocity for fast neutrons in such a reactor. The idea of reactor invariants will be expanded in Chap. 6, where a very general derivation of the reactor kinetics equation is given. The invariants discussed there hinge on whether or not the angular flux is separable into space and time factors. Another invariant of a reactor is its equilibrium noise spectrum, discussed in Chap. 3.

1-6 DERIVATION OF THE KINETICS EQUATIONS FROM THE MONOENERGETIC TRANSPORT EQUATION

There are a number of methods for deriving the reactor kinetics equations; they range from the rather complicated to the very simple. After some consideration it was felt that a somewhat involved, but pedagogically effective, method beginning with the time-dependent monoenergetic transport equation would best be used.

In Chap. 6 the reactor kinetics equations are rederived from the general time- and energy-dependent transport equation essentially without approximation, where the principal kinetics parameters are redefined in a very general manner. For example, a theoretical basis is given for the effective delayed neutron yield fraction of the ith delayed neutron precursor $\overline{\beta}_i$.

An initial reason for adopting the following avenue of derivation is that it exposes the reader to some elements of transport theory, which aid in providing a basis for many phases of nuclear reactor theory and design. Second, it reiterates the unfortunate plethora of approximations needed to justify neutron diffusion theory from which the kinetics equations spring in this development.

† Also, for large reactors $\exp(-\tau B_0^2) \cong 1$, so that leakage changes exert a negligible effect on Λ as opposed to small reactors, or reactors controlled by varying the leakage, where τB_0^2 can change appreciably.

The derivation begins by obtaining the one-velocity (thermal) time-dependent transport equation, initially for neutrons in a non-multiplying medium. From the above-mentioned approximations, the transport equation is finally reduced to the time-dependent one-velocity diffusion equation. From there, the reactor kinetics equations are a straightforward result. Later, the kinetics equations (seven coupled first-order differential equations) are aggregated first to an integro-differential equation in the neutron density and then to an integral equation of the Volterra type in the neutron density, whose kernel depends on the delayed neutron decay function discussed in Sec. 1-2.

Let $\rho(\mathbf{r}, \mathbf{\Omega}, t)\, d\mathbf{r}\, d\mathbf{\Omega}$ be the number density of neutrons, all at speed v, in volume element $d\mathbf{r}$ at $\mathbf{r}$ and time t and traveling in the incremental solid angle $d\mathbf{\Omega}$ about the direction given by the unit vector $\mathbf{\Omega}$; for example, $\mathbf{v} = v\mathbf{\Omega}$ (Fig. 1-4). Often, $\mathbf{\Omega}$ can be taken in the $+z$ direction without loss of generality. This simplifies some of the calculational labor involved.

The transport equation is a rather delicate balance equation that audits all of the neutrons that leave in a specified direction $\mathbf{\Omega}$ in $d\mathbf{\Omega}$, the infinitesimal solid angle, but *only* those that are contained in an infinitesimal volume $d\mathbf{r}$ at $\mathbf{r}$ and time t. This genre of equations is ubiquitous throughout the fields of fluid mechanics and hydrodynamics.

Further, let an angular flux, a scalar, be defined as

$$\phi(\mathbf{r}, \mathbf{\Omega}, t) = \rho(\mathbf{r}, \mathbf{\Omega}, t)v \tag{1-6}$$

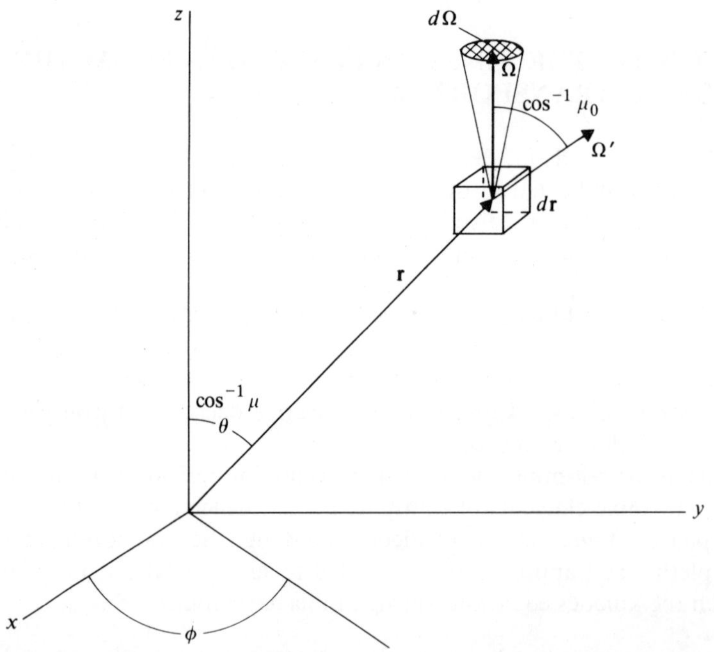

Figure 1-4 Coordinates for the derivation of the kinetics equations from the monoenergetic transport equation.

$\phi(\mathbf{r}, \mathbf{\Omega}, t)$ is the number of neutrons per square centimeter per second traveling with speed v in direction $\mathbf{\Omega}$ at $\mathbf{r}$ and time t. What is usually called the "flux" is the scalar

$$\phi_0(\mathbf{r}, t) = \int \phi(\mathbf{r}, \mathbf{\Omega}, t)\, d\mathbf{\Omega} \tag{1-7}$$

the integral of the angular flux over all solid angle. It can be defined as the total track length of neutrons of velocity v per cubic centimeter per second. It is a sought-after quantity, because multiplying it by the appropriate macroscopic cross section yields the total number of corresponding collisional processes per cubic centimeter per second.

The current, a vector, is given by the integral

$$\mathbf{j}(\mathbf{r}, t) = \int \phi(\mathbf{r}, \mathbf{\Omega}, t)\mathbf{\Omega}\, d\mathbf{\Omega} \tag{1-8}$$

It is the *net* number of neutrons traveling in direction $\mathbf{\Omega}$ per square centimeter per second at $\mathbf{r}$ and t.

Then, with these quantities, following Placzek,[9] one can write a monoenergetic transport equation in the angular flux, as

$$\frac{1}{v}\frac{\partial}{\partial t}\phi(\mathbf{r}, \mathbf{\Omega}, t) = -\mathbf{\Omega}\cdot\nabla\phi(\mathbf{r}, \mathbf{\Omega}, t) - \Sigma\phi(\mathbf{r}, \mathbf{\Omega}, t) + \int \Sigma_s(\mathbf{\Omega}, \mathbf{\Omega}')\phi(\mathbf{r}, \mathbf{\Omega}', t)\, d\mathbf{\Omega}' + S(\mathbf{r}, \mathbf{\Omega}, t) \tag{1-9}$$

Taking Eq. (1-9) apart piece by piece delineates all the contributions to the neutron inventory traveling in $d\mathbf{\Omega}$ about $\mathbf{\Omega}$ in $d\mathbf{r}$ at $\mathbf{r}$ and t. The term on the left in Eq. (1-9) is merely the change in the number of neutrons, $\partial\rho/\partial t$, local to the volume $d\mathbf{r}$ in $d\mathbf{\Omega}$ about $\mathbf{\Omega}$ at $\mathbf{r}$ and time t. Against this are balanced the influx and efflux of these particularly specified neutrons due to causes, considered in turn, existing in the medium.

The first term on the right-hand side of Eq. (1-9) is the net efflux due to the streaming of neutrons, per second, into and out of $d\mathbf{r}\, d\mathbf{\Omega}$ about $\mathbf{\Omega}$. This convection term can also be written as $-\nabla\cdot\bar{\mathbf{\Omega}}\phi(\mathbf{r}, \bar{\mathbf{\Omega}}, t)$, by employing a simple vector identity for the divergence of the product of a scalar by a vector.

The second term on the right-hand side of Eq. (1-9) is also a loss term. Σ is a total macroscopic removal cross section from $d\mathbf{r}\, d\mathbf{\Omega}$, for example,

$$\Sigma = \Sigma_{s0} + \Sigma_a \equiv \Sigma_{s0} + \Sigma_{a,\mathrm{mod}} + \Sigma_{a,\mathrm{fuel}} + \cdots \tag{1-10}$$

Further appropriate cross sections can be added in a straightforward manner later. Σ_{s0} is the total macroscopic scattering cross section, and Σ_a is the total macroscopic absorption cross section. The second term, then, is the number of neutrons that have been removed from the direction $d\mathbf{\Omega}$ about $\mathbf{\Omega}$ in $d\mathbf{r}$ at $\mathbf{r}$ and t owing to collisions of all kinds. "Collisions" is a generic term for all types of neutron encounters, so that absorption is here also counted as a collision.

The third term on the right-hand side of Eq. (1-9) represents neutrons that have been scattered into $d\mathbf{\Omega}$ about $\mathbf{\Omega}$ in $d\mathbf{r}$ at $\mathbf{r}$ and t from all other solid angles and directions $d\mathbf{\Omega}'$ about $\mathbf{\Omega}'$, also in $d\mathbf{r}$ at $\mathbf{r}$ and t. $\Sigma_s(\mathbf{\Omega}, \mathbf{\Omega}')\, d\mathbf{\Omega}$, the differential macroscopic cross section, is the probability per cubic centimeter that a neutron is scattered into $d\mathbf{\Omega}$ about $\mathbf{\Omega}$ if, prior to the collision, it was headed in direction $\mathbf{\Omega}'$. $\Sigma_{s0} = \int \Sigma_s(\mathbf{\Omega}, \mathbf{\Omega}')\, d\mathbf{\Omega}$, as defined above.

The medium is assumed to be homogeneous (and isotropic) so that cross sections are not a function of position. Neutron-neutron collisions, as well as triple collisions, are assumed to be so rare as to be negligible. Considering that the thermal neutron density in an ordinary reactor is about 10^8 to 10^{10} neutrons per cubic centimeter while the medium contains $\sim 10^{20}$ target nuclei per cubic centimeter, it is certain that neutron-neutron collisions can indeed be neglected compared with neutron-medium collisions. Neglecting the former type of collisions is not trivial but mathematically vital, in that the transport equation is thereby rendered linear. This equation is sufficiently difficult to solve generally without including terms proportional to $\phi^2(\mathbf{r}, \mathbf{\Omega}, t)$, which represent nonlinear neutron-neutron collisions.

The fourth term on the right-hand side of Eq. (1-9) is a source of neutrons in $d\mathbf{\Omega}$ about $\mathbf{\Omega}$ in $d\mathbf{r}$ at $\mathbf{r}$ and time t. One example of such a source could be the number of neutrons slowing down to the thermal speed v owing to being moderated from fission energies in $d\mathbf{\Omega}$ about $\mathbf{\Omega}, \ldots,$ etc. Such a source will be introduced into this derivation.

To derive the time-dependent diffusion equation in the flux $\phi_0(\mathbf{r}, t)$, multiply Eq. (1-9) by $d\mathbf{\Omega}$ and integrate over all solid angle to obtain, after transposing

$$\frac{1}{v}\frac{\partial}{\partial t}\phi_0(\mathbf{r}, t) + \nabla\cdot\mathbf{j}(\mathbf{r}, t) + \Sigma\phi_0(\mathbf{r}, t) = \iint \Sigma_s(\mathbf{\Omega}, \mathbf{\Omega}')\phi(\mathbf{r}, \mathbf{\Omega}', t)\, d\mathbf{\Omega}'\, d\mathbf{\Omega} + S_0 \qquad (1\text{-}11)$$

where $S_0 = \int S(\mathbf{r}, \mathbf{\Omega}, t)\, d\mathbf{\Omega}$ and $S(\mathbf{r}, \mathbf{\Omega}, t)$ is assumed to be isotropic. Inverting the order of integration on the right-hand side of Eq. (1-11) now makes it read as follows:

$$\frac{1}{v}\frac{\partial}{\partial t}\phi_0(\mathbf{r}, t) + \nabla\cdot\mathbf{j}(\mathbf{r}, t) + \Sigma_a\phi_0(\mathbf{r}, t) = S_0 \qquad (1\text{-}12)$$

Starting with Eq. (1-9) again, multiply it by $\mathbf{\Omega}\, d\mathbf{\Omega}$ and integrate again over all solid angle. This yields

$$\frac{1}{v}\frac{\partial}{\partial t}\mathbf{j}(\mathbf{r}, t) + \int \mathbf{\Omega}\cdot\nabla\phi(\mathbf{r}, \mathbf{\Omega}, t)\mathbf{\Omega}\, d\mathbf{\Omega} + \Sigma\mathbf{j}(\mathbf{r}, t) = \iint \Sigma_s(\mathbf{\Omega}, \mathbf{\Omega}')\phi(\mathbf{r}, \mathbf{\Omega}', t)\, d\mathbf{\Omega}'\, \mathbf{\Omega}\, d\mathbf{\Omega} \qquad (1\text{-}13)$$

where the integral term containing the source vanishes because of its assumed isotropy.

To make the integrals in Eq. (1-13) tractable, assume that the angular density $\phi(\mathbf{r}, \mathbf{\Omega}, t)$ and $\Sigma_s(\bar{\mathbf{\Omega}}, \bar{\mathbf{\Omega}}')$ are almost isotropic. This is one of the principal assumptions from which the time-dependent diffusion equation is obtained. It implies that:

1. The volume of interest is many mean free paths away from anisotropic sources or sinks, such as the interfaces of a heavily absorbent medium as well as voids or boundaries from which neutrons cannot scatter back into the medium.
2. $\Sigma_a \ll \Sigma_{s0}$, that is, the collisions are predominantly "scatters" so that almost isotropic diffusion is the main phenomenon occurring, as opposed to streaming and absorption, for example. This amounts to saying that not only is the angular flux almost isotropic but its magnitude changes only slightly in one mean free path.

Hence both $\varphi(\bar{r}, \bar{\mathbf{\Omega}}, t)$ and $\Sigma_s(\bar{\mathbf{\Omega}} \cdot \bar{\mathbf{\Omega}}')$ can be expanded in Legendre Polynomials, that is

$$\varphi(\bar{r}, \bar{\mathbf{\Omega}}, t) = \sum_{l=0}^{\infty} \frac{2l+1}{4\pi} C_l(\bar{r}, t) P_l(\bar{\mathbf{\Omega}}) \tag{1-14}$$

$$\Sigma_s(\bar{\mathbf{\Omega}} \cdot \bar{\mathbf{\Omega}}') = \sum_{l=0}^{\infty} \frac{2l+1}{4\pi} A_l P_l(\mu_0) \tag{1-15}$$

and only the first two terms in each expansion retained. This is called the P_1 or diffusion approximation in the hierarchy of P_n approximations for obtaining solutions to the transport equation. For an isotropic medium, $\Sigma_s(\bar{\mathbf{\Omega}}, \bar{\mathbf{\Omega}}') = \Sigma_s(\bar{\mathbf{\Omega}} \cdot \bar{\mathbf{\Omega}}')$, that is, the scattering cross section depends only on the scattering angle, or more conveniently on the cosine of this angle, namely, $\mu_0 = \bar{\mathbf{\Omega}} \cdot \bar{\mathbf{\Omega}}'$.

Since $P_0(x) = 1$, and $P_1(x) = x$, in the P_1 approximation Eqs (1-14) and (1-15), respectively, reduce to

$$\varphi(\bar{r}_1 \bar{\mathbf{\Omega}}_1 t) \cong \frac{1}{4\pi} [C_0(\bar{r}, t) + 3C_1(\bar{r}, t)\bar{\mathbf{\Omega}}] \tag{1-16}$$

$$\Sigma_s(\mu_0) \cong \frac{1}{4\pi} [A_0 + 3A_1 \mu_0] \tag{1-17}$$

The coefficients C_0, C_1, A_0, and A_1 are obtained by integrating moments of $\varphi(\bar{r}_1 \bar{\mathbf{\Omega}}, t)$ and $\Sigma_s(\bar{\mathbf{\Omega}} \cdot \bar{\mathbf{\Omega}}') \equiv \Sigma_s(\mu_0)$ with respect to $\bar{\mathbf{\Omega}}$ over all solid angle. For example, from Eq. (1-17), the zero moment integral is

$$\int_{\bar{\Omega}} \Sigma_s(\mu_0)\, d\bar{\mathbf{\Omega}} = \Sigma_{s0} = \frac{1}{4\pi} \left[4\pi A_0 + 3A_1 \cdot 2\pi \int_{-1}^{+1} \mu_0 \, d\mu_0 \right] = A_0 \tag{1-18}$$

and the first moment integral, since $d\bar{\mathbf{\Omega}} = -d\mu\, d\phi$, yields

$$\begin{aligned} 2\pi \int_{-1}^{+1} \Sigma_s(\mu_0)\mu_0 \, d\mu_0 &= \Sigma_{s0}\bar{\mu}_0 \\ &= \frac{1}{4\pi} \left[2\pi A_0 \int_{-1}^{+1} \mu_0 \, d\mu_0 + 3A_1 \cdot 2\pi \int_{-1}^{+1} \mu_0^2 \, d\mu_0 \right] = A_1 \end{aligned} \tag{1-19}$$

where the average cosine of the scattering angle $\bar{\mu}_0$ is given by

$$\bar{\mu}_0 = \frac{\int_{-1}^{+1} \Sigma_s(\mu_0)\mu_0 \, d\mu_0}{\Sigma_{s0}}$$

so that from Eqs (1-17)–(1-19)

$$\Sigma_s(\bar{\mathbf{\Omega}} \cdot \bar{\mathbf{\Omega}}') = \frac{1}{4\pi}[\Sigma_{s0} + 3\Sigma_{s0}\bar{\mu}_0 \bar{\mathbf{\Omega}} \cdot \bar{\mathbf{\Omega}}'] \tag{1-20}$$

The coefficients of $\varphi(\bar{r}, \bar{\mathbf{\Omega}}, t)$ are obtained similarly by first integrating Eq. (1-16) over all solid angles to obtain the zero moment integral as

$$\varphi_0(\bar{r}, t) = \frac{1}{4\pi}\left[4\pi C_0(\bar{r}, t) + 3C_1(\bar{r}, t)\int_{\bar{\Omega}} \bar{\mathbf{\Omega}}\, d\bar{\mathbf{\Omega}}\right] = C_0(\bar{r}, t) \tag{1-21}$$

Thus, it is seen that the coefficient C_0 is the total flux $\varphi_0(\bar{r}, t)$. The integral

$$\int_{\bar{\Omega}} \bar{\mathbf{\Omega}}\, d\bar{\mathbf{\Omega}} = \int_{\bar{\Omega}} (\bar{i}\mathbf{\Omega}_x + \bar{j}\mathbf{\Omega}_y + \bar{k}\mathbf{\Omega}_z)\, d\bar{\mathbf{\Omega}}$$

$$= \int_0^{2\pi}\int_{-1}^{+1} (\bar{i}\sqrt{1-\mu^2}\cos\phi + \bar{j}\sqrt{1-\mu^2}\sin\phi + \bar{k}\mu)\, d\mu\, d\phi = 0 \tag{1-22}$$

which causes the second term in Eq. (1-21) to vanish. $\bar{i}$, $\bar{j}$, and $\bar{k}$ are the unit vectors along the x, y, and z axes, respectively. The first moment integral of, say, the x component $\varphi(x, \mu, t)$ from Eq. (1-16) is

$$2\pi \int_{-1}^{+1} \varphi(x, \mu, t)\mu\, d\mu = j(x, t)$$

$$= \frac{2\pi}{4\pi}\left[C_0(x, t)\int_{-1}^{+1} \mu\, d\mu + 3C_1(x, t)\int_{-1}^{+1} \mu^2\, d\mu\right] = C_1(x, t) \tag{1-23}$$

so that from Eq. (1-16) and the above, the corresponding x components are given by

$$\varphi(x, \mu, t) = \frac{1}{4\pi}[\phi_0(x, t) + 3\mu j(x, t)] \tag{1-24}$$

Since $\mu j(x, t) = \bar{\mathbf{\Omega}} \cdot \bar{j}(\bar{r}, t)$, Eq. (1-16) can be generalized to

$$\varphi(\bar{r}, \bar{\mathbf{\Omega}}, t) = \frac{1}{4\pi}[\phi_0(\bar{r}, t) + 3\bar{\mathbf{\Omega}} \cdot \bar{j}(\bar{r}, t)] \tag{1-25}$$

Now substituting Eqs (1-20) and (1-25) into their respective integrals in Eq. (1-13) yields from the leftmost integral therein,

$$\int \frac{\bar{\mathbf{\Omega}} \cdot \nabla}{4\pi}[\phi_0(\bar{r}, t) + 3\bar{\mathbf{\Omega}} \cdot \bar{j}(\bar{r}, t)]\bar{\mathbf{\Omega}}\, d\bar{\mathbf{\Omega}}$$

$$= \frac{1}{4\pi}\nabla\varphi_0(\bar{r}, t) \cdot \int \bar{\mathbf{\Omega}}\bar{\mathbf{\Omega}}\, d\bar{\mathbf{\Omega}} + \frac{3}{4\pi}\int \nabla(\bar{\mathbf{\Omega}} \cdot \bar{j}) \cdot \bar{\mathbf{\Omega}}\bar{\mathbf{\Omega}}\, d\bar{\mathbf{\Omega}} \tag{1-26}$$

Using the preceding cartesian vector expressions, that is, $\bar{\mathbf{\Omega}} = \bar{i}\mathbf{\Omega}_x + \bar{j}\mathbf{\Omega}_y + \bar{k}\mathbf{\Omega}_z$ and

$$\nabla\phi_0 = \bar{i}\frac{\partial\varphi_0}{\partial x} + \bar{j}\frac{\partial\varphi_0}{\partial y} + \bar{k}\frac{\partial\varphi_0}{\partial z}$$

then

$$\frac{1}{4\pi}\nabla\varphi_0(\bar{r},t)\cdot\int \bar{\mathbf{\Omega}}\bar{\mathbf{\Omega}}\, d\bar{\mathbf{\Omega}} = \tfrac{1}{3}\nabla\varphi_0(\bar{r},t) \tag{1-27}$$

and $\int \nabla(\bar{\mathbf{\Omega}}\cdot\bar{j})\cdot\bar{\mathbf{\Omega}}\bar{\mathbf{\Omega}}\, d\bar{\mathbf{\Omega}} = 0$. Similarly, from Eq. (1-13) its rightmost integral yields, using Eq. (1-20) and the above vector relationships

$$\int \varphi(\bar{r},\bar{\mathbf{\Omega}}')\, d\bar{\mathbf{\Omega}}' \int \frac{1}{4\pi}[\Sigma_{s0} + 3\Sigma_{s0}\bar{\mu}_0\bar{\mathbf{\Omega}}\cdot\bar{\mathbf{\Omega}}']\bar{\mathbf{\Omega}}\, d\bar{\mathbf{\Omega}} = \Sigma_{s0}\bar{\mu}_0\bar{j}(\bar{r},t) \tag{1-28}$$

Then Eq. (1-13) becomes

$$\left(\frac{1}{v}\frac{\partial}{\partial t} + \Sigma_{tr}\right)\bar{j}(\bar{r},t) + \tfrac{1}{3}\nabla\varphi_0(\bar{r},t) = 0 \tag{1-29}$$

where the transport cross section $\Sigma_{tr} = \Sigma_{s0}(1-\bar{\mu}_0)$, and that $\Sigma_a \ll \Sigma_{s0}$, thereby allowing the $\Sigma_a\bar{j}(\bar{r},t)$ term that appears in Eq. (1-13) to be neglected. Equation (1-12) is now rewritten as

$$\left(\frac{1}{v}\frac{\partial}{\partial t} + \Sigma_a\right)\varphi_0(\bar{r},t) + \nabla\cdot\bar{j}(\bar{r},t) = S_0 \tag{1-30}$$

Taking the divergence of Eq. (1-29) allows the $\nabla\cdot\bar{j}(\bar{r},t)$ term to be eliminated using Eq. (1-30). The resulting equation is

$$\frac{3D}{v^2}\frac{\partial^2\varphi_0}{\partial t^2}(\bar{r},t) + \left(\frac{1+3D\Sigma_a}{v}\right)\frac{\partial\varphi_0}{\partial t}(\bar{r},t) = D\nabla^2\varphi_0(\bar{r},t) - \Sigma_a\varphi_0(\bar{r},t) + S_0 \tag{1-31}$$

where the diffusion coefficient $D = \lambda_{tr}/3$, and the transport mean free path $\lambda_{tr} = \Sigma_{tr}^{-1}$.

Equation (1-31) is called the telegrapher's equation. Its solutions exhibit the characteristic retarded neutron wave behavior which travels with speed v and takes time $t = |\mathbf{r}|/v$ to get to the position $|\mathbf{r}|$ if, for example, the waves emanate from the origin. The asymptotic form of this equation provides the time-dependent diffusion equation discussed below.

More will be said about neutron waves in Chap. 7. Now, however, the road to the diffusion equation from Eq. (1-31) is that of letting the speed v grow without bound and the transport mean free path approach zero in such a fashion that the products Dv and Σv remain finite. This forces all disturbances to be propagated instantaneously throughout the reactor. That is, following a disturbance, the reactor as a whole is perturbed, since all propagation phenomena (waves) are assumed absent. In this sense then, the asymptotic form of Eq. (1-31) becomes the time-dependent diffusion equation for thermal neutrons, which is thereby valid only for times greater than the time needed to propagate the disturbance throughout the reactor. It is, after the limits are taken

$$\frac{1}{v}\frac{\partial}{\partial t}\phi_0(\mathbf{r},t) = D\nabla^2\phi_0(\mathbf{r},t) - \Sigma_a\phi_0(\mathbf{r},t) + S_0 \tag{1-32}$$

For simplicity of derivation, assume that the neutron moderation (slowing down) is continuous so that the Fermi slowing-down kernel can be employed. The source term (excluding an artificial source) is then given by the number of neutrons slowing down to thermal energies per cubic centimeter per second.[10] It is

$$S_0 = (1 - \beta)k_\infty[\exp(-\tau B_0^2)]\Sigma_a \phi_0(\mathbf{r}, t) + [p \exp(-\tau B_0^2)] \sum_{i=1}^{6} \lambda_i C_i(\mathbf{r}, t) \tag{1-33}$$

where $C_i(\mathbf{r}, t)$ is the concentration of the ith group of delayed neutron precursors, p is the resonance escape probability, k_∞ is the reproduction factor for an infinitely large reactor, β is the fraction of fission neutrons that are delayed, and λ_i is the decay constant of the ith delayed neutron emitter.

The delayed neutron precursors satisfy

$$\frac{\partial C_i(\mathbf{r}, t)}{\partial t} = -\lambda_i C_i(\mathbf{r}, t) + \frac{\beta_i k_\infty}{p} \Sigma_a \phi_0(\mathbf{r}, t) \qquad i = 1, \ldots, 6 \tag{1-34}$$

The source term may be represented more accurately by the convolution of the thermal flux with the appropriate slowing-down kernel. The second fundamental pile theorem[10] is assumed to hold. This essentially states that the flux ϕ_0 that satisfies Eq. (1-32) with zero left-hand side (critical reactor) and with S_0 given by Eq. (1-33) also satisfies $\nabla^2 \phi_0 + B_0^2 \phi_0 = 0$ with $k_{\text{eff}} = 1$. Now assume that the reactor is not far from critical and that it is homogeneous and bare. This allows $\phi_0(\mathbf{r}, t)$ to be approximated by the lowest harmonic ($n = 0$) of the sequence of solutions of $\nabla^2 \phi_n + B_n^2 \phi_n = 0$. B_0 is the lowest eigenvalue, where the fluxes vanish on the extrapolated boundary. Assume also that the flux is separable in space and time, so that $\phi_0(\mathbf{r}, t) = \rho(\mathbf{r})n(t)v$ and $C_i(\mathbf{r}, t) = \rho(\mathbf{r})c_i(t)$. All of this is substituted into Eqs (1-32) to (1-34) from which the kinetics equation, in the neutron density n, results

$$\frac{dn(t)}{dt} = \frac{(1 - \beta)k_{\text{eff}} - 1}{l} n(t) + [p \exp(-\tau B_0^2)] \sum_{i=1}^{6} \lambda_i c_i(t) \tag{1-35}$$

From Eq. (1-34) a similar set of balance equations can be easily written for the time-dependent portion of the six delayed neutron precursors. They are

$$\frac{dc_i(t)}{dt} = -\lambda_i c_i(t) + \frac{\beta_i k_{\text{eff}} \exp(+\tau B_0^2)}{pl} n(t) \qquad i = 1, \ldots, 6 \tag{1-36}$$

When the above derivation is repeated with the approximation that all neutrons are thermal, even those born of fission, so that

$$\varepsilon = p = \exp(-\tau B_0^2) = 1$$

Equations (1-35) and (1-36) become, respectively,

$$\frac{dn(t)}{dt} = \frac{(1 - \beta)k_{\text{eff}} - 1}{l} n(t) + \sum_{i=1}^{6} \lambda_i c_i(t) \tag{1-37}$$

$$\frac{dc_i(t)}{dt} = -\lambda_i c_i(t) + \frac{\beta_i k_{\text{eff}}}{l} n(t) \qquad i = 1, \ldots, 6 \tag{1-38}$$

1-7 KINETICS EQUATIONS

Define an excess multiplication factor $k_{ex} = k_{eff} - 1$ and its fractional change, the reactivity, ρ as $(k_{eff} - 1)/k_{eff}$. If k is written without a subscript, it is meant as k_{eff}. There are three principal units of reactivity used with various degrees of acceptance at present:

1. Inhour, discussed in Chap. 2.
2. Percent reactivity or percent excess k, which is merely that these quantities are expressed in percent; for example, $k_{ex} = 0.005$ is $\frac{1}{2}$ percent excess k.
3. Dollars and cents. This unit has enjoyed a variegated popularity, being used extensively during the Manhattan Project years and somewhat fading since then, but regaining in popularity until (2) and (3) are used almost interchangeably. The reactivity scale between 0 and β is divided into 100 cents. For example, in a ^{235}U-fueled reactor, one dollar of reactivity corresponds to a $k_{ex} = \bar{\beta} \cong 0.0064$, or 0.64 percent. Since 0.0064 of the neutrons from fission are available as delayed neutrons, a reactor with $k_{ex} = \$1.00$, or a reactivity of the same amount, since k_{ex} and reactivity correspond closely for k_{ex} near one dollar, is supercritical in just exactly the amount needed to compensate for the immediate absence of delayed neutrons. Such a reactor is just critical on prompt neutrons alone; such a state is termed *prompt critical*. The reactor population can now grow very rapidly to dangerously large magnitudes, since the damping influence due to delayed neutrons is no longer present.

Let the reactivity as expressed in dollars (units of $\bar{\beta}$) be given by $K(t) = (k - 1)/\bar{\beta}k$ and prompt generation time $\Lambda = l/\bar{\beta}k$. If $K(t)$ is substituted into the Eqs (1-37) and (1-38), the result is the kinetics equations which form the basis of much of reactor kinetics.

$$\dot{n}(t) = \frac{K(t) - 1}{\Lambda} n(t) + \sum_{i=1}^{6} \lambda_i c_i(t) + \frac{R(t)}{\Lambda} \tag{1-39}$$

$$\dot{c}_i(t) = -\lambda_i c_i(t) + \frac{\bar{a}_i n(t)}{\Lambda} \qquad i = 1, \ldots, 6 \tag{1-40}$$

where an external source of $R(t)/\Lambda$ neutrons per second has been added and $\bar{a}_i = \bar{\beta}_i/\bar{\beta}$, the effective fractional yield of the ith delayed neutron precursor.†

There are two principal kinds of initial conditions. These are (1) startup in a neutronless reactor so that $n(0) = c_i(0) = 0$ for all i, then the above equations imply $\dot{n}(0) = R(0)/\Lambda$, (2) a critical reactor with $\dot{n}(0) = \dot{c}_i(0) = 0$ for all i implying

† The theoretical basis for the effective delayed neutron fraction $\bar{\beta}$ is discussed in detail in Chap. 6. $\bar{\beta} = \gamma\beta$; γ is slightly greater than unity ($\gamma \cong 1.05$ to 1.25) except for certain Pu fast reactors; cf. Keepin and Wimett, reference 4, Chap. 2. One way to determine $\bar{\beta}$ is to correlate experimentally measured reactivity fuel mass coefficients between critical and prompt critical with transport theory perturbation calculations; cf. Paxton et al., and Hansen and Maier, Further Reading, Chap. 2. Throughout the text, β should be understood as $\bar{\beta}$ unless otherwise indicated.

$K(0)n(0) + R(0) = 0$. For an arbitrary initial equilibrium (critical) neutron density, $K(0) = -R(0)/n(0)$, which means that the initial source strength must be compensated by the above negative reactivity. For most criticality situations of interest, the equilibrium neutron density dwarfs, by orders of magnitude, the integrated source intensity. Then R is deleted in Eq. (1-39) so that initially $K(0)n(0) = 0$, implying that $K(0) = 0$. Then the critical state is given by $\dot{n}(0) = K(0) = \dot{c}_i(0) = 0$. If no initial conditions are specified, these sourceless equilibrium initial conditions are meant to hold. Conditions of $t = -\infty$ are discussed in a problem format at the end of this chapter.

1-8 INTEGRAL EQUATION FORMULATION

The solution to Eq. (1-40) for an initial equilibrium state is

$$c_i(t) = c_i(0)e^{-\lambda_i t} + \frac{\bar{a}_i}{\Lambda}\int_0^t n(\tau)e^{-\lambda_i(t-\tau)}\,d\tau \tag{1-41}$$

Inserting Eq. (1-41) into Eq. (1-39), and after one integration by parts, results in a very useful aggregate form of the kinetics equations which holds for both the startup and equilibrium initial conditions. It is[12]

$$\Lambda\dot{n}(t) = K(t)n(t) - \int_0^t n'(\tau)f(t-\tau)\,d\tau + R(t) \tag{1-42}$$

where $f(t) = \sum_{i=1}^{6} \bar{a}_i e^{-\lambda_i t}$ is the delayed neutron yield function. This equation can be derived more directly from the kinetics Eqs. (1-35) and (1-36) under which the all thermal assumption discussed in Sec. 1-6 need not be made. The Laplace transform of Eq. (1-42) yields

$$\Lambda(s\bar{n}(s) - n_0) = L(K(t)n(t)) - \bar{f}(s)(s\bar{n}(s) - n_0) + \bar{R}(s) \tag{1-43}$$

The barred quantities denote the Laplace transforms of the corresponding functions. $L(K(t)n(t))$ represents the transform of the product $K(t)n(t)$. Equation (1-43) is transposed as

$$\bar{n}(s) = \frac{n_0}{s} + \bar{G}(s)[L(K(t)n(t)) + \bar{R}(s)] \qquad \bar{G}(s) = \{s[\Lambda + \bar{f}(s)]\}^{-1} \tag{1-44}$$

Using the convolution theorem of Laplace transforms, inverting Eq. (1-44) yields a Volterra-type integral equation, which is another aggregate form of the kinetics Eqs (1-39) and (1-40). It is[12]

$$n(t) = n_0 + \int_0^t G(t-\tau)[K(\tau)n(\tau) + R(\tau)]\,d\tau \tag{1-45}$$

As alluded to earlier, for startup where $n_0 = 0$ this same derivation gives

$$n(t) = \int_0^t G(t-\tau)[K(\tau)n(\tau) + R(\tau)]\,d\tau \tag{1-46}$$

while for a reactivity change following equilibrium,† $(Kn \gg R)$

$$n(t) = n_0 + \int_0^t G(t-\tau)K(\tau)n(\tau)\,d\tau \tag{1-47}$$

The kernel of the integral equation, $G(t)$, in turn satisfies the integral equation

$$\Lambda G(t) + \int_0^t f(t-\tau)G(\tau)\,d\tau = 1 \tag{1-48}$$

which is easily deduced by inverting $\bar{G}(s)$ as defined above. To gain some insight into the behavior of $G(t)$, its transform can be expanded in partial fractions; thus

$$\bar{G}(s) = \sum_{r=0}^{6} \frac{A_r}{s + s_r} \tag{1-49}$$

where the roots s_r (all distinct) are solutions of

$$s_r(\Lambda + \bar{f}(s_r)) \equiv s_r\left(\Lambda + \sum_{i=1}^{6} \frac{a_i}{s_r + \lambda_i}\right) = 0 \tag{1-50}$$

From the method of partial-fraction expansion, the coefficients A_r are obtained from

$$A_r = \left[s_r \bar{f}'(s_r) + \Lambda + \bar{f}(s_r)\right]^{-1} \qquad r = 0, \ldots, 6 \tag{1-51}$$

so that inverting Eq. (1-49) yields

$$G(t) = \sum_{r=0}^{6} A_r e^{-s_r t} \tag{1-52}$$

Tabulations of A_r and s_r for various values of Λ are given in Tables 1-3 and 1-4, cf. Keepin and Cox, reference 8, Chap. 2, computed from the table of delayed neutron parameters ($\gamma = 1$) of Keepin and Wimett, reference 4, Chap. 2.

With the present-day propinquity of EDPM equipment, the difference form of Eq. (1-45) is useful for programming purposes. ($R \equiv 0$ will be assumed in the remainder of the text, since this is the more usual situation, unless otherwise specified.) From Eq. (1-45) it is, with $R \equiv 0$,

$$n_r = 1 + \sum_{m=0}^{r-1} G_{r-m} K_m n_m \qquad n_0 = 1 \qquad r = 1, 2, \ldots \tag{1-53}$$

where n_r is the neutron density at the rth time increment. The time interval Δ is suppressed above but should be considered as multiplying $G_{r-m}K_m$. Equation (1-53) is a recurrence relation for obtaining the sequence n_r in terms of the earlier values n_m; $m < r$.

† When these equations are referred to in the remainder of the text, $R = 0$ unless otherwise indicated.

Table 1-3 Roots in the partial-fraction expansion of $\Lambda\bar{G}(s)$ for ^{235}U ($\pm 0m \equiv x \times 10^{\pm m}$)†

$\beta\Lambda$, sec	s_0	s_1	s_2	s_3	s_4	s_5	s_6
10^{-3}	0	1.5032 − 02	7.0394 − 02	1.9626 − 01	1.2099	3.6891	7.0596
5×10^{-4}	0	1.5042 − 02	7.0696 − 02	1.9739 − 01	1.2262	3.7491	1.3481 + 01
10^{-4}	0	1.5050 − 02	7.0939 − 02	1.9829 − 01	1.2377	3.7752	6.5443 + 01
5×10^{-5}	0	1.5051 − 02	7.0969 − 02	1.9841 − 01	1.2391	3.7777	1.3043 + 02
10^{-5}	0	1.5052 − 02	7.0993 − 02	1.9850 − 01	1.2401	3.7796	6.5043 + 02
5×10^{-6}	0	1.5052 − 02	7.0996 − 02	1.9851 − 01	1.2403	3.7798	1.3004 + 03
10^{-6}	0	1.5052 − 02	7.0999 − 02	1.9852 − 01	1.2404	3.7800	6.5004 + 03
5×10^{-7}	0	1.5052 − 02	7.0999 − 02	1.9852 − 01	1.2404	3.7801	1.3000 + 04
10^{-7}	0	1.5052 − 02	7.0999 − 02	1.9852 − 01	1.2404	3.7801	6.5000 + 04
5×10^{-8}	0	1.5052 − 02	7.0999 − 02	1.9852 − 01	1.2404	3.7801	1.3000 + 05
10^{-8}	0	1.5052 − 02	7.0999 − 02	1.9852 − 01	1.2404	3.7801	6.5000 + 05

† After Keepin and Cox, reference 8, Chap. 2.

Table 1-4 Coefficients in the partial-fraction expansion of $\Lambda\bar{G}(s)$ for ^{235}U ($\pm 0m \equiv x \times 10^{\pm m}$)†

$\beta\Lambda$, sec	A_0	A_1	A_2	A_3	A_4	A_5	A_6
10^{-3}	1.1919 − 02	1.3167 − 03	8.5614 − 03	1.1448 − 02	2.8987 − 02	4.6959 − 02	8.9080 − 01
5×10^{-4}	5.9956 − 03	6.6288 − 04	4.7292 − 03	5.7147 − 03	1.2404 − 02	1.0970 − 02	9.5997 − 01
10^{-4}	1.2049 − 03	1.3330 − 04	8.5544 − 04	1.1408 − 03	2.1953 − 03	1.3659 − 03	9.9310 − 01
5×10^{-5}	6.0281 − 04	6.6699 − 04	4.2769 − 04	5.7026 − 04	1.0812 − 03	6.4783 − 04	9.9660 − 01
10^{-5}	1.2062 − 04	1.3347 − 05	8.5533 − 05	1.1402 − 04	2.1366 − 04	1.2431 − 04	9.9932 − 01
5×10^{-6}	6.0314 − 05	6.6740 − 06	4.2766 − 05	5.7012 − 05	1.0667 − 04	6.1842 − 05	9.9966 − 01
10^{-6}	1.2063 − 05	1.3348 − 06	8.5532 − 06	1.1402 − 05	2.1308 − 05	1.2318 − 05	9.9993 − 01
5×10^{-7}	6.0317 − 06	6.6744 − 07	4.2766 − 06	5.7011 − 06	1.0652 − 05	6.1559 − 06	9.9996 − 01
10^{-7}	1.2063 − 06	1.3349 − 07	8.5532 − 07	1.1402 − 06	2.1303 − 06	1.2306 − 06	9.9999 − 01
5×10^{-8}	6.0317 − 07	6.6745 − 08	4.2766 − 07	5.7011 − 07	1.0651 − 06	6.1531 − 07	9.9999 − 01
10^{-8}	1.2063 − 07	1.3349 − 08	8.5532 − 08	1.1402 − 07	2.1302 − 07	1.2305 − 07	9.9999 − 01

† After Keepin and Cox, reference 8, Chap. 2.

PROBLEMS

1-1 Replot Fig. 1-1 to a reasonable scale on semilog paper by using reference 3. Obtain the exponential terms given in Eq. (1-1) by using the peeling method. Replot parts of the curve in detail, as the peeling proceeds, to obtain at least two-place accuracy.

1-2 Californium bullets are mentioned[11] as a possible technological "achievement," since californium fission releases more neutrons than uranium and plutonium release. Assume that ν for californium is 3.8. Calculate the prompt neutron lifetime for such a bullet for fast neutrons at the peak of the fission spectrum (0.72 MeV). Assume all cross sections are the same as those for ^{235}U.

1-3 Assume that the spontaneous fissions in natural ^{235}U, ~20 per gramme per hour, are caused by cosmic radiation. Calculate the magnitude of this cosmic-ray flux.

1-4 Consider the totality of the delayed neutron emitters and photoneutron emitters in a heavy-water moderator as a single group of delayed neutrons with a weighted mean life. Calculate this mean life.

1-5 Show that the number of neutron-neutron collisions per cubic centimeter is given by $\frac{1}{2}n^2(t)\sigma_{nn}$. Why is the factor $\frac{1}{2}$ necessary?

1-6 Using Fick's law of diffusion, $\mathbf{j} = -D\nabla\phi_0$, derive Eq. (1-32) directly from Eq. (1-12).

1-7 Retrace the derivation of the telegraphers' equation (1-31) from Eq. (1-18), filling in the steps in detail.

1-8 Using the second fundamental pile theorem, derive in detail the kinetics equations (1-35) and (1-36) from the time-dependent diffusion equation (1-32) by using the source term from Eq. (1-33).

1-9 Instead of equilibrium initial conditions, use the fact that the delayed neutron concentration products $c_i \exp \lambda_i t$ vanish at $t = -\infty$, to derive the modified form of the integro-differential kinetics equation as $\Lambda\dot{n} = Kn - \int_{-\infty}^{t} n'(\tau)f(t-\tau)\,d\tau = Kn - \int_0^\infty n'(t-\tau)f(\tau)\,d\tau$.

1-10 In nuclear reactor analysis, a principal quantity is the flux $\phi_0(\mathbf{r}, t)$. In the analysis of time-dependent nuclear weapon effects, why is the analogous quantity construed to be the fluence, or time integrated flux $\int_0^t \phi_0(\mathbf{r}, \tau)\,d\tau$, as opposed to the flux?

1-11 The neutron is unstable in that it decays with a half-life of about 11 minutes. What is the effect on the prompt neutron lifetime of a reactor so large that the neutron half-life must be taken into account? Estimate the critical radius of such a spherical reactor by simply relating this radius to the neutron lifetime.

REFERENCES

1. M. A. Schultz: "Control of Nuclear Reactors and Power Plants," 2d ed., McGraw-Hill, New York, 1961, pp. 21–22.
2. G. R. Keepin: *Prog. Nucl. Energy, Ser. I*, **1**: 191–225 (1956).
3. G. R. Keepin and T. F. Wimett: Reactor Kinetic Functions: A New Evaluation, *Nucleonics*, **16**(10): 86–90 (1958).
4. G. R. Keepin, T. F. Wimett, and R. K. Zeigler: Delayed Neutrons from Fissionable Isotopes of Uranium, Plutonium and Thorium, *Phys. Rev.*, **107**: 1044–1049 (1957).
5. S. Bernstein, W. M. Preston, G. Wolfe, and R. E. Slattery: *Phys. Rev.*, **71**: 573 (1947).
6. A. K. Krasin et al.: Physics of a Beryllium Moderated Reactor, *Proc. U.N. Intern. Conf. Peaceful Uses At. Energy, 2nd*, 1958, *Geneva*, **12**: 571–579 (1958).
7. S. Glasstone and M. C. Edlund: "The Elements of Nuclear Reactor Theory," Van Nostrand, Princeton, N.J., 1952, p. 64.
8. *Ibid.*, p. 203.
9. G. Placzek: Notes on Diffusion of Neutrons without Change in Energy, *Natl. Res. Council Can. Doc. MT4*, 1943.
10. A. Weinberg and E. P. Wigner: "The Physical Theory of Neutron Chain Reactors," University of Chicago Press, Chicago, 1958, chap. 12.
11. H. Kahn: "On Thermonuclear War," Princeton University Press, Princeton, N.J., 1961, p. 494.
12. M. Ash: Solutions of the Reactor Kinetics Equations for Time Varying Reactivities. *J. Appl. Phys.*, **27**: 1030–1031 (1956).

CHAPTER

TWO

THE POINT MODEL OF REACTOR KINETICS

2-1 INTRODUCTION

The derivation of the kinetics equations in the preceding chapter began with the one-velocity transport equation for neutrons. Certain assumptions were made there regarding flux isotropy, remoteness from heavy anisotropic absorbers, and predominantly scattering processes to affect the transition to the time-dependent diffusion equation via the telegrapher's equation. The most important further assumption more directly associated with the kinetics equations was that all but the fundamental spatial harmonics of the flux, following a flux disturbance, have been attenuated in a very short time (a few microseconds) compared to times of reactor kinetics interest. That is, the neutron density in the reactor is now assumed to rise or fall as a whole following a reactivity change. Neutron waves and higher spatial flux modes induced by reactivity changes are discussed in Chap. 6. Hence, all energy and spacial dependencies factor away, leaving the result that time is the only remaining independent variable. The resulting equations describe what is called the point kinetics model.

A second consideration is that in many instances the kinetics equations will be seen to lend themselves to solutions using the methods of Laplace transforms. It is assumed that a working knowledge of Laplace transforms is at hand—not only for direct cookbook solution of many of the problems but also as a means for recasting the kinetics equations into more suitable forms. Further, use is made of the various tauberian and abelian theorems which respectively relate large and small values of the Laplace transform variable s in the argument of a transform to small and large values of the time in the argument of its inverse. The verb "invert" used throughout this chapter means to find the function of time corresponding to the particular Laplace transform being discussed.

In the following, the kinetics equations will be reintroduced. Variations of the kinetics equations and their solutions will be constructed for physically interesting situations such as those that are encountered in the design and analysis of reactor control systems. The integral equation form of the kinetics equations will be discussed insofar as transfer function notions will follow from this formulation. The various feedback contributions to the reactivity, such as temperature, density, and poison effects, as well as the essentially nonlinear kinetic behavior of reactors will be discussed in Chap. 3.

2-2 SIMPLE SOLUTIONS

If delayed neutrons are neglected in Eq. (1-39), $\beta_i = c_i = 0$ for all i. This would, for example, be a good approximation to the situation where a very large amount of reactivity (many dollars over prompt critical) is quickly inserted into an equilibrium reactor. Neglecting R in comparison with $K(t)n(t)$, Eq. (1-39) becomes

$$\Lambda \dot{n}(t) = K(t)n(t); \qquad n(0) = 1 \tag{2-1}$$

with solution

$$n(t) = \exp\left[\int_0^t \frac{K(\tau)}{\Lambda}\,d\tau\right] = \exp\int_0^t \alpha(\tau)\,d\tau \tag{2-2}$$

where $\alpha = K/\Lambda$ is an important kinetics parameter discussed later.

However, if the change in reactivity following equilibrium is rapid, but not necessarily large, the delayed neutron emitters will remain at their equilibrium concentration for short times following such reactivity insertion. Then $\lambda_i c_i(0) = \bar{a}_i n(0)/\Lambda$, which when inserted in Eq. (1-39) with $R = 0$ results in

$$\Lambda \dot{n}(t) = [K(t) - 1]n(t) + 1; \qquad n(0) = 1 \tag{2-3}$$

The solution of this equation is often termed the constant source solution, and is given by

$$n(t) = \left[1 + \frac{1}{\Lambda}\int_0^t \left(\exp\left\{-\int_0^\tau \frac{[K(x) - 1]}{\Lambda}dx\right\}\right)d\tau\right] \cdot \left(\exp\left\{\int_0^t \frac{[K(\tau) - 1]}{\Lambda}d\tau\right\}\right) \tag{2-4}$$

which for constant K_0, that is, an initial positive step function reactivity, gives

$$n(t) = \frac{K_0(\exp(K_0 - 1)t/\Lambda) - 1}{K_0 - 1} \tag{2-5}$$

The above is meant to be suggestive of the many other simple solutions that can easily be derived for various initial reactor states.[1]

Even though solutions can be obtained in general by computation using Eq. (1-53), it is instructive to seek analytical solutions to the kinetics equations where possible. Also, not everyone is endowed with digital computing machinery.

An important characteristic of Eqs (1-42) and (1-47) is that the kernels in both cases have an initial sharp drop as the argument is increased. For example, $G(t) \cong e^{-t/\Lambda}/\Lambda$ for small times as obtained from $\bar{G}(s)$ for very large s as shown below. This fact can be exploited when searching for techniques to obtain approximate solutions of these equations. This is done by noting that for reasonably slow variations in the other factors of the convolution integrals of Eqs (1-42) and (1-47), the preponderant contribution to the integral will come from small values of the kernel argument, i.e., from small values of $t - \tau$. This allows one to expand the other factors in the convolution about $\tau = t$. When this is very sharply the case, as it certainly is for increasing reactivities and large times, good approximations are obtained when only the first term is retained in such an expansion. For example, Eq. (1-42) would then become the following first-order differential equation, with $R \equiv 0$

$$\left[\Lambda + \int_0^t f(\tau)\,d\tau\right]\dot{n}(t) = K(t)n(t) \qquad n(0) = 1 \tag{2-6}$$

with solution

$$n(t) = \exp\left[\int_0^t \frac{K(\tau)\,d\tau}{\Lambda + \int_0^\tau f(x)\,dx}\right] \tag{2-7}$$

To press this approximation further: since the main contribution of the integral $\int_0^\tau f(x)\,dx$ is obtained for small values of the argument of $f(x)$, for long times following equilibrium,

$$\int_0^t f(\tau)\,d\tau \cong \int_0^\infty f(\tau)\,d\tau = \bar{\tau} = \sum_{i=1}^{6} a_i/\lambda_i$$

where $\bar{\tau}$ is the mean life of the delayed neutrons. With $\Lambda \ll \bar{\tau}$, which is the usual case, Eq. (2-7) would then give an asymptotic form

$$n(t) \sim \exp\left[\int_0^t \frac{K(\tau)\,d\tau}{\bar{\tau}}\right] \tag{2-8}$$

Still another example of approximate solutions to Eq. (1-47) is the use of $G(t)$ above for short times. If it is substituted for the kernel in Eq. (1-47) and the resulting integral is eliminated by differentiation, the result is the differential equation

$$\Lambda\dot{n}(t) = [K(t) - 1]n(t) + 1 \qquad n(0) = 1 \tag{2-9}$$

which is Eq. (2-3) obtained by these means.

$G(t)$ for small times is obtained from $\bar{G}(s)$ for large s by using the appropriate properties of Laplace transforms.[2] That is, from Eq. (1-44) with $\bar{f}(s) = \sum_{i=1}^{6} a_i/(s + \lambda_i)$, letting s grow large gives

$$\bar{G}(s) \approx \frac{1}{s\Lambda + 1} \tag{2-10}$$

Inverting gives the above $G(t)$.

2-3 CONSTANT REACTIVITIES

The integral equation (1-45) of point model reactor kinetics is rewritten here for an equilibrium neutron density n_0 and zero source, as

$$n(t) = n_0 + \int_0^t G(t - \tau)K(\tau)n(\tau)\,d\tau \tag{2-11}$$

All of the "prompt" and delayed neutron parameters of the point model are embodied in the above integral equation, residing in its kernel $G(t)$ as defined in Eq. (1-52).

In general, closed form solutions of integral equations are difficult to obtain, and the above equation is no exception. To date, no general closed form solution of this equation for reasonably arbitrary reactivities, $K(t)$, has been obtained. For example, no exact solutions of the form $n(t) = F(G*K)$ for generally variable reactivity are known, where F is the sought for function.† All manner of approximations for particular reactor kinetic behavior are available, however. From the mathematical point of view, this chapter largely comprises a study of some of these approximations, plus a discussion of results for certain particularly specified reactivity variations. This section is such a discussion for constant reactivities, and the corresponding neutron densities are obtained in a straightforward manner.

The constant-reactivity case, step function reactivity at the origin, is the most straightforward of the exact solutions of the kinetics equations that include the delayed neutrons and is therefore discussed first. Equation (2-11) for the constant-reactivity case is rewritten as

$$n(t) = n_0 + K\int_0^t G(t - \tau)n(\tau)\,d\tau \qquad K = \text{const} \tag{2-12}$$

It is very natural to solve this equation by using Laplace transform methods because the Laplace transform convolution theorem is immediately applicable. It gives, on taking the Laplace transform

$$\bar{n}(s) = n_0\{s[1 - K\bar{G}(s)]\}^{-1} \tag{2-13}$$

† The symbols $G*K = \int_0^t G(t - \tau)K(\tau)\,d\tau$, where the asterisk signifies the convolution operator.

If $\bar{G}(s)$ from Eq. (1-44) is substituted in Eq. (2-13), the result is

$$\bar{n}(s) = \frac{n_0(\Lambda + \bar{f}(s))}{s(\Lambda + \bar{f}(s)) - K} \tag{2-14}$$

Assuming that both numerator and denominator can be expressed as polynomials in s with the degree of the numerator less than that of the denominator, then Eq. (2-14) can be expanded in partial fractions in the manner of $\bar{G}(s)$ earlier. That is

$$\bar{n}(s) = n_0 \sum_{m=0}^{6} \frac{A_m}{s - s_m} \tag{2-15}$$

where the s_m are the seven roots of the denominator in Eq. (2-14), that is, roots of $s_m[\Lambda + \bar{f}(s_m)] = K$.

For K positive, all but one of the seven roots are negative and real; the other root is positive and real. For K negative, all seven roots are negative and real.

To find the coefficients† A_m; $m = 0, 1, \ldots, 6$, multiply Eqs (2-14) and (2-15) through by $s - s_m$ and equate them to get

$$\frac{\Lambda + \bar{f}(s)}{s[\Lambda + \bar{f}(s)] - K}(s - s_m) = \sum_{j \neq m} A_j \frac{s - s_m}{s - s_j} + A_m \tag{2-16}$$

Now let s approach s_m to yield A_m on the right side of Eq. (2-16). But this procedure yields the indeterminate form 0/0 on the left because $(s - s_m)$ is a factor of both numerator and denominator. Use of l'Hopital's rule, however, yields from Eq. (2-16)

$$A_m = \frac{\Lambda + \bar{f}(s_m)}{\Lambda + \bar{f}(s_m) + s_m \bar{f}'(s_m)} \tag{2-17}$$

Knowing the coefficients, then,‡ Eq. (2-15) can be inverted as

$$n(t) = n_0 \sum_{m=0}^{6} A_m e^{s_m t} \tag{2-18}$$

Figure 2-1 is a plot of $s\bar{f}(s) = K - \Lambda s$ from which the roots s_m, with the exception of the most negative root, can be determined. To obtain the most negative root, which is usually very large in magnitude, it is seen that $s\bar{f}(s) \sim 1$ for large $|s|$. Then it is given by $s \sim -(1 - K)/\Lambda$. The corresponding $A_m \sim K/(K - 1)$.

A remark should be made at this point regarding the continuity of n and c_i for sudden and/or discontinuous reactivity changes[3] such as step function reactivities. Since the kinetics equations (1-39) and (1-40) are first-order differential equations, n and c_i can be simply taken as continuous at the instant of the

† The coefficients A_m are, of course, different from those in the partial-fraction expansion of $\bar{G}(s)$.

‡ Nomograms for the coefficients A_m as a function of constant K and Λ are given in Schultz, reference 1, Chap. 1.

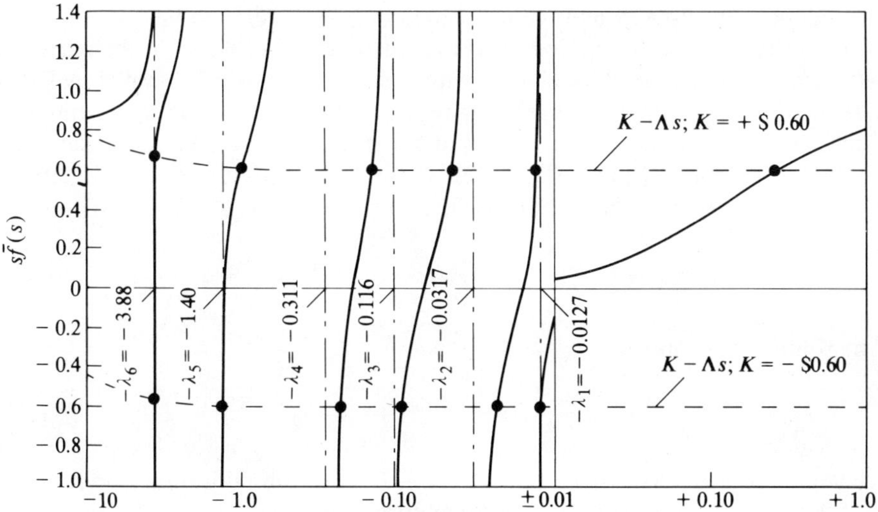

Figure 2-1 Plot $s\bar{f}(s) = \pm K$—versus s for ^{235}U using delayed neutron parameters given in Keepin and Wimett[4] for $\Lambda = 10^{-4}/\beta = 0.0156$. The most negative root is not obtained from this graph, but from simple symptotic considerations of $s\bar{f}(s)$ given in the text.

reactivity change. The solutions are then obtained for constant K as above, where K is the magnitude of the reactivity jump, with initial conditions $\dot{n}(0) = \dot{c}_i(0) = 0$. $\dot{n}(t)$ is then discontinuous at the instant of the jump as seen from the kinetics equations.

If $n(t)$ is assumed continuous at the time of the reactivity jump, or during the short interval over which the reactivity very rapidly changes to its new constant value, then from physical reasoning the reactivity change occurred over a time short compared to the prompt neutron lifetime. Such a constraint on the reactivity is much too strong, since reactivities in actual systems normally change over intervals much longer than the prompt neutron lifetime. A weaker restriction on the reactivity change is to assume that it occurs over a time only short compared to any of the delayed neutron periods (usually the smallest). For example, if this assumption is made for the equilibrium reactor subject to a sudden change at $t = 0$, then the c_i can be assumed continuous during the reactivity change (or at the instant of the jump) and as such be taken as constant at their equilibrium value $c_i(0) \cong a_i n(0)/\lambda_i \Lambda$. Then the kinetics equations reduce to the solutions discussed that resulted in Eq. (2-3). If the reactivity change occurs over a time not short compared to the delayed neutron periods, then the general kinetics equations (1-42) or (1-47) hold and must be solved accordingly. If the reactivity is slowly varying, certain quasi-constant reactivity solutions can be obtained.[6] They are discussed later.

Although obtaining solutions to the kinetics equations for constant reactivities is straightforward, the computations are tedious and time-consuming. For rapid calculation many approximations are available.[6,7] Most deal with how to approximate $f(t)$ by fewer representative delayed neutron groups chosen for the particular problem and time scale at hand. Herewith is a simple solution obtained by letting $f(t) = e^{-t/\bar{\tau}}$, which is a one representative delayed neutron group approximation. Such an approximation holds better and better as the reactivity is decreased ($\ll$\$1.00). Inserting $f(t)$ into Eq. (1-42), and differentiating once to eliminate the integral term, gives for constant reactivity and $R = 0$.

$$\Lambda \ddot{n}(t) + \dot{n}\left(1 - K + \frac{\Lambda}{\bar{\tau}}\right) - \frac{nK}{\bar{\tau}} = 0; \qquad n(0) = 1, \qquad \dot{n}(0) = \frac{K}{\Lambda} \tag{2-19}$$

The solution is simply

$$n(t) = \frac{e^{-at}}{2b}\left[(b + 1)\,e^{abt} + (b - 1)\,e^{-abt}\right] \tag{2-20}$$

where $a = (1 - K)/2\Lambda$, $b = \left[1 + 4\Lambda K/\bar{\tau}(1 - K)^2\right]^{1/2}$. Simplifications for $K \ll 1$ or $K \cong 1$ can be made by suitably approximating a and b.

With the advent of high-speed EDPM equipment, solutions to the kinetics equations can be obtained by programming the integral Eq. (1-47), or its variants, by first rendering it in difference form as in Eq. (1-53).[9] Once the kernel $G(t)$ is tabulated as a function of time (or generated in the computer as required) for a particular set of delayed neutron parameters and prompt neutron lifetime, the computer can very quickly produce families of solutions for various and diverse

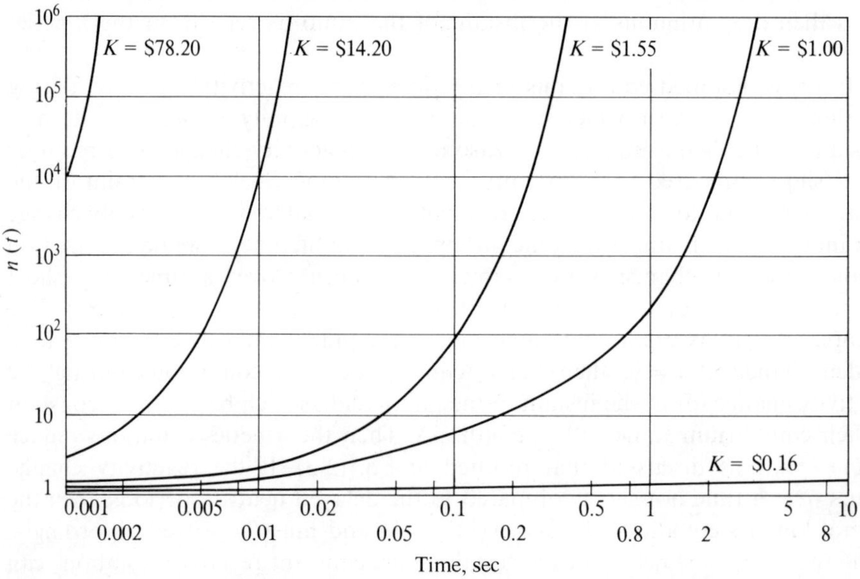

Figure 2-2 Neutron density versus time for positive step reactivities from initial equilibrium for ^{235}U, $\Lambda = 10^{-4}/\beta$.[8]

types of time-dependent reactivities. Such computations for constant reactivities (step functions at the time origin) are depicted in Fig. 2-2. The delayed neutron parameters range over very wide values, which makes the equations difficult to integrate without the use of special numerical mathematical devices.

2-4 INHOUR FORMULA

The inhour relationship was developed during the era of the Manhattan Project. It is a formula relating the constant reactivity inserted into a reactor at equilibrium to the asymptotic (stable) period. The "stable" period is the time constant of the neutron density for times long after insertion of a step reactivity. It supplies a simple method for calibrating the reactivity of a control rod. For example, the rod is withdrawn a small amount quickly from the equilibrium reactor and is assumed to represent a small positive step reactivity thereby. The asymptotic period is then noted and correlated with the reactivity of the rod as discussed below.

The inhour formula is developed by assuming the simple exponential asymptotic solution $n(t) \sim e^{\lambda t}$ in Eq. (2-11), which becomes with K constant

$$1 = e^{-\lambda t} + K \int_0^t G(\tau)\, e^{-\lambda \tau}\, d\tau \tag{2-21}$$

As time grows without bound, the asymptotic form ensues and the above yields the inhour relationship as

$$K\overline{G}(\lambda) - 1 = 0 \tag{2-22}$$

The asymptotic decay constant λ (inverse of asymptotic or stable reactor period) is the algebraically largest root of the above equation. $\overline{G}(\lambda)$ is the Laplace transform of $G(t)$ with respect to the parameter λ.

Using the Laplace transform of $G(t)$ defined in Eq. (1-44), Eq. (2-22) yields the more familiar inhour formula

$$K = \lambda\left(\Lambda + \sum_{i=1}^{6} \frac{a_i}{\lambda + \lambda_i}\right) \tag{2-23}$$

Letting $\lambda = (3600T)^{-1}$ defines the asymptotic stable period T in hours if λ has units of inverse seconds. The above is then rewritten to yield

$$K = \frac{\Lambda}{3600T} + \sum_{i=1}^{6} \frac{a_i}{1 + 3600\lambda_i T} \tag{2-24}$$

One inhour of reactivity is that magnitude which results in a stable period of one hour. For $T = 1$ then, Eq. (2-24) becomes

$$K = \frac{\Lambda}{3600} + \sum_{i=1}^{6} \frac{a_i}{1 + 3600\lambda_i} \tag{2-25}$$

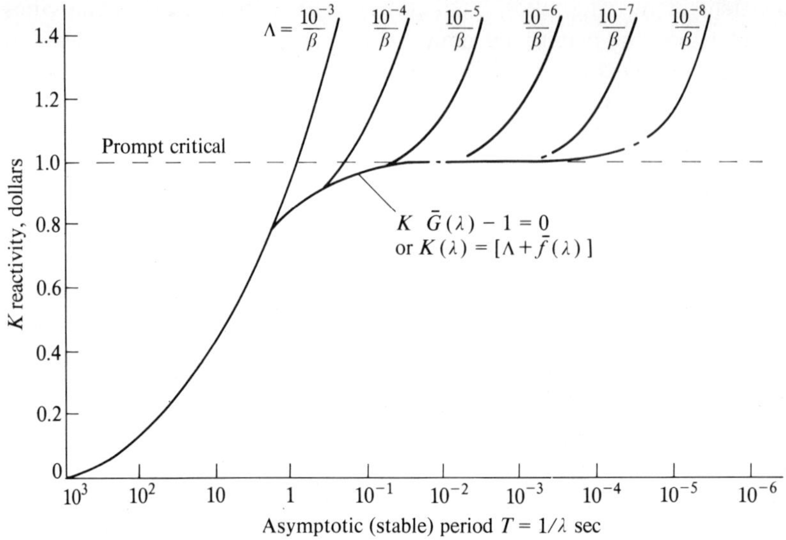

Figure 2-3 Plot of inhour relationship for ^{235}U as a function of reactivity for various Λ.[4]

For comparison, a prompt neutron lifetime of one millisecond ($\Lambda = 10^{-3}/\beta$) yields one inhour of reactivity from the above equation for ^{235}U as

$$K \cong \frac{\bar{\tau}}{3600} = 0.355\,¢ \tag{2-26}$$

since $3600\lambda_i \gg 1$ for all i and $\bar{\tau} \gg \Lambda$. $\bar{\tau} = 12.8$ sec for ^{235}U.

To express the reactivity in inhour units by using Eq. (2-24), it is necessary to divide the latter by Eq. (2-25), the inhour unit. This gives

$$K = \frac{\dfrac{\Lambda}{3600T} + \displaystyle\sum_{i=1}^{6} \frac{a_i}{1 + 3600\lambda_i T}}{\dfrac{\Lambda}{3600} + \displaystyle\sum_{i=1}^{6} \frac{a_i}{1 + 3600\lambda_i}} \quad \text{inhr} \tag{2-27}$$

For very large T, the inequalities resulting in Eq. (2-26) hold even more strongly, so that

$$K \cong \frac{1}{T} \quad \text{inhr} \qquad \text{or} \qquad \frac{3600}{T} \qquad \text{for } T \text{ sec} \tag{2-28}$$

ensues as an approximation. This was the rule of thumb used in the days of the first reactors. A plot of the stable reactor period $T = \lambda^{-1}$ (sec) versus the reactivity K in $K\bar{G}(1/T) - 1 = 0$ is shown in Fig. 2-3. Table 2-1 gives a few representative values of the stable period as a function of reactivity.

Table 2-1 Stable reactor period versus reactivity for ^{235}U† ($\beta = 0.0064$)

Positive reactivity inserted, $	Stable period, sec		
	$\Lambda = 10^{-3}/\beta$	$\Lambda = 10^{-4}/\beta$	$\Lambda = 10^{-5}/\beta$
0.16	60	60	60
0.47	10	10	10
0.62	3.5	2.5	2.5
1.09	0.8	0.2	0.14
1.56	0.3	0.04	0.0003

† After Glasstone and Sesonske.[10]

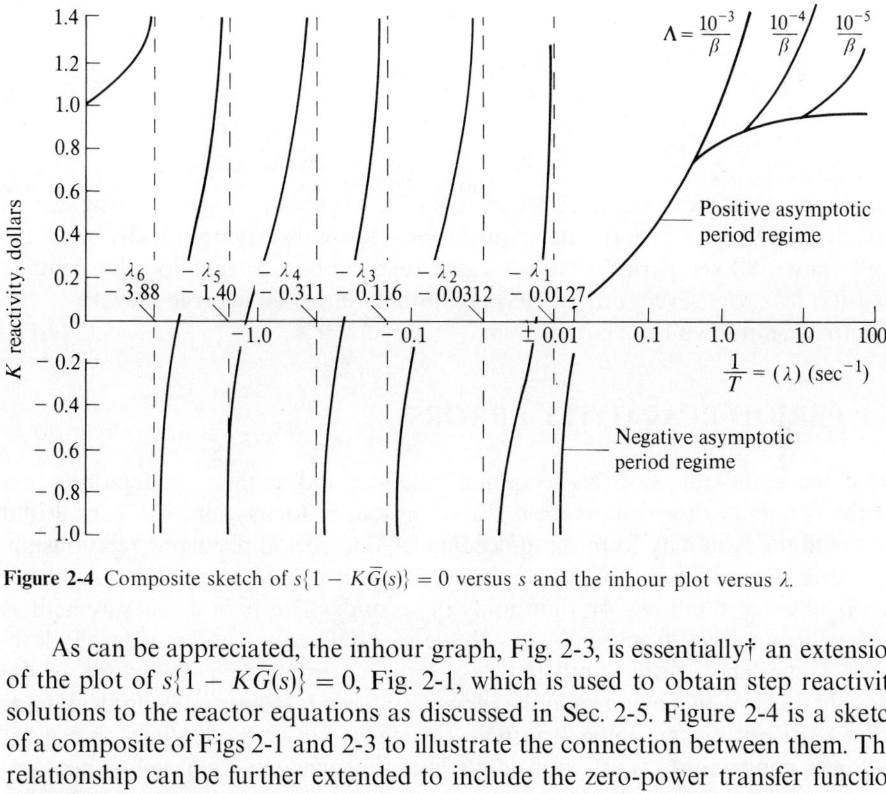

Figure 2-4 Composite sketch of $s\{1 - K\bar{G}(s)\} = 0$ versus s and the inhour plot versus λ.

As can be appreciated, the inhour graph, Fig. 2-3, is essentially† an extension of the plot of $s\{1 - K\bar{G}(s)\} = 0$, Fig. 2-1, which is used to obtain step reactivity solutions to the reactor equations as discussed in Sec. 2-5. Figure 2-4 is a sketch of a composite of Figs 2-1 and 2-3 to illustrate the connection between them. This relationship can be further extended to include the zero-power transfer function discussed in Chap. 3.

For large positive reactivities that are greater than prompt critical ($K > 1$);

† Figure 2-1 is a plot of both $K - \Lambda s$ and $s\bar{f}(s)$ versus s, whereas Fig. 2-3 is a plot of $K(\lambda) = \lambda[\Lambda + \bar{f}(\lambda)]$ versus $1/\lambda$. These plots are virtually identical except for very large s (or λ), since $\Lambda s \ll s\bar{f}(s)$ except for large s.

it is seen from Fig. 2-1 that this corresponds to large asymptotic decay constants, λ, in the inhour equation (2-23), namely $K = \lambda[\Lambda + \bar{f}(\lambda)]$. For very large λ, $\bar{f}(\lambda)$ approaches zero. Hence the asymptotic decay constant $\lambda \sim K/\Lambda$ as expected, since the delayed neutron effects are vitiated by the prompt period $(K/\Lambda)^{-1}$ obtained for large step reactivities.

For negative reactivities, the asymptotic decay constant (negative stable period), for both small and large reactivity magnitude, is obtained by investigating the inhour equation, that is, $K = \lambda[\Lambda + \bar{f}(\lambda)]$ in the vicinity of small negative λ near λ_1. That is, for $-\lambda_1 \leq \lambda \leq 0$, $\lambda_1 = 0.0125\ \text{sec}^{-1}$, it reduces to

$$|K_0| = |\lambda|[c - a_1/(|\lambda| - \lambda_1)]; \qquad K_0 < 0 \tag{2-29}$$

$c = \Lambda + \sum_{i=2}^{6} a_i/\lambda_i \simeq 10$, which is that part of $\Lambda + \bar{f}(\lambda)$ that remains essentially constant near the singularity at $\lambda = -\lambda_1$. Multiplying through by $|\lambda| - \lambda_1$ yields a quadratic equation in $|\lambda|$. Examining this for large and small reactivities, respectively, yields the corresponding asymptotic decay constants. They are

$$\lambda \sim \begin{cases} -\lambda_1; & |K_0| \gg 2\lambda_1 c \cong 25\,¢ \\ -\dfrac{\lambda_1 K_0}{c\lambda_1 + a_1}; & |K_0| \ll c\lambda_1 + a_1 = 16\,¢ \end{cases} \tag{2-30}$$

Hence, for a reactor shutdown by means of a large magnitude negative step of reactivity ($|K_0| \gg 25\,¢$), the neutron density decreases asymptotically with the well-known 80 sec period $T = -1/\lambda_1$. It takes about 20 min for the neutron density to drop seven orders of magnitude following a large negative step reactivity shutdown from equilibrium.

2-5 PERIOD-REACTIVITY ERRORS

As stated at the outset of this section, a time honored method for determination of the reactivity of a reactor on a stable period, is to measure the period and compute the reactivity from the preceding inhour period-reactivity relationship. However, for small reactivity magnitudes the corresponding periods are rather large, often in hundreds or thousands of seconds. The period measurement is normally computed from semi-logarithmic plots of reactor neutron density level versus time, and waiting until the data form a straight line. The slope of the straight line is then a measure of the reciprocal period. However, slight deviations from a straight line are difficult to detect using such a method. Hence, it at least becomes important to wait a sufficiently long time when making such a measurement to within a specified measurement error. The following, Toppel,[11] quantifies these relationships, including the waiting time to measure the asymptotic period to within an a priori specified error accuracy.

For constant reactivity, Eq. (2-18) is rewritten as

$$n(t)/n_0 = \sum_{m=0}^{6} A_m\, e^{s_m t} \tag{2-31}$$

with a derivative given by

$$\dot{n}(t)/n_0 = \sum_{m=0}^{6} A_m s_m e^{s_m t} \tag{2-32}$$

and the time-dependent period $T(t) = n/\dot{n}$. From above

$$T(t) = \sum_{m=0}^{6} A_m e^{s_m t} \sum_{m=0}^{6} A_m s_m e^{s_m t} \tag{2-33}$$

After a long time following the insertion of reactivity K_0 from an equilibrium state, the asymptotic or stable period T is characterized by the single persistent (algebraically largest) decay constant, labeled s_1. Then from the above equation, and where s_1 is equivalent to λ, of the previous section

$$\lim_{t \to \infty} T(t) = T_\infty = 1/s_1 \tag{2-34}$$

So that the deviation of the period from its asymptotic value, $\Delta(t)$, is

$$\Delta(t) = (T_\infty - T)/T_\infty = 1 - s_1 \frac{n(t)}{\dot{n}(t)} \tag{2-35}$$

If an artificial source plays a role in a particular type of kinetic behavior, such as that from a startup source, spontaneous fission, or photoneutron production from a beryllium or heavy water moderator coolant, then the reactivity behavior is altered. This is reflected in the period as well. Equation (1-45) for the case of a constant source R_0, and constant reactivity K_0, is given by

$$n(t) = n_0 + \int_0^t G(t - \tau)[K_0 n(\tau) + R_0]\, d\tau \tag{2-36}$$

The Laplace transform of the above gives for the transform of the neutron density $\bar{n}(s)$

$$\bar{n}(s) = \frac{n_0 + R_0/K_0}{s[1 - K_0 \bar{G}(s)]} - R_0/K_0 s \tag{2-37}$$

Inverting yields the neutron density as

$$n(t) = (n_0 + R_0/K_0) \sum_{m=0}^{6} A_m e^{s_m t} - R_0/K_0 \tag{2-38}$$

where the first term is that for the sourceless case, and that the above reduces to the sourceless neutron density for $R_0 = 0$. The corresponding asymptotic period $T = 1/s_1$ is the same as the sourceless case. However, from Eq. (2-38), the "near" asymptotic period for constant source is given by its $n/\dot{n}$, or

$$T(t) \sim \frac{1}{s_1}\left[1 - \frac{e^{-s_1 t}}{A_1(1 + K_0 n_0/R_0)}\right] \tag{2-39}$$

which shows that the deviation from the asymptotic period persists for a longer time than in the sourceless situation.

In any event, Toppel[11] computed the deviation of the period from the asymptotic period, as a function of time, for a priori specified errors. Some graphical results are depicted in Figs 2-5 and 2-6. This was done by digital machinery computation of $T(t)$ with time for various initial positive and negative step changes in reactivity. The waiting times following a step change in reactivity to observe a period within 1%, 5%, and 10% of its asymptotic value was also computed. From Figs 2-5 and 2-6 it is noted that the negative step reactivity waiting times are much longer than those for positive reactivities. Similarly, as per the previous discussion, the presence of a source increases the corresponding waiting times over those with no source. This is also borne out in the above computations, but not depicted here. As discussed in Sec. 2-4 for large reactivities, that is, $|K| \gg s_1\bar{\tau}$, the neutron density decreases with the well-known $1/s_1 = 80$ sec period as depicted in Fig. 2-6.

One alternative for determining the (negative) reactivity due to the introduction of a control rod, instead of using the inhour equation method, is to employ integral or total count data. That is, if the control rod is rapidly introduced initially, e.g. dropped part way into the core, then this well approximates an initial negative step function of reactivity, of amount K_0 for example. From Eq. (1-47), the corresponding neutron density satisfies

$$n(t) = n_0 - K_0 \int_0^t G(t - \tau)n(\tau)\,d\tau \tag{2-40}$$

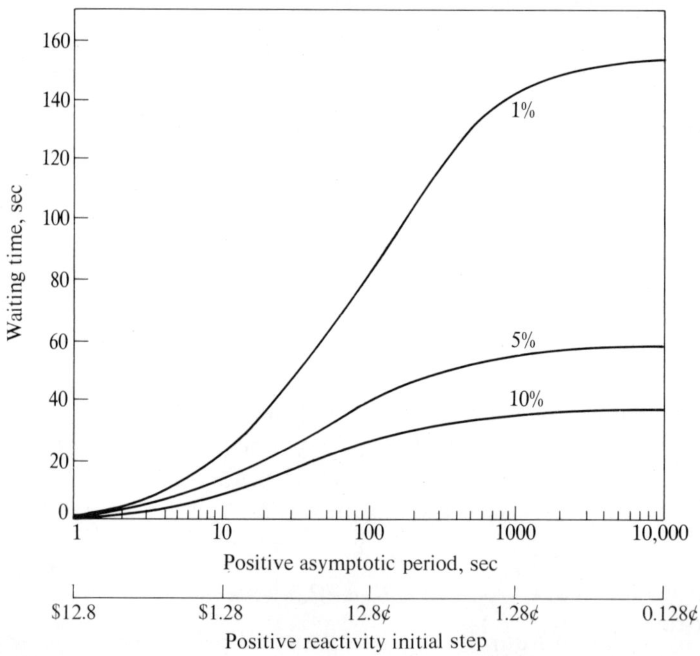

Figure 2-5 Positive reactivity initial step.

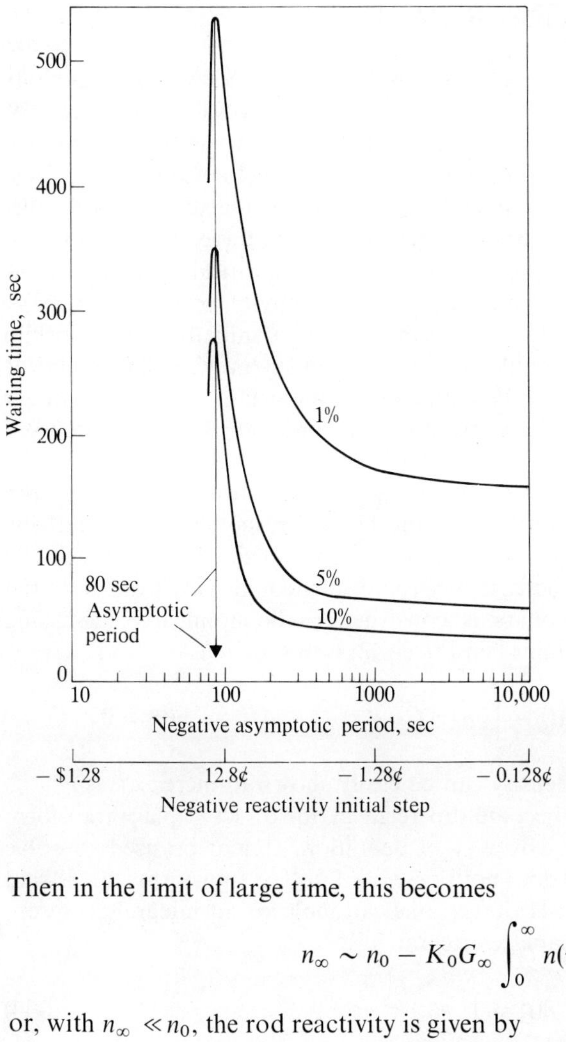

Figure 2-6 Negative reactivity initial step.[11]

Then in the limit of large time, this becomes

$$n_\infty \sim n_0 - K_0 G_\infty \int_0^\infty n(\tau)\, d\tau \tag{2-41}$$

or, with $n_\infty \ll n_0$, the rod reactivity is given by

$$K_0 = n_0 \bar{\tau} \bigg/ \int_0^\infty n(t)\, dt \tag{2-42}$$

The integral in the denominator corresponds to the total count rate from the time the rod was introduced into the core, until the neutron density falls to a negligible level, with the reactor on a stable period all the while. It turns out that the total count rate is insensitive to the precise time at which the rod was inserted at the outset. Hence only one final measurement is involved, assuming that the equilibrium neutron density, n_0, is known. This is as opposed to the inhour method where flux-time data needs to be recorded as the reactor decays on a stable period. Other methods for obtaining reactivity are discussed in the ensuing sections.

2-6 LINEAR TIME-VARYING REACTIVITIES

Following the constant-reactivity solutions is the linear time or ramp reactivity reactor behavior. Important examples of linear reactivity variations arise, for example, where a source is introduced into an essentially neutronless reactor ($n \cong 0$, but is actually seven to 10 orders of magnitude below equilibrium for a shutdown reactor) and the rods are pulled out to bring the reactor to equilibrium. The rod motion, at least over an appreciable range of its travel, can be approximated by a linear reactivity increase from an initial subcritical state. The latter state corresponds to a large amount of negative reactivity, measured in many dollars. A second important case is when the reactor is initially at equilibrium (critical) and the rods are pulled out (or pushed in) in such a way that the linear time reactivity approximation can be made. Other important examples of linear reactivity changes occur in the area of reactor accident kinetics analysis discussed in Chap. 7.

First, consider a startup problem where a source emitting R neutrons per second is inserted into a neutronless reactor. This corresponds to a brand-new reactor or one that has been previously completely shut down for a time. The rod motion (out) employed to increase the reactor neutron density from an initial negative reactivity of $-K_0$ dollars is construed to be a linearly increasing reactivity of m dollars per second. Then Eq. (1-42) is rewritten as

$$\Lambda \dot{n}(t) = (-K_0 + mt)n(t) - \int_0^t n'(\tau) f(t-\tau)\, d\tau + \Lambda R \qquad n(0) = 0 \tag{2-43}$$

Since the asymptotic neutron density can be easily shown to increase† as $e^{mt^2/2\Lambda}$, especially after the reactor has become supercritical, the direct Laplace transform method, and all that it implies by way of definition, cannot be used to solve Eq. (2-43). Such a function $n(t)$ does not possess a Laplace transform; $n(t)$ is said to be not of exponential order. However, one can look for an integral representation of $n(t)$ in the form of a Laplace integral[12-14]

$$n(t) = \int_C \bar{n}(z)\, e^{zt}\, dz \tag{2-44}$$

where $\bar{n}(z)$ and the contour C are so chosen that $n(t)$ is the particularly desired solution of Eq. (2-43). When this representation of $n(t)$ is substituted into Eq. (2-43), where the limit on the convolution integral has been extended to ∞ since $\dot{n}(t) = f(t) = 0$ for $t < 0$, the result is

$$\int_C dz\, \bar{n}(z)\, e^{zt} \left[\Lambda z + z\bar{f}(z) + K_0 - mt\right] = \Lambda R \qquad \bar{f}(z) = \sum_{i=1}^{6} \frac{a_i}{z + \lambda_i} \tag{2-45}$$

† This can be seen by noting that after a sufficient length of time, the neutron density has grown large enough (past prompt critical) to vitiate the effects of the delayed neutrons, the source R, and the initial reactivity $-K_0$. Then the asymptotic form of the kinetics equations is simply $\Lambda \dot{n} \sim mtn$ with solution $e^{mt^2/2\Lambda}$.

The last term on the left-hand side is integrated by parts so that

$$\int_C dz\, e^{zt}\,[\bar{n}(z)\{z[\Lambda + \bar{f}(z)] + K_0\} + m\bar{n}'(z)] = m\bar{n}(z)\, e^{zt}]_C + \Lambda R \qquad (2\text{-}46)$$

where $\ldots]_C$ means that the function is evaluated at the end points of the particular contour chosen. If the contour is closed, $\ldots]_C = 0$.

Equation (2-43) is one aggregate form of the kinetics equations, which are a system of seven first-order differential equations, or equivalently a seventh-order nonhomogeneous linear differential equation. As such, its general solution is a linear combination of seven linearly independent solutions of the corresponding homogeneous equation ($R = 0$) plus any solution of the complete equation ($R \neq 0$). These individual solutions are gotten by first letting the bracket in the integrand on the left-hand side of Eq. (2-46) vanish, so that $\bar{n}(z)$ satisfies

$$m\bar{n}'(z) + \{z[\Lambda + \bar{f}(z)] + K_0\}\bar{n}(z) = 0 \qquad (2\text{-}47)$$

with solution

$$\bar{n}_k(z) = A_k \prod_{i=1}^{6} (z + \lambda_i)^{r_i/m} \exp - \left(\frac{\Lambda}{2m} z^2 + \frac{1 + K_0}{m} z\right) \qquad r_i = a_i \lambda_i \qquad (2\text{-}48)$$

where A_k are arbitrary constants. Then the individual solutions are of the form

$$n_k(t) = A_k \int_{C_k} dz \prod_{i=1}^{6} (z + \lambda_i)^{r_i/m} \exp - \left(\frac{\Lambda}{2m} z^2 + \frac{1 + K_0}{m} z - zt\right) \qquad (2\text{-}49)$$

if the integral is convergent and the contours are so chosen that the bilinear concomitant,† from Eq. (2-46),

$$-m\bar{n}(z)\, e^{zt}\Big]_C \equiv -mA_k \prod_{i=1}^{6} (z + \lambda_i)^{r_i/m} \exp - \left(\frac{\Lambda}{2m} z^2 + \frac{1 + K_0}{m} z - zt\right)\Big]_C = \Lambda R \qquad (2\text{-}50)$$

The seven linearly independent solutions of the homogeneous counterpart of the seventh-order kinetics equation are obtained by choosing seven contours C_k: $k = 0, 1, \ldots, 6$, so that for all t and for each contour C_k

$$-m\bar{n}(z)\, e^{zt}\Big]_{C_k} = 0 \qquad (2\text{-}51)$$

With $\bar{n}(z)$ above, these contours are immediately seen to be along the real axis, respectively from $-\infty$ to $-\lambda_6$, $-\lambda_6$ to $-\lambda_5, \ldots, -\lambda_1$ to ∞ as in Fig. 2-7. For the corresponding particular solution to the complete nonhomogeneous equation, we choose the contour from 0 to ∞ so that

$$-m\bar{n}(z)\, e^{zt}\Big]_C \equiv -mA_k \prod_{i=1}^{6} (z + \lambda_i)^{r_i/m} \exp - \left(\frac{\Lambda}{2m} z^2 + \frac{1 + K_0}{m} z - zt\right)\Big]_C = \Lambda R \qquad (2\text{-}52)$$

† Cf. Ince.[13]

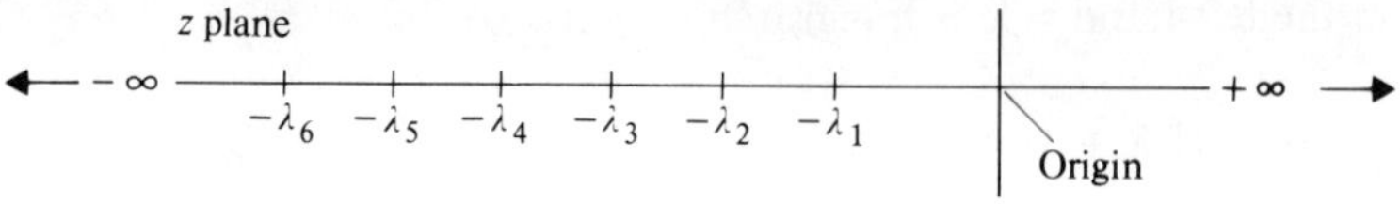

Figure 2-7 Contours for linear reactivity solutions.

Therefore, the solution to the homogeneous equation which gives the transient behavior of $n(t)$, as seen from the integration limits, is the linear combination

$$\left[A_6 \int_{-\infty}^{-\lambda_6} + A_5 \int_{-\lambda_6}^{-\lambda_5} + \cdots + A_0 \int_{-\lambda_1}^{\infty}\right]\left[dz \prod_{i=1}^{6} (z+\lambda_i)^{r_i/m} \cdot \exp -\left(\frac{\Lambda}{2m} z^2 + \frac{1+K_0}{m} z - zt\right)\right] \tag{2-53}$$

The particular solution to the complete equation which yields the asymptotic behavior, or steady state if there is one, is

$$\int_0^{\infty} dz \prod_{i=1}^{6} (z+\lambda_i)^{r_i/m} \exp -\left(\frac{\Lambda}{2m} z^2 + \frac{1+K_0}{m} z - zt\right) \tag{2-54}$$

The general solution is the sum of these solutions after determining the arbitrary constants $A_0, \ldots, A_6$ from seven initial conditions on the neutron density and its derivatives. Starting with $n(0) = 0$ and $\dot{n}(0) = R$, the remaining initial conditions can be obtained by successively differentiating Eq. (2-43) and evaluating at $t = 0$. Usually, only the asymptotic solution is desired, so that, from Eq. (2-52)

$$A = \frac{\Lambda R}{m} \prod_{i=1}^{6} (\lambda_i)^{-r_i/m}$$

giving

$$n(t) \sim \frac{R\Lambda}{m} \int_0^{\infty} dz \prod_{i=1}^{6} \left(1 + \frac{z}{\lambda_i}\right)^{r_i/m} \exp -\left(\frac{\Lambda}{2m} z^2 + \frac{1+K_0}{m} z - zt\right) \tag{2-55}$$

The second case, that of the linear reactivity increase from the equilibrium state, is gotten in the same way, but by first letting $K_0 = R = 0$. R is neglected because its contribution to the neutron density is overwhelmingly dwarfed by the equilibrium neutron population. Then the bilinear concomitant is

$$-m\bar{n}(z)\, e^{zt}\Big]_{C_k} = -mA_k \prod_{i=1}^{6} (z+\lambda_i)^{r_i/m} \exp -\left(\frac{\Lambda}{2m} z^2 + \frac{z}{m} - zt\right) = 0 \qquad k = 0, \ldots, 6 \tag{2-56}$$

The complete solution is again given by the linear combination of the seven solutions

$$\left(A_6 \int_{-\infty}^{-\lambda_6} + A_5 \int_{-\lambda_6}^{-\lambda_5} + \cdots + A_0 \int_{-\lambda_1}^{\infty}\right)\left[dz \prod_{i=1}^{6} (z + \lambda_i)^{r_i/m} \exp - \left(\frac{\Lambda}{2m} z^2 + \frac{z}{m} - zr\right)\right] \tag{2-57}$$

where the arbitrary constants $A_0, \ldots, A_6$ are obtained in the same way from the initial conditions, but beginning with $n(0) = 1$, $\dot{n}(0) = 0$, etc.

Again, the asymptotic solution is obtained as

$$n(t) \sim A_0 \int_0^{\infty} dz \prod_{i=1}^{6} (z + \lambda_i)^{r_i/m} \exp - \left(\frac{\Lambda}{2m} z^2 + \frac{z}{m} - zt\right) \tag{2-58}$$

where the portion of the integral from $-\lambda_1$ to 0 has been neglected. Since the asymptotic solution itself cannot satisfy all of the initial conditions, then, for example, let $n(0) = 1$ and obtain A_0 in Eq. (2-58) by evaluating

$$1 = A_0 \int_0^{\infty} dz \prod_{i=1}^{6} (z + \lambda_i)^{r_i/m} \exp - \left(\frac{\Lambda}{2m} z^2 + \frac{z}{m}\right) \tag{2-59}$$

Now consider the situation where negative reactivity is inserted at a constant rate of $-m$ dollars per second into a reactor initially at equilibrium. Assume that the prompt drop (Sec. 2-9) in neutron density immediately following equilibrium holds well. This then implies that, subsequent to the immediate (prompt) drop, $\Lambda\dot{n}(t) \ll |mtn(t)|$. The time origin will be taken at this point. Then the left-hand side of Eq. (2-43) is neglected by comparison, giving

$$0 = -(K_0 + mt)n(t) - \int_0^t n'(\tau) f(t - \tau)\, d\tau + \Lambda R \qquad n(0) = 1 \tag{2-60}$$

This will be assumed as the kinetics equation holding for all times, so that $-K_0$ is the negative reactivity that compensates the source to maintain initial equilibrium ($K_0 = \Lambda R$). This assumption is inaccurate for only a very short time following equilibrium, as explained above.

Since the solutions are decreasing with time, they are certainly of exponential order, which allows the Laplace transform method to be employed in obtaining them. Using the device of letting

$$n(t) = 1 + \int_0^t p\, d\tau \tag{2-61}$$

with Laplace transform $\bar{n}(s) = 1/s + \bar{p}(s)/s$, simplifies the development somewhat. With this inserted in Eq. (2-60), its Laplace transform yields the following differential equation in $\bar{p}(s)$:

$$\bar{p}'(s) - \bar{p}(s)\left[\frac{K_0}{m} + \frac{1}{s} + \frac{s\bar{f}(s)}{m}\right] = \frac{1}{s} \qquad \bar{p}(\infty) = 0 \qquad \bar{f}(s) = \sum_{i=1}^{6} \frac{a_i}{s + \lambda_i} \tag{2-62}$$

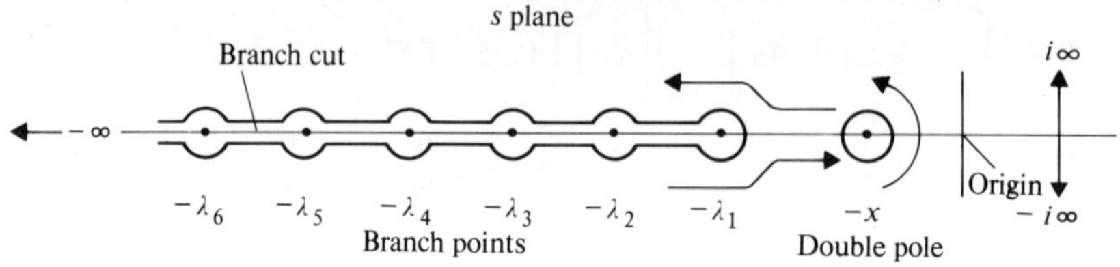

Figure 2-8 Contour in the s-plane for Laplace transform inversion integral.

Its solution is easily obtained and is inserted into the Laplace transform inversion integral thus:

$$n(t) = 1 - \frac{1}{2\pi i}\int_{\Gamma} \frac{ds \exp\{s[t + (K_0 + 1)/m]\}}{\prod_{i=1}^{6} (s + \lambda_i)^{r_i/m}} \cdot \int_{s}^{\infty} \frac{dz}{z^2} \prod_{i=1}^{6} (z + \lambda_i)^{r_i/m} \exp\left(-\frac{K_0 + 1}{m} z\right) \tag{2-63}$$

Interchanging the order of integration, and with $z - s = x$, yields

$$n(t) = 1 - \frac{1}{2\pi i}\int_0^{\infty} dx \exp\left(-\frac{K_0 + 1}{m} x\right) \int_{\Gamma} \frac{ds}{(s + x)^2} \prod_{i=1}^{6} \left(1 + \frac{x}{s + \lambda_i}\right)^{r_i/m} e^{st} \tag{2-64}$$

The contour Γ is essentially the usual semicircular one for the inversion integral except for a portion skirting a branch cut along part of the negative real axis, as shown in Fig. 2-8. The singularities in the s plane are seen to occur at the six branch points $s = -\lambda_i$, $i = 1, \ldots, 6$, plus a double pole at $s = -x$, where x is real. The residues at the branch points yield six transient solutions each of which consists of hypergeometric functions of complicated arguments.† The asymptotic solution comes from the residue at the pole. It is, after an integration by parts

$$n(t) \sim \frac{\Lambda R}{m} \int_0^{\infty} dx \frac{\exp[-x(K_0 + 1 + mt)]/m}{\prod_{i=1}^{6} (1 - x/\lambda_i)^{r_i/m}} \qquad m > r_i \tag{2-65}$$

Following the initial shutdown, the negative reactivity causes the neutron density to decrease on an increasing period as the delayed neutron effects become more and more manifest. That is, as the reactor is being shut down, the delayed neutron population becomes an increasingly larger fraction of the total population because of its response lag as compared with the prompt neutrons. For very long times after shutdown ($|mt| \gg 1$) the neutron density varies like

† The transient solutions are discussed in Smets.[6] Cf. H. Bateman, in A. Erdélyi (ed.), "Tables of Integral Transforms," vol. I, McGraw-Hill, New York, 1954, no. 9, p. 238.

$$n(t) \sim \frac{\Lambda R}{\Lambda R + mt} \qquad K_0 = \Lambda R \tag{2-66}$$

obtained from Eq. (2-65), where the exponential dominates the integrand so that

$$\prod_{i=1}^{6} (1 - x/\lambda_i)^{r_i/m}$$

approximates unity as mt approaches ∞. This expression can also be obtained from Eq. (2-60) after a long time, when the derivative of the neutron density is assumed to be so small that it is neglected in comparison with the remaining terms.

Therefore, after a ramp decrease in reactivity from equilibrium, the neutron density at first drops rapidly [prompt drop $n \cong \exp - mt^2/2\Lambda$] and then more gradually until the above asymptotic behavior results. In actual systems, the reactivity insertion is usually halted long before the asymptotic state ensues. In that case, the neutron density decreases gradually with the 80 sec period λ_1^{-1}, which corresponds to the delayed neutron precursor with the longest half-life. This is discussed further in Sec. 2-9. Here, the assumed relentlessly decreasing reactivity (for sufficiently large $|m|$) pushes the neutron density decrease so that delayed neutron effects are ultimately vitiated.

The preceding solutions form part of the basis for somewhat more approximate expressions for the neutron density under various linear rates of reactivity change. The following illustrates some of these expressions and the techniques for obtaining them.

For very large reactivity insertion rates ($|m| \gg 1$), the neutron density departs from equilibrium, for example, so rapidly that the delayed neutrons can be completely neglected ($\beta \equiv 0$) for positive insertion rates. This is because the reactor becomes prompt critical so quickly that any damping influence of the delayed neutrons is utterly lost. Then $f \equiv 0$, so that Eq. (1-42) yields simply

$$\Lambda \dot{n}(t) = mtn(t); \qquad n(0) = 1 \tag{2-67}$$

with solution

$$n(t) = e^{mt^2/2\Lambda} \tag{2-68}$$

For very large negative insertion rates, $n = e^{-mt^2/2\Lambda}$ holds only for the first small fraction of a second, after which the delayed neutron damping is manifest. Then the delayed neutrons must be taken into account, and for this Eq. (1-42), etc. are available. Short-time $m < 0$ solutions can be "faired" with asymptotic solutions for decreasing neutron level as discussed in Sec. 2-9.

For merely large insertion rates ($|m| \gtrsim 1$), the implication is that the delayed neutron precursor concentration remains at its equilibrium level during the (short) times of interest. Then $f \cong f(0) = 1$ in Eq. (1-42) so that

$$\Lambda \dot{n}(t) = (mt - 1)n(t) + 1 \qquad n(0) = 1 \tag{2-69}$$

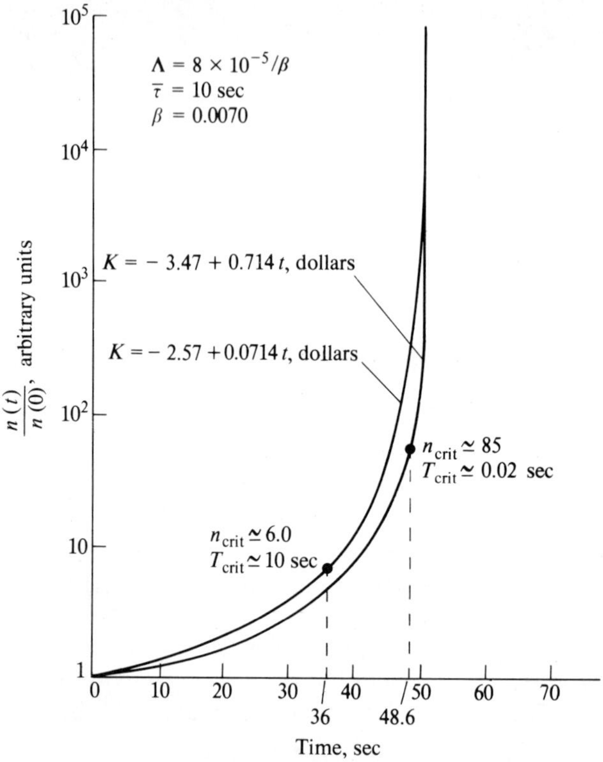

Figure 2-9 Neutron density versus time for startup with ramp reactivity increase for one delayed neutron group approximation to the Laplace integral representation of $n(t)/n(0)$.[15]

with solution

$$n(t) = \left(\frac{2}{m\Lambda}\right)^{1/2} \left\{\operatorname{erf}\left[\left(\frac{m}{2\Lambda}\right)^{1/2} t - (2m\Lambda)^{-1/2}\right] + \operatorname{erf}(2m\Lambda)^{-1/2}\right\}$$
$$\cdot \exp\left[\frac{m}{2\Lambda}\left(t - \frac{1}{m}\right)^2\right] + \exp\left\{\frac{m}{2\Lambda}\left[\left(t - \frac{1}{m}\right)^2 - \frac{1}{m^2}\right]\right\} \qquad (2\text{-}70)$$

which can be approximated in various ways[1] depending on the parameter $m/2\Lambda$ and whether the reactor is subcritical, supercritical, or prompt supercritical (prompt critical).

As shown above, large positive reactivity insertion rates quickly bring the reactor from equilibrium to the prompt critical state. Since the reactor is now critical on prompt neutrons only† ($K \cong 1$), then $\Lambda\dot{n} = n$; and in the neighborhood of this state its logarithmic derivative $\alpha = \dot{n}/n$, called inverse period or α, is essentially that of the inverse lifetime Λ^{-1}. The period is defined as α^{-1}.

If the reactor passes through critical and becomes prompt critical from the

† At prompt critical $k_{eff} = 1 + \beta$, so that $K = 1/1 + \beta \cong 1$ dollar.

neutronless $[n(0) = 0]$ state, the period and the neutron level at prompt critical are different from those attained through a large ramp reactivity from equilibrium where delayed neutrons are neglected. The latter implies $n = e^{mt^2/2\Lambda}$, n_{pc} (prompt critical) $= e^{1/2\Lambda m}$, and the corresponding inverse period $\alpha_{pc} = 1/\Lambda$. From Eq. (2-55), the asymptotic neutron density at prompt critical, from a ramp startup with $n(0) = 0$, is obtained by inserting the time at prompt critical $t_c = (1 + K_0)/m$. The result is

$$n_{pc} \approx \frac{R\Lambda}{m}\int_0^\infty dz \prod_{i=1}^{6}\left(1 + \frac{z}{\lambda_i}\right)^{r_i/m} \exp\left(-\frac{\Lambda}{2m} z^2\right)$$

$$\cong \frac{R\Lambda}{m}\int_0^\infty dz \prod_{i=1}^{6}\left(\frac{z}{\lambda_i}\right)^{r_i/m} \exp\left(-\frac{\Lambda}{2m} z^2\right) \tag{2-71}$$

where the rightmost expression is a good approximation, since for not too large Λ, or too small $|m|$, the preponderant contribution to the integral comes from values of $z/\lambda_i \gg 1$. Then the neutron level at prompt critical reduces easily to give

$$n_{pc} \approx \frac{\Lambda R}{2m}\prod_{i=1}^{6}(\lambda_i)^{-r_i/m}\left(\frac{2m}{\Lambda}\right)^{1/2+\lambda/2m}\Gamma\left(\frac{1}{2} + \frac{\lambda}{2m}\right) \tag{2-72}$$

where $\lambda = \sum_{i=1}^{6} a_i\lambda_i \neq 1/\bar{\tau}$. If Eq. (2-55) is differentiated and evaluated at t_{pc} by using similar approximations, the inverse period α at prompt critical is

$$\alpha_{pc} \approx \frac{(2m/\Lambda)^{1/2}\Gamma(1 + \lambda/2m)}{\Gamma(\frac{1}{2} + \lambda/2m)} \tag{2-73}$$

As m gets very large, $n_{pc} \approx R(\pi/2m\Lambda)^{1/2}$ and $\alpha_{pc} \approx (2m/\pi\Lambda)^{1/2}$. Similar expressions can be obtained for the neutron level and the behavior of the inverse period as the system passes through criticality ($K = 0$) from startup.

On the other hand, if the ramp reactivity insertion rate from equilibrium is very small, so that $|m|$ is about one cent per second, the neutron density does not deviate from equilibrium by an appreciable amount for the first few seconds. Since $n \cong 1$, one iteration of Eq. (1-47), which is now

$$n(t) = 1 + \int_0^t G(t - \tau)m\tau n(\tau)\, d\tau \tag{2-74}$$

yields

$$n(t) \cong 1 + m\int_0^t G(t - \tau)\tau\, d\tau \equiv 1 + m\int_0^t d\tau \int_0^\tau G(x)\, dx \tag{2-75}$$

which should suffice.

For small $|m|$, a method of Hurwitz[3] which holds for any gradual reactivity change gives handy solutions. This method assumes that the inhour equation is available and that the algebraically largest root therefrom is time varying slowly as seen in the following. The inhour equation is rederived in a way that is slightly different from its development earlier. The kinetics equations are rewritten here for convenience.

$$\dot{n} = \frac{K-1}{\Lambda} n + \sum_{i=1}^{6} \lambda_i c_i \qquad n(0) = 1 \tag{2-76}$$

$$\dot{c}_i = -\lambda_i c_i + \frac{a_i n}{\Lambda} \qquad \dot{c}_i(0) = 0 \qquad i = 1, \ldots, 6 \tag{2-77}$$

When the reactivity is constant (positive or negative step function at the origin), the inhour equation is rederived by assuming that the asymptotic form of $n(t)$ and $c_i(t)$ are, respectively,

$$n(t) \sim e^{\lambda t} \qquad c_i(t) \sim \frac{a_i\, e^{\lambda t}}{\Lambda(\lambda + \lambda_i)} \qquad i = 1, \ldots, 6 \tag{2-78}$$

where λ is the algebraically largest root of the inhour equation. Inserting these expressions into the kinetics equations above yields the inhour equation as

$$K\bar{G}(\lambda) = 1 \qquad \text{or} \qquad K = \lambda[\Lambda + \bar{f}(\lambda)] \tag{2-79}$$

$\bar{G}(\lambda)$ and $\bar{f}(\lambda)$ are the Laplace transforms of $G(t)$ and $f(t)$ with respect to the parameter λ.

Now assume that K is slowly increasing with time. For such a slowly increasing reactivity, the inhour equation is still assumed to hold. That is, there is still a persistent positive root, now $\lambda(t)$, which still satisfies

$$K(\lambda(t)) = \lambda(t)[\Lambda + \bar{f}(\lambda(t))] \tag{2-80}$$

Then let

$$n(t) = \xi(t) \exp\left[\int_0^t \lambda(x)\, dx\right] \tag{2-81}$$

and also

$$c_i(t) = \Lambda^{-1}\left[\frac{a_i \xi}{\lambda + \lambda_i} + \rho_i(t)\right] \exp\left[\int_0^t \lambda(x)\, dx\right] \qquad i = 1, \ldots, 6 \tag{2-82}$$

where $\xi(t)$ and $\rho_i(t)$ are to be obtained such that $n(t)$ satisfies the kinetics equations. If $n(t)$ and $c_i(t)$ above are substituted into the kinetics equation (2-76) for $n(t)$, together with the inhour equation the result yields a differential equation for $\xi(t)$. It is

$$\Lambda \frac{d\xi}{dt} = \sum_{i=1}^{6} \lambda_i \rho_i(t) \qquad \xi(0) = 1 \tag{2-83}$$

To obtain the corresponding differential equations for ρ_i, insert $n(t)$ and $c_i(t)$ into the kinetics equations (2-77) for $c_i(t)$ to get

$$\frac{d\rho_i}{dt} + (\lambda + \lambda_i)\rho_i(t) = -\frac{d}{dt}\left[\frac{a_i \xi(t)}{\lambda + \lambda_i}\right] \qquad i = 1, \ldots, 6 \tag{2-84}$$

Consider the right-hand side as a "negative forcing" function, such that in the limit of large time $d\rho_i/dt$ approaches zero. Thus, asymptotically,

$$\rho_i(t) \sim -\frac{1}{\lambda + \lambda_i}\frac{d}{dt}\left[\frac{a_i\xi(t)}{\lambda + \lambda_i}\right] \tag{2-85}$$

Substituting this into Eq. (2-83) gives

$$\Lambda\frac{d\xi}{dt} = -\sum_{i=1}^{6}\frac{1}{\lambda + \lambda_i}\frac{d}{dt}\left[\frac{a_i\lambda_i\xi(t)}{\lambda + \lambda_i}\right] \tag{2-86}$$

or

$$\frac{d}{dt}\ln\xi = -\frac{\dfrac{1}{2}\displaystyle\sum_{i=1}^{6} a_i\lambda_i\frac{d}{dt}\left(\frac{1}{\lambda + \lambda_i}\right)^2}{\Lambda + \displaystyle\sum_{i=1}^{6} a_i\lambda_i/(\lambda + \lambda_i)^2} \tag{2-87}$$

With initial conditions $\xi(0) = 1$, $\lambda(0) = 0$, integrating gives

$$\xi(t) = \left[\frac{\Lambda + \displaystyle\sum_{i=1}^{6} a_i/\lambda_i}{\Lambda + \displaystyle\sum_{i=1}^{6} a_i\lambda_i/(\lambda + \lambda_i)^2}\right]^{1/2} = \left[\frac{\Lambda + \bar{\tau}}{\Lambda + \bar{f}(\lambda) + \lambda\bar{f}'(\lambda)}\right]^{1/2} \tag{2-88}$$

so that the neutron density is

$$n(t) = \left[\frac{\Lambda + \bar{\tau}}{\Lambda + \bar{f}(\lambda) + \lambda\bar{f}'(\lambda)}\right]^{1/2}\exp\left[\int_0^t \lambda(\tau)\,d\tau\right] \tag{2-89}$$

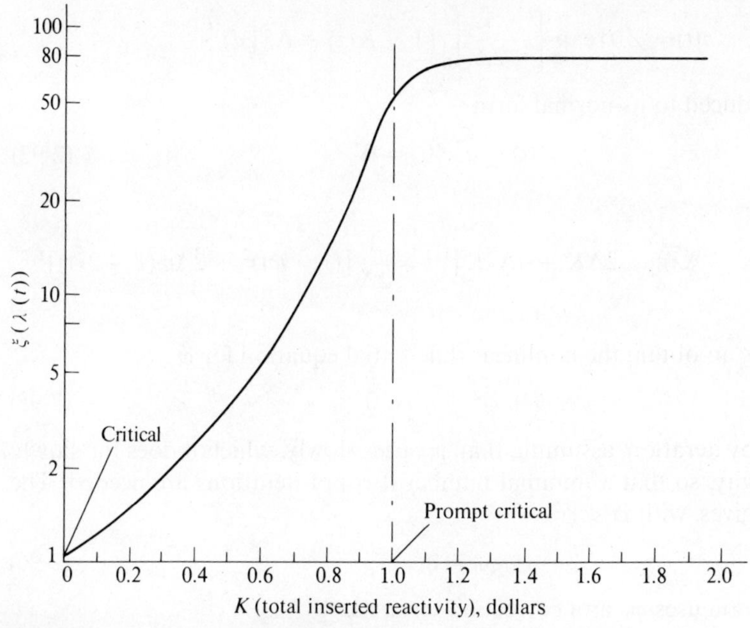

Figure 2-10 Hurwitz reactivity approximation function ξ for $\Lambda = 0.003$.[1]

For $K = mt$ for small $m \ll 1$, $\xi \cong [\bar{\tau}/K'(\lambda)]^{1/2} = [\bar{\tau}\dot{\lambda}(t)/m]^{1/2}$. $\xi(t)$ is a slowly varying function compared to the exponential factor as seen in Fig. 2-10.

The above development for obtaining solutions for gradual reactivities is essentially a variant of the Wentzel, Kramers, Brillouin (WKB) method. This method is used to obtain solutions to certain linear differential equations when their coefficients are slowly varying, corresponding here to slowly varying reactivities. An illustration of the method is given for the kinetics differential equation obtained when all of the delayed neutron groups are lumped into one representative group with a mean decay constant $\bar{\lambda} = 1/\bar{\tau}$. This corresponds to approximating the delayed neutron decay function $f(t)$ by a single exponential, that is, $f(t) = \exp(-\bar{\lambda}t)$. Equation (1-42) thus rewritten is

$$\Lambda\dot{n}(t) = K(t)n(t) - \int_0^t n'(\tau)\exp[-\bar{\lambda}(t-\tau)]\,d\tau \tag{2-90}$$

Eliminating the integral between this equation and its derivative results in the second-order differential equation, for a ramp reactivity

$$\Lambda\ddot{n}(t) + [1 - K(t) + \Lambda\bar{\lambda}]\dot{n}(t) - [\dot{K}(t) + \bar{\lambda}K(t)]n(t) = 0$$

$$n(0) = 1 \qquad \dot{n}(0) = 0 \tag{2-91}$$

For a ramp reactivity $K = mt$ and $\Lambda\bar{\lambda} \ll 1$, which is almost always the case,

$$\Lambda\ddot{n}(t) + (1 - mt)\dot{n}(t) - m(1 + \bar{\lambda}t)n(t) = 0 \tag{2-92}$$

By letting

$$n(t) = Z(t)\exp\left\{-\frac{1}{2\Lambda}\int_0^t [1 - K(\tau) + \Lambda\bar{\lambda}]\,d\tau\right\}$$

Eq. (2-91) is reduced to its normal form

$$\ddot{Z}(t) - q^2 Z(t) = 0 \tag{2-93}$$

where

$$q = \frac{1}{2\Lambda}[(1 - K + \Lambda\bar{\lambda})^2 + 2\Lambda\dot{K} + 4\Lambda\bar{\lambda}K]^{1/2} = \frac{1}{2\Lambda}[(1 - mt)^2 + 2\Lambda m(1 + 2\bar{\lambda}t)]^{1/2} \tag{2-94}$$

Now let $Z = e^{\phi}$ to obtain the nonlinear differential equation for ϕ

$$\ddot{\phi} + \dot{\phi}^2 - q^2 = 0 \tag{2-95}$$

This is solved by iteration, assuming that $\dot{\phi}$ varies slowly, which it does for slowly varying reactivity, so that a minimal number (two) of iterations are needed. The first iteration gives, with $\ddot{\phi} \ll \dot{\phi}^2$

$$\dot{\phi}_0^2 = q^2 \tag{2-96}$$

The next iteration uses $\ddot{\phi}_0$ as a correction term, i.e.,

$$\dot{\phi}_1^2 = q^2 - \ddot{\phi}_0 \cong q^2 - \dot{q} \tag{2-97}$$

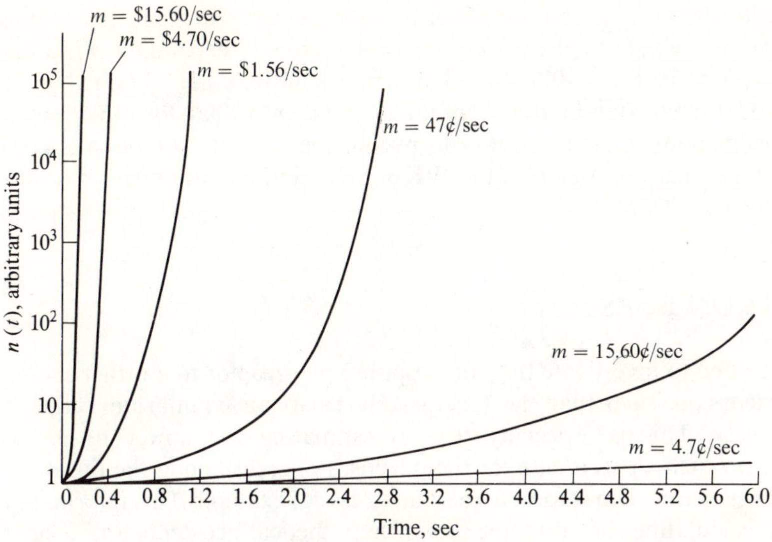

Figure 2-11 Neutron density versus time for positive ramp reactivities from initial equilibrium for ^{235}U, $\Lambda = 10^{-4}/\beta$.[8]

For the iterative process to converge with sufficient rapidity that two iterations suffice, $|\dot{q}| \ll q^2$ must hold. If such is the case

$$\dot{\phi}_1 \cong \pm(q^2 - \dot{q})^{1/2} \cong \pm\left[q - \frac{1}{2}\frac{\dot{q}}{q}\right] \tag{2-98}$$

Then the general solution to Eq. (2-93) is the linear combination†

$$Z(t) = \frac{1}{q^{1/2}}\left\{A \exp\left[\int_0^t q(\tau)\,d\tau\right] + B \exp\left[-\int_0^t q(\tau)\,d\tau\right]\right\} \tag{2-99}$$

assuming the validity of the approximation

$$\frac{|\dot{q}|}{q^2} \cong \frac{2|m(1 - mt)|}{[2m\Lambda^{1/3}(1 + 2\bar{\lambda}t) + (1 - mt)^2\Lambda^{-2/3}]^{3/2}} \ll 1 \tag{2-100}$$

which holds strongly for small $|m|$.

The method *per se* introduces artificial singularities in the solutions at times corresponding to roots of the denominator in $|\dot{q}|/q^2$, that is, in the roots of q^2. Such times, called turning points, correspond to where q changes from a real to an imaginary quantity, while $Z(t)$ and thereby $n(t)$ change from real exponential to imaginary exponential (sinusoidal) solutions.

† By comparing the neutron density obtained by the WKB method with that obtained from the one delayed neutron group equivalent of the Hurwitz method, using the corresponding inhour equation, it is seen that they are essentially equivalent. The major difference is in the coefficient multiplying the exponential factor. This coefficient is only slowly varying, however.

It is easily shown that $q(t)$, for small positive or negative ramp reactivities, has no roots for $t > 0$, so that no turning points exist. By expanding $q(t)$ about the turning point in Eq. (2-93), Bessel function solutions are obtained in its neighborhood. Linear combinations[16] of these are taken so that, for times remote from the turning point, their asymptotic form coincides with the solutions as given in Eq. (2-99). Further discussion of the WKB method applied to nuclear reactor kinetics is given by Tan.[17]

2-7 NEUTRON BURSTS

Often, it is desired to investigate the time response of a reactor to a rather sudden reactivity excursion, such that the reactor returns to an equilibrium value of neutron density. This is especially true in estimating the power or energy developed from such an excursion for the purposes of ascertaining the durability of the reactor structure with respect to thermal stress, for example. This information is needed in calculating the consequences of hypothetical accidents and general reactor hazard behavior. This aspect, as well as the space-time dependencies of neutron bursts, will be discussed in Chap. 7.

First, consider an idealized reactivity excursion represented by $K_0\delta(t)$. This is a delta function at the origin, of "amplitude" K_0. This might correspond approximately to quickly ejecting and reinserting a rod into an equilibrium system. Substituting for K in Eq. (1-47) gives, with $n_0 = 1$

$$n(t) = 1 + K_0 \int_0^t G(t - \tau)\delta(\tau)n(\tau)\,d\tau = 1 + K_0 G(t) \tag{2-101}$$

Noting the behavior of $G(t)$ from Chap. 1, it is seen that there is an initial jump of neutron density of $K_0 G(0) = K_0/\Lambda$ followed by a rapid drop and (by using the Laplace transform final-value theorem) thence to the asymptotic value $K_0 G(\infty) = K_0/[\Lambda + \bar{f}(0)] = K_0/(\Lambda + \bar{\tau}) \cong 0.078 K_0$ for ^{235}U and $\Lambda \ll \bar{\tau}$. That is, the new neutron density level is $n(\infty) = 1 + 0.078 K_0$. As an aside, such shock excitation can be construed as a "ringing experiment"† to obtain the reactor zero-power transfer function $\bar{G}(\omega)$, that is, the Fourier transform of its impulse response $G(t)$, gotten by appropriate instrumentation for decomposing $G(t)$ into frequency components. There are, however, accuracy difficulties with this method. This is more fully discussed in the section on transfer functions in Chap. 3.

Shock excitation is more accurately represented by an initial reactivity "spike" of amplitude K_0 and width t_0. The spike is defined as

$$K(t) = \begin{cases} K_0 & 0 \le t \le t_0 \\ 0 & \text{otherwise} \end{cases} \tag{2-102}$$

† This term is taken from telephone engineering jargon for the (electrical) shock excitation of electronic components to obtain an indication of their frequency response.

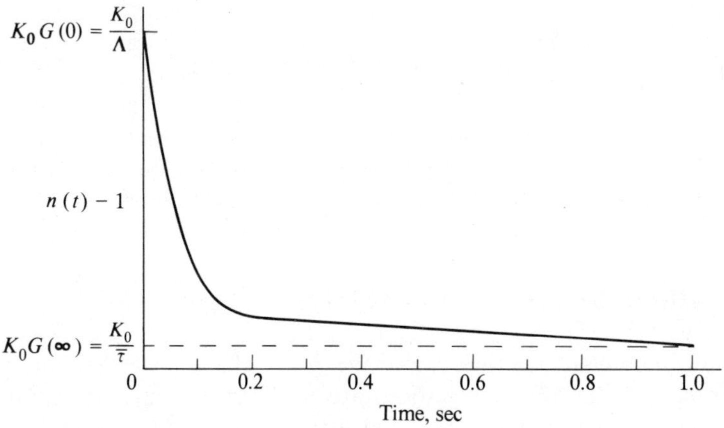

Figure 2-12 Neutron density behavior for delta function reactivity excursion.

With this spike substituted for the reactivity in Eq. (1-47), one obtains

$$n_<(t) = n_0 + K_0 \int_0^t G(t-\tau) n_<(\tau)\, d\tau \qquad t \leq t_0 \tag{2-103}$$

While for times beyond the duration of the spike

$$n_>(t) = n_0 + K_0 \int_0^{t_0} G(t-\tau) n_<(\tau)\, d\tau ; \qquad t > t_0 \tag{2-104}$$

where $n_<$ is the neutron density for times prior to t_0. If the spike is of sufficiently short duration such that it is dominated by prompt neutrons, then $n_< = n_0 \exp(K_0/\Lambda)t$. Hence, substituting the latter expression for $n_<$ above yields

$$n_>(t) = n_0 \left[1 + K_0 \int_0^{t_0} G(t-\tau) \exp(K_0\tau/\Lambda)\, d\tau\right] \tag{2-105}$$

or

$$n_>(t) = n_0 \left[1 + K_0 \sum_{r=0}^{6} \frac{A_r e^{-s_r t}}{s_r + K_0/\Lambda} \{\exp[(s_r + K_0/\Lambda)t_0] - 1\}\right] \tag{2-106}$$

Notice that for a very short spike, that is, in the limit of small t_0

$$n_>(t) \simeq n_0[1 + K_0 t_0 G(t)] \tag{2-107}$$

essentially the same behavior as the delta function spike of Eq. (2-101). Other approximations for $n_<$ to be inserted into the integral on the right-hand side of Eq. (2-103) suggest themselves. Equation (2-103) can be generalized to the case of variable reactivities that cut off at t_0, and are described by the pair of equations

$$n_>(t) = n_0 + \int_0^{t_0} G(t-\tau) K(\tau) n_<(\tau)\, d\tau ; \qquad t > t_0 \tag{2-108}$$

where

$$n_<(t) = n_0 + \int_0^t G(t-\tau)K(\tau)n_<(\tau)\,d\tau; \qquad t \le t_0 \tag{2-109}$$

Note that Eq. (2-108) is simply a formula for $n_>(t)$ for a given $n_<(t)$, obtained from Eq. (1-47), and not an integral equation for $n_>(t)$. However, Eq. (2-109) is an integral equation for $n_<(t)$, the fundamental integral equation of the point model.

2-8 PROMPT DROP/JUMP REACTIVITIES

For negative step function reactivities, all of the roots of the denominator of Eq. (2-14) (equivalent to the inhour equation) are negative, which corresponds to a solution given by a linear combination of exponential terms all with negative decay constants. The asymptotic solution is, of course, given by the single exponential term with the algebraically largest root. If the Laplace transform of Eq. (1-47) ($R \neq 0$) is obtained and the $\lim_{s\to 0} s\bar{n}(s)$ employed for negative step reactivity K, the neutron density after a very long time following the negative step insertion is given by $n(\infty) \sim \Lambda R/|K|$.

Immediately after the insertion of a negative step function of reactivity K, there is a sharp drop in the neutron density followed by a much more gradual and decreasing neutron density. The immediate sharp drop is investigated by looking at Eq. (2-5) for negative step reactivity changes. It is

$$\frac{n(t)}{n_0} = \frac{1 + K\,e^{-(K+1)t/\Lambda}}{1+K} \tag{2-110}$$

so that for short times, the neutron density decreases rapidly on the "prompt group" of neutrons.

The neutron level immediately following the prompt drop can be obtained by letting Λ approach zero in Eq. (2-110). This is equivalent to extrapolating the long time asymptote of this equation back to the origin. Rigorously, this holds only for times appreciably long after the occurrence of the prompt change in neutron level where $|\Lambda\dot{n}| \ll |(K-1)n|$. From Eq. (2-1), it is seen that $|\Lambda\dot{n}| \ll |(K-1)n|$ holds increasingly better as the delayed neutrons slow the response of $\dot{n}(t)$ after appreciable time has elapsed. The inequality, of course, approaches an equality as the system approaches prompt critical, since the prompt neutron population overcomes that of the delayed neutrons. At any rate, for $K < 0$, the asymptote of Eq. (2-110) gives the prompt neutron drop level as

$$n_{\text{prompt drop}} = n_0(1+K)^{-1}, \qquad K < 0 \tag{2-111}$$

Similar arguments supply a general expression for the prompt jump following a positive step function reactivity change. Equation (2-111) is rewritten in general to combine both as

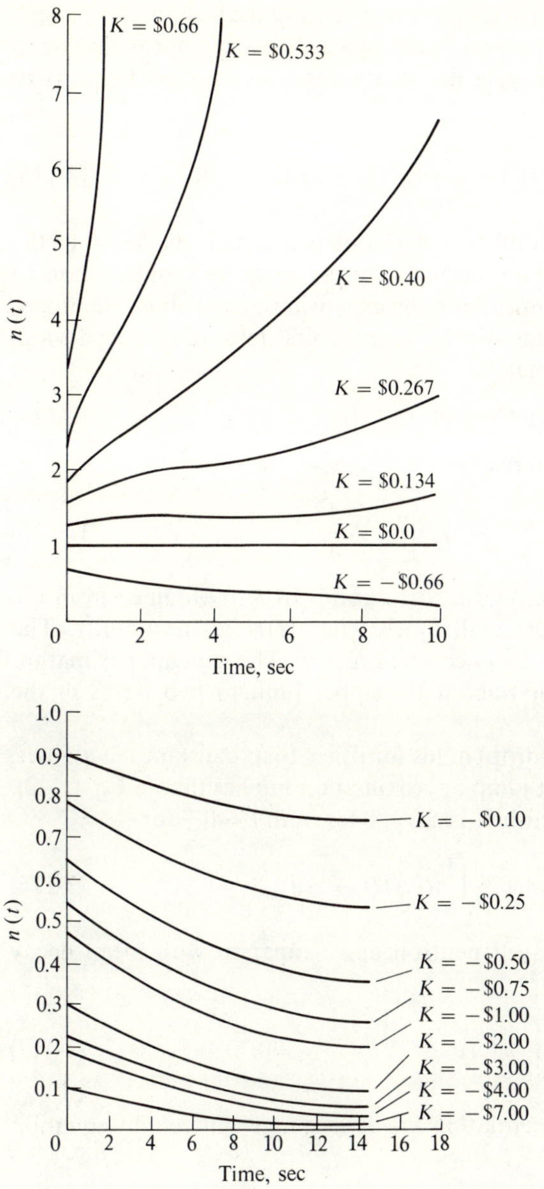

Figure 2-13 Representative curves of $n(t)$ for positive and negative constant (step function) reactivity, $\Lambda = 10^{-4}/\beta$. These curves illustrate the "prompt" jump (or drop) in neutron density occurring for very short times. They are well approximated by using the prompt jump (drop) to establish the $n(0+)$ value. Then the remainder of the curve can be approximated by the exponential corresponding to the algebraically largest root of the inhour equation.

$$n_{\text{prompt}} = (1 \mp |K_\pm|)^{-1} n_0, \qquad \text{or} \qquad |K_\pm| = \mp\left(\frac{n_0 - n_p}{n_p}\right) \tag{2-112}$$

Figure 2-13 illustrates the prompt jump approximation employed in calculating $n(t)$ for step function reactivities.

To calibrate the reactivity "worth" of control rods by quickly inserting them into a previously critical reactor (rod drop method), it is desirable to obtain an explicit expression for K, the reactivity, in terms of $n(t)$ and the reactor parameters. For accurate calibration, Eq. (2-111) will not suffice. A better approximation is obtained from Eq. (1-42) by integrating the second term on the right by parts to obtain

$$\Lambda\dot{n}(t) = (K-1)n(t) + f(t) - \int_0^t n(\tau)f'(t-\tau)\,d\tau \qquad \dot{n}(0) = 0 \tag{2-113}$$

Since $f'(t)$ has an even sharper cutoff than $f(t)$, one can exploit the fact that the sharp cutoff characteristic of $f'(t)$ implies that the preponderant contribution to the integral term in Eq. (2-113) comes from the expansion of $n(\tau)$ about the upper limit of the integral term. Replacing $n(\tau)$ by $n(t)$, the first term in this expansion, Eq. (2-113) yields to this approximation

$$\Lambda\dot{n} \cong Kn + (1-n)f(t) \tag{2-114}$$

The explicit expression (for K negative) is

$$|K| = \left(\frac{1-n}{n}\right)f(t) - \Lambda\frac{\dot{n}}{n} \tag{2-115}$$

It is seen that neglecting the second term corresponds to $K(t)$ obtained from the prompt drop equation (2-112) for small times, where $f(t)$ lies near unity. The second term corresponds to a reactivity correction term. The next approximation can be obtained by expanding $n(\tau)$ about the upper limit to two terms in the integrand of Eq. (2-113), etc.

Of course, the prompt jump (drop) holds for other than constant reactivities. In general, employing the prompt jump approximation implies that in Eq. (1-42), with $K = 0$, the left-hand side vanishes, i.e., $\dot{n} = 0$ for both $t = 0\pm$, or

$$0 = K(t)n(t) - \int_0^t n'(\tau)f(t-\tau)\,d\tau \tag{2-116}$$

For example, the one group delayed neutron approximation with mean decay constant $\bar{\lambda}$ yields

$$0 = K(t)n(t) - \int_0^t n'(\tau)\,e^{-\bar{\lambda}(t-\tau)}\,d\tau; \qquad n(0) = 1 \tag{2-117}$$

Eliminating the integral by differentiation yields a simple differential equation whose solution, for $K < 1$, is

$$n(t) = \frac{\exp\left\{\bar{\lambda}\int_0^t [K/(1-K)]\,d\tau\right\}}{1-K} \tag{2-118}$$

For example, if the ramp reactivity $K = mt < 1$, then

$$n(t) = \frac{e^{-\bar{\lambda}t}}{(1-mt)^{1+\bar{\lambda}/m}} \tag{2-119}$$

For an equilibrium reactor with $n(0-) = 1$, the new level immediately following the prompt reactivity change is $n(0+) = (1 - K)^{-1}$. The magnitude of the jump is $\Delta n(0)_{\text{prompt jump}} = n(0+) - n(0-) = K/(1 - K)$. Assuming that the prompt jump approximation $n(t) = [1 - K(t)]^{-1}$ is valid, the corresponding inverse period[1] $\alpha = \dot{n}/n = \dot{K}(t)/(K - 1)$. The magnitude of the jump in the inverse period from equilibrium $\Delta\alpha = \alpha(0+) - \alpha(0-) = -\dot{K}(0)$, since $K(0) = \alpha(0-) = 0$. Also

$$\alpha^{-1} = \frac{(K - 1)}{\dot{K}}$$

is the time to reach prompt critical ($K = 1$) at the ramp rate $\dot{K}(t)$. The prompt jump in α can be used as a reactor safety indicator by measuring the resultant $\dot{\alpha}$ as it reflects the period rate of change.

Similar to the rod drop method is the source jerk method for the determination of control rod reactivity. If the reactor level is such that the reactor can be held in equilibrium (critical) with its source, then the control rod configuration must be such as to constitute a negative reactivity of absolute magnitude equal to the source reactivity equivalent.

If the source is then suddenly removed (jerked) from the core, the reactor power level will begin to fall, since the net reactivity is now negative. Consider Eq. (1-42) which includes the constant source rate R_0, that is

$$\Lambda\dot{n} = Kn - \int_0^t n'(\tau) f(t - \tau)\, d\tau + R_0 \tag{2-120}$$

Near initial equilibrium (small times), the delayed neutron concentration is assumed to persist to good approximation, especially in the context of the prompt drop approximation. So that $f \equiv 1$ can be substituted into the above, giving

$$\Lambda\dot{n} = (K - 1)n + n_0 + R_0 \tag{2-121}$$

For negative constant reactivity, $K_0 < 0$, and with the prompt drop approximation, that is, $\Lambda\dot{n} \ll (K - 1)n$, so that at initial equilibrium, that is, at $t = 0_-$

$$0 = -(|K_0| + 1)n_0 + n_0 + R_0; \qquad K_0 n_0 = R_0 \tag{2-122}$$

Now assume that the source is suddenly removed from the core at $t = 0_+$. Then from Eq. (2-121)

$$0 = -(|K_0| + 1)n_1 + n_0 \tag{2-123}$$

So that the rod reactivity that was balanced by the source is given by

$$|K_0| = (n_0 - n_1)/n_1 \tag{2-124}$$

identical to the rod drop result of Eq. (2-112).

The principal advantage of the source jerk method is that the source mass is usually very much smaller than that of the control rods under investigation. Hence, the source sudden removal can approximate a step reactivity equivalent change much better than control rod motion, because of the usually greater mechanical inertia of the latter.

2-9 INVERSE KINETICS

As is appreciated by now, insofar as the point model is concerned, one of the major problems is the determination of the neutron density, $n(t)$, for a given reactivity $K(t)$. In general, the integral equation (1-47) must be solved, or equivalently the integro-differential equation (1-42). However, if the latter is rewritten as

$$K(t) = \left(\Lambda\dot{n} + \int_0^t n'(\tau) f(t-\tau)\, d\tau\right)\bigg/ n \tag{2-125}$$

it becomes simply an explicit expression for reactivity $K(t)$, given a neutron density $n(t)$. That is, the inverse problem of the determination of the reactivity for a given neutron density is simply obtained as a formula from Eq. (1-42).

For example, if it is desired that the equilibrium neutron density n_0 jump initially in zero time to a new larger value n_1, the required reactivity is given from Eq. (2-125), where $n = n_0 + (n_1 - n_0)H(t)$, and $\dot{n} = (n_1 - n_0)\delta(t)$

$$K(t) = \left[\Lambda(n_1 - n_0)\delta(t) + (n_1 - n_0)\int_0^t \delta(\tau) f(t-\tau)\, d\tau\right]\bigg/(n_0 + (n_1 - n_0)H(t)) \tag{2-126}$$

where $H(t)$ and $\delta(t)$ are, respectively, the heaviside step and delta function at the origin, so that

$$K(t) = \left(\frac{n_1 - n_0}{n_1}\right)(\Lambda\delta(t) + f(t)) \tag{2-127}$$

Perhaps a more straightforward way to appreciate Eq. (2-127) is to obtain the reactivity using the integral equation (1-47) written for this case, for $t \geq 0_+$, as

$$n_1 = n_0 + n_1 \int_0^t G(t-\tau) K(\tau)\, d\tau \tag{2-128}$$

which, on using the Laplace transform, immediately gives the above expression for the reactivity $K(t)$. In general, either the integro-differential or integral equation can be used to determine reactivity herein, whichever is more expedient.

A second illustration is to determine the required reactivity for a ramp increase of the neutron density, that is, $n = n_0 + mt$. Inserting this into Eq. (2-125) gives

$$K(t) = m\left(\Lambda + \int_0^t f(\tau)\, d\tau\right)\bigg/(n_0 + mt) \tag{2-129}$$

A third illustration is to determine the reactivity required for a periodic neutron density given by $n = n_0(1 + m \sin \omega_0 t)$, a sinusoidal variation of modulation index m, superposed on an equilibrium level n_0. Rewriting Eq. (2-125) with this neutron density results in

$$K(t) = m\omega_0\left(\Lambda \cos \omega_0 t + \int_0^t \cos \omega_0(t-\tau) f(\tau)\, d\tau\right)\bigg/(1 + m \sin \omega_0 t) \tag{2-130}$$

This is revised to a more compact form through the use of the Laplace transform of the above, which is

$$\mathscr{L}(Kn) = \frac{m\omega_0}{(s^2 + \omega_0^2)\bar{G}(s)} \tag{2-131}$$

where the Laplace transform of the integral equation kernel is $\bar{G}(s) = [s(\Lambda + f(s)]^{-1}$. Since $\bar{G}(s) = |\bar{G}(s)| \exp i\phi(s)$, where $\bar{G}(s)$ is the modulus and $\phi(s)$ is the phase of $\bar{G}(s)$ written in polar form, the above becomes, using the Laplace transform inversion integral

$$K(t)n(t) = \frac{m\omega_0}{2\pi i} \int_{c-i\infty}^{c+i\infty} e^{st}\, ds/\bar{G}(s)(s^2 + \omega_0^2) \tag{2-132}$$

For no singularities of the integrand except on the imaginary axis, letting $s = i\omega$ in the above with $c = 0$, results in the real (fourier transform inversion) integral

$$K(t)n(t) = \frac{m\omega_0}{2\pi} \int_{-\infty}^{+\infty} e^{i\omega t}\, d\omega/\bar{G}(i\omega)(\omega_0^2 - \omega^2) \tag{2-133}$$

Evaluating the integral, with poles $\omega = \pm\omega_0$ results in

$$K(t) = \frac{m \sin(\omega_0 t - \phi)}{|\bar{G}(\omega_0)|(1 + m \sin \omega_0 t)} \tag{2-134}$$

Although the above reactivity is seen to be periodic, it is not a pure sinusoid. The relationship between the periodicities of the reactivity $K(t)$ and neutron density $n(t)$, will be discussed in Sec. (2-10) where periodic reactivities are examined in general.

2-10 PERIODIC REACTIVITIES

Periodic reactivity variation is quite important in the study of reactor reactivity excitation for a number of reasons. One principal reason is that, since the reactor and control system comprise a feedback loop, such a system can exhibit oscillatory behavior. As a matter of fact, the reactor *per se* comprises an inherent feedback system because of the structural and physical reactivities, both positive and negative, that it possesses. With the advent of more exotic reactor designs and applications, oscillatory behavior will become a ubiquitous aspect of reactor kinetics.

First, periodic reactivity solutions are obtained for a number of reactivity amplitudes by digital computation using the discrete version of the integral Eq. (1-47) for $K = K_0 \sin \omega_0 t$ and $K = K_0 \cos \omega_0 t$. That is

$$n_r = 1 + K_0\Delta \sum_{m=0}^{r-1} G_{r-m} \begin{Bmatrix} \sin m\omega_0\Delta \\ \cos m\omega_0\Delta \end{Bmatrix} n_m, \qquad n_0 = 1 \qquad r = 1, 2, \ldots \tag{2-135}$$

To obtain solutions in short computer running times, recurrence relations are used in these programs whenever feasible. These require only simple addition and replacement computing operations which are completed in a very few microseconds on present-day large-scale machines. For example, the sine and cosine functions are related by the two-term recurrence relation

$$\cos(m\omega_0\Delta) = 2\cos\omega_0\Delta\cos(m-1)\omega_0\Delta - \cos(m-2)\omega_0\Delta \qquad m = 3, 4, \ldots \tag{2-136}$$

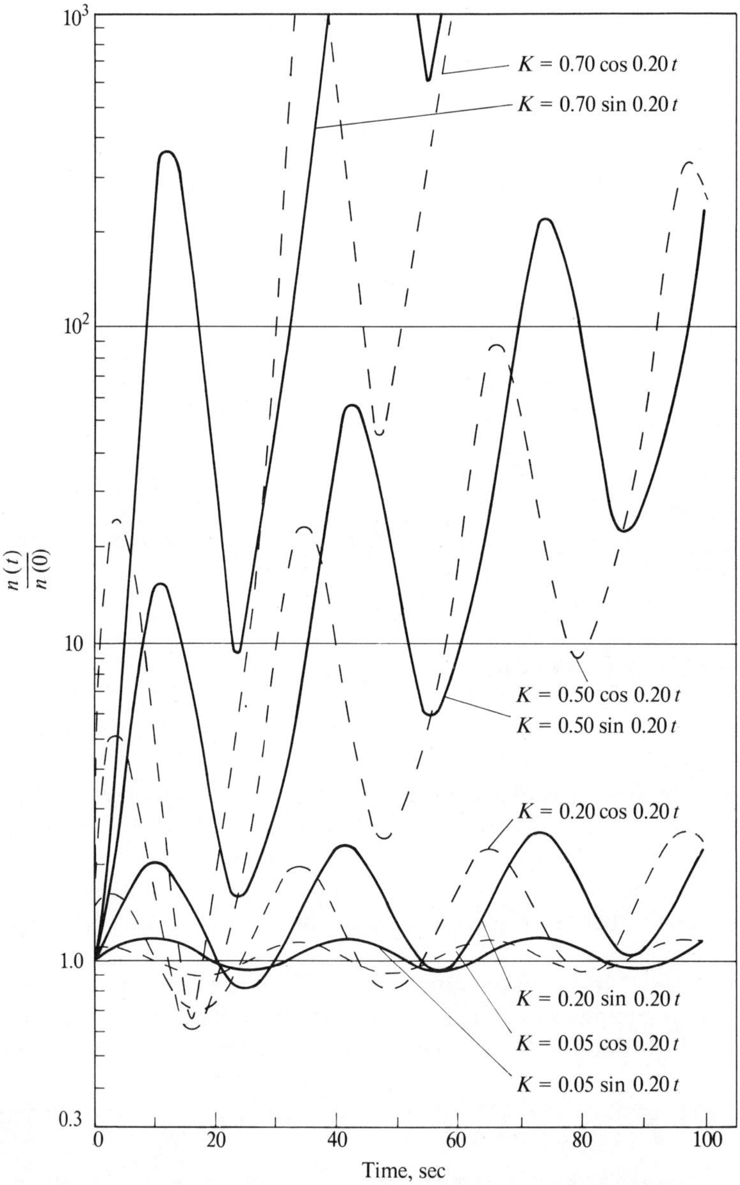

Figure 2-14 Sinusoidal reactivity variations for various reactivity amplitudes.

$$\sin(m\omega_0\Delta) = 2\cos\omega_0\Delta \sin(m-1)\omega_0\Delta - \sin(m-2)\omega_0\Delta \qquad m = 3, 4, \ldots \tag{2-137}$$

with appropriate initial conditions.

Similarly, the recurrence relation for the neutron density at the rth time increment can be written as

$$n_r \to n_r + K_0 \Delta G_{r-m} \begin{Bmatrix} \sin m\omega_0\Delta \\ \cos m\omega_0\Delta \end{Bmatrix} n_m \qquad n_0 = 1 \qquad 0 \le m < r \qquad r = 1, 2, \ldots \tag{2-138}$$

where the arrow implies that n_r is computed by summing iteratively on m.

With $\Delta = 0.1$, 1000 incremental values for n_r are obtained in a very short time on a large-scale digital computer per each value of reactivity amplitude K_0. The results are plotted in Fig. 2-14.

It is seen from Fig. 2-14 that for very small reactivity amplitudes, the neutron density corresponds closely to the response of a linear system to sinusoidal forcing. This will be discussed later in connection with the derivation of the reactor transfer function.

For all but the very small reactivity amplitudes, it is also seen that the inherent nonlinear behavior of the reactor is evident. Even for a reactivity amplitude of only 5 cents, the neutron density is seen to be asymmetric about equilibrium, being slightly "top heavy." It can be shown in general that an oscillating reactivity about a zero mean implies a diverging power. Smith[18] and Hetrick[19] obtained the approximate negative reactivity bias needed to maintain a non-divergent power using the kinetics description consisting of the prompt-jump approximation and one delayed neutron group.

For reactivity amplitudes much above 50 cents, the neutron density increases sufficiently fast so that delayed neutron effects can be neglected in approximation for large $\omega_0 > 1$. This results in a simple exponential solution. From Eq. (2-2) for a cosine reactivity variation

$$n(t) \cong \exp\left(\frac{K_0}{\Lambda}\right) \int_0^t \cos\omega_0 x \, dx \tag{2-139}$$

or

$$n(t) \cong \exp\left(\frac{K_0}{\omega_0 \Lambda}\right) \sin\omega_0 t \tag{2-140}$$

which, however, only holds for small times following equilibrium since a bounded neutron density always ensues when delayed neutrons are neglected. The curves are easy to sketch, since the extrema occur at $\omega_0 t_0 = (2r+1)\pi/2$, $r = 0, 1, \ldots$, where $\dot{n}(t_0) = 0$. Maxima occur at even r, while minima occur at odd r; that is

$$n\left[\frac{(2r+1)\pi}{2\omega_0}\right] = \exp(-1)^r \frac{K_0}{\Lambda\omega_0} \tag{2-141}$$

It should be noted that if K_0 is sufficiently large, the reactor becomes prompt critical during the very first undulation of the neutron density. However, neglecting delayed neutrons produces a periodic neutron density regardless of the reactivity amplitude. This reflects an unconditionally bound neutron density which is, of course, erroneous. This is an interesting "counterexample" to the stability arguments discussed in Chap. 4, which assert that the inclusion of delayed neutrons increases stability margins. Here, neglecting delayed neutrons gives stable (bounded) neutron densities, while considering them gives unstable (unbounded) neutron densities, as seen from Fig. 2-14. How delayed neutrons influence reactor stability is discussed in Chap. 4.

For small amplitude sinusoidal reactivity and small ω_0 (so that the reactivity changes only slowly), the approximation can be made that $n'(\tau) = n'(t)$ in the integrand of Eq. (1-42), ($R = 0$), resulting in

$$\Lambda\dot{n}(t) = K_0 n(t) \sin \omega_0 t - \dot{n}(t) \int_0^{\infty} f(x)\, dx \tag{2-142}$$

A further approximation has been made by extending the upper limit of the integral, since $f(t)$ cuts off quickly, except at very short times. Since $\int_0^\infty f(x)\, dx = \bar{\tau}$, the mean delayed neutron lifetime, Eq. (2-142), becomes, with $\bar{\tau} \gg \Lambda$

$$\bar{\tau}\dot{n} \cong K_0 n(t) \sin \omega_0 t \tag{2-143}$$

also yielding an exponential solution, but with a smaller "decay constant" than Eq. (2-140),

$$n = \exp\left(\frac{K_0}{\omega_0 \bar{\tau}}\right)(1 - \cos \omega_0 t) \tag{2-144}$$

This solution depicts the fact that the time behavior is governed by $\bar{\tau}$ when the reactivity varies slowly. For such reactivities the Wentzel–Kramers–Brillouin method is also applicable.

For periodically varying reactivities, the WKB method is straightforward, especially if no turning points occur, as discussed in Sec. 2-6. This is the case if the amplitude K_0 of the periodic reactivity is small ($K_0 <$ \$1.00). Consider a

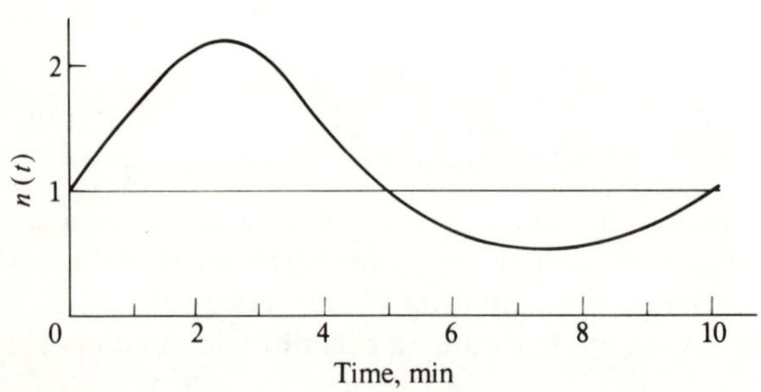

Figure 2-15 Neutron density for periodic reactivity using WKB method with equilibrium conditions. $K = K_0 \sin wt$, $K_0 = 0.10$, $w = 0.01$, $A = 0.10$.

slowly varying periodic reactivity $K = K_0 \sin \omega t$ with $K_0 = 0.10$ and $\omega = 0.01$. Assume that one group of delayed neutrons with mean decay constant $\bar{\lambda}$ adequately represents the delayed neutron behavior. Then the kinetics equation is given by Eq. (2-91). For the set of parameters $\bar{\lambda} = \Lambda = K_0 = 0.10$ and $\omega = 0.01$

$$q(t) \cong \frac{1}{2\Lambda} [(1 - K)^2 + 2\Lambda\dot{K} + 4\Lambda\bar{\lambda}K]^{1/2} \qquad (2\text{-}145)$$

where $\Lambda\bar{\lambda} \ll 1$ and is therefore neglected. The first term under the radical dominates the second and third for the parameters selected. The first two terms of the binomial expansion should then suffice to give

$$q(t) \cong \frac{1}{2\Lambda}(1 - K_0 \sin \omega t) - \frac{d}{dt} \ln (1 - K_0 \sin \omega t) + \bar{\lambda} K_0 \sin \omega t \qquad (2\text{-}146)$$

From Sec. 2-6 the neutron density is given by the linear combination

$$n(t) = q^{-1/2}\left\{A\left[\exp\left(+\int_0^t q\,d\tau\right)\right] + B\left[\exp\left(-\int_0^t q\,d\tau\right)\right]\right\} \cdot \exp\left(-\frac{1}{2\Lambda}\int_0^t (1 - K)\,d\tau\right) \qquad (2\text{-}147)$$

or

$$n(t) = q^{-1/2}[A(1 - K_0 \sin \omega t)^{-1/2} + B(1 - K_0 \sin \omega t)^{1/2}] \exp\left(\frac{2K_0}{\Lambda\omega} \sin^2 \frac{\omega t}{2} - \frac{t}{\Lambda}\right) \qquad (2\text{-}148)$$

Another method for determining the neutron density for certain periodic reactivities is to consider Eq. (1-42) rewritten as

$$\Lambda\dot{n} = Kn - \int_0^t n(\tau)\left[\frac{d}{d\tau} \log n(\tau)\right] f(t - \tau)\,d\tau \qquad (2\text{-}149)$$

Let the reactivity be given by $K = K_0 \sin \omega t$ with ω not small. This means that ω is sufficiently large that the inverse period, $d/dt \log n(t)$, changes rapidly compared with $n(t)$ itself. With the rapid cutoff characteristics of $f(t)$, $n(\tau)$ is assumed to contribute to the integral substantially only for τ near t, since it is slowly varying. Then, factoring $n(\tau)$ from the integral gives to good approximation

$$\Lambda\dot{n} \cong Kn - n(t) \int_0^t \left[\frac{\dot{n}(\tau)}{n(\tau)}\right] f(t - \tau)\,d\tau \qquad (2\text{-}150)$$

This approximation holds for general reactivities as well, when the assumption that n is slowly varying compared to $\dot{n}/n$ can be made. With $Q = \log n$ inserted

$$\Lambda\dot{Q} = K - \int_0^t Q'(\tau) f(t - \tau)\,d\tau \qquad Q(0) = 0 \qquad (2\text{-}151)$$

This is easily solved by using Laplace transforms to obtain

$$Q(t) = \int_0^t G(t - \tau)K(\tau)\,d\tau \tag{2-152}$$

so that

$$n(t) \cong \exp\left[\int_0^t G(t - \tau)K(\tau)\,d\tau\right] \tag{2-153}$$

For the above periodic reactivity, where $G(t) = \sum_{r=0}^{6} A_r e^{-s_r t}$

$$n(t) \cong \exp\left[K_0 \sum_{r=0}^{6} \frac{\omega A_r}{\omega^2 + s_r^2}\left(1 - \cos\omega t + \frac{s_r}{\omega}\sin\omega t\right)\right] \tag{2-154}$$

As ω gets larger, the effects of the delayed neutrons are lessened because of the high-frequency reactivity variations compared with the delayed neutron periods. For large ω, the above reduces to

$$n(t) \sim \exp\left(\frac{2K_0}{\omega\Lambda}\sin^2\frac{\omega t}{2}\right) \tag{2-155}$$

which is equivalent to beginning with the kinetics equations, with sinusoidal reactivity and neglecting delayed neutrons. For very large ω, the neutron density above approaches unity—its equilibrium value—as it should, since the reactivity corresponds more closely to its average value of zero in terms of its effect on the reactor.

2-11 STARTUP AND SHUTDOWN

When a reactor has been shut down for a long time, or is to be started for the first time, besides there being an essentially zero neutron population there are no delayed neutron emitters from which neutrons can be derived. To start such a reactor, a neutron source must be introduced to provide an original neutron population. Then a startup procedure to rated power, by withdrawing control rods, is used that is safe and does not consume disproportionate amounts of time. If the rod withdrawal is too rapid, the reactor can pass through critical and proceed on a dangerous supercritical fast period. It is usually desired to bring the reactor up to near critical rapidly and then expand the period so that criticality is passed through on a slow safe period facilitating easy control, especially for the human operator. Such a period is maintained until the desired operating power is reached, at which time the reactivity, or inverse period, is reduced to essentially zero. That is, the reactor becomes, and is maintained, critical at the desired power. Equation (1-42), the startup form of the aggregate kinetics equation, is rewritten here for convenience.

$$\Lambda\dot{n} = Kn - \int_0^t n'(\tau)f(t - \tau)\,d\tau + \Lambda R \qquad n(0) = 0 \tag{2-156}$$

where a constant neutron source of strength R neutrons per second is introduced initially.

If the startup reactivity rate is not large, so that both $K(t)$ and $n(t)$ are slowly increasing, then the sharp cutoff characteristic of $f(t)$, the delayed neutron decay function, can be exploited. Again, this is done by assuming that $n'(\tau) \approx \dot{n}(t)$ within the integral in Eq. (2-156), and also here

$$\int_0^t f(t-\tau)\,d\tau \cong \int_0^\infty f(\tau)\,d\tau = \bar{\tau}$$

This yields a reduction of Eq. (2-156) to

$$\bar{\tau}\dot{n}(t) \cong K(t)n(t) + \Lambda R \qquad n(0) = 0 \tag{2-157}$$

where $\Lambda \ll \bar{\tau}$, on the left-hand side of Eq. (2-157), which is the usual situation, so that Λ is neglected. As mentioned previously, a "neutronless" reactor is very subcritical, with the magnitude of the subcritical reactivity being reckoned in tens of dollars. For example, if the total reactivity

$$K(t) = mt - K_0 \tag{2-158}$$

is made up of a linearly increasing startup reactivity mt and an initial subcritical reactivity $|K_0|$, then $m \ll |K_0|$ should hold for safe startup procedures. $|K_0|$ could be about \$20, while $m \cong 10^{-3}$ dollars per second. If this is the startup procedure for a reactor, the corresponding solution to Eq. (2-157) is, for a constant source R and magnitude K_0

$$\frac{n(t)}{\Lambda R} = \left(\frac{2}{m\bar{\tau}}\right)^{1/2} \left\{\exp\left[\frac{m}{2\bar{\tau}}\left(t - \frac{K_0}{m}\right)^2\right]\right\}\left[\operatorname{erf}\left(\frac{m}{2\bar{\tau}}\right)^{1/2}\left(t - \frac{K_0}{m}\right) + \operatorname{erf}\frac{K_0}{(2\bar{\tau}m)^{1/2}}\right] \tag{2-159}$$

At criticality, $K = 0$, so that it takes $t_c = K_0/m$ sec to bring the reactor to initial criticality. From this point on, the reactor can be made slightly supercritical to bring it up to design power, where it is again made critical. At initial criticality the neutron level becomes, from Eq. (2-159) at t_c

$$\frac{n(t_c)}{\Lambda R} = \left(\frac{2}{m\bar{\tau}}\right)^{1/2} \operatorname{erf}\frac{K_0}{(2m\bar{\tau})^{1/2}} \approx \left(\frac{\pi}{m\bar{\tau}}\right)^{1/2} \tag{2-160}$$

since $K_0/(2m\bar{\tau})^{1/2}$ is very large so that $\operatorname{erf} K_0/(2m\bar{\tau})^{1/2} \cong \operatorname{erf}(\infty) = (\pi/2)^{1/2}$.

As seen in such a reactor startup situation, the delayed neutrons are taken into account in a gross manner. For the case of large reactivity insertions during a startup, delayed neutrons can be neglected. This can be construed in two ways. First, to actually assume that the phenomenon of delayed neutrons is nonexistent, where the reactor level is changing rapidly according to prompt neutron influences only, the delayed neutron decay function $f(t) \equiv 0$. If this is inserted in the startup integral equation (1-47), the corresponding kernel is given by $G(t) = 1/\Lambda$, since $\bar{G}(s) = 1/s\Lambda$. From Eq. (1-47), one then obtains

$$\Lambda\dot{n}(t) = K(t)n(t) + \Lambda R \tag{2-161}$$

which is formally identical to Eq. (2-157) except that now the prompt neutron lifetime dominates the neutron density rates.

The second manner of construing the delayed neutrons is to consider that for short times following the introduction of a reactivity disturbance, the delayed neutrons remain at their equilibrium concentration if such is applicable. This corresponds to† $f(t) \equiv 1$. Therefore, $\bar{f}(s) = 1/s$ with $\bar{G}(s) = 1/(s\Lambda + 1)$ and $G(t) = e^{-t/\Lambda}/\Lambda$, so that from Eq. (1-47)

$$\Lambda \dot{n}(t) = [K(t) - 1]n(t) + \Lambda R \tag{2-162}$$

When $K \gg 1$, Eqs (2-161) and (2-162) coincide. This would be the case where the reactivity is many dollars over prompt critical. For neutron levels above critical, the neutron population dwarfs the source contribution, so that R can be neglected. Then for large reactivities at supercritical neutron densities, Eqs (2-161) and (2-162) both are given by

$$\Lambda \dot{n}(t) \sim K(t)n(t) \tag{2-163}$$

Now consider a nascent reactor with a source of R neutrons per second introduced initially. The subcritical reactivity corresponds to a particular effective multiplication factor $0 < k_{\text{eff}} < 1$. For the first prompt neutron generation time interval l, after the source is "turned on," it provides a neutron population Rl which produces $k_{\text{eff}}Rl$ progeny. During the second prompt generation time interval, the $k_{\text{eff}}Rl$ neutrons produce in turn $k_{\text{eff}}^2 Rl$ progeny simultaneously with the production of additional $k_{\text{eff}}Rl$ neutrons from the source population Rl provided in the second prompt generation time interval. After two such intervals, the total subcritical population has increased to $Rl(1 + k_{\text{eff}} + k_{\text{eff}}^2)$. After many prompt generations, the original source population has been multiplied by $1 + k_{\text{eff}} + k_{\text{eff}}^2 + \cdots = (1 - k_{\text{eff}})^{-1}$, called the *subcritical multiplication.* The subcritical population becomes $n_{\text{sub}} = Rl/(1 - k_{\text{eff}}) = R\Lambda/|K|$; $K < 0$ is the subcritical reactivity in dollars. A subcritical period T_{sub} can be obtained from $T_{\text{sub}} = n_{\text{sub}}/\dot{n}_{\text{sub}} = |K|/\dot{K}$. This relation can also be obtained by examining the kinetics equation (1-42) rewritten for the subcritical situation, where R and $\dot{n}(t)$ have the same units and $K < 0$, as

$$\Lambda \dot{n}(t) = K(t)n(t) - \int_0^t n'(\tau) f(t - \tau)\, d\tau + \Lambda R \tag{2-164}$$

When the reactor becomes more and more subcritical after a long time, terms involving $\dot{n}(t)$ are small compared with the remaining terms, so that they are neglected above and we have simply

$$n_{\text{sub}} \sim \frac{R\Lambda}{K(t)} \tag{2-165}$$

† However, for the startup case, which corresponds to a nascent reactor, their equilibrium concentration is zero.

An inequality involving the period (or inverse period) can be obtained by integrating the integral in Eq. (2-164) by parts to obtain, now for $K > 0$,

$$\Lambda \dot{n} = (K - 1)n(t) + f(t) + \int_0^t n(\tau)\,|\,f'(t-\tau)\,|\,d\tau + \Lambda R \tag{2-166}$$

Equations (2-164) and (2-166), respectively, give $\Lambda \dot{n} \le Kn$ and $\Lambda \dot{n} \ge (K - 1)n$, so that

$$\frac{K-1}{\Lambda} \le \frac{\dot{n}}{n} \le \frac{K}{\Lambda} \tag{2-167}$$

Then for large positive reactivity, the inequality narrows to

$$\alpha \equiv \frac{\dot{n}}{n} = \frac{K}{\Lambda} \tag{2-168}$$

which is the supercritical inverse period. Therefore, both an upper (supercritical) bound for the period (Λ/K) as well as a lower bound $(|K|/\dot{K})$ have been obtained. Figure 2-16 depicts the manner by which the bounds delimit the neutron period as a function of a ramp reactivity increase of 1.6 cents per second

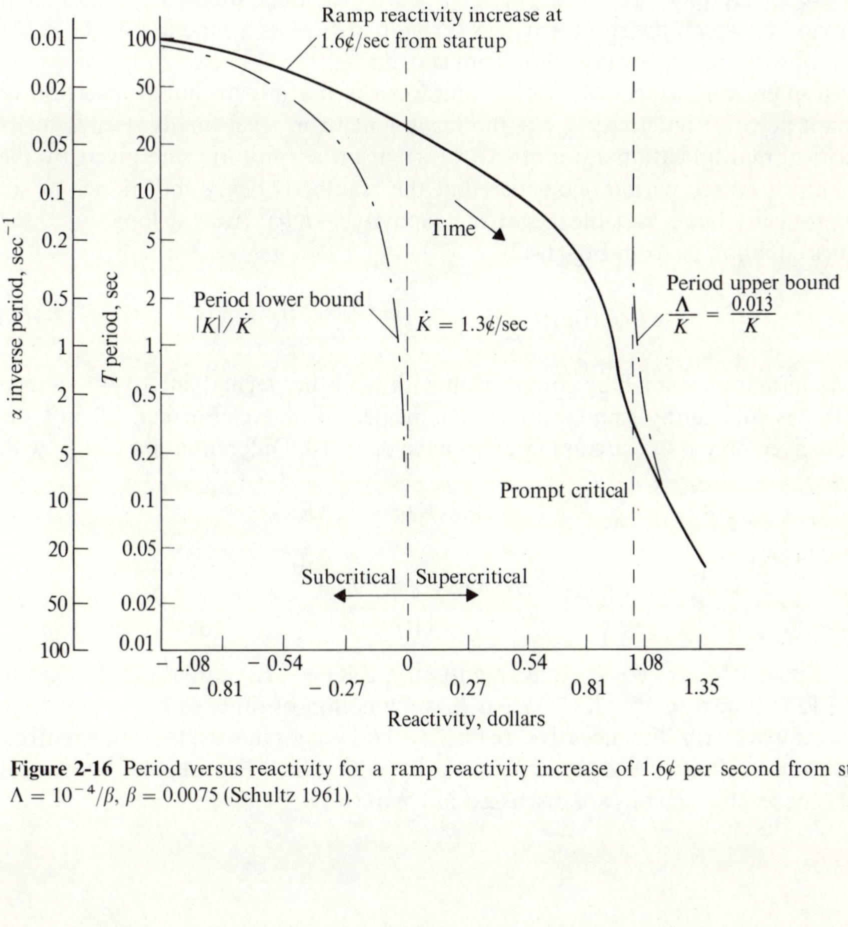

Figure 2-16 Period versus reactivity for a ramp reactivity increase of 1.6¢ per second from startup, $\Lambda = 10^{-4}/\beta$, $\beta = 0.0075$ (Schultz 1961).

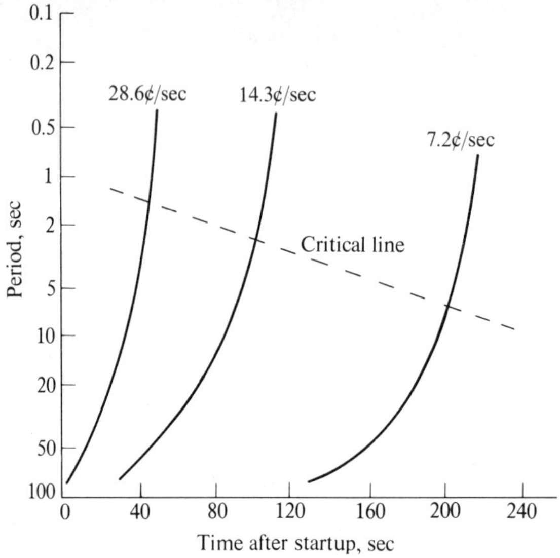

Figure 2-17 Period as a function of time for various ramp reactivities from startup.

from the startup ($n = 0$) state. Figure 2-17 shows the large differences attained in the period by which the reactor passes through critical as a function of the ramp rate of positive reactivity insertion from startup.

An interesting aspect of reactor shutdown is that for an initial insertion of constant negative reactivity $-K_0$, the reactor neutron level finally decays to its subcritical multiplication amount ($\Lambda R/|K_0|$) at an asymptotic rate given by the well-known 80 sec period. Consider that the reactor is being shut down by an asymptotically large variable negative reactivity $-K(t)$. After a long time, the kinetics equation is, from Eq. (1-42),

$$0 \cong -K(t)n(t) - a_1 \int_0^t n'(\tau)\, e^{-\lambda_1(t-\tau)}\, d\tau + \Lambda R \tag{2-169}$$

The $\Lambda \dot{n}$ term is now small by comparison with the other terms, since it is assumed that $\dot{n}(t)$ is sufficiently small, and so it is neglected above. Further, all but the longest-lived precursors are assumed to have decayed. Differentiation yields, with $|K| \gg a_1$

$$K\dot{n} + (\dot{K} + \lambda_1 K)n \sim \lambda_1 \Lambda R \tag{2-170}$$

with solution

$$n \sim \frac{\Lambda R + c\, e^{-\lambda_1 t}}{K(t)} \tag{2-171}$$

where c is an arbitrary constant. As mentioned, if $K = -K_0$ (initial step function), after a long time $n \approx \Lambda R/|K_0|$. When K is a function of time, so is $n \approx \Lambda R/K(t)$. For example, with the negative reactivity $|K| = mt$, the asymptotic neutron population ultimately decays hyperbolically with time. The detailed statistical aspects of reactor startup are discussed in Chap. 10.

PROBLEMS

2-1 Derive the kinetics equation for the case when the delayed neutron emitters remain at their equilibrium concentration. Solve it for linearly varying reactivity (ramp reactivity) of $K = mt$, for $m = 0.001, 0.005, 0.01$, and plot the results.

2-2 Use Laplace transforms to obtain the integral equation formulation of the kinetics equations (1-45) directly from Eqs (1-39) and (1-40).

2-3 Obtain expressions for $G(0)$ and $G(\infty)$ from $\bar{G}(s)$ by using the appropriate initial- and final-value theorems of Laplace transform theory.

2-4 Derive the integrodifferential form of the kinetics equation (1-42) from the kinetics equations (1-39) and (1-40), in which the all thermal neutron approximation has not been made.

2-5 Write out the first few members of the sequence of approximate solutions of the recurrence relation, Eq. (1-53), for constant reactivity.

2-6 Show that the solution to Eq. (2-58) is indeed the asymptotic neutron density for a ramp reactivity increase from startup by substituting into the kinetics equation (2-46) and letting time grow without bound.

2-7 Discuss how to obtain the WKB-type solution to the kinetics equations

$$n(t) \cong \exp\left[\int_0^t G(t-\tau)K(\tau)\,d\tau\right]$$

from the iteration (Neumann series) of the kinetics integral equation

$$n(t) = 1 + \int_0^t G(t-\tau)K(\tau)n(\tau)\,d\tau$$

Determine the restrictions on $G(t)$ and $K(t)$ required to make the iteration yield the above exponential approximation to the neutron density.

2-8 For a general periodic reactivity variation, let the reactivity be $K = K_0 \sin \omega_0 t$ in the integral equation formulation of the kinetics equations. Derive the corresponding difference equation for the transform variable $\bar{n}(s)$. Compare the first two terms of its solution by the iteration procedure with the linearized equation

$$\varepsilon(t) = K_0 \int_0^t G(t-\tau)\sin \omega_0 \tau\,d\tau$$

and show the equivalence. Why does not the substitution of $K = K_0 e^{i\omega_0 t}$, instead of the above, into the integral equation give the solution for general sinusoidal reactivities (in the same way that the forcing function of a linear differential equation can be written as a complex exponential)?

2-9 Show by considering the cutoff properties of $G(t)$ that

$$\left[1 - K_0 \int_0^t G(t-\tau)\sin \omega_0 \tau\,d\tau\right]^{-1}$$

provides upper and lower bounds for the neutron density.

2-10 The source multiplication method,[20] an alternative for measuring subcritical reactivities near critical, involves an experimental determination of the subcritical multiplication, M, and the asymptotic measured linear rise rate, R, of neutron density following the initial introduction of a source ΛS into the critical reactor. From the integral equation (1-45) kinetics formulation with the introduction of such a source, show that this asymptotic rise is given by $n \sim n_0 + \Lambda St/(\Lambda + \bar{\tau})$, and that the corresponding sought for reactivity in dollars is given by $K = R/M(\Lambda + \bar{\tau})$, where the analytical expression for the linear rise rate $R = \Lambda S/(\Lambda + \bar{\tau})$.

2-11 Suppose that the reactivity can be well approximated by a series of equal magnitude spikes that can be represented by periodic delta functions, that is, $K = K_0 \sum_m \delta(t - mt_0)$ where t_0 is the period.† Using the kinetics integral equation, show that the resulting neutron density satisfies a recurrence relation identical to that obtained for approximating the integral equation as given in Sec. 1-8, but for constant reactivity K_0. Further, that the corresponding exact constant reactivity solutions obtained in Sec. 1-8 are envelopes of the above spike reactivity neutron density.

2-12 Using the integro-differential form of the kinetics equations given in Prob. 1-9, show that if the neutron density is periodic, that is, $n(t - T) = n(t)$, with period T, then the "driving" reactivity must be periodic as well. Is the converse true?

2-13 For the periodic reactivity $K(t) = K_0 + K_1 \sin \omega_0 t$, show (*a*) by substitution into the kinetics integral equation (1-47) that the neutron density now satisfies a modified version, namely, $n(t) = n_c(t) + K_1 \int_0^t Y(t - \tau)n(\tau) \sin \omega_0 \tau \, d\tau$, where $n_c(t)$ is the neutron density for a constant reactivity K_0, and the Laplace transform of $Y(t)$ is $\bar{Y}(s) = [s(\Lambda + \bar{f}(s)) - K_0]^{-1}$. Also show (*b*), since $\bar{Y}(s)$ and $\bar{G}(s)$ possess the same roots $\{s_r\}$ in their partial fraction expansion, that when $n(t)$ is replaced by n_0 in the right-hand side of the above integral equation, the resulting neutron density is then given by

$$n(t)/n_0 \simeq \sum_r \left(n_r + \frac{K_1 \omega_0}{s_r^2 + \omega_0^2} Y_r \right) e^{s_r t} + K_1 \sum_r \frac{Y_r}{\sqrt{s_r^2 + \omega_0^2}} \sin\left(\omega_0 t + \tan^{-1} \frac{\omega_0}{s_r} \right)$$

where n_r and Y_r are the corresponding (Laplace transform) expansion coefficients of n_c and Y, both expressed as linear combinations of exponentials in time.

What is (*c*) the physical interpretation when n is replaced by n_0 in the above equation? Show (*d*) that after a long time, the asymptotic relationship for $n(t)$ is given by

$$n(t)/n_0 \sim \left(1 + \frac{K_1 \omega_0 \bar{\tau}}{K_0^2 + (\omega_0 \bar{\tau})^2} \right) e^{K_0 t/\bar{\tau}} + \frac{K_1}{\sqrt{K_0^2 + (\omega_0 \bar{\tau})^2}} \sin\left(\omega_0 t + \tan^{-1} \omega_0 \bar{\tau}/K_0 \right)$$

where $\bar{\tau} = \sum_{i=1}^{6} a_i/\lambda_i$, the mean delayed neutron decay time. Find (*e*), for the three cases of constant reactivity "bias" $K_0 > 0$, $K_0 = 0$, $K_0 < 0$, the corresponding parameter relationships involving K_0, K_1, etc., to insure that only periodic (and not increasing) neutron density ensues asymptotically.

2-14 Consider a type of breeder consisting of fissile and fissionable material and controlled by a pulsed neutron source at its center. The system is held very subcritical corresponding to a negative reactivity $-K_0$. Using the reactor kinetic integral equation (1-46), where the pulsed source is represented by a series of periodic delta functions in time, of source strength R_0 and period T_0, derive the corresponding relationship for the neutron density as

$$n(t) = \int_0^t G(t - \tau) \left[-K_0 n + \Lambda R_0 \sum_{r=0}^{\infty} \delta(\tau - rT_0) \right] d\tau$$

(*a*) Use the Laplace transform on the above to obtain $\bar{n}(s) = \Lambda R_0/(s(\Lambda + \bar{f}(s)) + K_0)(1 - e^{-sT_0})$ and take the proper limit to yield the asymptotic neutron density n_∞, after a long time as $n_\infty = \Lambda R_0/K_0 T_0$. If the breeder subcritical multiplication is M, show that $M > T_0/l$ in order that the system multiply the original neutron source strength R_0.

(*b*) Show that for times long after the pulsed source is turned on, that the above neutron density can be approximated by

$$n(t) \sim \frac{\Lambda R_0}{K_0 T_0} (1 - e^{-K_0 t/\bar{\tau}})$$

where $\bar{\tau} = \sum_{i=1}^{6} a_i/\lambda_i$ is the mean delayed neutron period.

Hint: Invert the above Laplace transform for small values of the transform variable s.

† Such as the reactivity in a periodically pulsed dragon assembly, cf. D. L. Hetrick, "Dynamics of Nuclear Reactors," University of Chicago Press, 1971, pp. 51, 52.

REFERENCES

1. H. Soodak (ed.): "Reactor Handbook, Part A, Physics," Interscience, New York, 1962, chap. 5.
2. I. Sneddon: "Fourier Transforms," Oxford, New York, 1955, appendix IV.
3. H. Hurwitz: Derivation and Integration of the Pile Kinetics Equations, *Nucleonics*, **5**(1): 61 (1949). (*AECD* 2438, January, 1949.)
4. G. R. Keepin and T. F. Wimett: Reactor Kinetics Functions: A New Evaluation, *Nucleonics*, **16**(10): 86–90 (1958).
5. G. R. Keepin, T. F. Wimett, and R. E. Zeigler: Delayed Neutrons from Fissionable Isotopes, *Phys. Rev.*, **107**: 1044 (1957).
6. H. B. Smets: Reactor Dynamics at Low Power, *Proc. U.N. Intern. Conf. Peaceful Uses At. Energy, 2nd, Geneva*, 1958, **11**: 237–244 (1958).
7. A. Kirchenmayer: Delayed Neutron Groups Number Reduction, *Nucl. Sci. Eng.*, **8**: 720–721 (1960).
8. G. R. Keepin and C. W. Cox: General Solution of Reactor Kinetics Equations, *Nucl. Sci. Eng.*, **8**: 670–690 (1960).
9. M. Ash: Solutions of the Reactor Kinetics Equations for Time Varying Reactivities, *J. Appl. Phys.*, **27**: 1030–1031 (1956).
10. S. Glasstone and A. Sesonske: Nuclear Reactor Engineering," Van Nostrand, Princeton, N.J., 1963, p. 246.
11. B. J. Toppel: Sources of Error in Reactivity Determination by Means of Asymptotic Period Measurements, *Nucl. Sci. Eng.*, **5**: 88–98 (1959).
12. E. A. Coddington and N. Levenson: "Theory of Ordinary Differential Equations," McGraw-Hill, 1955, p. 172.
13. E. L. Ince: "Ordinary Differential Equations," Dover, New York, 1956, chaps. 18, 19. (Paperback.)
14. P. M. Morse and H. Feshbach: "Methods of Theoretical Physics," vol. I, McGraw-Hill, 1953, chap. 5.
15. S. Wallach: "Solutions of the Pile Kinetics Equations when the Reactivity is a Linear Function of Time," Westinghouse Electric Corporation *WAPD*-13, 1950.
16. Morse and Feshbach: *op cit.*, chap. 9.
17. S. Tan: Solution of Reactor Kinetics Equations by B.W.K. Approximation, *Nucleonik*, **8**: 480, 1966; Sinusoidal Analysis of a Low Power Nuclear Reactor by the Wentzel-Kramers-Brillouin Approach, *Nucl. Sci. Eng.*, **30**: 436 (1967).
18. H. P. Smith JR.: Divergence of the Mean Power Level During an Oscillation Experiment, *Nucl. Sci. Eng.*, **22**: 253 (1965).
19. D. L. Hetrick: "Dynamics of Nuclear Reactors," University of Chicago Press, 1971, pp. 77–78.
20. E. Greenspan: The Source Multiplication Method for Absolute Reactivity Determination, *J. Nucl. Energy*, **27**: 129–138 (1973).

FURTHER READING

G. E. Hansen and C. Maier: Perturbation Theory of Reactivity Coefficients for Fast Neutron Critical Systems, *Nucl. Sci. Eng.*, **8**: 532–542 (1960).

G. A. Jarvis, G. A. Linenberger, J. D. Orndoff, and H. C. Paxton: Two Plutonium Metal Critical Assemblies, *Nucl. Sci. Eng.*, **8**: 525–531 (1960).

P. Schmid: Derivation of a Basic Integral Equation of Reactor Kinetics, *Proc. U.N. Intern. Conf. Peaceful Uses At. Energy, 2nd, Geneva*, 1958, **11**: 277–288 (1958).

M. A. Schultz: "Control of Nuclear Reactors and Power Plants," 2d ed., McGraw-Hill, New York, 1961.

CHAPTER

THREE

REACTOR TRANSFER FUNCTIONS

3-1 INTRODUCTION

During the era immediately following World War II, it was realized that, as an outgrowth of using oscillatory reactivity perturbations of the neutron density to measure absorption cross sections, the reactor and control system could be considered as a feedback loop. As such, it could be analyzed under certain linearizing conditions as a servomechanism, for which the methods of analysis and synthesis already existed and were being continually improved upon.

The method of analyzing servomechanisms by using transfer function theory rests on the fact that the equations characterizing the system are linear. Because these dynamical equations—usually ordinary differential equations with constant coefficients—are linear, the superposition theorem (linear combinations of particular solutions are also solutions) holds for their homogeneous counterparts. Assuming the reader has some familiarity with the main ideas of transfer functions, we need say only that in Laplace transform space the output (response) function of a linear system is given by the product of the transfer function and the input (stimulus) function. Transforming this statement to the time domain implies that the output time function is given as a convolution of a kernel function with the input time function. The transfer function is the Laplace transform of the kernel function. The transfer function characterizes the system under discussion, since all of the pertinent system parameters are contained in it. The convolution can be thought of, in the limiting sense of an integral, as a superposition of inputs each displaced in time and each modified by the kernel in accordance with its time displacement.

A nuclear reactor is basically a complex nonlinear entity from the temporal point of view. Even the aggregate formulation of the zero-power kinetics equations, as given by Eq. (1-47), results in a nonhomogeneous linear integral equation. As seen in Chap. 2, it is essentially intractable by analytic means for arbitrary time variations of reactivity. In general, the reactor kinetic behavior at operating power is given by

$$n(t) = n_0 + \int_{-\infty}^{t} G(t-\tau)K\{\tau, n(\tau)\}\, d\tau$$

which is a nonlinear integral equation. This representation of the kinetics includes the many and diverse feedback mechanisms, alluded to in the functional $K\{t, n(t)\}$, which are manifest at the high neutron density levels corresponding to the reactor operating power. The reactivity feedback mechanisms, such as temperature, density, doppler broadening, voids, and thermal expansion, are considered non-existent for the zero-power situation. The zero-power kinetics corresponds to reactor power sometimes as much as 10 orders of magnitude below operating power.

3-2 PERTURBATION THEORY AND TRANSFER FUNCTIONS

To bring home the idea that the transfer function stems from the first order level in the hierarchy of the perturbation theory solution to the integral equation (1-47) of reactor kinetics, let the neutron density be given as

$$n(t) = n^{(0)} + n^{(1)} + n^{(2)} + \cdots \tag{3-1}$$

and the reactivity similarly as

$$K(t) = K^{(0)} + K^{(1)} + K^{(2)} + \cdots \tag{3-2}$$

where the superscripts correspond to the order of the perturbation, and that both

$$n^{(0)} \gg n^{(1)} \gg n^{(2)} \gg \cdots \qquad \text{and} \qquad K^{(0)} \gg K^{(1)} \gg K^{(2)} \gg \cdots$$

If the neutron density and reactivity cannot be written in this above manner, then the perturbation theory approach is not applicable. However, for most physical situations of interest, they can be so compartmentalized. Inserting Eqs (3-1) and (3-2) into the integral equation (1-47) yields

$$\begin{aligned} n^{(0)} &+ n^{(1)} + n^{(2)} + \cdots \\ &= n_0 + \int_0^t G(t-\tau)(K^{(0)} + K^{(1)} + K^{(2)} + \cdots)(n^{(0)} + n^{(1)} + n^{(2)} + \cdots)\, d\tau \end{aligned} \tag{3-3}$$

The perturbation decomposition proceeds simply by equating the sum of like orders on each side of Eq. (3-3). The orders on the right-hand side are identified by first making all of the two factor multiplications, and then adding the order superscripts of the two factors in each product (like exponents). That sum is the order of that term. The zeroth order result, assumed to include n_0 is given by

$$n^{(0)}(t) = n_0 + \int_0^t G(t - \tau)K^{(0)}(\tau)n^{(0)}(\tau)\, d\tau \tag{3-4}$$

while the first order result is

$$n^{(1)}(t) = \int_0^t G(t - \tau)(K^{(0)}(\tau)n^{(1)}(\tau) + K^{(1)}(\tau)n^{(0)}(\tau))\, d\tau \tag{3-5}$$

and the second order result is

$$n^{(2)}(t) = \int_0^t G(t - \tau)(K^{(0)}(\tau)n^{(2)}(\tau) + K^{(1)}(\tau)n^{(1)}(\tau) + K^{(2)}(\tau)n^{(0)}(\tau)\, d\tau \tag{3-6}$$

and so on for higher orders.

For a near equilibrium reactor, $K^{(0)} = 0$ to zeroth order, implying from Eq. (3-4) that $n^{(0)} = n_0$. Then the first order neutron density from Eq. (3-5) satisfies the linear equation

$$n^{(1)}(t) = n_0 \int_0^t G(t - \tau)K(\tau)\, d\tau \tag{3-7}$$

where the superscript on $K^{(1)}$ has been dropped. Then the Laplace transform of the above yields the well-known transfer function relationship between the stimulus reactivity and the neutron density (deviation from equilibrium) response $\bar{n}^{(1)}$, namely

$$\bar{n}^{(1)}(s) = n_0\bar{G}(s)\bar{K}(s) \tag{3-8}$$

so that, within the constant n_0, the Laplace transform of the kernel of the integral equation, $\bar{G}(s)$, becomes the transfer function by definition. As is evident from the foregoing, it is seen that the central "output equals transfer function times input" relationship of the transfer idea has roots in the concept of the Greens function.

3-3 PILE OSCILLATORS

The idea of using periodic reactivity fluctuations to modulate the neutron density level was proposed during the Manhattan Project days by E. P. Wigner. It was suggested as a more sensitive method for measuring absorption cross sections than the danger coefficient method.† The latter consisted of merely introducing an amount of the absorber into a critical reactor. The reactor was then restored to criticality by using a calibrated control rod. From the amount of reactivity needed

† One of the first pile oscillators built was used to oscillate absorbers in the Oak Ridge reactors.[1]

to restore the reactor to critical, the absorption cross section could be determined. The difficulty with this method was not only that large amounts of absorbers must be used (they were not readily available in the case of microgram amounts of isotopes from the separation plants) but that the large amount of absorber affects the local diffusion properties, i.e., the absorber is usually a good scatterer as well. To correct for the scattering is an involved calculation.

If the oscillations are such that large reactivity amplitudes ensue, the resulting nonlinear flux behavior in both time and space produces such phenomena as neutron waves emanating from the oscillator. This is investigated in Chap. 7, where spatially dependent kinetics is discussed. These investigations were furthered at Argonne National Laboratories by oscillating a control rod with small amplitude in the CP-2 reactor.[2] The CP-2 reactor was the original University of Chicago graphite-moderated natural uranium reactor.

The next epoch in the acceptance of the use of the transfer function as a fundamental descriptor and tool of reactor kinetics, occurred *circa* 1955 with the EBR-I excursion incident. In a classic paper H. A. Bethe (1956) recognized that the sought after EBR-I excursion information could have been obtained through the use of a pile oscillator. This helped to pave the way for the use of transfer functions at various power levels to discover the relationships between power-dependent transfer functions in a given reactor, and thus to yield insight into reactor stability as it depends on power. Later, the statistics of reactor derived noise was proven to be a substantive method from which to obtain, at least, the gain-frequency component of the transfer function, without the necessity of perturbing the reactor due to the presence of the pile oscillator within. Details of this, as well as the preceding transfer function ramifications are discussed in the following sections.

3-4 TRANSFER FUNCTIONS

A second approach to the transfer function is through the integral equation (1-47)

$$n(t) = n(0) + \int_0^t G(t-\tau)K(\tau)n(\tau)\,d\tau \tag{3-9}$$

with $\varepsilon(t) = n(t) - n(0)$ the deviation about steady state, the above is rewritten to give

$$\varepsilon(t) = n(0)\int_0^t G(t-\tau)K(\tau)\,d\tau + \int_0^t G(t-\tau)K(\tau)\varepsilon(\tau)\,d\tau \tag{3-10}$$

This is iterated to give

$$\varepsilon(t) = n(0)\left[\int_0^t G(t-\tau)K(\tau)\,d\tau + \int_0^t G(t-\tau)K(\tau)\,d\tau \int_0^\tau G(\tau-x)K(x)\,dx + \cdots\right] \tag{3-11}$$

For sufficiently small reactivities causing only small fluctuations of neutron density about equilibrium, the first term is sufficient, or

$$\varepsilon(t) = n(0) \int_0^t G(t - \tau) K(\tau)\, d\tau \tag{3-12}$$

and, as with Eq. (3-8), the Laplace transform of Eq. (3-12) gives

$$\bar{\varepsilon}(s) = n(0) \bar{G}(s) \bar{K}(s) \tag{3-13}$$

$n(0)\bar{G}(s)$ is the transfer function, and this is stressed as being different from that given in Eq. (3-8). This is because the $n(0)$ amplitude factor introduces an equilibrium power-dependent gain in the amplitude response $|\bar{G}(s)|$, in the same sense that a pendulum undergoing large-amplitude oscillations causes the frequency of oscillations to depend on the amplitude. This dependence in the pendulum disappears when the amplitude is made sufficiently small. There is, however, the more deep-seated nonlinearity which was discussed previously and which will be further investigated in subsequent sections. $\bar{G}(s)$ or $\bar{G}(\omega)$ is termed the zero-power transfer function. This means simply that the reactor is operating at a nominally low power.

To show that generally s can be replaced by $i\omega$ in a transfer function† the zero-power transfer function $\bar{G}(s)$ [or $\bar{G}_0(s)$, as it is often termed] is first rewritten from Eq. (1-49) in terms of its poles and zeros by examining the relationship

$$\bar{G}(s) = \sum_{r=0}^{6} \frac{A_r}{s + s_r} = [s(\Lambda + \bar{f}(s))]^{-1} = \frac{\prod_{i=1}^{6} (s + \lambda_i)}{s\left[\Lambda \prod_{i=1}^{6} (s + \lambda_i) + \sum_{i=1}^{6} a_i \prod_{j \neq i}^{5} (s + \lambda_j)\right]} \tag{3-14}$$

where s_r are the roots of $s_r(\Lambda + \bar{f}(s_r)) = 0$ and $s_0 = 0$. The s_r are given in Table 1-3. Then from the above equation, $\bar{G}(s)$ can be rewritten in product form, to within a multiplicative constant, as the quotient of its poles and zeros, that is

$$\bar{G}(s) = \frac{\prod_{i=1}^{6} (s + \lambda_i)}{Cs \prod_{r=1}^{6} (s + s_r)}$$

The constant $C = \Lambda$ is determined by using the initial-value theorem of Laplace transforms ($\lim_{s \to \infty} s\bar{G}(s) = 1/\Lambda$), so that finally

$$\bar{G}(s) = \frac{\prod_{i=1}^{6} (s + \lambda_i)}{\Lambda s \prod_{r=1}^{6} (s + s_r)} \tag{3-15}$$

† Because, physically, ω corresponds to excitation frequency, since $\bar{G}(i\omega) \equiv \bar{G}(\omega)$ is the quantity determined experimentally from transfer function measurements.

From the theory of Laplace transforms, the inversion integral for $\bar{G}(s)$ is

$$G(t) = \frac{1}{2\pi i} \int_{c-i\infty}^{c+i\infty} ds\, \bar{G}(s)\, e^{st}$$

This is a line integral in the complex s plane along the straight line L depicted in Fig. 3-1. It should be noted that any point on the line L has coordinates $s = c + i\omega$, where $-\infty \leq \omega \leq +\infty$. In order to use the theorems in the theory of functions of a complex variable (Cauchy's integral theorems) to evaluate the inversion line integral, the closed contour C is formed. The contour consists, as in Fig. 3-1, of the line L, the two smaller lines L_1 and L_2, and the semicircle of infinite radius R in the left half plane in the sense (arrows) shown. The contour must enclose all the singularities of $\bar{G}(s)$, that is, its poles. The poles of $\bar{G}(s)$ are the roots s_r, all of which lie in the left half plane, plus one at the origin ($s_0 = 0$), since all the s_r but one are negative. For a closed contour enclosing all of the poles of $\bar{G}(s)$, the above theorems apply, so that the integral can be evaluated.

The parameter c in the inversion integral denotes the line L in the complex s plane, which must lie to the right of all of the singularities of $\bar{G}(s)$. Because the roots s_r are negative, c can be made zero so that the line L coincides with the imaginary axis and $L_1 = L_2 = 0$. The contour is still closed by the semicircle of infinite radius in the left half plane plus a semicircular nodule at the origin to accommodate the pole there, which vanishes in the limit of zero radius about the origin. It is easily shown that, for $\bar{G}(s)$ above, the line integral along this semicircle yields a zero contribution. Since $c = 0$, the contour integral becomes merely a line integral along the infinite imaginary axis. Therefore, let $s = i\omega$ in the integrand of the above integral to obtain the real integral

$$G(t) = \frac{1}{2\pi} \int_{-\infty}^{+\infty} \bar{G}(i\omega)\, e^{i\omega t}\, d\omega \tag{3-16}$$

which is the Fourier transform inversion integral, so that $\bar{G}(i\omega) \equiv \bar{G}(\omega)$ is the Fourier transform of $G(t)$ and is also the transfer function with s replaced by $i\omega$.

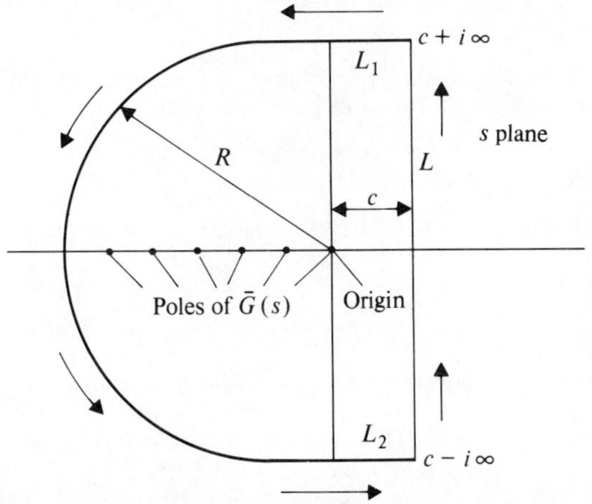

Figure 3-1 Contour of inversion integral of transfer function $\bar{G}(s)$.

Therefore, Eq. (3-15) can be rewritten as

$$\bar{G}(\omega) = \frac{\prod_{j=1}^{6} (i\omega + \lambda_j)}{i\omega\Lambda \prod_{r=1}^{6} (i\omega + s_r)} \tag{3-17}$$

Or in polar form

$$\bar{G}(\omega) = \frac{\prod_{j=1}^{6} \left[1 + \left(\frac{\omega}{\lambda_j}\right)^2\right]^{1/2}}{\omega\bar{\tau} \prod_{r=1}^{6} \left[1 + \left(\frac{\omega}{s_r}\right)^2\right]^{1/2}} \exp\left[i\left(\sum_{j=1}^{6} \tan^{-1}\frac{\omega}{\lambda_j} - \sum_{r=1}^{6} \tan^{-1}\frac{\omega}{s_r} - \frac{\pi}{2}\right)\right] \tag{3-18}$$

where the final-value theorem of Laplace transforms is used to evaluate the constant $\prod_{j=1}^{6} (\lambda_j)/\Lambda \prod_{r=1}^{6} (s_r)$, which is factored from $\bar{G}(\omega)$ and is equal to $(\bar{\tau})^{-1}$, since $\Lambda \ll \bar{\tau}$ holds to good approximation. $\bar{G}(\omega)$ characterizes any given reactor through the roots s_r. The roots depend on the delayed neutron parameters a_i, λ_i, and Λ, the prompt lifetime.† a_i and λ_i depend on the fuel used, that is, ^{233}U or ^{235}U. Λ depends on the physical constituents as well as on the size and shape of the particular reactor. The amplitude response is the modulus of Eq. (3-18)

$$|\bar{G}(\omega)| = \frac{\prod_{j=1}^{6} \left[1 + \left(\frac{\omega}{\lambda_j}\right)^2\right]^{1/2}}{\omega\bar{\tau} \prod_{r=1}^{6} \left[1 + \left(\frac{\omega}{s_r}\right)^2\right]^{1/2}} \tag{3-19}$$

while the phase response is the angular portion $\delta(\omega) = \arg \bar{G}(\omega)$ where

$$\exp\,[i\delta(\omega)] = \exp\left[i\left(\sum_{j=1}^{6} \tan^{-1}\frac{\omega}{\lambda_j} - \sum_{r=1}^{6} \tan^{-1}\frac{\omega}{s_r} - \frac{\pi}{2}\right)\right] \tag{3-20}$$

It is customary to use the decibel as the unit of amplitude, so that Eq. (3-19) becomes

$$N_{\text{db}}(\omega) = 20 \log_{10} |\bar{G}(\omega)| \tag{3-21}$$

Then

$$N_{\text{db}}(\omega) = 20\left\{\sum_{j=1}^{6} \log_{10}\left[1 + \left(\frac{\omega}{\lambda_j}\right)^2\right]^{1/2} - \sum_{r=1}^{6} \log_{10}\left[1 + \left(\frac{\omega}{s_r}\right)^2\right]^{1/2} - \log_{10} \omega\bar{\tau}\right\} \tag{3-22}$$

† a_i and λ_i actually depend on the physical composition of the reactor through the importance function, which also depends on the reactor size and shape. The importance function determines the effective fraction of delayed neutrons $\bar{\beta}$, which in turn yields $\bar{a}_i$. This dependence is only slowly varying over the spectrum of reactor shapes and compositions; cf. Sec. 6-3.

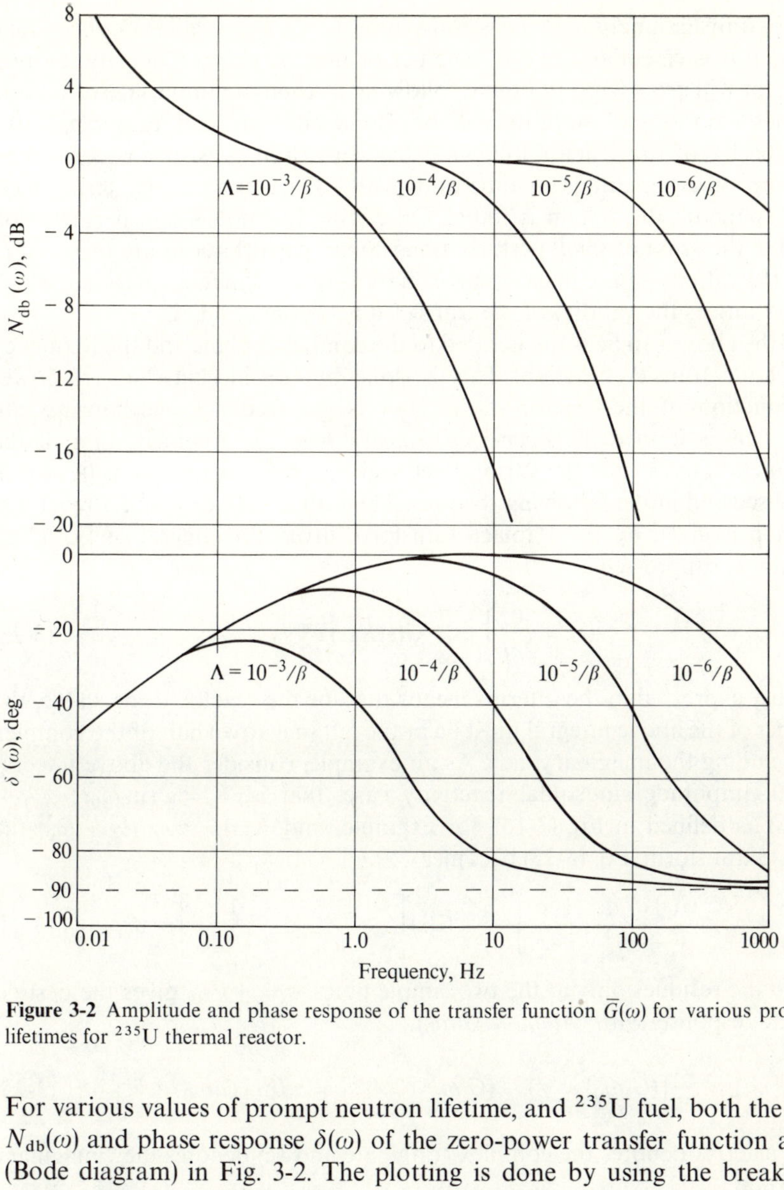

Figure 3-2 Amplitude and phase response of the transfer function $\overline{G}(\omega)$ for various prompt neutron lifetimes for ^{235}U thermal reactor.

For various values of prompt neutron lifetime, and ^{235}U fuel, both the amplitude $N_{db}(\omega)$ and phase response $\delta(\omega)$ of the zero-power transfer function are plotted (Bode diagram) in Fig. 3-2. The plotting is done by using the break-frequency method on the amplitude response, which is a labor-saving technique developed for such plots.[3]

3-5 TRANSFER FUNCTION STABILITY

To investigate a reactor control system by using the methods of transfer functions, the broad central criterion ordinarily employed is that of stability. Stability in the sense of transfer function analysis and synthesis has a particular mathematical connotation. It is essentially that of whether or not the neutron density (output) of the reactor will grow without bound following a reactivity (input) perturbation. More general notions of stability will be discussed later. For example, if the neutron density of the reactor following the perturbation is given as a linear combination of exponentials in time and if all of the exponential terms have negative exponents, the system is stable. Otherwise, it is not. Again perturbation is implied in the sense of *small* perturbations. Small perturbations are required to maintain the validity of the linear approximation to the kinetic equations, which in turn guarantees the validity of the transfer function method.

All of the above can be transliterated to the complex s plane and the frequency domain of the transfer function. This is done by considering the zero-power transfer function of the reactor itself. That is, no feedback mechanisms are included. This is termed the *open-loop transfer function.* Augmentation of the transfer function to include the various feedback reactivity mechanisms (feedback loops) is discussed in the following sections. The output $\varepsilon(t)$ for a small reactivity perturbation is given by the Laplace transform inversion integral of Eq. (3-8), that is, with n_0 suppressed

$$\varepsilon(t) = \frac{1}{2\pi i}\int_{c-i\infty}^{c+i\infty} \bar{G}(s)\bar{K}(s)\, e^{st}\, ds \qquad (3\text{-}23)$$

Stability expressed in these terms means that for the reactor to be stable, the singularities of the above integral must lie in the left (negative) half of the complex s plane including the imaginary axis. As an example, consider the above integral for a small-amplitude sinusoidal reactivity case, that is, $K = \alpha \sin \omega_0 t$; $\alpha \ll 1$. Using $\bar{G}(s)$ as defined in Eq. (3-15), for example, and $\bar{K}(s) = \alpha\omega_0/(s^2 + \omega_0^2)$ the appropriate transform, Eq. (3-23) becomes

$$\varepsilon(t) = \frac{\alpha\omega_0}{2\pi i}\int_{c-i\infty}^{c+i\infty} \frac{\bar{G}(s)\, e^{st}\, ds}{s^2+\omega_0^2} = \frac{\alpha}{4\pi}\int_{c-i\infty}^{c+i\infty} \bar{G}(s)\left[\frac{1}{s+i\omega_0} - \frac{1}{s-i\omega_0}\right] e^{st}\, ds \qquad (3\text{-}24)$$

Evaluating the residues only at the two simple poles, $s = \pm i\omega_0$, gives the desired steady-state response, with $\bar{G}(i\omega_0) \equiv \bar{G}(\omega_0)$

$$\varepsilon(t) = \frac{\alpha}{2i}\left[(\bar{G}(\omega_0)\, e^{i\omega_0 t}) - (\bar{G}(\omega_0)\, e^{i\omega_0 t})^*\right] = \alpha\, \mathrm{Im}\left[G(\omega_0)\, e^{i\omega_0 t}\right] \qquad (3\text{-}25)$$

where the asterisk denotes the complex conjugate and Im denotes the "imaginary part of." Then

$$\varepsilon(t) = \alpha\left|\bar{G}(\omega_0)\right| \sin\left[\omega_0 t + \delta(\omega_0)\right] \qquad \delta(\omega_0) = \arg \bar{G}(\omega_0) \qquad (3\text{-}26)$$

so that the input reactivity sinusoid has been recovered in the output neutron density deviation $\varepsilon(t)$ but with a shift in phase $\delta(\omega_0)$ which depends on the excitation frequency. Similarly, the output amplitude $\alpha\left|\bar{G}(\omega_0)\right|$, the amplitude response of the transfer function, is also dependent on the excitation frequency. Zero-power instability is immediately evident at low frequencies, since the definition of $\bar{G}(\omega_0)$ implies that, as ω_0 approaches zero

$$\bar{G}(\omega_0) \cong \frac{1}{\omega_0 \bar{\tau}} e^{-i\pi/2} \to \infty \qquad \bar{\tau} \gg \Lambda \tag{3-27}$$

with

$$\varepsilon(t) \sim \frac{\alpha}{\omega_0 \bar{\tau}} \sin\left(\omega_0 t - \frac{\pi}{2}\right) \tag{3-28}$$

growing without bound and with an asymptotic phase lag of $\pi/2$ radians. For high frequencies $\bar{G}(\omega_0) \sim (1/\omega_0 \Lambda)\, e^{-i\pi/2}$ and

$$\varepsilon(t) \sim \frac{\alpha}{\omega_0 \Lambda} \sin\left(\omega_0 t - \frac{\pi}{2}\right) \tag{3-29}$$

which decreases to zero as ω_0 approaches very large values, also with the same asymptotic phase lag of $\pi/2$ radians.

The above open-loop system is therefore unstable at low frequencies, since the output amplitude increases hyperbolically without bound. It decreases to zero at sufficiently high frequencies, with a characteristic roll-off or corner frequency defined such that $\omega_0 \Lambda = 1$. The above result regarding stability must be taken with a grain of salt, since the above linearizing approximations provide a model valid only for small deviations of the neutron density about equilibrium and, more importantly, only for the zero-power situation. Since the amplitude $|\bar{G}(\omega_0)|$ at low frequencies is intolerably large, other arguments, which are discussed later, must be used to ascertain whether or not the system possesses stability there.

Now a simple physical argument for instability due to low-frequency sinusoidal reactivity excitation can be given. If the reactivity is increasing slowly enough (very low ω_0), the reactor can approach prompt critical before the sinusoidal reactivity downturn at a time corresponding to $\omega t = \pi/2$ radians is reached. Then the neutron level increases very rapidly in an unbounded fashion before the reactivity downturn can become effective in restoring the level to equilibrium. A factor militating against easy restoration of equilibrium neutron density is that the neutron level deviation is not symmetrical with regard to given amounts of reactivity excitation. It is seen, by inspecting the appropriate curves of $n(t)$ versus time (Fig. 2-13), that much greater step reactivity magnitudes must be used to achieve a given subcritical asymptotic neutron level rate than would be required for its supercritical counterpart. Or alternatively, for example, a positive step function of only 16 cents increases the neutron level from equilibrium by about a factor of 20 in 140 sec, while a negative step approximately four times as large (-63 cents) is needed to drop the level by a factor of 20 in the same time. This is what is meant when it is stated that "it is harder to shut a reactor down than it is to increase its level."

Again, the difficulty with the zero-power open-loop development, besides ignoring nonlinearities in general, is that no feedback mechanisms are taken into account. Even for not very large deviations of neutron level about equilibrium, stability can in fact be provided by such feedback mechanisms as thermal

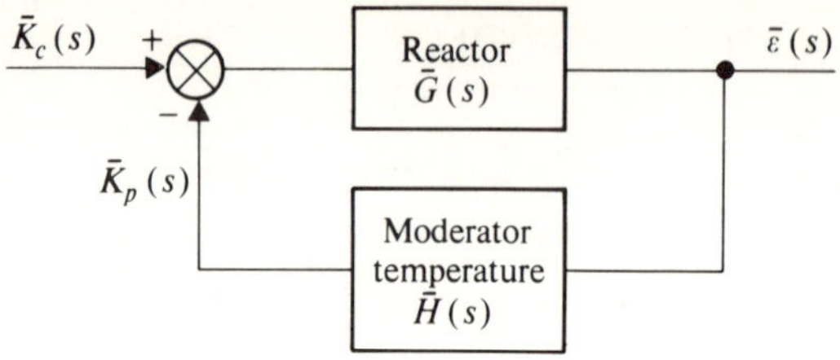

Figure 3-3 Schematic diagram of reactor with temperature reactivity feedback.

expansion of the core in certain reactors. The methods and ideas of feedback (closed-loop) control and all of its ramifications are here only introduced. Comprehensive discussions of linear feedback control analysis and synthesis using transfer functions are available, and references are given at the end of the chapter.[4-6] However, certain principal ideas, comprising an abbreviated formulation of reactor linear stability, are now discussed.

For example, it is well known that after a liquid-moderated reactor is started, the heat generated decreases the density of the moderator. This decrease in density impairs the diffusion and moderating properties of the moderator so that the neutron density begins to decrease, with the reactor tending to shut itself off. This is one example of a feedback mechanism manifest in a reactor. There are others including xenon poisoning, discussed in Chap. 5. As to the above moderator temperature-density effect, a given change in moderator temperature is reflected in a given small change in reactivity. However, there is a time lag due to the heat transfer from the fuel to the moderator, as well as that due to thermodynamic properties of the reactor core structure.

Assume that the relationship between this temperature density-induced reactivity and neutron density can be expressed by the ordinary linear convolution equation. The lag memory function is given by a kernel $H(t)$ [reactivity per unit $\varepsilon(t)$] which reflects the reactor physical and thermodynamical characteristics. Then this reactivity $K_p(t)$ can be written as a superposition of past reactivities, $H(t-\tau)\varepsilon(\tau)$, to give the corresponding reactivity now, that is

$$K_p(t) = \int_0^t H(t-\tau)\varepsilon(\tau)\,d\tau \tag{3-30}$$

where $\varepsilon(t)$ is the neutron density deviation about equilibrium.

From the definition of transfer function—the Laplace transform of the output is equal to the product of the transfer function and the Laplace transform of the input—an inspection of Fig. 3-3 yields

$$\bar{\varepsilon}(s) = \bar{G}(s)[\bar{K}_c(s) - \bar{K}_p(s)] \tag{3-31}$$

The subscript c refers to the controllable rod reactivity. From Eq. (3-30)

$$\bar{K}_p(s) = \bar{H}(s)\bar{\varepsilon}(s) \tag{3-32}$$

Substitution of this into Eq. (3-27) yields the combined "transfer function," $\bar{T}(s)$, that is

$$\bar{\varepsilon}(s) = \bar{T}(s)\bar{K}_c(s) \qquad \bar{T}(s) = \frac{\bar{G}(s)}{1 + \bar{G}(s)\bar{H}(s)} \tag{3-33}$$

Since we know that all of the singularities of $\bar{K}_c(s)$, and $\bar{G}(s)$ in the numerator of $\bar{T}(s)$, lie in the left half plane (including the origin), we need be concerned only with those due to the denominator to ascertain stability. That is, it is desired to know whether or not any roots of $1 + \bar{G}(s)\bar{H}(s) = 0$ lie in the right half of the s plane.

The stability of such a feedback system can be stated in a somewhat more rigorous fashion called the Nyquist criterion. To formulate this criterion, consider two complex planes, the s plane and the $\bar{G}(s)\bar{H}(s)$ plane. Consider s and the polar form of $\bar{G}(s)\bar{H}(s)$, namely

$$\bar{G}(s)\bar{H}(s) = |\bar{G}(s)\bar{H}(s)|\, e^{i\theta(s)} \qquad \theta(s) = \arg \bar{G}(s)\bar{H}(s) \tag{3-34}$$

as two polar vectors, each in its own plane. As the real and imaginary parts of s are varied, each of these vectors traces a path in its respective plane. The Nyquist criterion is simply that if there are no roots of $1 + \bar{G}(s)\bar{H}(s) = 0$ in the right half of the s plane, the system is stable, since all the roots must lie in the left half of the s plane. The system is stable because the residues of the inversion integral, from Eq. (3-33)

$$\varepsilon(t) = \frac{1}{2\pi i}\int_C \frac{\bar{G}(s)\bar{K}(s)}{1 + \bar{G}(s)\bar{H}(s)}\, e^{st}\, ds \tag{3-35}$$

over the usual inverse Laplace transform semicircular contour C in the s plane now enclosing all of the singularities of the integrand will give a sum of exponential terms all with negative real exponents. This assumes that the residues consist of simple poles, which is usually the case. Negative real exponents implies that the neutron density dies out with time, as opposed to it growing without bound.

In order to determine whether there exist roots of $1 + \bar{G}(s)\bar{H}(s) = 0$ when s traverses its positive real part semicircular contour of infinite radius in its right half plane, the Nyquist criterion is reformulated in terms of a well-known result from the theory of complex variables. It is thence easy to show that the number of encirclements N of $1 + \bar{G}(s)\bar{H}(s)$ about the origin, or $\bar{G}(s)\bar{H}(s)$ about $(-1, 0)$ on its closed contour in its plane, is given by $Z - P$. P is the number of poles of $1 + \bar{G}(s)\bar{H}(s)$ with positive real parts encircled by the s contour in the s plane, while Z is the number of zeros with positive real parts also enclosed by the s contour.† Since the zeros of $1 + \bar{G}(s)\bar{H}(s)$ with positive real parts correspond to poles of $\bar{T}(s)$ with positive real parts, $Z = 0$ must hold to ensure stability. Therefore, if the number of revolutions of $\bar{G}(s)\bar{H}(s)$ about $(-1, 0)$ in its plane is equal to the number of its poles with positive real parts, the system is stable, since $Z = 0$ is then implied. If $P = 0$ also, then the number of revolutions of $\bar{G}(s)\bar{H}(s)$ about $(-1, 0)$ must be zero for stability, because for $P = 0$ the number of revolutions

† The multiplicity of the poles and zeros must be included, e.g., for $(s - a)^{-2}$ and $(s - b)^3$, $P = 2$ and $Z = 3$, respectively.

then counts the number of zeros with positive real parts of $\bar{G}(s)\bar{H}(s)$ about $(-1,0)$. If $P=0$, then, to ensure stability, $\bar{G}(s)\bar{H}(s)$ must not encircle $(-1,0)$ to form a closed contour. Actually, from Eq. (3-33), if $1+\bar{G}(s)\bar{H}(s)=0$, that is, if $\bar{G}(s)\bar{H}(s)$ passes through $(-1,0)$, the system is unstable as well.

Since $i\omega$ is usually substituted for s, as discussed earlier, one investigates the circumlocutions of $\bar{G}(i\omega)\bar{H}(i\omega)$ about $(-1,0)$. As $\bar{G}(s)\bar{H}(s)$ describes its closed contour, $\bar{G}(i\omega)\bar{H}(i\omega)$ describes a corresponding closed contour from $\bar{G}(0)\bar{H}(0)$ to

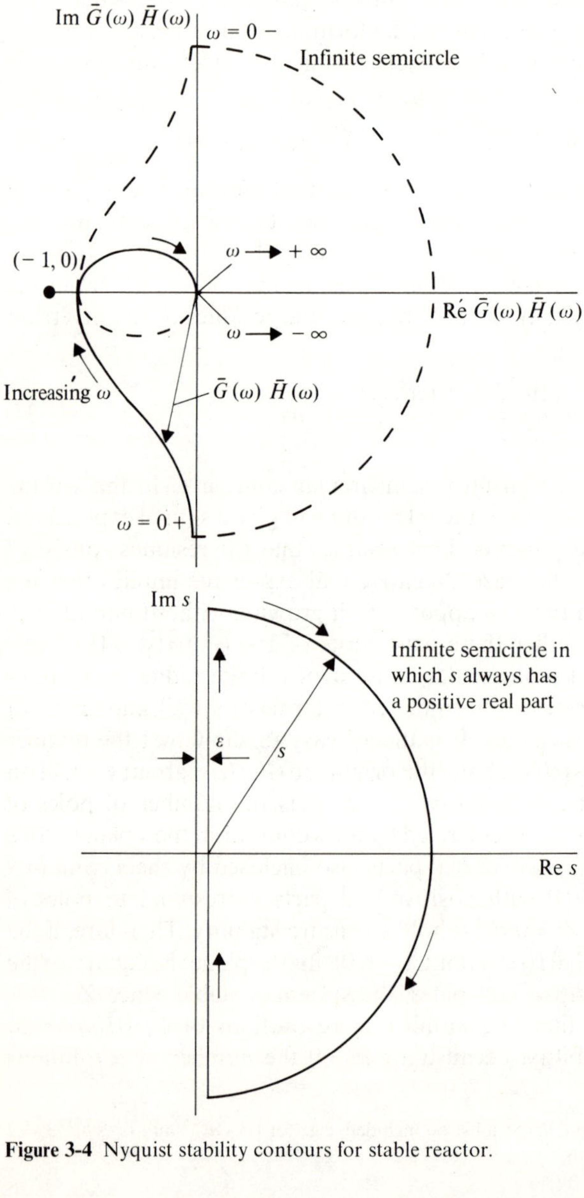

Figure 3-4 Nyquist stability contours for stable reactor.

$\bar{G}(i\infty)\bar{H}(i\infty)$ joining its mirror-image contour $\bar{G}(-i\omega)\bar{H}(-i\omega)$ and back around an infinite semicircle in the right half plane to $\bar{G}(0)\bar{H}(0)$, as shown in Fig. 3-4. That is, $\bar{G}(i\omega)\bar{H}(i\omega)$ completes a closed contour as $-\infty \le i\omega \le \infty$. In applying this technique, one must exercise a certain amount of care when tracing the $\bar{G}(i\omega)\bar{H}(i\omega)$ contour to correctly count the poles and zeros.[7] The locus (contour) of the $\bar{G}(s)\bar{H}(s)$ is called the Nyquist plot. Examples are discussed in Chap. 4, and an application is given for the experimental boiling-water reactor (EBWR) in Chap. 9.

Stability control synthesis in this context amounts to building in additional physical control elements with their individual transfer functions to be included in the total transfer function $\bar{T}(s)$. These are so designed as to distort the $\bar{G}(s)\bar{H}(s)$ vector locus to not enclose the point $(-1, 0)$ as the s vector locus traverses its contour in its right half plane. These individual transfer functions are physically realized by additional electrical or electromechanical components that are added to the reactor control system.

3-6 REACTIVITY COEFFICIENTS

When the temperature of the moderator changes owing to general reactivity excursions, as mentioned in the preceding section, it lags behind the reactivity excursion as characterized by the function $H(t)$ defined in Eq. (3-30). When the situation is examined in more detail, it is realized that the fuel temperature changes almost immediately with reactivity changes, since the fuel itself is the heat source. The moderator becomes hotter owing to the increased heat power transferred from the fuel, assuming a positive reactivity insertion. The moderator is now less dense, so that its moderating properties are impaired owing to the increase of the thermal diffusion area L^2, which thereby increases the thermal neutron leakage probability $1 - (1 + L^2B^2)^{-1}$. Therefore, the reactivity is decreased because of the decrease in multiplication factor k_{eff} and so acts as a negative-feedback influence. This is only part of the picture, because there are positive-feedback effects such as an increase in size due to thermal expansion which reduces (buckling) leakage and thereby increases k_{eff}. For not too large reactivity changes, the moderator-induced reactivity can be assumed to be proportional to the moderator temperature with essentially zero lag. Then a moderator reactivity temperature coefficient α_m can be defined as

$$K_m = \alpha_m \, \Delta T_{\text{mod}} \tag{3-36}$$

The reactivity temperature coefficient has the dimensions of, say, dollars per degree. Other reactivity coefficients, such as reactivity dependence on certain fission product poisons, voids in boiling-liquid moderators, and reflector composition, can be defined. Again, these reactivity changes are considered small and applicable over restricted ranges of the parameters involved because of nonlinearities that arise when these limitations are not maintained.

Feedback due to heating of the reactor constituents increases with the operating power level of the reactor. One type of positive feedback in high-flux thermal reactors, discussed in Chaps 6 and 7 and alluded to above, is that due to the existence of ^{135}Xe poison as a concomitant of fission. That the xenon feedback is positive is apparent, since an increase in flux level due to an increase of reactivity drops the xenon concentration because more neutrons are available for the xenon to absorb.† Therefore, the negative reactivity represented by the presence of xenon is decreased, so that the total reactivity is further increased, causing an increase in power level. The time constants involved are measured in fractions of a day, so that these effects are easily anticipated and thereby controlled.

Besides temperature reactivity coefficients, power reactivity coefficients can be defined.[8] Consider a reactor with a liquid-moderator coolant with a fixed inlet temperature T_{in}. The effect of the operating power level on the fuel and moderator temperatures, T_f and T_m, respectively, is depicted by temperature rate balance equations of the following type.[9]

$$\dot{T}_f = a_1 n(t) - b_1(T_f - T_m) \qquad \dot{T}_f(0) = 0 \tag{3-37}$$

$$\dot{T}_m = c_1(T_f - T_m) - e_1(T_m - T_{\text{in}}) \qquad \dot{T}_m(0) = 0 \tag{3-38}$$

Equation (3-37) represents the fuel temperature–time rate of change as that due to fission heating rate less moderator cooling rate. Equation (3-38) represents the moderator temperature rate due to the fuel heating the moderator, less the heat removed by moderator transport out of the reactor to another part of the coolant loop. The constants a_1, b_1, c_1, and e_1 represent heat capacities, flow rates, etc.

At equilibrium, the derivatives in Eqs (3-37) and (3-38) vanish, giving

$$T_f(0) = T_{\text{in}} + \left(1 + \frac{c_1}{e_1}\right)\frac{a_1}{b_1} n(0) \tag{3-39}$$

and

$$T_m(0) = T_{\text{in}} + \frac{a_1 c_1}{b_1 e_1} n(0) \tag{3-40}$$

As before, for small temperature changes about equilibrium

$$K_m = \alpha_m \, \Delta T_m \tag{3-41}$$

$$K_f = \alpha_f \, \Delta T_f \tag{3-42}$$

where $\Delta T_m = T_m - T_m(0)$ and $\Delta T_f = T_f - T_f(0)$. α_m and α_f are the temperature reactivity coefficients of the moderator coolant and fuel, respectively.

If the power is only slowly varying, there will be essentially no time lag between the temperatures and the corresponding power level at a given time. Then from Eqs (3-39) and (3-40), respectively, for small changes about equilibrium and fixed inlet temperature

† An increase of flux level also increases the amount of xenon present because of the increased fission rate. However, this source of xenon is very small compared with that due to the decay of ^{135}I, which largely governs the equilibrium concentration of ^{135}Xe. Cf. Secs 5-8 and 7-4.

$$\Delta T_f(0) \cong \Delta T_f = \left(1 + \frac{c_1}{e_1}\right)\frac{a_1}{b_1}\,\Delta n \tag{3-43}$$

$$\Delta T_m(0) \cong \Delta T_m = \frac{a_1 c_1}{b_1 e_1}\,\Delta n \tag{3-44}$$

with $\Delta n = n - n(0)$. Therefore, a slow power variation about equilibrium causes a total reactivity change, from Eqs (3-41) to (3-44), given by

$$K_{\text{tot}} = K_m + K_f = \left[\frac{a_1 c_1}{b_1 e_1}(\alpha_f + \alpha_m) + \frac{a_1}{b_1}\,\alpha_f\right]\Delta n \tag{3-45}$$

A power coefficient of reactivity can now be defined as the reactivity change per fractional change in power level about equilibrium, that is

$$\frac{K_{\text{tot}}}{\Delta n/n(0)} = \left[\frac{a_1 c_1}{b_1 e_1}(\alpha_f + \alpha_m) + \frac{a_1}{b_1}\,\alpha_f\right] n(0) \tag{3-46}$$

Also, a gross power coefficient of reactivity can be defined by including all other applicable feedback influences and summing the individual coefficients as above.

It is of interest to investigate in some detail certain of the reactivity coefficients including the doppler broadening coefficient. First, temperature reactivity coefficients of large thermal homogeneous reactors will be derived after El-Wakil.[10] Fast and boiling reactor reactivity coefficients are discussed in their respective sections in Chap. 8. For the two-energy-group (fast and slow) slowing-down model and a large reactor (very small B^2)

$$k_{\text{eff}} = \frac{k_\infty}{1 + M^2 B^2} \tag{3-47}$$

where the migration area $M^2 = L^2 + L_f^2$. L is the thermal diffusion length, and L_f is the fast diffusion length. Then the reactivity in absolute units is

$$K = 1 - k_{\text{eff}}^{-1} = 1 - \frac{1 + M^2 B^2}{k_\infty} = 1 - \frac{1 + (L^2 + L_f^2)B^2}{k_\infty} \tag{3-48}$$

In order to delineate the various partial temperature coefficients that sum to yield the total temperature coefficient, the total derivative of K with respect to T is calculated:

$$\frac{dK}{dT} = \frac{\partial K}{\partial k_\infty}\frac{dk_\infty}{dT} + \frac{\partial K}{\partial L^2}\frac{dL^2}{dT} + \frac{\partial K}{\partial L_f^2}\frac{dL_f^2}{dT} + \frac{\partial K}{\partial B^2}\frac{dB^2}{dT} \tag{3-49}$$

First the change in $k_\infty = \eta f \varepsilon p$ with temperature is examined. The fast fission effect is a function only of the fast neutrons. It should be remembered that fast neutrons have energies corresponding to temperatures of millions of degrees absolute. The thermodynamic temperatures involved in thermal reactivity coefficients are almost infinitesimal by comparison. ε is then considered to be independent of temperature, although it does change slightly owing to fuel density changes. This change is small, however, compared to the changes in L^2 and L_f^2, due to changes in density, and is therefore neglected. Similar arguments hold for the resonance escape probability p, which decreases slightly because of doppler broadening (discussed later).

All microscopic capture cross sections are assumed to vary as $1/v$, while all microscopic scattering cross sections are considered constant in their energy group range. Since v is proportional to $T^{1/2}$, the ratio

$$\eta = \frac{\nu\sigma_{\text{fiss}}v}{\sigma_{a,\text{fuel}}v}$$

is then independent of temperature in a thermal reactor. Also, the thermal utilization $f = (N\sigma_a)_{\text{fuel}}/(N_a\sigma_a)_{\text{tot}}$ is likewise independent of temperature except implicitly through the material density changes. These are also small compared to the changes in L^2 and L_f^2 due to density variations, since they partially cancel because N appears both in the numerator and denominator of f. Therefore k_∞ is considered to be independent of temperature, so that $dk_\infty/dT \cong 0$.

Now

$$L^2(T, \rho(T)) = \frac{1}{3N^2\sigma_{\text{tr}}\sigma_a} = \frac{\text{const } T^{1/2}}{3\rho^2\sigma_{\text{tr}}(\sigma_a v)} \tag{3-50}$$

and

$$L_f^2(\rho(T)) = \frac{1}{3\xi N^2\sigma_{\text{tr}}\sigma_s}\ln\frac{E_f}{E_t} = \text{const}/\rho^2(T) \tag{3-51}$$

where ρ is the ordinary density, N is the number of target nuclei per cubic centimeter, and E_f and E_t are the fission neutron energy and the lower bound of the fast neutron group energy, respectively. The following expressions are needed in Eq. (3-49)

$$\frac{dL^2}{dT} = \frac{\partial L^2}{\partial T} + \frac{\partial L^2}{\partial \rho}\frac{d\rho}{dT} \qquad \frac{dL_f^2}{dT} = \frac{\partial L_f^2}{\partial \rho}\frac{d\rho}{dT} \tag{3-52}$$

where there is only one term in the second expression above, because it is assumed that there is no explicit dependence of L_f^2 on the temperature T.

For illustrative purposes consider that the reactor is spherical, so that the buckling $B^2 = (\pi/R)^2$, where R is the extrapolated core radius. Then

$$\frac{dB^2}{dT} = \frac{dB^2}{dR}\frac{dR}{dT} = -2B^2\frac{d}{dT}\ln R \tag{3-53}$$

The fourth term in Eq. (3-49) is then, using Eqs (3-48) and (3-53)

$$\frac{\partial K}{\partial B^2}\frac{dB^2}{dT} = \frac{2M^2B^2}{k_\infty}\frac{d}{dT}\ln R \tag{3-54}$$

The third term in Eq. (3-49) is, using Eqs (3-48), (3-51), and (3-52)

$$\frac{\partial K}{\partial L_f^2}\frac{dL_f^2}{dT} = \frac{\partial K}{\partial L_f^2}\frac{\partial L_f^2}{\partial \rho}\frac{d\rho}{dT} = \frac{2B^2}{k_\infty}L_f^2\frac{d}{dT}\ln\rho \tag{3-55}$$

The second term in Eq. (3-49) is, using Eqs (3-48), (3-50), and (3-52)

$$\frac{\partial K}{\partial L^2}\frac{dL^2}{dT} = \frac{\partial K}{\partial L^2}\left(\frac{\partial L^2}{\partial T} + \frac{\partial L^2}{\partial \rho}\frac{d\rho}{dT}\right) = \frac{2B^2L^2}{k_\infty}\left(\frac{d}{dT}\ln\rho - \frac{1}{4T}\right) \tag{3-56}$$

The first term in Eq. (3-49) vanishes by arguments already given. Then Eq. (3-49) becomes

$$\frac{dK}{dT} = -\frac{B^2L^2}{2k_\infty T} + \frac{2M^2B^2}{k_\infty}\left(\frac{d}{dT}\ln\rho + \frac{d}{dT}\ln R\right) \tag{3-57}$$

Let L_0 be a principal linear dimension of the critical core, such as the fuel element length. Then upon heating and expansion

$$\rho \approx \frac{1}{L_0^3(1+\alpha_{\text{core}}\Delta T)^3} \cong \frac{1}{L_0^3(1+3\alpha_{\text{core}}\Delta T)} \qquad \alpha_{\text{core}}\Delta T \ll 1 \tag{3-58}$$

where α_{core} is the corresponding coefficient of linear expansion.

$\Delta T = T - T_0$ is the temperature difference between the heated core and the ambient core. Then

$$\frac{d}{dT}\ln\rho = \frac{-3\alpha_{\text{core}}}{1+\alpha_{\text{core}}\Delta T} \cong -3\alpha_{\text{core}} \qquad \text{since } \alpha_{\text{core}}\Delta T \ll 1 \tag{3-59}$$

Similarly, $R = R_0(1 + \alpha_{\text{rad}}\Delta T)$, where α_{rad} is the radial coefficient of expansion, that is

$$\frac{d}{dT}\ln R \cong \alpha_{\text{rad}} \qquad \alpha_{\text{rad}}\Delta T \ll 1 \tag{3-60}$$

Since all of the above changes are considered small in the sense that the reactor is not far from being critical, $K \cong 0$ in Eq. (3-49), implying $M^2B^2 \cong k_\infty - 1$. Then the temperature reactivity coefficient is, from Eq. (3-57),

$$\frac{dK}{dT} = -\left[\frac{B^2L^2}{2k_\infty T} + \left(\frac{k_\infty - 1}{k_\infty}\right)(6\alpha_{\text{core}} - 2\alpha_{\text{rad}})\right] < 0 \qquad \alpha_{\text{core}} > \alpha_{\text{rad}} \tag{3-61}$$

that is, the temperature reactivity coefficient is negative as it should be for physical reasons. Over a spectrum of reactors, from pressurized-water to gas-cooled ones, the range of the temperature reactivity coefficients is roughly from -0.3 to -40 cents per degree F.[10]

An additional and physically interesting reactivity coefficient is that due to the doppler temperature broadening of, for example, the epithermal nonfission (^{238}U) capture resonances of the fuel in a heterogeneous thermal reactor. Even though it was given scant attention above in connection with the resonance escape probability, during very large accident-type power surges where temperatures can rise quickly to very high values, the doppler effect can become a dominant factor in the over-all reactivity temperature coefficient.

In a heterogeneous reactor the fuel is lumped to enhance the resonance escape probability. Since the first (thermal) reactors had to be fueled with natural uranium because of the relative unavailability of "pure" ^{235}U, the fuel was lumped to keep the resonance escape probability sufficiently high that criticality could be achieved. Agglomerating the fuel in lumps keeps the over-all ratio of fuel surface to volume low, as opposed to finely divided fuel as in a homogeneous reactor where this ratio would be high. If this ratio is minimized, relatively more fuel nuclei within the lumps are shielded by their respective fuel nuclei on the surface, which act as resonance absorption filters and thus save the interior nuclei for fission. This is called the *self-shielding effect*. The lumping also provides relatively more nonfuel volume within the moderator for the neutrons to slow down past the resonance energy region to thermal energies before encountering fuel.

As the fuel becomes heated owing to a power surge, the fuel nuclei agitate into thermal equilibrium at the fuel temperature.† Then the moderated neutrons which happen to be at resonance energies, incident on the fuel, encounter temperature-broadened resonances. This is due to the relative velocity of the neutrons and fuel nuclei, which now gives rise to a probability distribution of (velocities) energies about the resonances. The reaction rates are now proportional to σv, where v is a relative velocity governed by the above distribution function. These doppler-broadened resonances increase the mean epithermal neutron absorption and thereby decrease the resonance escape probability, resulting in a negative doppler coefficient of reactivity for the thermal reactor. The situation is more complicated for fast reactors because the doppler reactivity coefficient can be large and both positive and negative depending on the fissionable-fertile fuel mix.

Let $k_{\text{eff}} = \eta \varepsilon f p P_{nl}$, where P_{nl} is the nonleakage probability which includes both the thermal and slowing down nonleakage probabilities. Since the doppler reactivity coefficient is rather small in the sense that $k_{\text{eff}} \cong 1$, it is given in absolute units by

$$\left(\frac{\partial K}{\partial T}\right)_{\text{dopp}} = \frac{1}{k_{\text{eff}}}\frac{d}{dT}\ln p \cong \frac{d}{dT}\ln p \tag{3-62}$$

The resonance escape probability is given by

$$p = \exp\left[-\frac{N_{\text{fuel}}}{N_{\text{mod}}}\frac{I}{(\xi\sigma_s)}\right] \tag{3-63}$$

and the effective resonance integral is[11]

$$I_{\text{res}} = \int (\sigma_{a,\text{fuel}})_{\text{eff}}\frac{dE}{E} \tag{3-64}$$

Then from above, the doppler coefficient is, where $\alpha_0 = d/dT \ln I$, following Pearce[12]

$$\left(\frac{\partial K}{\partial T}\right)_{\text{dopp}} = \alpha_0 \ln p \tag{3-65}$$

† The fuel also is in thermal equilibrium with the neutrons, which can now be moderated upward.

Now I must be modified by the effect of doppler temperature broadening.† The well known Breit–Wigner formula for the cross section near a given resonance is

$$\sigma(E) = \frac{\sigma_{a,\text{fuel}}^{0}\Gamma^2/4}{(E - E_0)^2 + \Gamma^2/4} \qquad E_0 \gg \Gamma \tag{3-66}$$

where Γ is the natural energy width of the resonance and $\sigma_{a,\text{fuel}}^{0}$ is the amplitude of the capture cross section at the resonance peak occurring at E_0. It is well known, Bethe and Placzek,[13] that the thermally agitated fuel acts like a perfect gas at temperature T to good approximation where $E = kT$ corresponds to the energies under consideration.‡ The thermal energies of the fuel nuclei are then distributed with the Maxwell–Boltzmann spectrum, that is, if $p(E)\,dE$ is the probability distribution function

$$p(E)\,dE = \frac{1}{\pi^{1/2}\Delta}\exp - [(E - E_0)/\Delta]^2\,dE \tag{3-67}$$

where

$$\Delta = \left(\frac{4EkT}{A}\right)^{1/2} \tag{3-68}$$

is the ratio of doppler width to total width, A is the mass number of the fuel nuclei, k is the Boltzmann constant, and T is the fuel temperature. The doppler-broadened cross section is then

$$\sigma(E, T) = \frac{\displaystyle\int_{-\infty}^{+\infty} \sigma(E)p(E)\,dE}{\displaystyle\int_{-\infty}^{+\infty} p(E)\,dE} \tag{3-69}$$

Inserting Eqs (3-66) and (3-67) into Eq. (3-69) yields

$$\sigma(E, T) = \sigma_{a,\text{fuel}}^{0}\psi\left(\frac{E - E_0}{\Gamma/2}, \frac{\Delta^2}{\Gamma^2}\right) \tag{3-70}$$

with

$$\psi(x, z) = \frac{1}{2(\pi z)^{1/2}}\int_{-\infty}^{+\infty} \frac{dy}{1 + y^2}\exp - [(x - y)^2/4z] \tag{3-71}$$

which is a well-tabulated integral.[14] Essentially, then $\sigma(E, T)$ is reinserted into the resonance integral in Eq. (3-64) to calculate its new value due to the temperature-broadened epithermal resonances. A resulting thermal reactor typical value of $\alpha_0 \cong 2 \times 10^{-4}$ per degree C. For $p \cong 0.9$, the doppler coefficient $(\partial K/\partial T)_{\text{dopp}} \cong -0.3$ cents per degree C.

† I can be written more accurately to account for the disadvantage factor in the fuel, mean chord length of lump, etc.[11,12]

‡ The approximation becomes poorer with decreasing T.

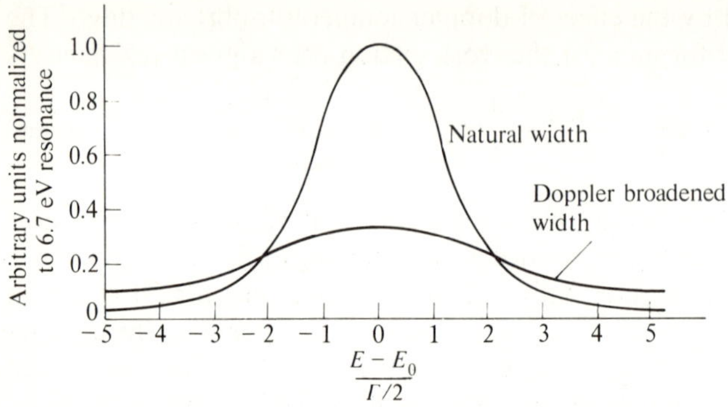

Figure 3-5 Doppler broadened resonance at 6.7 eV in ^{235}U.

3-7 TRANSFER FUNCTIONS FROM REACTOR NOISE

A reactor does not operate at a strictly steady-state neutron density level. What is implied is an expected or mean level, with random fluctuations about this mean. It should be realized not only that the neutron density is dancing about some mean value because the diffusion process *per se* is a random one but that reactor theory and neutronics, not to mention quantum mechanics, rest on a statistical basis. All cross sections, that is, mean free paths, imply certain expected values over probability distribution functions of corresponding random variables. It is just these stochastic aspects that are exploited in the techniques discussed in Chap. 10.

Therefore, an operating reactor has, to use the parlance of the radar systems engineer, a noise signature. All of the random fluctuations in all of the stochastic parameters in a reactor, such as the fission, scattering and absorption cross sections and age, act as a collection of reactivity coefficients causing the neutron density level to fluctuate in a random manner with a certain grand probability distribution. Once certain statistics of this probability distribution are obtained, the transfer function can in theory and usually in practice be found from them.

Some of the obvious advantages in obtaining the transfer function from reactor noise are (1) the data can often be obtained painlessly without perturbing the reactor with oscillating control rods, (2) it is much less expensive, with regard to instrumentation, setup time, and man-months of data taking, than oscillators, (3) in fact, it is so cheap that panoramic frequency instrumentation can be manufactured to obtain a visual real-time display of the transfer function which could be monitored along with the other instruments on the operator's console.

A principal disadvantage is that it is less accurate than the oscillator methods. However, with sufficient ingenuity this could be improved to the point of comparing favorably with the pile oscillator. For example, in the EBWR the oscillator method of obtaining the transfer function gave error limits of ± 0.5 db, whereas the noise method limits were ± 2.5 db.[16] In the instance that the autocorrelation function of the neutron level noise is required to obtain the transfer function, the principal disadvantage is that all phase information is lost by the very nature of the autocorrelation method. However, cross-correlation methods used to accomplish the same task retain phase information as well.

At this point, an introduction to at least the highlights of the reactor noise method for obtaining the transfer function is in order. Consider that the neutron density fluctuates about the mean as a random variable in time and that samples of the neutron density fluctuations about the mean are taken at small discrete (0.05 sec, say) time intervals. The collection of these samples ordered in time constitutes a random time series. No particular value in this set can be predicted with complete accuracy (with probability unity). Only certain statistical properties of this random series exist. It is desired that such a random time series possess both the property of stationarity and ergodicity defined below.

Assume a random time series of neutron density fluctuations partitioned into N strips each T sec in length. What a suitable length is will become apparent. If the strips are examined, it will be found that they are certainly not replicas of each other. However, if the series is a stationary time series, then they are statistical replicas of each other. The stationarity property implies that each of the N strips has the same statistical properties, such as the same mean (over the length of the sample) and variance. That is, their statistics are time-invariant. This type of invariance may be thought of as another reactor invariant as discussed in Chap. 1.

Now consider the ensemble (collection) of the N time strips, one under the other, to form a two-dimensional array of data (Fig. 3-13). Since each strip, because of its stationarity property, contains identical statistical information, it is as if the neutron density data were taken from an ensemble of N identical reactors and each strip were a record from each reactor. First consider the time average of some property of the series taken over any one of the strips in time—e.g., the time average of the square of the deviation from the mean neutron density level, that is, the variance over the time. Second, consider the two-dimensional array of amplitude data from the N strips (Fig. 3-13) and calculate the variance, but take each datum from a given column, say the 34th, of the array. This calculation gives the variance over the ensemble of N hypothetical reactors. If the individual strips possess the stationarity property, the ensemble variance, or any other ensemble average (like the ensemble mean), will be independent of which column of the array is chosen to obtain the data. Further, that the ensemble average is equal to the time average enunciates the *ergodicity property*. The ergodic hypothesis states that if the process is stationary, then time averages of any function of the data (like the mean or variance) are equal to corresponding ensemble averages. A proof of the ergodic hypothesis has been found for only a limited number of statistical processes. Mathematically speaking, processes which are of physical interest are only quasiergodic.[17] This distinction can be ignored for our needs.

The ergodicity property plays an important role in obtaining the correlation function of random neutron density noise outputs which are excited by random reactivity noise inputs. These are not readily characterized in terms of continuous well-behaved functions. The correlation function is then used to obtain the output spectral density and finally the transfer function. If the output can be obtained directly in terms of continuous functions of time, such as in an oscillator experiment, its Fourier transform divided by that of the input yields the transfer function without the adjunct of the correlation function. Since, in principle, random noise cannot be couched in terms of continuous functions but only in probabilistic terms, correlation techniques are a necessity to determine transfer functions of systems by means of noise excitation. However, the frequency range of interest is often low enough for reactor transfer functions that a reasonably smooth output ensues, even though the reactor is excited by its noise reactivity.

The frequency response of the usual output pen recorder is generally high enough that it adequately smooths the reactor noise output without sacrificing too much of its high-frequency content. The output then appears as a continuous function of time. Both methods (with or without the correlation function) can be used, and complementary information can be derived from each.

Let $dP\{n(t)\}$ be the probability distribution function of $n(t)$, the neutron density. That is, $dP\{n(t)\}$ is the probability that the neutron density level lies between n and $n + dn$, so that the ensemble average is (wavy overbar denotes this average)

$$\tilde{n} = \int n(t_i)\, dP\{n(t_i)\} \tag{3-72}$$

which is independent of the time interval t_i corresponding to the ith column of the array mentioned above. The above integral is over the ensemble of $n(t_i)$ for a given i, whereas the time average of $n(t)$ over a given strip is

$$\bar{n} = \lim_{T\to\infty} \frac{1}{T}\int_0^T n(t)\, dt \cong \frac{1}{T}\int_0^T n(t)\, dt \tag{3-73}$$

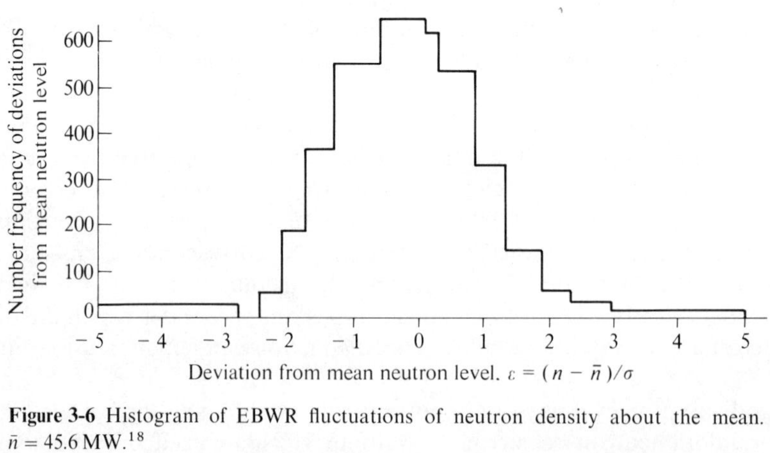

Figure 3-6 Histogram of EBWR fluctuations of neutron density about the mean. $\sigma = 0.83$ MW, $\bar{n} = 45.6$ MW.[18]

where T is the finite time duration of the strip. Ergodicity implies that if the collection $\{n(t_i)\}$, for any column i in the array, is a stationary process, then

$$\bar{n} = \tilde{n} \tag{3-74}$$

with probability 1. As an example of the above, consider the calculation of the mean deviation of the neutron density n, where the deviation is termed $\varepsilon(t)$, calculated from its "ensemble" probability distribution as depicted in Fig. 3-6. For the purpose of illustration, the histogram of Fig. 3-6 is fitted by

$$dp(\varepsilon) = \frac{A}{(2\pi\sigma^2)^{1/2}} \exp\left(-\varepsilon^2/2\sigma^2\right) d\varepsilon \tag{3-75}$$

where A is a normalization constant. The ordinates of Fig. 3-6 are, as is usual, the number of particular deviations from the mean for a given ε abscissa. The data used to plot Fig. 3-6 can be thought of as taken from any given column of a two-dimensional array made up of data from neutron density time strips from an EBWR run. Therefore, by using Eq. (3-75), the ensemble mean deviation is

$$\tilde{\varepsilon} = \int_{-\infty}^{+\infty} \varepsilon \, dp(\varepsilon) = 0 \tag{3-76}$$

since the integrand is an odd function over the infinite interval.

Now, the calculation of the time average of $\varepsilon(t)$, given by

$$\bar{\varepsilon} = \frac{1}{T} \int_0^T \varepsilon(t) \, dt \cong 0 \tag{3-77}$$

could be verified by numerical integration of the curves in Fig. 3-7. Here it will be estimated by inspection of the curves. That is, using the finite sum equivalent of Eq. (3-77)

$$\bar{\varepsilon} = \frac{1}{N} \sum_{j=1}^{N} \varepsilon_j \Delta \cong 0 \qquad T = N\Delta \tag{3-78}$$

inspection of Fig. 3-7 shows that there are, grossly, as many positive as negative ε_j so that the above zero mean deviation ensues. It is also trivially apparent that

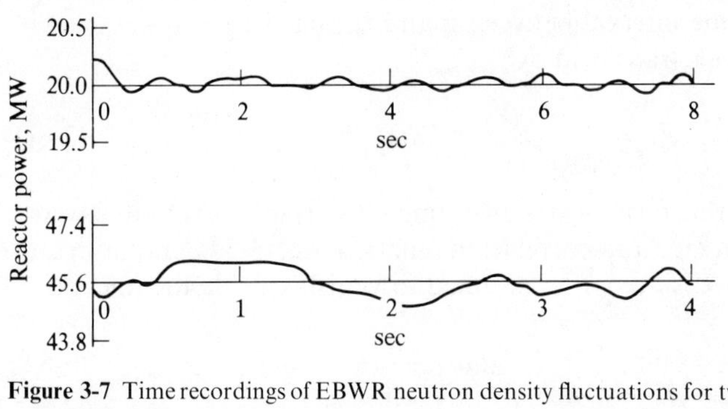

Figure 3-7 Time recordings of EBWR neutron density fluctuations for two separate mean power levels.

$\bar{\varepsilon} \cong \tilde{\varepsilon} = 0$ for physical reasons, since the EBWR was essentially at equilibrium (critical) during the runs and thus constituted a system with the required stochastic properties.

It should be noted that in the electrical engineering world, the quantity corresponding to $n(t)$ or $\varepsilon(t)$ has the dimensions of voltage, so that $\bar{P}_\varepsilon(\omega)$ has the dimensions of power. This is the reason it is called the power spectral density. However, the power output of a reactor is proportional to $n(t)$ or $|\bar{n}(\omega)|$, not $|\bar{n}|^2$. This is, of course, because $nv\Sigma_f$ gives the expected number of fissions per cubic centimeter per second and each fission results in the release of about 200 MeV of energy. The corresponding $\bar{P}(\omega)$ for reactor noise has units of reactor power squared. However, this dimensionality quirk acts as no deterrent to the use of the method.

As mentioned earlier, to obtain the transfer function of a system from its random noise output, correlation function techniques are employed. This is because the noise output is not available as a smooth function of time from which its Fourier transform can be computed to obtain the transfer function directly. If the noise output is obtainable only as an ensemble of stationary time series, then the correlation function must be obtained from ensemble averaging as opposed to time averaging.

The bridging relationship by which the correlation function is used to obtain the transfer function from the system noise output is Wiener's theorem. It relates the spectral density of a noise output with its correlation function in the following way. The autocorrelation function of the neutron density fluctuation about the mean, $\varepsilon(t)$, is defined as

$$\phi_{\varepsilon\varepsilon}(\tau) = \lim_{T\to\infty} \frac{1}{T} \int_{-T/2}^{+T/2} \varepsilon(t)\varepsilon(t+\tau)\, dt \tag{3-79}$$

Since the reactor noise is assumed to be ergodic, an alternative definition is the ensemble average

$$\phi_{\varepsilon\varepsilon}(\tau) = \int_{-\infty}^{+\infty}\int_{-\infty}^{+\infty} \varepsilon_i \varepsilon_j \, dP(\varepsilon_i, \varepsilon_j; \tau) \tag{3-80}$$

where $dP(\varepsilon_i, \varepsilon_j; \tau)$ is the joint probability that the instantaneous neutron density fluctuation at the ith time lies between ε_i and $\varepsilon_i + d\varepsilon_i$, and between ε_j and $\varepsilon_j + d\varepsilon_j$ at the jth time. τ is the time interval between t_i and t_j, that is, $\tau = t_i - t_j$. $\varepsilon(t)$ is related to its inverse Fourier transform by

$$\varepsilon(t) = \int_{-\infty}^{+\infty} \bar{\varepsilon}(\omega)\, e^{i\omega t}\, d\omega \tag{3-81}$$

where $\bar{\varepsilon}(\omega)$ is the Fourier transform of $\varepsilon(t)$. Inserting $\varepsilon(t+\tau)$ in terms of its inverse transform into the time average autocorrelation function, Eq. (3-79), and inverting the order of integration give, since $\varepsilon(t)$ is assumed to vanish outside the interval T

$$\phi_{\varepsilon\varepsilon}(\tau) = \lim_{T\to\infty} \frac{1}{T} \int_{-\infty}^{+\infty} \bar{\varepsilon}(\omega)\bar{\varepsilon}(-\omega)\, e^{i\omega\tau}\, d\omega \tag{3-82}$$

Since $\bar{\varepsilon}(\omega)$ is an even function, i.e., from physical reasoning $\bar{\varepsilon}(\omega) = \bar{\varepsilon}(-\omega)$, then

$$\phi_{\varepsilon\varepsilon}(\tau) = \lim_{T\to\infty} \frac{1}{T} \int_0^T |\bar{\varepsilon}(\omega)|^2 \cos \omega\tau \, d\omega \tag{3-83}$$

In the limit of large T

$$\phi_{\varepsilon\varepsilon}(\tau) = 2 \int_0^\infty \bar{P}_\varepsilon(\omega) \cos \omega\tau \, d\omega \tag{3-84}$$

where the spectral density is defined as

$$\bar{P}_\varepsilon(\omega) = \lim_{T\to\infty} \frac{1}{T} \left| \int_{-T/2}^{+T/2} \varepsilon(t)\, e^{i\omega t}\, dt \right|^2 \cong \lim_{T\to\infty} \frac{1}{T} |\bar{\varepsilon}(\omega)|^2 \tag{3-85}$$

The Fourier transform mate to $\phi_{\varepsilon\varepsilon}(\tau)$ is

$$\bar{P}_\varepsilon(\omega) = \int_0^\infty \phi_{\varepsilon\varepsilon}(\tau) \cos \omega\tau \, d\tau \tag{3-86}$$

Equations (3-84) and (3-86) comprise Wiener's theorem, which asserts that the output noise fluctuation autocorrelation function and the output noise fluctuation power spectral density are Fourier cosine transform mates. This holds for the cross-correlation function as well. That is, suppose

$$\phi_{\varepsilon K}(\tau) = \lim_{T\to\infty} \frac{1}{T} \int_{-T/2}^{+T/2} \varepsilon(t) K(t+\tau)\, dt \tag{3-87}$$

is the cross-correlation function between the reactivity input and the neutron density fluctuation output; then the corresponding Wiener's theorem is that

$$\phi_{\varepsilon K}(\tau) = 2 \int_0^\infty \bar{P}_{\varepsilon K}(\omega) \cos \omega\tau \, d\omega \qquad \bar{P}_{\varepsilon K}(\omega) = \int_0^\infty \phi_{\varepsilon K}(\tau) \cos \omega\tau \, d\tau \tag{3-88}$$

and the cross-spectral density function

$$\bar{P}_{\varepsilon K}(\omega) = \lim_{T\to\infty} \frac{1}{T} |\bar{\varepsilon}(\omega)\bar{K}(\omega)|^2$$

For a linear system—in other words, the transfer function relationship $\bar{\varepsilon}(\omega) = \bar{G}(\omega)\bar{K}(\omega)$ holds, where $\bar{G}(\omega)$ is the transfer function† and $\bar{K}(\omega)$ is the Fourier transform of the reactivity—the definition of the neutron level fluctuation output spectral density provides

$$\bar{P}_\varepsilon(\omega) = \lim_{T\to\infty} \frac{1}{T} |\bar{\varepsilon}(\omega)|^2 = \lim_{T\to\infty} \frac{1}{T} |\bar{G}(\omega)\bar{K}(\omega)|^2 \tag{3-89}$$

or

$$\bar{P}_\varepsilon(\omega) = |\bar{G}(\omega)|^2\, \bar{P}_K(\omega) \tag{3-90}$$

† $\bar{G}(\omega)$ here is the total, or closed-loop, transfer function.

with the reactivity input spectral density given by

$$\bar{P}_K(\omega) = \lim_{T\to\infty} \frac{1}{T} |\bar{K}(\omega)|^2$$

Then the transfer function amplitude response can be obtained from

$$|\bar{G}(\omega)| = \left[\frac{\bar{P}_\varepsilon(\omega)}{\bar{P}_K(\omega)}\right]^{1/2} \tag{3-91}$$

Therefore, if one obtains the spectral density of the input reactivity and the spectral density of the output neutron density fluctuations from, say, their respective autocorrelation functions, the square root of their quotient gives the amplitude response of the transfer function, as above. Usually it can be assumed that the input reactivity noise is at least slowly varying over the relatively low frequency region for which $\bar{G}(\omega)$ changes appreciably. That is, if $\bar{K}(\omega)$ is reasonably "flat" or "white" over most of the frequency range of $\bar{G}(\omega)$ less than the corner frequency given by $\omega = 1/\Lambda$, then $\bar{P}_K(\omega) \cong 1$ and the amplitude response is simply

$$|\bar{G}(\omega)| = [\bar{P}_\varepsilon(\omega)]^{1/2} \approx |\bar{\varepsilon}(\omega)| \tag{3-92}$$

Then $|\bar{G}(\omega)|$ is the square root of the fluctuation power output spectral density, and it is also equal to the modulus of the Fourier transform of the neutron density fluctuation if available.

The autocorrelation method above yields only the amplitude response of the transfer function—all phase information is lost. Cross-correlation methods, though entailing more experimental effort as explained below, will yield both the phase and amplitude response of the transfer function. From Wiener's theorem in Eq. (3-58)

$$\bar{\phi}_{\varepsilon K}(\omega) = \bar{P}_{\varepsilon K}(\omega) \tag{3-93}$$

where the barred quantities are the Fourier transforms of the corresponding time functions, which in this case are the input-output cross-correlation function $\bar{\phi}_{\varepsilon K}(\omega)$ and the cross-spectral power density $\bar{P}_{\varepsilon K}(\omega)$. Since the cross-correlation function is

$$\phi_{\varepsilon K}(\tau) = \lim_{T\to\infty} \frac{1}{T} \int_{-T/2}^{+T/2} \varepsilon(t) K(t+\tau)\, dt \tag{3-94}$$

inserting the inverse transform of $K(t+\tau)$ gives

$$\phi_{\varepsilon K}(\tau) = \lim_{T\to\infty} \frac{1}{T} \int_{-T/2}^{+T/2} \bar{\varepsilon}(-\omega) \bar{K}(\omega)\, e^{i\omega\tau}\, d\omega \tag{3-95}$$

$\bar{\varepsilon}(\omega)$ is even and $\bar{\varepsilon}(\omega) = \bar{G}(\omega)\bar{K}(\omega)$, so that

$$\phi_{\varepsilon K}(\tau) = \int_{-\infty}^{+\infty} \bar{G}(\omega) \bar{P}_K(\omega)\, e^{i\omega\tau}\, d\omega \tag{3-96}$$

where $\bar{P}_K(\omega) = \lim_{T\to\infty} (1/T)|\bar{K}(\omega)|^2$. The transform mate to $\phi_{\varepsilon K}(\tau)$ is $\bar{\phi}_{\varepsilon K}(\omega)$, and

$$\bar{G}(\omega)\bar{P}_K(\omega) = \bar{\phi}_{\varepsilon K}(\omega) = \bar{P}_{\varepsilon K}(\omega) \tag{3-97}$$

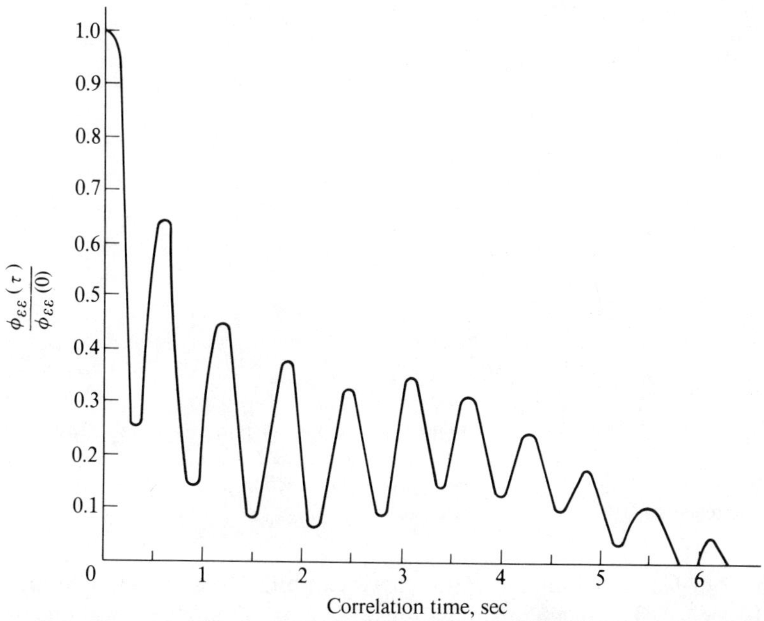

Figure 3-8 Autocorrelation function obtained from EBWR power traces (Fig. 3-6) at 45.6 MW, 600 psig.[18]

the last equality being available from Eq. (3-93). Then the transfer function in both phase and amplitude is, with $\bar{\phi}_{KK}(\omega)$ being the transform of the autocorrelation function of the reactivity input

$$\bar{G}(\omega) = \frac{\bar{P}_{\varepsilon K}(\omega)}{\bar{P}_K(\omega)} = \frac{\bar{\phi}_{\varepsilon K}(\omega)}{\bar{\phi}_{KK}(\omega)} \tag{3-98}$$

or the transfer function is given by dividing the input-output cross spectral density by the reactivity input spectral density. This technique for obtaining the transfer function gives good results and is elaborated on later, but first it is well to give another illustration of the use of autocorrelation functions.

From the EBWR power traces in Fig. 3-7, an autocorrelation function is computed by using Eq. (3-79). It is shown in Fig. 3-8. With the autocorrelation function of Fig. 3-8, its Fourier cosine transform is obtained [Eq. (3-86)]; it is plotted in Fig. 3-9.

It is of interest to compare $|\bar{\varepsilon}(\omega)| \cong [\bar{P}_\varepsilon(\omega)]^{1/2}$, where $[\bar{P}_\varepsilon(\omega)]^{1/2}$ is obtained from autocorrelation (Fig. 3-9), with $|\bar{\varepsilon}(\omega)|$ obtained directly from the Fourier transform of neutron density fluctuations, after Deshong.[19] The comparison is plotted in Fig. 3-10.

The curves in Figs 3-9 and 3-10 are smoothed from a dispersion of points obtained from the calculation of the corresponding quantities. For example, all of the integrals in the computation are replaced by summations whose addends are sampled from experimental data in discrete time increments.[18]

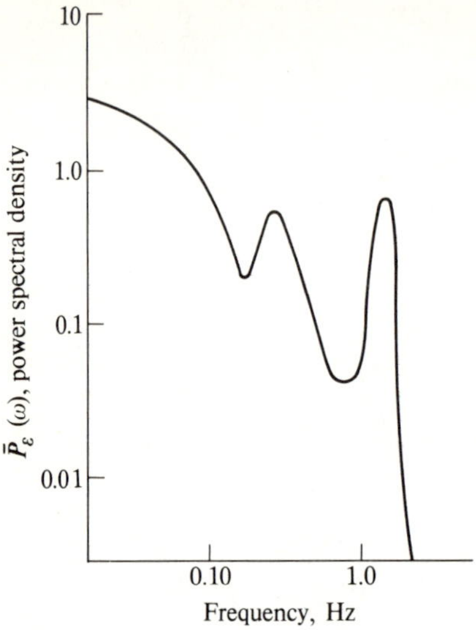

Figure 3-9 Smoothed power spectral density of EBWR noise obtained from cosine transform of autocorrelation function (Fig. 3-7) at 45.6 MW, 600 psig.

It is seen from Fig. 3-9 that the $\bar{P}_\varepsilon(\omega)$ curve exhibits a resonance at about 1.6 Hz ($\omega = 10\ \text{sec}^{-1}$). The mechanism assumed responsible for the resonance is discussed in Sec. 9-3 on boiling reactors. By definition, $[\bar{P}_\varepsilon(\omega)]^{1/2} \cong |\bar{\varepsilon}(\omega)|$: the square root of the neutron fluctuation spectral density is equivalent to its amplitude response within dimensionality considerations. As mentioned above, $|\bar{\varepsilon}(\omega)|$ can be obtained directly from the cosine transform of the pen recorder power output fluctuations because of the low frequencies involved. Figure 3-10 compares this $|\bar{\varepsilon}(\omega)|$ with that obtained by taking the square root $[\bar{P}_\varepsilon(\omega)]^{1/2}$, where $\bar{P}_\varepsilon(\omega)$ is gotten from the cosine transform of the autocorrelation function $\phi_{\varepsilon\varepsilon}(\tau)$ of the same power output fluctuations.[18,19] The curves compare rather well, as they should. Since the reactor is a low-frequency device, as opposed to radar systems whose amplifier noise outputs are in the hundred-kilohertz range, an important point with respect to reactor safety can be made. It is that even though a reactor might operate in the neighborhood of a resonance, a behavior that presages instability, its response is normally so slow that there is sufficient time for the controller to appraise the situation and move the reactor state point away from the instability by normal control rod action. This, of course, assumes that the reactor is operating below the prompt critical regime.

Also discussed earlier were cross-correlation techniques which yield both the phase and amplitude portions of the transfer function. Then $\bar{G}(\omega)$ is available as opposed to merely the amplitude $|\bar{G}(\omega)|$. The additional experimental difficulties entail the introduction of an external noise reactivity into the reactor. However, this is usually much less elaborate than the installation of a reactor oscillating rod system. Besides compensating for reactor power drift, cross-correlation methods eliminate extraneous instrument noise. The noise reactivity input can be obtained from an absorber placed within the reactor and driven in a randomly reciprocating manner. The absorber is electromechanically driven by a solenoid

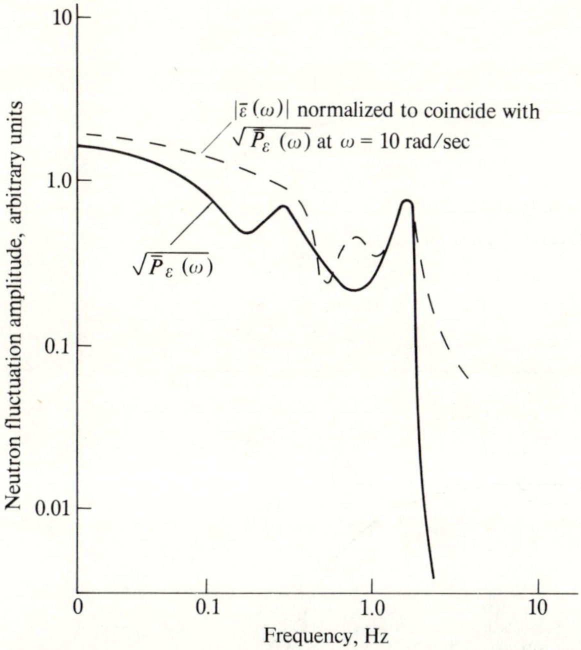

Figure 3-10 Comparison of EBWR smoothed $\{\bar{P}_\varepsilon(\omega)\}^{1/2} - \{\bar{\varepsilon}(\omega)\}$ obtained directly from Fourier transform of power fluctuations.[18,19]

in turn energized by pulses randomly triggered by the noise of a radioactive source; cf. Rajagopal.[20] The correlation functions are obtained from associated electronic playback equipment.

Instead of using the random emanations of a radioactive source from which to derive the random noise stimulus, pseudo-random binary sequences can be employed. Such sequences can be manufactured and outputted for this use from suitable computer program software. They can be tailored to fit the corresponding frequency spectral needs of the reactor system under investigation. That is, the required frequency bandwidth from the pseudo-random input can be adjusted through changes in its temporal parameters, even though such signals are deterministic, binary, and periodic. A prime example of their use was that of obtaining the transfer function of the KIWI-A′ nuclear rocket reactor prototype, where other means for its determination would have been well nigh impossible because of physical constraints peculiar to this type of reactor. (Balcomb, et al.[21])

From Eq. (3-98), for a reactor at zero power,

$$\bar{G}(\omega) = \frac{\bar{\phi}_{\varepsilon K}(\omega)}{\bar{\phi}_{KK}(\omega)} = \{i\omega[\Lambda + \bar{f}(i\omega)]\}^{-1} \tag{3-99}$$

Then $\bar{G}(\omega)$ can be obtained by cross-correlating the neutron density fluctuations ε with the random reactivity noise input K to get $\bar{\phi}_{\varepsilon K}(\omega)$. Autocorrelating the reactivity input gives $\bar{\phi}_{KK}(\omega)$, and the quotient yields $\bar{G}(\omega)$. The comparison of

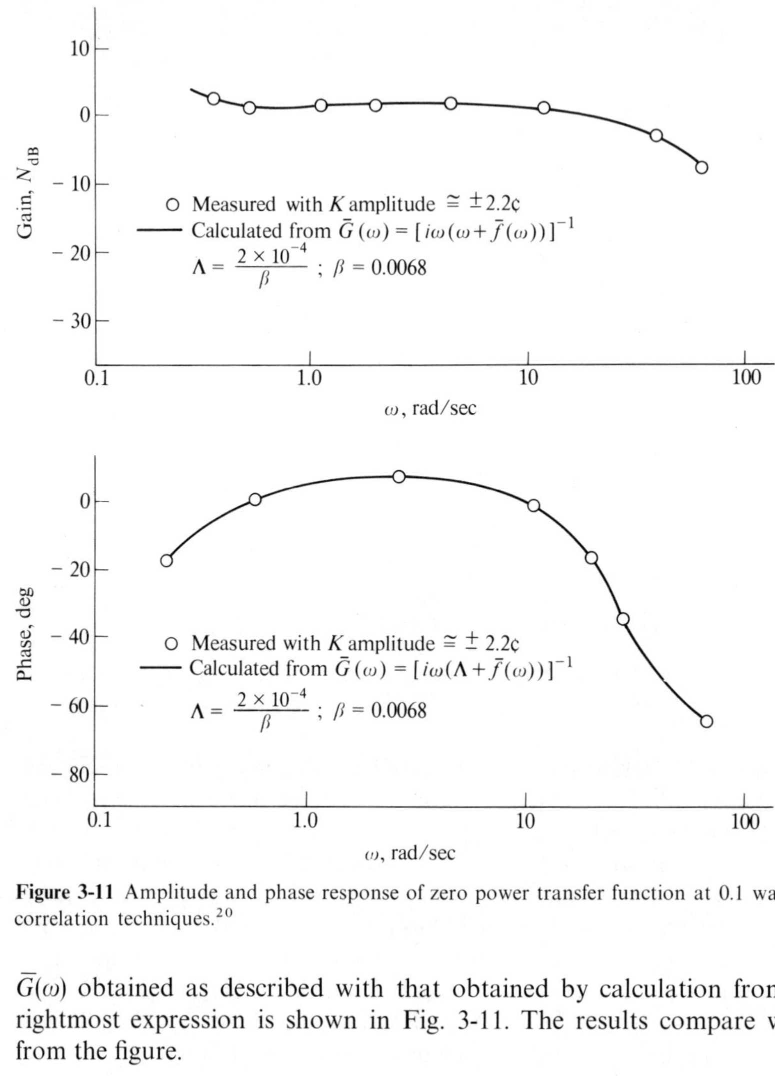

Figure 3-11 Amplitude and phase response of zero power transfer function at 0.1 watt, using cross-correlation techniques.[20]

$\bar{G}(\omega)$ obtained as described with that obtained by calculation from the above rightmost expression is shown in Fig. 3-11. The results compare well, as seen from the figure.

3-8 REACTIVITY NOISE

The techniques of the preceding section allow the investigation of at least the statistical nature of the reactivity noise input that causes the random neutron level fluctuations about equilibrium. The noise spectrum is easily obtained from the fact that

$$|\bar{K}(\omega)| = \frac{|\bar{\varepsilon}(\omega)|}{|\bar{G}(\omega)|} \tag{3-100}$$

which assumes that the fluctuation amplitudes are limited, so that system linearity is preserved. Equivalently

$$|\bar{K}(\omega)| \cong \frac{\bar{P}_\varepsilon(\omega)}{|\bar{G}(\omega)|} \tag{3-101}$$

where $\bar{P}_\varepsilon(\omega)$ is obtained from the autocorrelation of $\varepsilon(t)$, as described earlier, and $\bar{G}(\omega)$ is obtained from rod oscillator experiments. The experimental results from the EBWR at 45.6 MW are used to obtain such a $|\bar{K}(\omega)|$ as is plotted in Fig. 3-14.

To illustrate the ensemble averaging used to obtain correlation functions and gain some insight into the character of the reactivity noise, a feasible noise reactivity will be inferred from $|\bar{K}(\omega)|$. It will be synthesized by assuming random positive and negative reactivity waveforms of a shape to be determined. The waveforms have random initial points, assumed to be Poisson distributed in time, with a mean of k pulses per second. Their amplitude is also random with a given probability density $p(K)$. The specific underlying mechanism generating the pulses will not be examined here. Such investigations are made by postulating a reactivity feedback model as discussed in the sections in Chap. 9 on boiling-water reactors. The reactivity pulses (waveforms) might appear as drawn in Fig. 3-12. Let $dP(K_i, K_j; \tau)$ be the joint probability that the random reactivity magnitude at the ith time lies in dK_i about K_i and that at the jth time lies in dK_j about K_j where $\tau = t_i - t_j$. Then the reactivity autocorrelation function is

$$\phi_{KK}(\tau) = \int_{-\infty}^{+\infty} \int_{-\infty}^{+\infty} K_i K_j \, dP(K_i, K_j; \tau) \tag{3-102}$$

The above integration is over the ensemble of reactivity stationary time series for a given τ as shown in Fig. 3-13.

First the autocorrelation function of Poisson-distributed impulses (Dirac delta functions of a given area) will be obtained. Then the autocorrelation function for Poisson-distributed random waveforms whose collective time shape yields a spectrum corresponding to $\bar{K}(\omega)$, easily follows. From Eq. (3-102)

$$\phi_{KK}(\tau) = \int_{-\infty}^{+\infty} \int_{-\infty}^{+\infty} K_i K_j p(K_i) \, dP(K_i | K_j; \tau) \tag{3-103}$$

where $dP(K_i | K_j; \tau)$ is the conditional probability of reactivity impulse amplitude K_j, given K_i, where K_i and K_j are τ units of time apart. To calculate the autocorrelation function of Poisson-distributed Dirac impulses, consider an increment $d\tau$ about t_i, where a reactivity impulse of expected amplitude $K_i p(K_i)$ can occur. The probability of finding such an impulse in $d\tau$ is $k\, d\tau/2$, where k is the

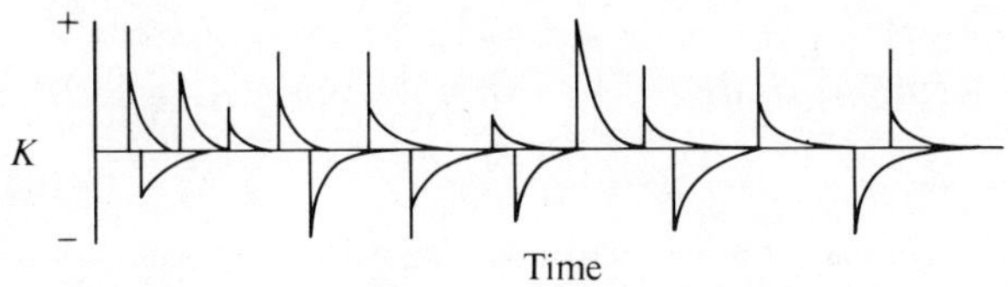

Figure 3-12 Feasible random reactivity noise pulses.

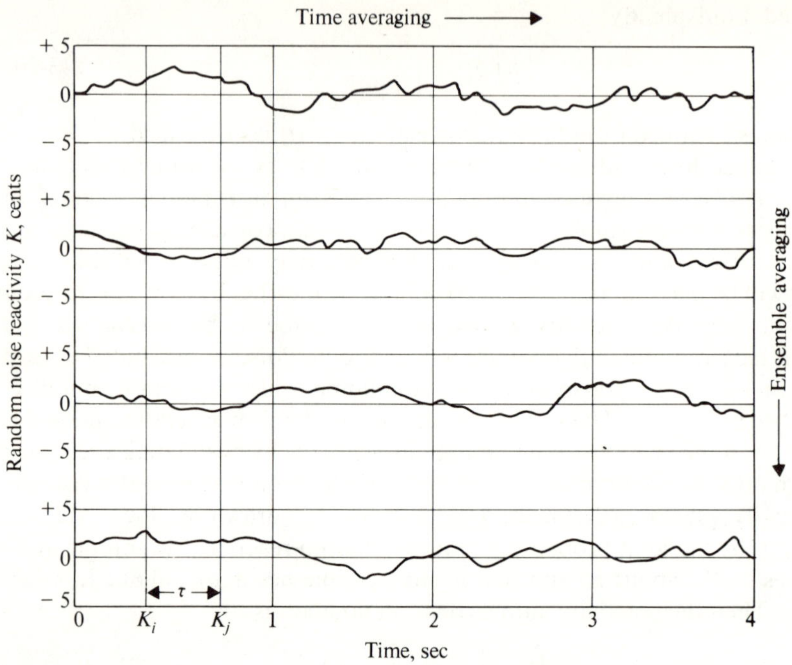

Figure 3-13 Ensemble of reactivity noise stationary time series.

mean number of impulses per second.† That is, the Poisson probability of finding n pulses in time τ is $(k\tau)^n e^{-k\tau}/n!$; so that for one pulse in time $d\tau$, it is $k\,d\tau$ to first order in $d\tau$. The factor $\frac{1}{2}$ is due to the fact that positive and negative reactivity impulses occur with equal probability. Equation (3-103) is now partitioned as

$$\phi_{KK}(\tau) = \int_{-\infty}^{+\infty}\int_{-\infty}^{+\infty} K_i K_i p(K_i)\, dP(K_i \mid K_i; 0) + \int_{-\infty}^{+\infty}\int_{-\infty}^{+\infty} K_i K_j p(K_i)\, dP(K_i \mid K_j; \tau) \qquad (3\text{-}104)$$

where the first term is the ensemble average of reactivity amplitudes in the same interval $d\tau$ ($\tau = 0$) and the second term is the ensemble average of those in different intervals ($\tau \neq 0$). The first term gives

$$\phi_{KK}(0) = \int_{-\infty}^{+\infty}\int_{-\infty}^{+\infty} K_i K_i p(K_i)\, dP(K_i \mid K_i; 0)$$
$$= \lim_{\substack{K \to \infty \\ K d\tau = 1}} \left[\int_{-\infty}^{+\infty}\int_{-\infty}^{+\infty} K_i^2 p(K_i) \frac{k}{2}\, d\tau\, dK_i + \int_{-\infty}^{+\infty}\int_{-\infty}^{+\infty} (-K_i)^2 p(K_i) \frac{k}{2}\, d\tau\, dK_i \right] \qquad (3\text{-}105)$$

† Thie[22] discusses the autocorrelation time τ^*, which marks the effective autocorrelation cutoff, as a function of essentially the number of zero crossings per second of the power fluctuations. Assuming that the reactivity zero crossings mirror the power zero crossings, with $\tau^* \cong 1/2\pi$ sec for EBWR at 45.6 MW, $k \cong 1/\pi\tau^* = 2$ crossings per second.

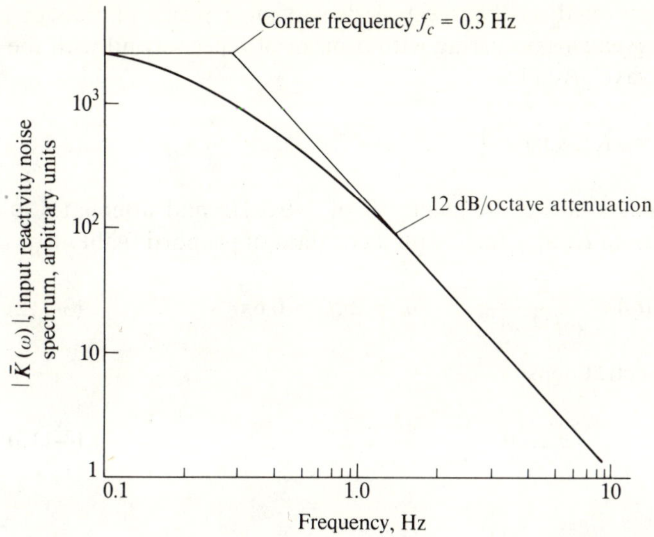

Figure 3-14 EBWR reactivity noise spectrum obtained experimentally from $\{\bar{P}_\varepsilon(\omega)\}^{1/2}/\{\bar{G}(\omega)\}$ at 45.6 MW.[16]

or

$$\phi_{KK}(0) = \sigma^2 k\delta(0) \tag{3-106}$$

where the variance $\sigma^2 = \int_{-\infty}^{+\infty} K_i^2 p(K_i)\, dK_i$ and $\delta(0)$ is the unit impulse at $\tau = 0$.

The random reactivity amplitudes which happen to be spaced τ units of time apart are assumed independent for $\tau \neq 0$. Then the second term in Eq. (3-104) vanishes with $dP(K_i | K_j; \tau) = p(K_i)p(K_j)\, dK_i\, dK_j$, so that

$$\int_{-\infty}^{+\infty} \int_{-\infty}^{+\infty} K_i K_j p^2(K_i) p(K_j)\, dK_i\, dK_j = 0 \tag{3-107}$$

since the mean reactivity K_i vanishes as the system is assumed in equilibrium in the mean. Then the autocorrelation function of the random impulses is

$$\phi_{KK}(\tau) = \sigma^2 k\delta(\tau) \tag{3-108}$$

with a corresponding spectral density $\bar{P}_\delta(\omega) = \int_{-\infty}^{+\infty} \sigma^2 k\delta(\tau) \cos \omega\tau\, d\tau$, that is

$$\bar{P}_\delta(\omega) = \sigma^2 \frac{k}{2\pi} \tag{3-109}$$

The autocorrelation function of Poisson-distributed impulses of area $\sigma^2 k$ is given above; it is equivalent to multiplying the autocorrelation of the individual waveshape (an impulse here) by k, the mean number of pulses per second. If the above impulses are fed to a hypothetical pulse-shaping network whose transfer function is $\bar{K}(\omega)$, the spectral density output is

$$\bar{P}_K(\omega) = \frac{k}{2\pi} \sigma^2 |\bar{K}(\omega)|^2 \tag{3-110}$$

$\bar{P}_K(\omega)$ can now be interpreted as the spectral density of a series of Poisson-distributed reactivity waveforms occurring with a mean of k per second with the form of the individual wave given by

$$K_{\text{pulse}}(t) = \int_{-\infty}^{+\infty} \bar{K}(\omega)\, e^{i\omega t}\, d\omega \tag{3-111}$$

From Fig. 3-13 $\bar{K}(\omega)$ has a corner frequency f_c of $\cong 0.3$ Hz and attenuates at 12 db per octave. A feasible $\bar{K}(\omega)$ is then, with a constant of proportionality $\bar{K}_0$

$$\bar{K}(\omega) = \frac{\bar{K}_0}{\omega_c^2 + \omega^2}; \qquad \omega_c = 2\pi f_c = 0.6\pi \tag{3-112}$$

and a corresponding spectral density

$$\bar{P}_K(\omega) = \frac{(k/2\pi)\bar{K}_0\sigma^2}{(\omega_c^2 + \omega^2)^2} \tag{3-113}$$

Then

$$K_{\text{pulse}}(t) = \begin{cases} \bar{K}_0 \sigma t\, e^{-\omega_c t} & t \geq 0 \\ 0 & t < 0 \end{cases} \tag{3-114}$$

Therefore, the reactivity noise $\bar{K}(\omega)$ can be realized by a Poisson-distributed series of random pulses with the above waveform. A reactivity generation model can be inferred, for example, from the fact that $K_{\text{pulse}}(t)$ satisfies a second-order differential equation whose constant coefficients are made up of the pulse parameters.

3-9 PERIOD TRANSFER FUNCTION

As seen from the perturbation theory development of the transfer function in Sec. 3-2, the reactor was assumed to be in equilibrium to zeroth order, that is, $K^{(0)} = 0$. In reality, reactors are seldom in actual equilibrium, but are varying on positive or negative stable periods, $T = 1/\lambda$, such that the neutron density is slowly waxing or waning with time. This means that to zeroth order $K^{(0)} = K_0$, a constant reactivity satisfying the inhour equation $K_0\bar{G}(\lambda) = 1$ instead of zero $K^{(0)}$. Then the zeroth and first order perturbation equations are taken from the discussion of Sec. 3-2, respectively,

$$n^{(0)} = n_0 + \int_0^t G(t-\tau)K^{(0)}n^{(0)}\, d\tau \tag{3-115}$$

$$n^{(1)} = \int_0^t G(t-\tau)(K^{(0)}n^{(1)} + K^{(1)}n^{(0)})\, d\tau \tag{3-116}$$

With the reactor on a stable period, Eq. (3-115) becomes

$$n^{(0)} = n_0 + K_0 \int_0^t G(t-\tau)n^{(0)}(\tau)\, d\tau \tag{3-117}$$

and using the Laplace transform gives $\bar{n}^{(0)}(s)$, the transform of the zeroth order neutron density is

$$\bar{n}^{(0)}(s) = n_0/s(1 - K_0\bar{G}(s)) \tag{3-118}$$

Then the Laplace transform of Eq. (3-116) yields

$$\bar{n}^{(1)}(s) = \frac{\bar{G}(s)}{2\pi i}\int_{\Gamma} \bar{K}^{(1)}(s - z)\bar{n}^{(0)}(z)\,dz/(1 - K_0\bar{G}(s)) \tag{3-119}$$

where Γ is the Laplace transform inversion contour, and the complex variable integral representation is that for the Laplace transform of the product $K^{(1)}(t)n^{(0)}(t)$. For a stable period then $\bar{n}^{(0)}(s) \sim n_0/(s - \lambda)$. Inserting this into the right-hand side of Eq. (3-119) allows an easy evaluation of the integral due to the pole at $s = \lambda$, resulting in

$$\bar{n}^{(1)}(s) = n_0\bar{G}(s)\bar{K}^{(1)}(s - \lambda)/(1 - K_0\bar{G}(s)) \tag{3-120}$$

Letting $s \to s + \lambda$, and using $K_0 = \bar{G}^{-1}(\lambda)$ results in

$$\bar{n}^{(1)}(s + \lambda) = n_0\bar{K}^{(1)}(s)/(\bar{G}^{-1}(s + \lambda) - \bar{G}^{-1}(\lambda)) \tag{3-121}$$

The factor above multiplying $K^{(1)}(s)$ is the period transfer function, that is, $\bar{G}_p^{-1}(s) = \bar{G}^{-1}(s + \lambda) - \bar{G}^{-1}(\lambda)$, or

$$\bar{G}_p(s) = \frac{n_0}{s\left(\Lambda + \sum_{i=1}^{6} a_i\lambda_i/(s + \lambda + \lambda_i)(\lambda + \lambda_i)\right)} \tag{3-122}$$

It is apparent that the integral in Eq. (3-119) could be avoided by substituting $n^{(0)} \sim n_0\,e^{\lambda t}$ into Eq. (3-116), and taking the Laplace transform of the result, ultimately yielding $\bar{G}_p(s)$ above. Also, notice that for actual equilibrium, that is in the limit of vanishing λ, $\bar{G}_p(s) = \bar{G}(s)$, as it should.

Manifestations of the period transfer function have been found in nuclear rocket reactors, discussed in Chap. 9, which developed instabilities during certain types of startup regimes, cf. Singer.[23]

3-10 DESCRIBING FUNCTIONS

As has been discussed, linear stability analysis using transfer functions has a limited application. Transfer function methods analyze only stability in the small, i.e., stability behavior for small fluctuations about an equilibrium power level. What is often more important is to investigate stability and control in the large. Is the system stable for large fluctuations about an equilibrium power level? There are a number of particular methods for making inroads into this question; they are discussed in later chapters. The describing function technique is a step in this direction; it corresponds to the next level above the transfer function method for analyzing the stability of quasi-linear systems. One aspect of its efficacy rests on the fact that resonances appearing in system transfer functions usually correspond to frequencies of potential instability. That is, if a periodic reactivity of nominal amplitude is used to oscillate the reactor power level at its dominant resonant frequency, then the power fluctuations can become dangerously large and stability margins will be reduced to intolerable levels.

The point is that in the neighborhood of the resonant frequency, the reactor acts like a bandpass filter. Then all input frequencies that are harmonics of the neutron level output resonance frequency are severely attentuated because they are displaced far outside the resonance skirt. The resonance thereby essentially passes only the fundamental, or first harmonic of the input, so that the neutron level output can be considered to consist solely of the input fundamental frequency. Then the stability of the system can be investigated by construing the fundamental frequency component of the neutron level as a quasi-linear response of the system. However, this response will be dependent on the amplitude of the reactivity input in the same way that the next level of approximation to the frequency of a simple pendulum is amplitude-dependent (cf. Prob. 3-1). The describing function technique is not limited to the resonance characteristics of systems. Most systems, including reactors, have good low-frequency characteristics, i.e., they are also good low-pass filters as well, and thereby attenuate the harmonics of fundamental frequencies that lie in the passband. Then the previous assumption of the output consisting almost entirely of the fundamental frequency of the input can be made.

Consider a nonlinear device whose response $n_d(t)$ to a sinusoidal input $A \sin \omega t$ is the imaginary part of

$$n_d(t) = \sum_{r=-\infty}^{+\infty} N_r(A, \omega)\, e^{ir\omega t} \tag{3-123}$$

with Fourier coefficients

$$N_r(A, \omega) = \frac{\omega}{2\pi} \int_{-\pi/\omega}^{\pi/\omega} n_d(t)\, e^{-ir\omega t}\, dt \tag{3-124}$$

If this nonlinear output can be represented by its describing function, all harmonics of its output but the first are neglected. Then

$$n_d(t) \cong \operatorname{Im} N_1(A, \omega)\, e^{i\omega t} = |N_1(A, \omega)| \sin(\omega t + \phi) \tag{3-125}$$

where $\phi(A, \omega) = \tan^{-1}[\operatorname{Im} N_1(A, \omega)/\operatorname{Re} N_1(A, \omega)]$. The describing function $\bar{D}(A, \omega)$ is then defined as

$$\bar{D}(A, \omega) = \frac{|N_1(A, \omega)|}{A} e^{i\phi(A, \omega)} \tag{3-126}$$

where it is seen that the describing function depends on the amplitude of the input. Equation (3-125) can now be written to good approximation as

$$n_d(t) \cong |\bar{D}(A, \omega)|\, A \sin(\omega t + \phi) \tag{3-127}$$

neglecting all harmonics higher than the first. If the device were truly linear, the output response would be an exact frequency replica of the input but amplified by $|\bar{D}(A, \omega)|$ and shifted in phase by ϕ radians. No harmonics but the first would be manifest in the output, and the describing function would revert to the input amplitude-independent transfer function $\bar{D}(\omega) = \bar{G}(\omega)$ and the two sides of Eq. (3-127) would be exactly equal.

In the present context, it is important to notice that the zero-power kinetics equations are linear. That is, the integral equation form of these kinetics equations

$$n(t) = 1 + \int_0^t G(t-\tau)K(\tau)n(\tau)\,d\tau \tag{3-128}$$

is a linear nonhomogeneous integral equation as discussed in Chap. 2. Let $n(t)$ be expanded in a Fourier sine series and inserted into the asymptotic (steady state) form of Eq. (3-128), that is, with infinite upper limit. However, for a sinusoidal reactivity $K = K_0 \sin \omega t$, the modified equation (3-128) gives

$$n(t) = 1 + K_0 \int_0^\infty G(\tau) \sin \omega(t-\tau) n(t-\tau)\,d\tau \tag{3-129}$$

The first-harmonic amplitude of the sine series is then given by

$$|N_1(K_0, \omega)| = K_0 |\bar{G}(\omega)| \tag{3-130}$$

where

$$\bar{G}(\omega) = \int_0^\infty G(t)\, e^{-i\omega t}\,dt$$

the Fourier transform of $G(t)$. Therefore, the amplitude of the describing function

$$|\bar{D}(K_0, \omega)| = \frac{|N_1(K_0, \omega)|}{K_0} = |\bar{G}(\omega)| \tag{3-131}$$

is, trivially, the zero-power amplitude transfer function.

It is when the general nonlinear kinetics integral equation, which would, for example, include nonlinear feedback effects, that is, when

$$n(t) = 1 + \int_0^t G(t-\tau)K(\tau, n(\tau))\,d\tau \tag{3-132}$$

is investigated, that the describing function, which is not the zero-power amplitude response, can be obtained. Then depending on the functional $K(t, n(t))$, the describing function can have any number of forms. For example, certain nonlinear approximate solutions to the zero-power kinetics equation (3-128) yield low power describing functions. Let us investigate, in the manner of Smets,[24] the describing function of a reactor, beginning with Eq. (2-153), the approximate neutron level for periodic reactivities whose amplitude is not too large but is larger than that required to linearize adequately the kinetics equations as discussed in Sec. 2-10. This approximate solution to the kinetics equations is rewritten for a sinusoidal reactivity $K = K_0 \sin \omega t$ to give

$$n(t) = \exp\left[K_0 \int_0^t G(t-\tau) \sin \omega\tau\,d\tau\right] \tag{3-133}$$

As before, assume that the asymptotic form of the above $n(t)$ is the nonlinear reactor response to the above sinusoidal reactivity. Then after a long time

$$n(t) = \exp\left[K_0 \int_0^\infty G(\tau) \sin \omega(t-\tau)\, d\tau\right] \tag{3-134}$$

or

$$\begin{aligned} n(t) &= \exp\left(\frac{K_0}{2i}\{[e^{i\omega t}\bar{G}(\omega)] - [e^{i\omega t}\bar{G}(\omega)]^*\}\right) \\ &= \exp\left[K_0 |\bar{G}(\omega)| \sin(\omega t + \delta)\right] \end{aligned} \tag{3-135}$$

where the asterisk denotes the complex conjugate and $\delta = \arg \bar{G}(\omega)$. The amplitude of the first harmonic, or fundamental, divided by the input reactivity amplitude K_0 gives the describing function amplitude as

$$\begin{aligned} |\bar{D}(K_0,\omega)| &= \frac{|N_1(K_0,\omega)|}{K_0} \\ &= \frac{\omega}{\pi K_0}\int_0^{2\pi/\omega} \exp\left[K_0|\bar{G}(\omega)|\sin\omega t\right]\sin\omega t\, dt = \frac{2}{K_0} I_1(K_0|\bar{G}(\omega)|) \end{aligned} \tag{3-136}$$

where $I_1(x)$ is the Bessel function of imaginary argument. For small K_0, $I_1(x)$ is expanded for small x, yielding

$$|\bar{D}(K_0,\omega)| \cong |\bar{G}(\omega)|[1 + \tfrac{1}{8}K_0^2|\bar{G}(\omega)|^2 + \tfrac{1}{192}K_0^4|\bar{G}(\omega)|^4 + \cdots] \tag{3-137}$$

Then the neutron level is, from Eq. (3-137)

$$n(t) \cong K_0|\bar{G}(\omega)|[1 + \tfrac{1}{8}K_0^2|\bar{G}(\omega)|^2 + \cdots]\sin(\omega t + \phi) \tag{3-138}$$

Again, if K_0 is small enough, the magnitude of the describing function becomes the gain or amplitude response of the transfer function as the reactor reverts to a linear system.

Note that the first iterant of the kinetics equation (3-128)

$$n_r(t) = 1 + \int_0^t G(t-\tau)K(\tau)n_{r-1}(\tau)\, d\tau \qquad n_0(t) = 1 \qquad r = 1, 2, \ldots \tag{3-139}$$

is

$$n_1 = 1 + \int_0^t G(t-\tau)K(\tau)\, d\tau \tag{3-140}$$

A better zero-order iterant to start the above iteration is $n(t)$ given from Eq. (3-135). Then for $K = K_0 \sin \omega t$, and after a long time so that equilibrium is reached, the new first-order iterant is obtained from Eq. (3-135) inserted in the right-hand side of Eq. (3-139) to yield

$$\varepsilon_1(t) = n_1(t) - 1 \approx \int_0^\infty G(t-\tau)K_0 \sin\omega\tau\,\{\exp\left[K_0|\bar{G}(\omega)|\sin(\omega\tau + \delta)\right]\}\, d\tau \tag{3-141}$$

It is seen that the equivalent open-loop system for the above equation, which will be considered as the system describing function, consists of a linear amplifier whose transfer function is $\bar{G}(\omega)$ followed by a nonlinear amplifier whose transfer function is the transform of Eq. (3-135). This is depicted in Fig. 3-15, where $\bar{G}(\omega) = \{i\omega[\Lambda + \bar{f}(\omega)]\}^{-1}$.

As mentioned earlier, there is a certain arbitrariness in deriving the describing function for the kinetics equations. As another example of a low power describing function, consider the iteration in Eq. (3-139) to be continued. Taking into account the cutoff properties of $G(t)$, an approximate expression for $n(t)$ ensues after summing the appropriate geometric series (Prob. 2-9). It is the lower bound

$$n(t) \geq \left[1 - \int_0^t G(t-\tau)K(\tau)\,d\tau\right]^{-1} \tag{3-142}$$

assuming that the magnitude of the integral is less than unity. Near the resonance, its asymptotic form for a periodic reactivity narrows the inequality to the approximation

$$\begin{aligned} n(t) &\cong \left[1 - K_0 \int_0^\infty G(\tau)\sin\omega(t-\tau)\,d\tau\right]^{-1} \\ &= [1 - K_0|\bar{G}(\omega)|\sin(\omega t + \delta)]^{-1} \end{aligned} \tag{3-143}$$

The amplitude of the first harmonic now becomes

$$\begin{aligned} |N_1(K_0, \omega)| &= \frac{\omega}{\pi}\int_0^{2\pi/\omega} [1 - K_0|\bar{G}(\omega)|\sin\omega t]^{-1}\sin\omega t\,dt \\ &= \frac{2}{K_0|\bar{G}(\omega)|}\left[\frac{1}{[1-(K_0|\bar{G}(\omega)|)^2]^{1/2}} - 1\right] \end{aligned} \tag{3-144}$$

The corresponding describing function amplitude is

$$|\bar{D}_2(K_0, \omega)| = \frac{|N_1(K_0, \omega)|}{K_0} = \frac{2}{K_0^2|\bar{G}(\omega)|}\left[\frac{1}{[1-(K_0|\bar{G}(\omega)|)^2]^{1/2}} - 1\right] \tag{3-145}$$

expanding the above result for small $K_0|\bar{G}(\omega)|$ yields

$$|\bar{D}_2(K_0, \omega)| \cong |\bar{G}(\omega)|[1 + \tfrac{3}{4}K_0^2|\bar{G}(\omega)|^2 + \tfrac{5}{8}K_0^4|\bar{G}(\omega)|^4 + \cdots] \tag{3-146}$$

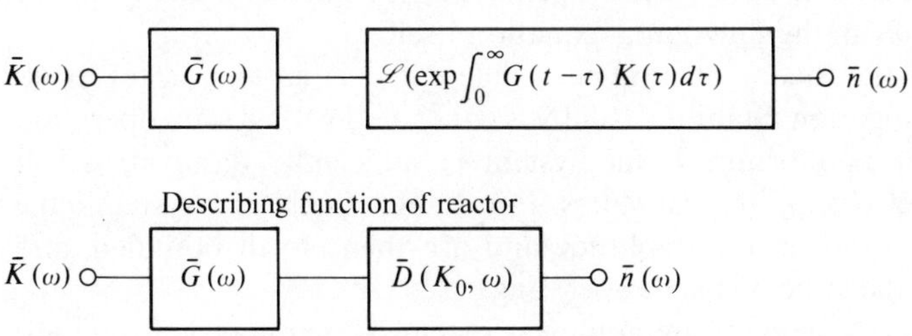

Figure 3-15 Schematic diagram of reactor open-loop system.[24]

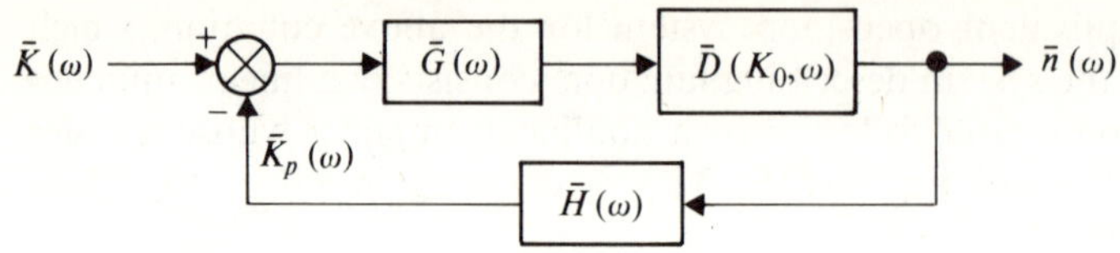

Figure 3-16 Closed loop reactor system.

At increased power, the coupling between the various reactivities in the reactor and the power level becomes much tighter, being represented by a power feedback reactivity $\bar{H}(\omega)$, that is, $\bar{K}_p(\omega) = \bar{H}(\omega)\bar{n}(\omega)$, where $\bar{K}_p(\omega)$ and $\bar{n}(\omega)$ are the Fourier transforms of the feedback reactivity and neutron density level, respectively. The equivalent closed-loop schematic diagram is depicted in Fig. 3-16. Then the over-all describing function is

$$\bar{n}(\omega)/\bar{K}_c(\omega) = \frac{\bar{D}(K_0, \omega)\bar{G}(\omega)}{1 + \bar{D}(K, \omega)\bar{G}(\omega)\bar{H}(\omega)} \tag{3-147}$$

Since all of the quantities in the describing function have good low-pass characteristics, the describing function model will render a better model of the system than the closed-loop transfer function (especially near a resonance) for the case of appreciably larger reactivities about an equilibrium level than can be handled by the transfer function. However, such reactivities are still much less than one dollar.

3-11 NONLINEAR KINETICS

To reemphasize, the kinetic behavior of a nuclear reactor is essentially nonlinear. Present-day reactors require sophisticated control and kinetics analysis because of their complex neutronics, structure, and exotic moderators, not to mention certain novel homogeneous types whose kinetic behavior is reflected in all kinds of nonlinear oscillations. A principal avenue of such reactor kinetic analysis seems to lie in the area of nonlinear differential equations.

Nonlinear methods are concerned in large part with stability. In most instances the nonlinear differential equations describing the reactor kinetics behavior cannot be solved in closed form. Then the approximate kinetic behavior must be investigated from the differential equation itself.

A good share of the theory of nonlinear differential equations is taken up with this type of investigation. Stability usually is concerned with system operation when disturbed from equilibrium. If the system is sufficiently damped, it will return to one of a set of equilibrium states. For stable oscillatory systems, the oscillations build up to a certain amplitude and are thenceforth bounded, and sustained, as long as the system is kept operating.

One of the difficulties of analyzing nonlinear reactor systems, aside from their mere complexity, is that the state of the mathematics of nonlinear differential equations is only partially developed. Some headway in this direction has been made in the USA, but it has not kept pace with developments made in other areas of applied mathematics such as probability theory, for example. Therefore, generalizations from results derived from one reactor system are difficult to make, so that each system must be analyzed individually.

Herein, investigations of nonlinear reactor systems will be discussed by using the so-called classical methods of nonlinear mathematical analysis. Questions of stability will be considered of first-priority importance.

Discontinuous nonlinear control behavior will not be discussed in the classical sense but will be attacked by using one of the newer, more powerful tools of analysis, which is dynamic programming. This methodology is sufficiently broad in application that parts of Chap. 6 are devoted to its espousal.

What we shall do is investigate a prototype nonlinear reactor kinetics problem, which is that of coupling the reactor temperature-density reactivity changes with the general kinetics equations. This is first attempted with no delayed neutrons and then followed up with an approximate method of taking them into account. Even these two seemingly simple examples will aptly demonstrate the vagaries and difficulties involved.

3-12 TEMPERATURE EFFECTS WITH NO DELAYED NEUTRONS

The following investigation is that of a reactor generating heat with a negative temperature coefficient of reactivity. Let the time rate of change of heat in the reactor be proportional to the deviation of the neutron density from equilibrium. This is the same as making the temperature rate proportional to the power deviation from equilibrium, that is

$$\dot{T} = \frac{n-1}{C} \qquad T(0) = 0 \tag{3-148}$$

where T = temperature of reactor
n = neutron density
C = heat capacity of reactor

With no delayed neutrons considered, the kinetics equation is simply

$$\dot{n} = (K - aT)\frac{n}{\Lambda} \qquad n(0) = 1 \tag{3-149}$$

where a is the temperature coefficient of reactivity due to the heated reactor. Λ is usually calculated on the basis of a mean lifetime in order to take into account the delayed neutrons to a crude approximation.† With $\dot{n} = \dot{T}\, dn/dT$ and Eqs

† The mean neutron lifetime over both delayed and prompt neutrons is given by

$$\sum_{i=1}^{6} \frac{a_i}{\lambda_i} + (1 - \beta)\Lambda \cong \bar{\tau}$$

since the prompt neutron lifetime is usually very much smaller than any delayed neutron period.

(3-148) and (3-149), the time is eliminated, resulting in

$$\frac{dn}{dT} = \frac{C}{\Lambda}(K - aT)\frac{n}{n-1} \qquad n(0) = 1 \tag{3-150}$$

If the initial reactivity change is a step function of magnitude K, the integral of the above is

$$n - \log n = \frac{C}{\Lambda}\left(KT - \frac{a}{2}T^2\right) + 1 \tag{3-151}$$

The plot of n as a function of T is depicted in Fig. 3-17. For each value of reactivity, the diagram yields a closed curve, called a limit cycle, implying stable bounded oscillations. One aspect of the reactor operation that can be gleaned from the above analysis is the extremes of temperature that are endured. These occur where $\dot{T} = 0$. From Eq. (3-148), $\dot{T} = 0$ when $n = 1$. For $n = 1$, Eq. (3-151) gives the extremum temperatures as

$$T_{\max} = \frac{2K}{a} \qquad T_{\min} = 0 \tag{3-152}$$

which gives an indication of the maximum temperature which the reactor structure must stand. It is also of interest to study this situation when time is the abscissa. Consider the case where Eq. (3-149) describes the neutron density for an initial positive step function insertion of reactivity of amount K. However, assume a cooling mechanism to be the reactor environment considered as an infinite heat sink. Then the temperature is governed by

$$C\dot{T} = n - \mu T \qquad T(0) = 0 \tag{3-153}$$

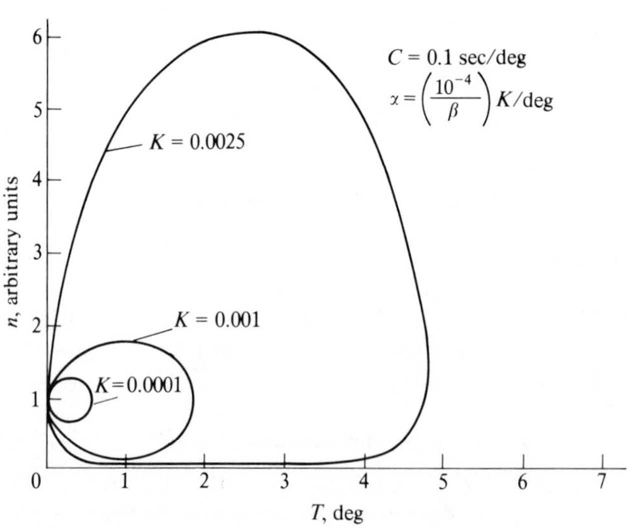

Figure 3-17 Phase plane plot of temperature-compensated system with no delayed neutrons.

where μ is an appropriate heat conduction coefficient. This particular model is studied in detail in Chap. 4 in the context of a family of such models on which stability investigations are made.

Letting $Q = \log n(t)$, and eliminating the temperature between Eqs (3-153) and (3-149), yields the following differential equation in Q.

$$\ddot{Q} + \frac{\mu}{C}\dot{Q} + \frac{a}{\Lambda C}e^{Q} = \frac{\mu}{\Lambda C}K \qquad Q(0) = 0 \qquad \dot{Q}(0) = \frac{K}{\Lambda} \tag{3-154}$$

This is almost akin to the standard second-order differential equation with a forcing function proportional to the "reactivity force." To make the analogy complete, let us investigate the situation for small changes of neutron density about equilibrium so that $e^{Q} \cong 1 + Q$. Then

$$\ddot{Q} + \frac{\mu}{C}\dot{Q} + \frac{a}{\Lambda C}Q = \frac{\mu K - a}{\Lambda C} \qquad K \ll 1 \tag{3-155}$$

Its solution exhibits the usual damped periodic behavior and is given as

$$\log n(t) = Q = \left(\frac{\mu K}{a} - 1\right) + e^{-\mu t/2C}\left\{\left(1 - \frac{\mu K}{a}\right)\cos \omega_0 t + \left[\frac{K}{\Lambda} + \frac{\mu}{2C}\left(1 - \frac{\mu K}{a}\right)\right]\frac{\sin \omega_0 t}{\omega_0}\right\} \tag{3-156}$$

with

$$\omega_0 = \frac{\mu}{2C}\left(\frac{4aC}{\Lambda\mu^2} - 1\right)^{1/2}$$

After a long time following the insertion of the reactivity step, the neutron density attains a new equilibrium state given by

$$\frac{n(\infty)}{n(0)} = e^{(\mu K/a) - 1} \tag{3-157}$$

3-13 APPROXIMATE TREATMENT OF DELAYED NEUTRONS IN THE PHASE PLANE

Equation (1-42), with the reactivity augmented by temperature dependence through a temperature reactivity coefficient, has an asymptotic form for long times after an initial step function excitation from equilibrium given by

$$\Lambda\dot{n}(t) = (K - aT)n - \int_0^{\infty} n'(\tau) f(t - \tau)\, d\tau \tag{3-158}$$

Again exploiting the sharp cutoff characteristic of $f(t)$ by removing $\dot{n}(\tau) \cong \dot{n}(t)$ from within the above integral gives, since $f(t) \equiv 0$ for $t < 0$

$$(\Lambda + \bar{\tau})\dot{n}(t) = (K - aT)n \tag{3-159}$$

$\Lambda \ll \bar{\tau}$ is assumed, which holds for most reactors, so that

$$\bar{\tau}\dot{n}(t) = (K - aT)n(t) \qquad n(0) = 1 \qquad K(0-) = 0 \qquad K(0+) = K \tag{3-160}$$

Equation (3-160) can be obtained, as well, from the integral equation (1-47) by using therein the asymptotic form of the kernel, $G(t) \sim (\bar{\tau})^{-1}$. The temperature dependence is given by

$$\dot{T} = \frac{n-1}{C} - \mu T \qquad T(0) = 0 \tag{3-161}$$

The second term is the heat loss to the reactor environment considered as an infinite heat sink. Equations (3-160) and (3-161) can be combined to yield

$$\frac{dn}{dT} = \frac{m(K - aT)n}{n - 1 - \mu CT} \tag{3-162}$$

where the following values for the parameters will be used in the illustration below:

$a = 10^{-4}$ dollars per deg; reactivity temperature coefficient
$C = 0.10$ sec per deg; reactor heat capacity
$m = 8 \times 10^{-3}\ \text{deg}^{-1}$; $m = C/\bar{\tau}$
$\mu = 0.01\ \text{sec}^{-1}$; heat transfer coefficient

Equation (3-162) is a first-order nonlinear differential equation of algebraic type amenable to the standard phase-plane treatment for such equations.† It has singular points (where $dn/dT = 0/0$) as given in Table 3-1.

Equation (3-162) is one of the following class of equations

$$\frac{dn}{dT} = \frac{Q(n, T)}{P(n, T)} \tag{3-163}$$

The singular points are the (n, T) coordinates that satisfy

$$P(n, T) = Q(n, T) = 0 \text{ simultaneously}$$

The particular type of singularity is determined by examining the solution behavior of Eq. (3-162) in the neighborhood of the singular points in turn. This is done by expanding Q and P separately in a Taylor series in n and T about the particular singular point and retaining only linear terms.

For the singular point (1) about $(0, -1/\mu C)$ in the (n, T) plane, the expansion linearizes Eq. (3-162) to give

$$\frac{dn}{dT} \cong \frac{m(K + a/\mu C)n}{n - 1 - \mu CT} \tag{3-164}$$

Near $(0, |1/\mu C)$ the equation is equivalent to the following pair of first-order

† This equation is the phase-plane equivalent of the nonlinear kinetic equation (3-154).

Table 3-1 Singular points of Eq. (3-162)

n	T	Singularity type
(1) 0	$-1/\mu C$	Saddle point (unstable)
(2) $1 + (\mu C K/a)$	K/a	Spiral point (stable)

differential equations. As written in vector matrix form, they are

$$\frac{d}{dt}\begin{bmatrix} n \\ 1 + \mu C T \end{bmatrix} = \begin{bmatrix} \frac{m}{\mu C}\left(K + \frac{a}{\mu C}\right) & 0 \\ 1 & -1 \end{bmatrix} \begin{bmatrix} n \\ 1 + \mu C T \end{bmatrix} \tag{3-165}$$

Since this system of equations is linear, exponential solutions used to ascertain the nature of the singular points are found by letting

$$\frac{d}{dt}\begin{bmatrix} n \\ 1 + \mu C T \end{bmatrix} = \begin{bmatrix} \lambda & 0 \\ 0 & \lambda \end{bmatrix} \begin{bmatrix} n \\ 1 + \mu C T \end{bmatrix} \tag{3-166}$$

Eliminating the derivatives between Eqs (3-165) and (3-166) gives a pair of algebraic equations

$$\begin{bmatrix} \lambda - \frac{m}{\mu C}\left(K + \frac{a}{\mu C}\right) & 0 \\ -1 & 1 + \lambda \end{bmatrix} \begin{bmatrix} n \\ 1 + \mu C T \end{bmatrix} = 0 \tag{3-167}$$

For nontrivial solutions of these, the determinant of their coefficients must vanish. This results in the usual characteristic equation for λ:

$$(\lambda + 1)\left[\lambda - \frac{m}{\mu C}\left(K + \frac{a}{\mu C}\right)\right] = 0 \tag{3-168}$$

or

$$\lambda = -1, \frac{m}{\mu C}\left(K + \frac{a}{\mu C}\right) \tag{3-169}$$

Then immediately

$$\begin{aligned} n &= A \exp\left[\frac{m}{\mu C}\left(K + \frac{a}{\mu C}\right)t\right] + B e^{-t} \\ 1 + \mu C T &= F \exp\left[\frac{m}{\mu C}\left(K + \frac{a}{\mu C}\right)t\right] + D e^{-t} \end{aligned} \tag{3-170}$$

Notice first that if none of the constants A, B, F, and D vanish, n and $1 + \mu C T$ do not intersect the origin (singular point) of the $(n, 1 + \mu C T)$ plane for any value of time. The only solutions intersecting the origin are obtained when $A = F = 0$

or $B = D = 0$. If $A = F = 0$, then $n = B\,e^{-t}$, $1 + \mu CT = D\,e^{-t}$ correspond to the straight lines

$$n = \pm(B/D)(1 + \mu CT)$$

which, for a chosen sign of B/D, are rays *approaching* the origin from opposite sides as time increases. Similarly, if $B = D = 0$, then

$$n = A\exp\left[\frac{m}{\mu C}\left(K + \frac{a}{\mu C}\right)t\right] \qquad 1 + \mu CT = F\left[\exp\frac{m}{\mu C}\left(K + \frac{a}{\mu C}\right)t\right]$$

are the straight lines $n = \pm(A/F)(1 + \mu CT)$ which are rays *emanating* from the origin with increasing time. All solutions, with the above exception $(A = F = 0)$, approach the origin and then recede as shown in Fig. 3-18, forming an unstable saddle point in the $(n, 1 + \mu CT)$ plane.

For the singular point (2) about $(1 + \mu CK/a, K/a)$ in the (n, T) plane, the expansion linearizes Eq. (3-162) so that

$$\frac{dn}{dT} = \frac{-m(a + \mu CK)(T - K/a)}{(n - 1 - \mu CK/a) - \mu C(T - K/a)} \tag{3-171}$$

or

$$\frac{d}{dt}\begin{bmatrix}\eta\\ \tau\end{bmatrix} = \begin{bmatrix}0 & -m(a + \mu CK)\\ 1 & -\mu C\end{bmatrix}\begin{bmatrix}\eta\\ \tau\end{bmatrix} \tag{3-172}$$

where $\eta = n - 1 - \mu CK/a$ and $\tau = T - K/a$. Again, to get exponential solutions, set

$$\frac{d}{dt}\begin{bmatrix}\eta\\ \tau\end{bmatrix} = \begin{bmatrix}\lambda & 0\\ 0 & \lambda\end{bmatrix}\begin{bmatrix}\eta\\ \tau\end{bmatrix} \tag{3-173}$$

and combine with Eq. (3-172) to give, with $\mu^2 \ll \mu$

$$\lambda_{\pm} = -\frac{\mu C}{2}\left\{1 \mp i\left[\frac{4m(a + \mu CK)}{\mu^2 C^2 - 1}\right]^{1/2}\right\} \tag{3-174}$$

Substituting into Eq. (3-173) gives linear combinations of oscillatory solutions as

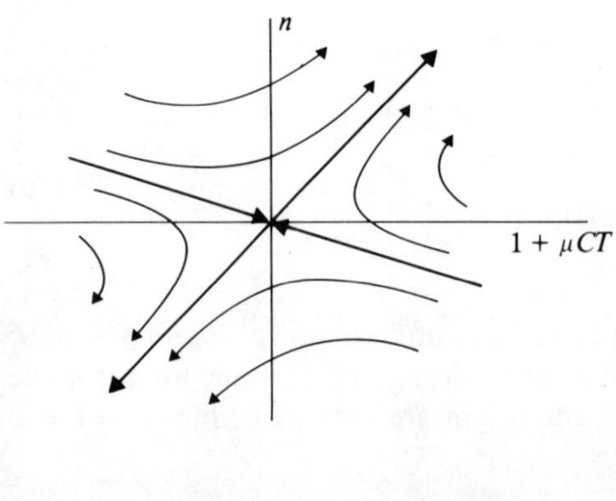

Figure 3-18 Unstable saddle point singularity in the $(n, 1 + \mu CT)$ plane $(K > 0)$.

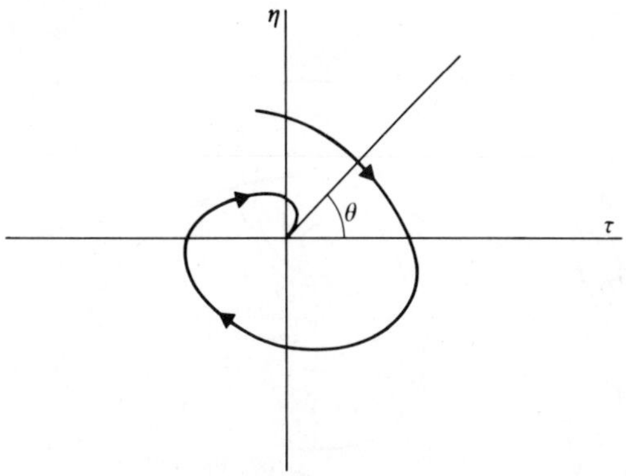

Figure 3-19 Stable spiral point singularity.

$$\begin{aligned}\eta &= e^{-\mu Ct/2}\left[A\cos\beta t + B\sin\beta t\right]\\ \tau &= e^{-\mu Ct/2}\left[C_1\cos\beta t + D\sin\beta t\right];\end{aligned} \tag{3-175}$$

where

$$\beta = \tfrac{1}{2}[4m(a+\mu CK) - \mu^2C^2]^{1/2}$$

In the (η, τ) plane, introduce polar coordinates; that is, let $\tan\theta = \eta/\tau$. Then, with Eq. (3-172)

$$\dot{\theta} = \frac{\tau\dot{\eta} - \eta\dot{\tau}}{\tau^2+\eta^2} = \frac{\mu C\eta\tau - m(a+\mu CK)\tau^2 - \eta^2}{\tau^2+\eta^2} < 0 \tag{3-176}$$

Using Eq. (3-172), $\dot{\theta}$ is negative since $-\eta^2$ dominates the numerator above.† As $\dot{\theta}$, and thereby θ, is decreasing, while η and τ are changing sign periodically, it is seen that the solutions in the neighborhood of singular point (2) spiral clockwise into the origin with increasing time. This is shown in Figs 3-19 and 3-20.

3-14 LIMIT CYCLES

It is of paramount importance for stability determination to ascertain the existence of limit cycle solutions in the phase plane. A limit cycle is a stable closed solution path usually surrounding a singular point. However, there seems to be no general way to determine whether or not a limit cycle exists. Only certain theorems that aid in the corroboration of the existence of limit cycles have been proved. One such is Bendixson's first theorem.[26] It states that in a region enclosed by a limit cycle, $\partial Q/\partial n + \partial P/\partial T$, where Q and P are defined in Eq. (3-163), must change sign. Or conversely, for no change in sign in the region, there is no limit cycle therein. From Eq. (3-162)

$$\frac{\partial Q}{\partial n} + \frac{\partial P}{\partial T} = m(K - aT) - \mu C \tag{3-177}$$

† Using the usual values for the parameters involved, e.g., those given in Fig. 3-20.

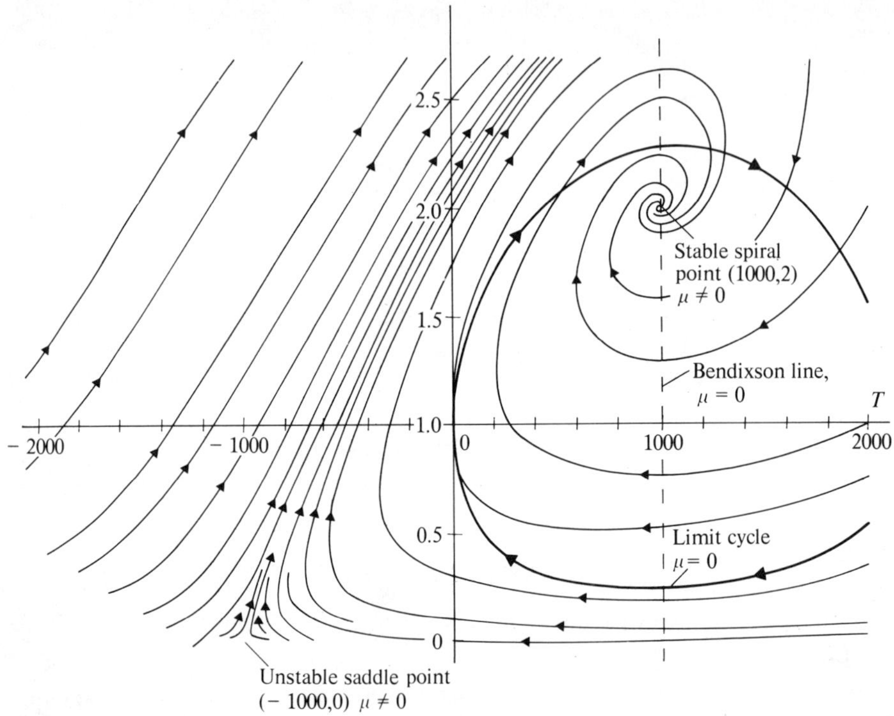

Figure 3-20 (n, T) phase plane behavior of Eq. (3-162): $K = 0.10$, $u = 0.01$, $a = 10^{-4}$, $m = 8 \times 10^{-3}$, $C = 0.10$. Arrows indicate increasing time.

Letting the right-hand side vanish yields the vertical line (Fig. 3-20)

$$aT = K - \frac{\mu C}{m} \tag{3-178}$$

which divides the region interior to the limit cycle, if there is one, since $\partial Q/\partial n + \partial P/\partial T$ changes sign from one side of this line to the other.

One scheme to aid in ascertaining the existence of a limit cycle is to calculate the time derivative of $R^2 = n^2 + T^2$, the square of the distance to the origin in the (n, T) plane, and note its behavior as a function of time along solutions of the differential equation (3-162) as discussed below. Using Eqs (3-160) and (3-161)

$$\frac{d}{dt}(R^2) = 2[n\dot{n} + T\dot{T}] = \frac{2}{C}[m(K - aT)n^2 + (n-1)T - \mu C T^2] \tag{3-179}$$

For very large n and T, $T > 0$, $(d/dt)(R^2) \sim -(2a/C)mn^2T < 0$, which means that solutions converge toward the origin for increasing time. For n near unity and $\mu = 0$, and T near zero

$$\frac{d}{dt}(R^2) \cong \frac{2mKn^2}{C} > 0$$

which means that solutions diverge near the origin. The latter is physically evident following a small positive step function reactivity change from equilibrium. Since the sign of $(d/dt)(R^2)$ changes as above, the existence of a limit cycle is indicated, but by no means proved. In this example, for $\mu = 0$, it is known that a limit cycle exists. From Eq. (3-162) for $\mu = 0$

$$\frac{dn}{dT} = \frac{m(K - aT)n}{n - 1} \qquad n(0) = 1 \tag{3-180}$$

This equation has been studied in Sec. 3-12. If μ is not zero, the reactor cools, dissipating its power, and so the solution spirals into singularity (2) and achieves a stable equilibrium. For $\mu = 0$, Eq. (3-180) is integrated to give the limit cycle ellipsoidal figure as its solution; i.e., for $n(0) = 1$

$$n\text{-}\log n = m\left(KT - \frac{aT^2}{2}\right) + 1 \tag{3-181}$$

All of the above is depicted in Fig. 3-20.

PROBLEMS

3-1 For a pendulum undergoing large-amplitude oscillations, the angular displacement is governed by $\ddot{\theta} + \omega^2 \sin\theta = 0$. Show that, as in the zero-power transfer function, the frequency of oscillation depends on the amplitude of oscillations θ_0. Carry out the integration of the differential equation to show the above as well as that the above dependence disappears as the oscillations are restricted to small amplitudes (i.e., when $\sin\theta \cong \theta$). Cf. L. A. Pipes: Applied Mathematics for Engineers and Physicists, McGraw-Hill, New York, 1946, chap. 22.

3-2 Plot the reactor transfer function amplitude open-loop response for a prompt neutron lifetime of 10^{-4} sec, using the "break-frequency" method described in Ref. 3.

3-3 Consider the reactor itself with negative feedback by letting the temperature kernel function $H(t) \equiv a$, a constant. What are the bounds on a with regard to the reactor stability?

3-4 A series of Poisson-distributed reactivity pulse forms has a spectrum given by $\overline{P}_K(\omega) = (k/2\pi)\sigma^2/(\omega_c^2 + \omega^2)^2$. Find the corresponding autocorrelation function. Relate the autocorrelation function of the individual reactivity pulse to its waveform in time. Plot this waveform given that the corner frequency is 0.3 Hz and that k, the mean number of pulses initiated, is 2 per second.

3-5 Compute the autocorrelation function of a time series of stock quotations, assuming stationarity, from the daily newspaper from the present to enough time in the past to satisfy a 10 percent accuracy requirement.

3-6 Carry out the derivation of the determination of the two singularities of the nonlinear equation (3-162).

3-7 For the case of small temperature fluctuations, using Laplace transforms, obtain the formal solution to the linearized integral equation (3-133).

REFERENCES

1. M. A. Schultz: "Control of Nuclear Reactors and Power Plants," 2d ed., McGraw-Hill, New York, 1961, p. 115.
2. J. M. Harrer, R. C. Boyer, and D. Krucoff: Transfer Function of Argonne CP-2 Reactor, *Nucleonics*, **10**(8): 32 (1950).

3. H. M. James et al.: "Theory of Servomechanisms," McGraw-Hill, New York, 1947.
4. J. G. Truxal: "Automatic Feedback Control Synthesis," McGraw-Hill, New York, 1955.
5. W. W. Seifert and C. W. Steeg (eds.): "Control Systems Engineering," McGraw-Hill, New York, 1960.
6. G. J. Thaler and R. G. Brown: "Analysis and Design of Feedback Control Systems," 2d ed., McGraw-Hill, New York, 1960.
7. James: *op. cit.*, chap. 2.
8. R. L. Murray: "Nuclear Reactor Physics," Prentice-Hall, Englewood Cliffs, N.J., 1957, chap. 7.
9. H. Soodak (ed.): "Reactor Handbook, Part A, Physics," Interscience, New York, 1962, chap. 5.
10. M. M. El-Wakil: "Nuclear Power Engineering," McGraw-Hill, New York, 1962, chap. 6.
11. S. Glasstone and A. Sesonske: "Nuclear Reactor Engineering," Van Nostrand, Princeton, N.J., 1963, chap. 4.
12. R. M. Pearce: The Doppler Effect in Thermal Reactors, *J. Nucl. Energy, Pt. A, Reactor Sci.*, **13**: 150–175 (1961).
13. H. Bethe and G. Placzek: *Phys. Rev.*, **51**: 450 (1937).
14. M. E. Rose et al.: Tables of $\psi(x, t)$, *WAPD*-506, Westinghouse Electric Corporation, October, 1954.
15. R. L. Murray: "Nuclear Reactor Physics," Prentice-Hall, Englewood Cliffs, N. J., 1957, p. 180.
16. J. A. Thie: Statistical Analysis of Nuclear Reactor Noise, *Nucleonics*, **17**(10): 102–111 (1959).
17. Y. W. Lee: "Statistical Theory of Communication," Wiley, New York, 1960, chap. 7.
18. Thie: *op. cit.*, p. 105.
19. J. A. Deshong: Flux, Reactivity, Steam Void, Steam Water Interlayer Noise Spectra in EBWR, *J. Nucl. Energy, Pt. A, Reactor Sci.*, **10**: 147–156 (1959).
20. V. Rajagopal: Correlation Methods for Transfer Functions, *Nucl. Sci. Eng.*, **12**: 218–224 (1962).
21. J. D. Balcomb, H. B. Demuth, and E. P. Gyftopoulos: A Crosscorrelation Method for Measuring the Impulse Response of Reactor Systems, *Nucl. Sci. Eng.*, **11**: 159–166 (1961).
22. J. Thie: Elementary Methods of Reactor Noise Analysis, *Nucl. Sci. Eng.*, **15**: 109–114 (1963).
23. S. Singer: "The Period Equilibrium Effect in Reactor Dynamics," Los Alamos Scientific Labs, LA-2654, 1962.
24. H. B. Smets: The Describing Function of Nuclear Reactors, *IRE Trans. Nucl. Sci.*, **NS-6**: 8–12 (1959).
25. Ergen and Weinberg: *Physica*, **20**: 413 (1954).
26. Y. H. Ku: "Analysis and Control of Non-linear Systems," Ronald, New York, 1958.

FURTHER READING

S. Barabaschi and E. Gatti: A Transfer Function for the Nuclear Reactor Parametric Excitation During a Transient, *Energia Nucl.*, **3**: 119–126 (April, 1956).
J. C. Carter, D. W. Sparks, and J. H. Tessier: "Period Effect in Reactor Dynamics," Proc. Symp. Reactor Kinetics and Control, Univ. Ariz. Mar. 25–27, 1963 (TID7662).
W. K. Ergen and A. M. Weinberg: *Physica*, **20**: 418 (1954).
P. Franklin: "A Treatise on Advanced Calculus," Wiley, New York, 1940, pp. 289–290.
James: *op. cit.*, p. 288.
W. Kaplan: "Ordinary Differential Equations," Addison-Wesley, Reading, Mass., 1959, chap. 6.
Lee: *op. cit.*, chap. 2.
Lee: *op. cit.*, p. 286.
J. Thie: "Reactor Noise," Rowman & Littlefield, New York, 1963, chap. 11.
R. Potter: The Transfer Function of a Steadily Diverging Reactor, *J. Nucl. Energy*, **6**: 291–299 (1958).

CHAPTER

FOUR

STABILITY

4-1 INTRODUCTION

The question of reactor stability is one of prime importance for the effective understanding of reactor dynamic behavior. Stability in various meanings and contexts will be defined and discussed. Some ideas of stability were discussed in the previous chapter, in connection with the transfer function and its implications.

First, let us introduce some rather rudimentary ideas of reactor stability, and proceed from that point. For example, if delayed neutrons are assumed non-existent, the neutron density following an introduction of positive reactivity will grow exponentially without bound, that is, as $n(t) \sim \exp \int_0^t K/\Lambda \, d\tau$ as appreciated from Sec. 1-8. This has some practical meaning with respect to neutron weapon like reactor behavior, and when exponential growth is applicable. This will be discussed further in Sec. 7-5 with regard to accident analysis, and in Sec. 9-16 where micro-reactors are discussed. Needless to say, the above is an example of unstable reactor behavior.

A second rudimentary example is demonstrated where the neutron density is represented by the integral equation of the point model, that is, from Sec. 1-7; it is rewritten here, where the kernel is partitioned, so that

$$n(t) = n_0 + A_0 \int_0^t K(\tau)n(\tau)\, d\tau + \sum_{r=1}^{6} A_r e^{-s_r t} \int_0^t e^{s_r t} K(\tau)n(\tau)\, d\tau \qquad (4\text{-}1)$$

From Eq. 1-52 the kernel is given as $G(t) = \sum_{r=0}^{6} A_r e^{-s_r t}$ and $s_0 = 0$. From Sec. 1-8 it is recalled that $A_0 = [\Lambda + \bar{f}(0)]^{-1} \simeq (\bar{\tau})^{-1}$ since $\Lambda \ll \bar{\tau}$ is the usual situation.

In the limit of large time, the neutron density is then given by

$$n_\infty = n_0 + (\bar{\tau})^{-1} \int_0^\infty K(\tau)n(\tau)\,d\tau + \lim_{t\to\infty} \sum_{r=1}^{6} A_r \int_0^\infty e^{s_r \tau} K(\tau)n(\tau)\,d\tau / e^{s_r t} \tag{4-2}$$

There are two cases:

1. If it is known that the excess neutron population density, $K(t)n(t)$, is varying only slowly, such that the numerator of the third term above will remain finite for large times, then from above

$$n_\infty = n_0 + (\bar{\tau})^{-1} \int_0^\infty K(\tau)n(\tau)\,d\tau \tag{4-3}$$

Hence, if the second term remains finite, that is, $\int_0^\infty K(\tau)n(\tau)\,d\tau < \infty$, then stability is assured insofar as this model mirrors the kinetics. This integral is seen to be essentially the total excess neutron production.

2. If it is known that $K(t)n(t)$ is varying so that the numerator of the third term in Eq. (4-2) grows without bound, then the limit must be taken (e.g. using L'Hopital's Rule). This results in

$$n_\infty = \left\{ n_0 + (\bar{\tau})^{-1} \int_0^\infty K(\tau)n(\tau)\,d\tau \right\} \Big/ \{1 - \bar{\tau}_2 K_\infty/(\bar{\tau})^2\} \tag{4-4}$$

where† $\bar{\tau}_2 = \sum_{i=1}^{6} a_i/\lambda_i^2$. For ^{235}U, $\bar{\tau}_2/(\bar{\tau})^2 \simeq 2.58$, so that for stability to be insured, in that $n(t)$ remains finite, not only must $\int_0^\infty K(\tau)n(\tau)\,d\tau < \infty$ as in 1, but that $K_\infty < 1/2.58 \simeq$ (39 cents) as well.

In the preceding chapters, the kinetics equations were investigated under manifold situations of reactor kinetic behavior. The idea of stability was introduced through the linear device of the transfer function. This was briefly extended to the more general nonlinear kinetics behavior through the phase-plane geometric analysis of simple versions of the kinetics equations coupled with simple feedback concepts.

In this chapter, the attempt is in part to further extend and generalize these ideas to the broader class of feedback phenomena that occur in reactors of various designs. The feedback reactivity component (as opposed to externally induced reactivity due to control rods, for example) is introduced by means of a very general notion. It is that of a linear integral operator. This comprises a linear superposition of past delayed reactivities which combine to constitute the reactivity in the reactor now. These delayed reactivities are given by the quantities $H(t - \tau)n(\tau)\,d\tau$, where $n(\tau)$ is the neutron density (essentially the power) at some time τ in the past. $H(t - \tau)\,d\tau$ is the negative reactivity increment which is manifest now, per unit neutron density (unit power) level τ sec ago. Superposing all of these past effects to find the total reactivity now gives, as in Eq. (3-30)

† The limit value is left as a problem at the end of this chapter.

$$K_p(t) = \int_{-\infty}^{t} H(t - \tau)n(\tau)\,d\tau \tag{4-5}$$

the present feedback reactivity component. $H(t)$, the power reactivity kernel, is a function of the kinetics characteristics of the particular reactor under investigation. It plays the major part in translating these characteristics into reactivity and its effect on the system.

4-2 LINEAR STABILITY—ROUTH'S METHOD

Besides the linear stability methods of Nyquist and Bode discussed in Chap. 3 in connection with the development of the implications of the transfer function, the following describes two other prominent methods used in linear stability theory. The first is Routh's method and the second is the root locus method. Routh's method is strictly analytical and bears little relation to any of the other stability methods extant. The root locus method, however, bears a somewhat close relation to that of Nyquist as it also combines both geometrical and analytical approaches to the determination of linear stability, as will be seen later. Routh's method is a scheme for determining the number of roots of a characteristic equation that have positive real parts. In this case, the characteristic equation is the polynomial as obtained from the Laplace transform of the linear differential equation whose solutions are being investigated for stability, that is, positive or negative going exponential terms that represent the form of its solution for long times. Routh's method can, of course, be used to determine the positive real part of the roots of any well-behaved polynomial in s. The method is an extension of Budan's theorem for complex roots together with Descartes rule of signs. Details of the proof can be found in most texts on the theory of equations. The following treatment is a paraphrasal of that in Gardner and Barnes (1943).

Let the characteristic equation be of the form

$$b_n s^n + b_{n-1}s^{n-1} + b_{n-2}s^{n-2} + \cdots b_1 s + b_0 = 0 \tag{4-6}$$

where the b_n are real, $b_n > 0$, and n is a positive real integer. Then begin to construct a Routh table by arranging the coefficients of the characteristic equation in the first two rows, which are labeled by descending powers of s

$$\begin{array}{lccccc} s^n: & b_n & b_{n-2} & b_{n-4} & b_{n-6} & \cdots \\ s^{n-1}: & b_{n-1} & b_{n-3} & b_{n-5} & b_{n-7} & \cdots \end{array}$$

The elements of the third row are found by computing the determinants

$$s^{n-2}: \quad \frac{\begin{vmatrix} b_{n-1} & b_{n-3} \\ b_n & b_{n-2} \end{vmatrix}}{b_{n-1}}, \quad \frac{\begin{vmatrix} b_{n-1} & b_{n-5} \\ b_n & b_{n-4} \end{vmatrix}}{b_{n-1}}, \quad \frac{\begin{vmatrix} b_{n-1} & b_{n-7} \\ b_n & b_{n-6} \end{vmatrix}}{b_{n-1}}, \quad \cdots$$

The elements of the fourth row are found in the same way by forming 2 by 2 determinants using elements from rows 2 and 3. As each succeeding row will become shorter than the one above, keep constructing rows until the elements are used up, that is, until no terms remain to make succeeding rows. To simplify the arithmetic the elements of a particular row may be divided by a common factor.

The Routh stability criteria begin with the fact that if all the elements in the first column of the completed Routh table have the same sign, then the corresponding characteristic equation has no roots with positive real parts, hence Routh stability is assured. Counting the number of changes in sign of the elements of the first column counts the number of roots with positive real parts.

As an example, for the characteristic equation $s^5 + 2s^4 + 2s^3 + 46s^2 + 89s + 260 = 0$ the Routh table is

s^5:	1	2	89	
s^4:	2	46	260	
s^3:	-1	-1.95		after division by 21
s^2:	1	6.18		after division by 42.1
s^1:	4.23			
s^0:	6.18			

In the first column there is seen to be two changes of sign, that is, + to −, and then − to +. This indicates that not only is the system that is represented by the above characteristic equation not stable, but that there are two roots with positive real parts. Since all of the coefficients of the characteristic equation are real, any complex roots must occur in conjugate pairs. Routh's table verifies this, in that the above characteristic equation can be factored as

$$(s + 4)(s - 2 + 3i)(s - 2 - 3i)(s + 1 + 2i)(s + 1 - 2i) = 0$$

so that the unstable conjugate pair of roots is seen to be $s = 2 \pm 3i$. Of course, all the other roots would give stable exponentials in the ultimate inversion of the characteristic equation to the time domain. The corresponding stable terms produced are, of course, e^{-4t}, $e^{-(1+2i)t}$, $e^{-(1+2i)t}$.

There are two special cases in the above treatment. The first is that if in the row by row construction of the Routh table, it is found that a particular row has a zero for the element in its first column. To continue to manufacture succeeding rows is futile, because their elements will all be infinite, because of division by this zero element. To circumvent this (*a*) merely replace the zero by a small constant ε and proceed from there, or (*b*) multiply the characteristic equation by an additional factor $(s + s_0)$ where s_0 is positive and real. The second special case is where all the elements of a row vanish. This indicates that there are roots opposite each other about the origin, real, imaginary, or complex. The

Routh procedure is to merely replace the row of zeros by the coefficients of the first derivative of an even power polynomial in s, whose coefficients are in turn formed from the elements in the last non-vanishing row of the Routh table. The roots of this polynomial coincide with some of those of the characteristic equation. These roots occur pairwise, being equal in magnitude and opposite in sign. Again, counting the changes in sign of the elements of the first column of the now augmented Routh table, counts the number of roots with positive real parts. The remaining roots have negative or zero real parts. The latter coincide with some of those of the even power polynomial.

As an example, consider the characteristic equation $s^5 + 3s^4 + 3s^3 + 9s^2 - 4s - 12 = 0$. The Routh table begins to look like

$$\begin{array}{llll} s^5: & 1 & 3 & -4 \\ s^4: & 1 & 3 & -4 \quad \text{after division by 3} \\ s^3: & 0 & 0 & \end{array}$$

The s^3 row is now replaced by the coefficients of the derivative of $s^4 + 3s^2 - 4$, the even power polynomial formed from elements of the last non-vanishing row, namely the s^4 row. So that the new s^3 row, and the remainder of the table looks like:

$$\begin{array}{lll} s^3: & 1 & 1.5 \quad \text{after dividing by 4} \\ s^2: & 1.5 & -4 \\ s^1: & 4.17 & \\ s^0: & -4 & \end{array}$$

As is seen, there is but one change in sign in the first column (+ to − then end of table). So that there is but one root with a positive real part. The remaining roots have either negative or zero real parts. The latter roots are among the roots of the even power polynomial $s^4 + 3s^2 - 4 = 0$, which has the four roots ± 1, $\pm 2i$, and all are roots of the original polynomial characteristic equation.

Recapitulating; for the system to be stable, and at most maintain an oscillatory steady state, then

1. all the coefficients of the characteristic equation are positive real, and
2. all the elements of the first column of the Routh table are positive which implies $b_n b_{n-3} < b_{n-1} b_{n-2}$, $b_n b_{n-3}^2 + b_{n-1}^2 b_{n-4} < b_{n-1} b_{n-2} b_{n-3} + b_n b_{n-1} b_{n-5}$, etc.

For oscillatory constant amplitude steady state behavior, the characteristic equation coefficients must be adjusted to make all the elements of one row of the Routh table vanish, and the resultant even power polynomial possess a pair of conjugate pure imaginary roots.

As an example of the above, consider the characteristic equation $s^3 + b_2 s^2 +$

$b_1 s + b_0 = 0$, with a Routh table given as

$$\begin{array}{lcc} s^3: & 1 & b_1 \\ s^2: & b_2 & b_0 \\ s^1: & (b_2 b_1 - b_0)/b_2 & \\ s^0: & b_0 & \end{array}$$

For no roots with positive real parts $b_2 > 0$, and $b_2 b_1 > b_0$. If a constant amplitude (undamped) oscillatory steady state is desired, then the s^1 row can be made to vanish by putting $b_2 b_1 = b_0$. The even power polynomial is then $b_2 s^2 + b_0 = 0$, with roots $\pm \sqrt{b_0/b_2}\, i$.

4-3 LINEAR STABILITY—ROOT LOCUS METHOD

Another method for the determination of linear stability using the Laplace transform s space is that of the root locus. Briefly, it is a method for depicting how the poles of the closed loop transfer function move through the s space as the gain of the open loop transfer function varies from zero to infinity. The gain, that is, the constant multiplier of the open loop transfer function is an important parameter with respect to the stability behavior in the transfer function sense.

If the loci of the above poles, that is, the loci of the roots of the closed loop transfer function denominator extend into the right half of the s plane for a range of values of the gain, then the system is unstable for these gain values. Root loci that skirt near the right-half plane presage system instability, in that the system becomes hypersensitive to spurious disturbances.

The root locus method will be illustrated by an example of a feedback transfer function analogous to that of the Experimental Breeder, EBR II, reactor, which is also discussed in Sec. 8-4 on fast reactors. The integral equation form of the point kinetics, with equilibrium power level n_0 is

$$n(t) = n_0 + \int_0^t G(t - \tau) K(\tau) n(\tau)\, d\tau \tag{4-7}$$

Assume that there is a power-reactivity feedback relationship between the neutron density, $n(t)$, and $K_p(t)$, the reactivity feedback due to the power excursion given by

$$K_p(t) = \int_0^t H(t - \tau)\, \varepsilon(\tau)\, d\tau \tag{4-8}$$

where the incremental deviation from equilibrium neutron density $\varepsilon = n - n_0$. Small excursions about power equilibrium levels n_0, allow Eq. (4-7) to be written as

$$\varepsilon(t) \cong n_0 \int_0^t G(t - \tau) K(\tau)\, d\tau \tag{4-9}$$

or in the Laplace transform space

$$\bar{\varepsilon}(s) \simeq n_0 \bar{G}(s) \bar{K}(s) \tag{4-10}$$

Let the reactivity K consist of the difference between the controllable reactivity (e.g. rod motion) K_c, and the feedback reactivity K_p, so that now

$$\bar{\varepsilon}(s) = n_0 \bar{G}(s) \{ \bar{K}_c(s) - \bar{K}_p(s) \} \tag{4-11}$$

Inserting the Laplace transform $\bar{K}_p(s)$ from Eq. (4-8), the result is

$$\bar{\varepsilon}(s) = \frac{n_0 \bar{G}(s)}{1 + n_0 \bar{G}(s) \bar{H}(s)} \cdot \bar{K}_c(s) \tag{4-12}$$

where the closed loop transfer function $G_c(s)$ is given by

$$\bar{G}_c(s) = n_0 \bar{G}(s) / (1 + n_0 \bar{G}(s) \bar{H}(s)) \tag{4-13}$$

To determine the stability characteristics of this system, the poles of $\bar{G}_c(s)$, that is, the roots s_r, of $1 + n_0 \bar{G}(s_r) \bar{H}(s_r) = 0$ are examined as to whether or not they have positive real parts. If they have, that is, the roots lie in the right half of the complex s plane, so that the system is unstable. This, of course, is meant in the sense that positive real parts of s_r will yield increasing positive time exponentials in the ultimate inversion of $\bar{G}_c(s)$ to $G_c(t)$.

The numerator of $\bar{G}_c(s)$, that is $n_0 \bar{G}(s)$, always has roots that have negative (or zero) real parts, so that $n_0 \bar{G}(s)$ does not need to be taken directly into consideration in this context. Now, the root locus method tracks the loci of the roots of $1 + n_0 \bar{G}(s_r) \bar{H}(s_r) = 0$ as a function of the power level gain constant n_0, where $0 \leq n_0 < \infty$. If there is a range of n_0 that does yield positive real parts of the above roots, then a system design modification is called for to devise measures to keep the system from attaining such (high) values of gain (n_0). The latter must also be consistent with normal system operation as well.

As discussed in Sec. 8-4 on fast reactors the EBR II experienced instabilities due to fuel element bowing. One model for reactivity feedback, through temperature, for the EBR, is that the fuel heats itself and the core coolant. The coolant is circulated out of the core proper to heat the immediate peripheral structure. Let T_c and T_p be the mean temperature difference between the coolant inlet temperature and the core coolant, and peripheral structure, respectively. A simple heat balance for the core coolant and periphery can be written as

$$\dot{T}_c = a n_0 - h_c T_c \tag{4-14}$$

$$\dot{T}_p = b T_c(t - T_0) - h_p T_p \tag{4-15}$$

Equation (4-14) asserts that the core rate of temperature rise (fuel plus coolant) is given by the fission power generated less the heat loss rate from the coolant. Equation (4-15) avers that the peripheral temperature rise is due to the rate of transport of heat out of the core via the coolant to the periphery, taking a delay time T_0, minus the heat loss from the latter. Eliminating T_c between these two equations through the Laplace transform gives the postulated temperature feedback equation

$$\bar{H}(s; n_0) \equiv \tilde{T}_p(s; n_0) = n_0 ab\, e^{-sT_0} \Big/ \left(s + \frac{1}{\tau_p}\right)\left(s + \frac{1}{\tau_c}\right) \tag{4-16}$$

where the thermal time constants are given by $\tau_p = h_p^{-1}$, and $\tau_c = h_c^{-1}$. The EBR time constants are relatively long, so that the low frequency approximation of $G(s) \simeq 1/s\bar{\tau}$, as shown in Sec. 2-2, is then used in

$$0 = 1 + n_0 \bar{G}(s)\bar{H}(s) = 1 + \frac{(abn_0/\bar{\tau})\, e^{-i\omega T_0}}{s\{s + (1/\tau_p)\}\{s + (1/\tau_c)\}} \tag{4-17}$$

Letting $abn_0/\bar{\tau} = K$, an augmented gain constant, then

$$K\bar{G}(s)\bar{H}(s) = -1 \tag{4-18}$$

implies two simultaneous relations, one for the modulus and one for the angle of a complex function. That for the modulus is

$$|\bar{G}(s)\bar{H}(s)| = 1/K \tag{4-19}$$

while that for the angle, arg GH, is

$$\arg GH = -\pi + 2l\pi; \qquad l = 0, 1, 2, \ldots \tag{4-20}$$

The latter relation means that for the above $\bar{G}(s)\bar{H}(s)$ and for $s = \alpha + i\omega$

$$\exp - i\left[\tan^{-1}\frac{\omega}{\alpha} + \tan^{-1}\frac{\omega}{\alpha + (1/\tau_p)} + \tan^{-1}\frac{\omega}{\alpha + (1/\tau_c)} + \omega T_0\right]$$

$$= \exp i(\pi + 2l\pi); \qquad l = 0, 1, \ldots \tag{4-21}$$

or

$$\tan^{-1}\frac{\omega}{\alpha} + \tan^{-1}\frac{\omega}{\alpha + (1/\tau_p)} + \tan^{-1}\frac{\omega}{\alpha + (1/\tau_c)} + \omega T_0 = \pm(\pi + 2l\pi) \tag{4-22}$$

This is the key root locus equation, in that it satisfies the locus of the intersection of the rays of the individual complex plane vectors $1/s$, $1/s + \tau_p$, $1/s + \tau_c$, and $e^{i\omega T_0}$ (the product makes up the GH function as seen from above, and is equated to -1) whose angles sum to $-\pi$ in the upper half of the s plane, or $+\pi$ in the lower half of the s plane. The modulus, Eq. (4-19), yields the corresponding values of the gain constant K for any pair (α, ω) on the locus. From Eq. (4-19)

$$\sqrt{(\alpha^2 + \omega^2)\left(\left(\alpha + \frac{1}{\tau_p}\right)^2 + \omega^2\right)\left(\left(\alpha + \frac{1}{\tau_c}\right)^2 + \omega^2\right)} = K \tag{4-23}$$

thus allowing the calibration of the locus in terms of the gain constant K.

Let us now introduce the tools used to qualitatively sketch the root locus for a general feedback function, without any lags such as $e^{-i\omega T_0}$

$$\frac{\prod_{i=1}^{m} (s - z_i)}{\prod_{j=1}^{n} (s - p_i)} = -\frac{1}{K} \tag{4-24}$$

and then particularize to the EBR-like feedback function. The corresponding root locus equation is of course

$$\sum_{i=1}^{m} \tan^{-1}\frac{\omega}{\alpha + z_i} - \sum_{j=1}^{n} \tan^{-1}\frac{\omega}{\alpha + p_i} = \pi + 2l\pi; \qquad l = 0, 1, 2 \tag{4-25}$$

Sometimes this is written as

$$\sum_{i=1}^{m} \text{phase}\,(s - z_i) - \sum_{j=1}^{n} \text{phase}\,(s - p_j) = \pm\pi + 2l\pi \tag{4-26}$$

Enumerating the means for sketching the root loci, paraphrasing Truxal[1] yields the following.

1. *As the gain K approaches zero*, from Eq. (4-24), the individual pole vectors $s - p_j$ must approach zero as well. The convention for the direction of the loci is from small to large K, so that all loci begin at $K = 0$, and so that all loci begin at the poles p_j of $\bar{G}(s)$ above.
2. *As the gain K approaches infinity*, that is, near the end of the loci, $\bar{G}(s)$ asserts that the loci must end on its zeros z_i. For the case where there are no zeros, then all the loci tend to infinity.
3. *The number of separate loci branches* is clearly given by the number of poles, or zeros, of $\bar{G}(s)$ whether they are finite or infinite, and if of multiple order, then each order is counted, for example, $1/s^k$ would contribute an additional k branches. For $\bar{G}(s)$ above, there are $n + 1$ branches.
4. *The locus is always symmetrical about* the real axis of the s plane, because the imaginary ($i\omega$) parts of the locus are always contributed in complex conjugate parts. The latter is due to the fact that the coefficients of $\bar{G}_c(s)$ in its expansion of its above product form

$$\bar{G}_c(s) = \frac{s^m + a_1 s^{m-1} + \cdots a_m}{s^{m+n} + b_1 s^{m+n-1} + \cdots b_{m+n}} = -\frac{1}{K} \tag{4-27}$$

   are customarily real.
5. *Asymptotes* to the loci are obtained by examining $\bar{G}(s)$ for very large s. From above

$$\bar{G}_c(s) \sim s^m/s^{m+n} = \frac{1}{s^n} = -\frac{1}{K} \tag{4-28}$$

   with n of the $m + n$ loci going to infinity. The corresponding phase is given by $|s|^{-1} e^{-in\theta} = e^{i\pi}$, so that the angle of the asymptotes with the real axis is

$$\theta = \pi/n \pm 2l\pi/n \tag{4-29}$$

   The point s_1 at which the asymptote(s) intersect the real axis can be determined in a number of ways. One is that if the numerator of $\bar{G}_c(s)$ in Eq. (4-27) is divided into its denominator, the result can be written as

$$\bar{f}(s) = s^n + (b_1 - a_1)s^{n-1} + \cdots = -K \tag{4-30}$$

where a_1 is the negative of the sum of the zero values, and b_1 is the negative of the sum of the pole values, since they each are the coefficient of the next to highest order term in their respective polynomial expansion, as seen in Eq. (4-27). Now for very large s, only the two highest terms are retained above in the calculation of its logarithmic derivative

$$\frac{\bar{f}'(s)}{\bar{f}(s)} \sim \frac{1}{s}\left[n - \frac{b_1 - a_1}{s}\right] \tag{4-31}$$

Now for very large s, this is essentially the logarithmic derivative of the asymptote as well. Also, for very large s, s_1, the origin, and all the poles and zeros appear closely clustered together. Then, as s in the right-hand side of Eq. (4-31) approaches the origin, with the above logarithmic derivative of the asymptote remaining finite, then

$$n - \frac{b_1 - a_1}{s_1} \simeq 0$$

or

$$s_1 = (b_1 - a_1)/n \tag{4-32}$$

so that the intersection on the real axis is simply

$$s_1 = \frac{\text{Sum of the pole values} - \text{sum of the zero values}}{\text{Number of finite poles} - \text{number of finite zeros}} \tag{4-33}$$

6. *Real axis branches* of the loci are easily found from Eq. (4-27). All poles or zeros whose coordinate values are complex, cancel each other out on the real axis, because they always appear in complex conjugate pairs. Hence the locus of $\bar{G}(\alpha)$, that is, $\bar{G}$ along the real axis is made up of contributions from real poles and zeros only. At any given point on the real axis, the angle contributed to $\bar{G}(\alpha)$ from a pole or zero to the left of the point is zero, because the "arg" of that pole or zero to the point is zero. If the pole or zero is to the right of the given point, then its contribution is $-\pi$. Since all contributions at the given point must sum to an odd multiple of $\pm\pi$ as in Eq. (4-26), then the occupied section of the real axis between real poles and zeros are only those that are to the left of an odd number of them. That is, loci branches will appear on sections of the real axis to the left of one, three, five, etc. poles and/or zeros.
7. *Imaginary axis intersection* point of the locus at a given K value is the laconic answer to the system stability question. That is, what is the maximum gain setting of the system which keeps it stable? The gain must always be set to a value less than that which allows the locus to cross the imaginary axis into the right-hand plane. The value of K at the intersection of the locus with the imaginary axis can be determined by using the Routh stability test discussed in the previous section. Such intersections correspond to a K (gain) value that yields pure imaginary conjugate roots of $\bar{G}_c(s)$.
8. *The immediate departure angles* of the loci from the complex poles and their

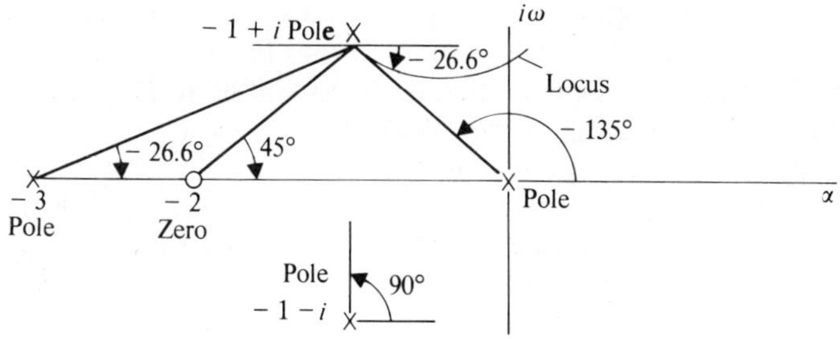

Figure 4-1 Angle of departure from pole at $-1+i$ for $K(s+2)/s(s+3)(s+1+i)(s+1-i)$.

arrival angles at the zeros are easily obtained. As an example, consider an $\overline{M}(s) = K(s+2)/s(s+3)(s+1+i)(s+1-i) = \pi + 2l\pi$, and it is desired to determine the immediate departure angle of the locus from the pole at $s = -1+i$. Since the sum of the angles from the zeros to $s = -1+i$, minus those from the poles to $s = -1+i$ must add to π, then it can be seen from Fig. 4-1, that

$$\text{phase}\,(s+2) - \text{phase}\,(s) - \text{phase}\,(s+1+i) - \text{phase}\,(s+3) - \text{phase}\,(s+1+i) = \pi \tag{4-34}$$

or

$$+45° - 135° - 90° - 26.6° - \theta = 180° \tag{4-35}$$

so that it is seen that $\theta = -26.6$ degrees, or that the locus departs from the pole $s = -1+i$ at an angle of -26.6 degrees.

9. *The point of departure from the real axis* of the locus is determined as follows. Consider a point on the locus a small distance ε away from α_0, the point where the locus departs from the real axis. For a point sufficiently close to α_0, the segment is so small that there is essentially a zero net change in $\overline{G}_c(s)$ along it to α_0. Since the arg $\overline{G}_c(\alpha_0) = 0$, because all pole and zero vectors must combine to yield zero at α_0, then

$$\sum_{i=1}^{m} \tan^{-1}\frac{\varepsilon}{\alpha+z_i} - \sum_{j=1}^{n} \tan^{-1}\frac{\varepsilon}{\alpha+p_j} \simeq \sum_{i=1}^{m}\frac{\varepsilon}{\alpha+z_i} - \sum_{j=1}^{n}\frac{\varepsilon}{\alpha+p_j} = 0 \tag{4-36}$$

The rightmost equation can be written because the angles are so small in the vicinity of α_0, that the angles can be replaced by their tangents, so that "$\tan^{-1}$ of tan" yields the arc itself. Dividing through by ε then, yields an equation for α. If $\overline{G}_c$ has complex conjugate singularities, the determination of α_0 is somewhat more complicated, as discussed by Truxal.[1]

10. *The sum of the poles and zeros* of $\overline{G}_c(s) = -1/K$ is constant. This is seen from Eq. (4-27) where $1/\overline{G}_c(s) = \overline{f}(s) = s^n + (b_1 - a_1)s^{n-1} + \cdots = -K$. For very large s, it is seen that $\overline{f}(s) \sim s^n + (b_1 - a_1)s^{n-1}$ and so the coefficient $(b_1 - a_1)$ is independent of K. Hence, this coefficient independence holds for all values of s. Since the coefficient is the negative of the sum of the pole values minus

the sum of the zero values, this means that as certain loci branches, say, turn to the right in the complex plane, other loci must turn to the left, so that this coefficient remains constant, that is, the difference between the totality of the pole values and the zero values remains constant.

11. *The product of the poles and zeros* is also constant. The product of the zeros of $\overline{f}(s) = 1/\overline{G}_c(s)$ is given by its lowest (constant) term. If $\overline{f}(s)$ has a pole at the origin, then it is seen that the constant term is directly proportional to K, hence the product of the poles and zeros is a constant.

The above rules will be illustrated by the EBR-like example above. Initially the pure lag factor corresponding to the delay time T_0 will be made zero. This will provide an apt example, whose characteristics can be determined using the above rules in the standard way. Then the pure lag factor will be included, as a separate example, because a radical departure from the rules will be made. This is mainly because, with the advent of the readily available programmable hand calculator a more analytical and less geometrical approach can be taken to determine the root loci of a closed loop transfer function containing a lag term.

The EBR-like† closed loop transfer function from Eq. (8-17) is given by

$$\overline{G}_c(s) = \frac{e^{-i\omega T_0}}{s\{s + (1/\tau_p)\}\{s + (1/\tau_c)\}} \tag{4-37}$$

The time constants used are $\tau_p = 1$ sec, $\tau_c = 0.5$ sec, and $T_0 = 10$ sec. Without the lag

$$\overline{G}_c(s) = \frac{K}{s(s+1)(s+2)} \tag{4-38}$$

Now it is desired to determine the root loci of the above $\overline{G}_c(s)$ using the foregoing rules that are applicable. Here then

$$f(s) = s(s+1)(s+2) \equiv s^3 + 3s^2 + 2s = -K \tag{4-39}$$

The following enumeration will correspond to the rules taken in order.

1. The loci start from $0, -1, -2$, where $K = 0$.
2. The loci all terminate at infinity, since $\overline{G}_c(s)$ has no finite zeros for the loci to terminate on.
3. There are three loci, since there are three singularities (poles).
4. By symmetry, for three loci, there is one on some part of the real axis, and the other two loci can be real or conjugate complex, but symmetrically displaced with respect to each other about the real axis.
5. Since $\overline{G}_c(s)$ has no finite zeros, its three loci approach infinity for K approaching

† Actually, the EBR II feedback portion of the transfer function contains two terms. The first is positive and is associated with a positive temperature reactivity coefficient. The second is negative and is associated with the negative temperature reactivity coefficient. The latter term is the one used as the example for root locus analysis, cf. Thalgott.[2]

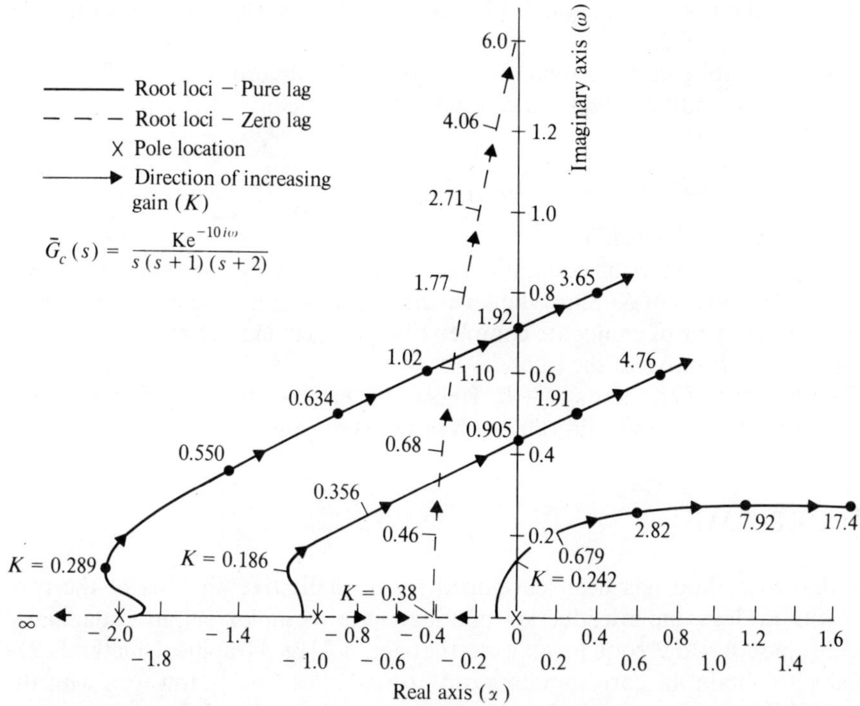

Figure 4-2 Root locus plot for EBR-like transfer function.

infinity. They approach infinite values on asymptotes at 60 degrees and 180 degrees. The asymptotes coincide on the real axis at $\alpha = -1$, using Eq. (4-38).

6. Branches of the loci on the real axis are in the real coordinate domain for $-\infty < \alpha < -2$, and $-1 < \alpha < 0$, increasing with increasing K as shown by arrows in Fig. 4-2.
7. The intersection of the loci with the imaginary axis are determined by suitably modifying the Routh table for $\bar{G}_c(s)$ as discussed in the previous section. Here, the characteristic equation is $s^3 + 3s^2 + 2s + K = 0$. The corresponding Routh table is

$$
\begin{array}{lll}
s^3: & 1 & 2 \\
s^2: & 3 & K \\
s^1: & (6-K)/3 & \\
s^0: & K &
\end{array}
$$

To obtain oscillatory stability, which is tantamount to determining K for a locus branch that intersects the imaginary axis, the third row can be made to vanish by choosing $K = 6$. Then the even order polynomial is $3s^2 + K = 0$, with a first derivative of $6s = 0$. The new third row is then $s^1: 6$, etc. The even order polynomial yields the two oscillatory roots, $s = \pm\sqrt{K/3}\,i = \pm\sqrt{2}\,i$ for $K = 6$.

8. Not applicable, since $\bar{G}_c(s)$ has no complex poles or zeros.
9. The corresponding $\bar{G}_c(\alpha_0) = 0$ equation is the cubic

$$\frac{1}{\alpha_0} - \frac{1}{\alpha_0 + 2} - \frac{1}{\alpha_0 + 1} = 0 \tag{4-40}$$

from which $|\alpha_0| = 0.423$.

10. From the characteristic equation above it is seen that the sum of the zeros is -3 independent of K. Since the locus branch from the pole at -2 moves to the left, the pair of conjugate complex branches leaving the real axis at $\alpha_0 = -0.423$ must move to the right.
11. The product of the zeros is $-K$. Hence this fact can be used to evaluate K along the loci, and so calibrate them in terms of gain.

4-4 PURE LAG

Now that a method has been developed for a qualitative analysis of the root locus without lags, the introduction of a lag factor (complex or pure imaginary) radically modifies the root locus over that for no lag. Graphical methods are available to obtain the corresponding root locus Y. See Chu.[3] However, with the advent of the programmable hand calculator, it is a straightforward task to compute the root locus including the lag, which can be then plotted.

For the EBR-like feedback transfer function, consider the angle root locus equation with pure lag which modifies the phase only, leaving the modulus as that without lag, as

$$\tan^{-1}\frac{\omega}{\alpha} + \tan^{-1}\frac{\omega}{\alpha + (1/\tau_p)} + \tan^{-1}\frac{\omega}{\alpha + (1/\tau_c)} + \omega T_0 = \pm\pi + 2l\pi \tag{4-41}$$

Combining the above arctangent terms will finally reduce the above to the cubic equation in α, in the (α, ω) space

$$\alpha^3 + (r + 3\omega\cot\omega T_0)\alpha^2 + (m + 2r\omega\cot\omega T_0 - 3\omega^2)\alpha + \{(m - \omega^2)\omega\cot\omega T_0 - r\omega^2\} = 0 \tag{4-42}$$

where $r = 1/\tau_p + 1/\tau_c$ and $m = 1/\tau_p\tau_c$. Using the time constants, that is, $T_0 = 10$ sec, $\tau_c = 1$ sec, and $\tau_p = 0.50$ sec, results in $r = 3$ and $m = 2$, so that

$$\alpha^3 + 3(1 + \omega\cot 10\omega)\alpha^2 + (2 - 3\omega^2 + 6\omega\cot 10\omega)\alpha + \{(2 - \omega^2)\omega\cot 10\omega - 3\omega^2\} = 0 \tag{4-43}$$

This is plotted in Fig. 4-2 where the zero lag root loci are superimposed.† The latter are plotted using the discussed procedure. Note that if the lag term (ωT_0)

† HP-65 Stored Program Hand Calculator Math Pac 1-06A, 1-07A.

is disregarded in Eq. (4-42) to obtain the no-lag locus,† the result is the hyperbola

$$\left(\alpha + \frac{r}{3}\right)^2 - \frac{\omega^2}{3} = \frac{1}{3}\left(\frac{r^2}{3} - m\right) \tag{4-44}$$

or for the above values of r and m

$$(\alpha + 1)^2 - \frac{\omega^2}{3} = \frac{1}{3} \tag{4-45}$$

whose rightmost branch is the no-lag root locus branch, that results from the merging of the two loci that start at the two poles (0, 0) and (−1, 0).

In Fig. 4-2 it is seen that the pure lag loci veer toward the right half of the complex plane. They are seen to cross into that half plane at lower values of gain than for the no-lag case. This, of course, implies that the pure lag system is less stable, in the root locus linear sense, than that without the lag—pure or not.

As is the case for the Nyquist stability investigation discussed in the preceding chapter, linear system stability is insured by designing compensating networks and inserting them into the system to keep the root loci in the left-half plane, or at most tangent to the imaginary axis. The above is tantamount to keeping the Nyquist path from enveloping the point (−1, 0) in the Nyquist diagram.

The system gain values along the root loci are simply obtained from the modulus portion of Eq. (4-41), that is

$$K(\alpha, \omega) = [(\alpha^2 + \omega^2)[\{\alpha + (1/\tau_p)\}^2 + \omega^2][\{\alpha + (1/\tau_c)\}^2 + \omega^2]]^{1/2} \tag{4-46}$$

These are shown in the root locus plots in Fig. 4-2. For a pure lag feedback as above, there is no real exponential attenuation multiplier in the right-hand side of the above $K(\alpha, \omega)$. The values of gain for incipient instability, that is, steady state oscillatory behavior are seen to be obtained from

$$K(0, \omega) = [\omega^2\{\omega^2 + (1/\tau_p^2)\}\{\omega^2 + (1/\tau_c^2)\}]^{1/2} \tag{4-47}$$

Figure 4-2 is plotted for the upper half plane only, since root loci are symmetrical about the real axis.

4-5 GENERALIZED EQUATIONS OF MOTION FOR REACTIVITY FEEDBACK

The next step in examining reactor stability is that due to Welton.[4]

Consider a reactor with a control system and a means for extracting power by circulating the moderator coolant from the reactor through a heat exchanger. One aggregate form of the kinetics equation is rewritten from Eq. (1-42), with zero source, as

$$\Lambda \dot{n}(t) = (K_c - K_p)n(t) - \int_{-\infty}^{t} n'(\tau) f(t - \tau)\, d\tau \tag{4-48}$$

† This is not the same as $\lim_{\omega \to 0} \omega \cot \omega T_0$ to get the no-lag locus, which is incorrect.

where K_c = controllable reactivity component through, e.g., rods controlled externally to the reactor.

K_p = the reactivity induced in the system by virtue of the neutron density, or power level, of the operating reactor. This is manifest through temperature and/or density effects.

K_p is the reactivity component that is fed back into the system as described in Sec. 4-1. It is rewritten from Eq. (4-5) as

$$K_p(t) = \int_0^\infty H(\tau)n(t-\tau)\,d\tau \tag{4-49}$$

and, as mentioned earlier, provides the salient feature of the discussion through the kernel $H(t)$, which is important in determining reactor stability.[4] For a reactor initially in equilibrium, that is, $n(t) \equiv 1$, $\dot{n}(t) \equiv 0$ for $t \leq 0$; then from Eqs (4-48) and (4-49)

$$K_c(0) = K_p(0) = \int_0^\infty H(\tau)\,d\tau \tag{4-50}$$

That is, the controllable component and the feedback component of reactivity balance each other at the initial reactivity level $\int_0^\infty H(\tau)\,d\tau$. The latter is termed the power coefficient of negative reactivity, since it multiplies $n(t)$ in the kinetics equations. The unit power coefficient of reactivity $H(t)$ can be positive or negative, but $\int_0^\infty H(\tau)\,d\tau$ is positive for a well-behaved system. Here $H(0) \geq 0$ and $H(\infty) = 0$ are assumed as well.

Let

$$Q = \log n(t) \qquad Q(0) = 0 \tag{4-51}$$

be a generalized coordinate of the system. Substitute this into Eq. (4-48) after differentiating the latter. This gives, paraphrasing Welton,[4]

$$\ddot{Q} = \frac{\dot{K}_c - \dot{K}_p}{\Lambda} - \frac{1}{\Lambda}\frac{d}{dt}\left[e^{-Q}\int_{-\infty}^{t} \dot{Q}(\tau)\,e^{Q} f(t-\tau)\,d\tau\right] \tag{4-52}$$

Now K_p, in Eq. (4-49), is differentiated to give, after a subsequent integration by parts

$$\frac{1}{\Lambda}\dot{K}_p = \omega^2\left[n(t) - \int_0^\infty n(t-\tau)R(\tau)\,d\tau\right] \tag{4-53}$$

where

$$R(t) \equiv -\frac{\dot{H}(t)}{H(0)} \qquad \int_0^\infty R(t)\,dt = 1 \qquad \text{and} \qquad \omega^2 = \frac{H(0)}{\Lambda} \tag{4-54}$$

With Eq. (4-53) inserted into Eq. (4-52) for $K_p(t)$, the result can be manipulated to finally give

$$\ddot{Q} + \omega^2(e^{Q} - 1) = \frac{\dot{K}_c}{\Lambda} + \omega^2 \int_0^\infty \{e^{Q(t-\tau)} - 1\}R(\tau)\,d\tau$$
$$- \frac{1}{\Lambda}\frac{d}{dt}\left[e^{-Q}\int_0^\infty e^{Q(t-\tau)}\,|f'(\tau)|\,d\tau\right] \tag{4-55}$$

This is an equation for a one-dimensional nonlinear oscillator where the displacement motion is given by the generalized coordinate Q. The right-hand side of the equation consists of forcing functions. The first term is the generalized applied force $\dot{K}_c/\Lambda$, and the second and third terms are memory effects due to past operating history. Specifically, the second term is the delayed reactivity feedback effect. The third is that due to delayed neutrons. Equation (4-55) is the basic equation which will be investigated in a number of ramifications.

Example 1 This is an approximation made by neglecting delayed neutrons, i.e., $f(t) \equiv 0$, and constant power is extracted from the reactor at the equilibrium level. The latter implies that H, the per unit power reactivity, is constant, making $R = 0$ as well. Then Eq. (4-55) becomes

$$\ddot{Q} + \omega^2(e^Q - 1) = \frac{\dot{K}_c}{\Lambda} \tag{4-56}$$

which has been investigated extensively.[6] Two simple cases, that for constant and that for linearly varying K_c, are discussed in Prob. 4-3. A first integral of the homogeneous part of Eq. (4-56), obtained by multiplying by $\dot{Q}$ and then integrating, is

$$\tfrac{1}{2}\dot{Q}^2 + \omega^2(e^Q - Q) = E \qquad \text{constant} \tag{4-57}$$

which corresponds to the conservation of energy for this system. Since the generalized kinetic energy is $\frac{1}{2}\dot{Q}^2$, a homogeneous quadratic function of the generalized velocities Q, the hamiltonian $H = T + V$, where T is the kinetic energy-like term, and V is the potential energy-like term, is†

$$H = E = \tfrac{1}{2}\dot{Q}^2 + \omega^2(e^Q - Q) = \frac{P^2}{2} + \omega^2(e^Q - Q) \tag{4-58}$$

where $P = \partial L/\partial \dot{Q} = \dot{Q}$, and E is the total energy. That is, given the above hamiltonian

$$H = H(Q, P)$$

the set of $(2n = 2)$ first-order differential equations are from Eq. (4-58)

$$\dot{Q} = \frac{\partial H}{\partial P} = P \qquad \dot{P} = -\frac{\partial H}{\partial Q} = -\omega^2(e^Q - 1) \tag{4-59}$$

which combine to give

$$\ddot{Q} + \omega^2(e^Q - 1) = 0 \tag{4-60}$$

The lagrangian can be defined as

$$L = 2T - H = \tfrac{1}{2}\dot{Q}^2 - \omega^2(e^Q - Q) \tag{4-61}$$

† The hamiltonian should not be confused with the per unit power reactivity which has the same symbol. Which quantity referred to is clear from the context.

which can be inserted in the Euler–Lagrange equations to obtain Eq. (4-60) as well. The phase-plane analysis of Eq. (4-56) or (4-57), where a first integral gives for linear reactivity variation

$$\frac{\dot{Q}^2}{2} + \omega^2 \left\{ e^Q - \left[1 + \left(\frac{\omega_0}{\omega} \right)^2 \right] Q \right\} = E_1 \qquad \omega^2 = \frac{\dot{K}_c}{\Lambda} \tag{4-62}$$

is precisely that analyzed in Sec. 3-10, which shows the Q motion to be stable.

Example 2 This is the same as Example 1 but with the delayed neutron concentration assumed to be constant at its equilibrium value. This corresponds to $f \equiv 1$, with $\dot{K}_c = 0$. This results in, from Eq. (4-55)

$$\ddot{Q} + \omega^2(e^Q - 1) = -\frac{1}{\Lambda}\dot{Q}\,e^{-Q} \tag{4-63}$$

Further discussion is reserved for Prob. 4-4.

Example 3 This is the case where no delayed neutrons are assumed. The power is extracted by means of a circulating moderator coolant pumped into the reactor at constant temperature. The moderator temperature is assumed constant throughout the reactor volume. This model could be used as well for a circulating fuel reactor where a fuel slurry is being pumped into the core at constant temperature and delayed neutrons are neglected. Let

C = heat capacity of the reactor.

λ = fractional flow replacement rate of liquid moderator coolant through the reactor.

α = temperature coefficient of reactivity, $-\alpha$ dollars per degree.

A heat balance equation is readily

$$\frac{d}{dt}(CT) = n - \lambda CT \tag{4-64}$$

which merely equates the rate of change of heat content to that generated by the reactor minus the heat outflow rate. This is rewritten as

$$\dot{T} = \frac{n}{C} - \lambda T \qquad \dot{T}(0) = 0 \tag{4-65}$$

Since $K_p = -\alpha T$, the above becomes an equation in the power reactivity:

$$\dot{K}_p + \lambda K_p = \frac{\alpha}{C} n \qquad K_p(0) = \frac{\alpha}{\lambda} C \tag{4-66}$$

with solution

$$K_p = \frac{\alpha}{C} \int_0^\infty n(t - \tau)\, e^{-\lambda\tau}\, d\tau \tag{4-67}$$

Upon making the identification of the above with Eq. (4-49), one notes that

$$H(t) = e^{-\lambda t} \tag{4-68}$$

giving

$$R(t) = -\frac{\dot{H}(t)}{H(0)} = \lambda\, e^{-\lambda t} \tag{4-69}$$

We substitute this into Eq. (4-55), with no delayed neutrons, to obtain

$$\ddot{Q} + \omega^2(e^Q - 1) = \omega^2\lambda \int_0^\infty (e^{Q(t-\tau)} - 1)\, e^{-\lambda\tau}\, d\tau \tag{4-70}$$

The integral can be eliminated, by differentiating, to yield

$$\dddot{Q} + \lambda\ddot{Q} + \omega^2\, e^Q\, \dot{Q} = 0 \tag{4-71}$$

with a "first" integral

$$\ddot{Q} + \lambda\dot{Q} + \omega^2\, e^Q = C_1^2 \tag{4-72}$$

This second-order equation can be replaced by the following pair of first-order equations

$$\dot{v} = p \tag{4-73}$$

$$\dot{p} = C_1^2 - \lambda p - \omega^2\, e^v \tag{4-74}$$

where

$$v = Q \qquad p = \dot{Q} \tag{4-75}$$

This system of first-order differential equations has been investigated[6] and proved to be stable by using conventional methods of nonlinear differential equations.† Equation (4-72) can be reduced in the same manner as Eq. (4-63) (cf. Prob. 4-4).

4-6 STABILITY OF THE POWER REACTIVITY FEEDBACK SYSTEM

There are other examples of the basic equation (4-55) that can be linked to various kinds of reactors, but now let us turn to a more general aspect of stability, after Welton.[4] First, the linear situation, usually resolved by using the Nyquist criterion of stability, will be revisited in the context of the analogy to classical mechanics. If in Eq. (4-55) only terms through first order in Q are kept, that is, $e^Q \cong 1 + Q$, then the equation reduces to

$$\ddot{Q} + \omega^2 Q = \frac{\dot{K}_c}{\Lambda} + \omega^2 \int_0^\infty Q(t-\tau)R(\tau)\, d\tau - \frac{1}{\Lambda}\frac{d}{dt}\left[Q + \int_0^\infty Q(t-\tau) f'(\tau)\, d\tau\right] \tag{4-76}$$

† These are essentially the equations discussed in Secs 3-12 and 3-13.

Now, multiply the above by $\dot{Q}$, with $F = \dot{K}_c/\Lambda$, the generalized applied force, to get

$$F\dot{Q} = \frac{d}{dt}\left(\frac{\dot{Q}^2}{2} + \frac{\omega^2}{2}Q^2\right) - \omega^2 \int_0^\infty Q(t-\tau)\dot{Q}(t)R(\tau)\,d\tau + \frac{1}{\Lambda}\left[\dot{Q}^2 + \int_0^\infty f'(\tau)Q(t-\tau)Q(t)\,d\tau\right] \tag{4-77}$$

The mean time integral of $F\dot{Q}$

$$\lim_{T\to\infty} \frac{1}{T}\int_0^T F\dot{Q}\,dt \equiv \langle F\dot{Q}\rangle$$

has the dimensions of work, with respect to the Q motion. If $\langle F\dot{Q}\rangle$ is always positive, this implies that work is continually being done by the force F to maintain the Q motion against the energy-dissipating mechanisms (the losses) of the Q system oscillator. In these terms, then, if $\langle F\dot{Q}\rangle$ is always positive *for all functions* F, then no self-sustained ($F = 0$) oscillations can persist. In other words, the system cannot sustain self oscillations.

Further,[7] if the generalized force is periodic, and positive work per cycle is always being done on the system, then any self-sustained motion (when $F = 0$) will be dissipated by the losses in the system. In this sense, then, the system will enjoy stability. This, of course, holds only for the situation where the power output $n(t)$ is linearly related to the periodic input force. For the nonlinear (large Q amplitude) case, the work per cycle must be positive for all types of F.

Another way of expressing the stability for linear Q motion is that, for the system to be unstable, there must exist a certain set of values for the parameters ω^2, Λ, etc. for which the stability will at least be neutral (that is, Q motion with $F = 0$) where the Q motion is simply harmonic. For such Q motion $(d/dt)\langle F\dot{Q}\rangle$, the rate of doing work on the system will vanish for the above set of parameters. Therefore, if it can be shown that $\langle F\dot{Q}\rangle$ is positive for all values of the parameters $(\omega^2, \Lambda, \ldots)$ and all functions F, then this comprises a sufficient condition for stability. The necessary condition follows from the Nyquist criterion investigation of poles in the frequency domain of Eq. (4-76).

To continue with the "linearized" investigation, insert

$$Q_p = A \sin \omega t \tag{4-78}$$

into Eq. (4-77) and integrate over one period. This yields, since

$$\frac{d}{dt}\left(\frac{\dot{Q}_p^2 + \omega^2 Q_p^2}{2}\right) = 0$$

$$\langle F\dot{Q}\rangle = \frac{A}{2}\left[\omega^2 \int_0^\infty R(\tau)\omega \sin \omega\tau\,d\tau + \frac{\omega^2}{\Lambda}\int_0^\infty |f'(\tau)|(1 - \cos \omega\tau)\,d\tau\right] \tag{4-79}$$

where

$$\langle F\dot{Q}\rangle = \frac{\omega}{2\pi}\int_0^{2\pi/\omega} F\dot{Q}\,dt$$

for periodic motion. Now, $\langle F\dot{Q}\rangle$ is certainly positive if both

$$\int_0^\infty R(\tau)\omega \sin \omega\tau \, d\tau \geq 0 \tag{4-80}$$

and

$$\int_0^\infty |f'(\tau)|(1 - \cos \omega\tau) \, d\tau \geq 0 \tag{4-81}$$

Equation (4-81) is never negative, since $|f'(\tau)|$ and $(1 - \cos \omega\tau)$ are always positive. Since Eq. (4-81) reflects the role of delayed neutrons, the effect of delayed neutrons always aids in attaining stability (at least in the linearized case) because the addition of this term makes $\langle F\dot{Q}\rangle$ "more positive." This will be shown to hold for the general nonlinear case as well.

Here, then, a sufficient condition for stability is that

$$\int_0^\infty R(\tau)\omega \sin \omega\tau \, d\tau \geq 0 \tag{4-82}$$

that is, if the Fourier sine transform of $R(t)$ is nonnegative for $0 \leq \omega \leq \infty$. This comprises the criterion for *unconditional* linear stability. Conditional stability is that $\langle F\dot{Q}\rangle \geq 0$ for all but a certain set of values of the parameters $(\omega^2, \Lambda, \ldots)$. Welton's stability criterion can be derived along strictly mathematical lines using Pitt's form of Wiener's theorem in the theory of Fourier transforms, cf. Hetrick.[8]

To check the preceding examples for stability, it is only necessary to calculate the Fourier sine transform of the kernel $R(t)$ in each case above.

Example 1 $R(t) = 0$. This is the case of neutral stability as evidenced by the linearized homogeneous part of Eq. (4-56), or

$$\ddot{Q} + \omega^2 Q = 0 \tag{4-83}$$

Example 2 $R(t) = 0$ also. The linearized equation of motion corresponding to Eq. (4-63) for this example is

$$\ddot{Q} + \frac{1}{\Lambda}\dot{Q} + \omega^2 Q = 0 \tag{4-84}$$

with a solution

$$Q(t) = C_1 e^{-H(0)t} + C_2 e^{-[1/2\Lambda - H(0)]t} \tag{4-85}$$

for $\Lambda H(0) \ll 1$. This is the usual situation for most reactors of orthodox design. Also, $1/\Lambda > H(0)$, so that the motion is overdamped.

Example 3 $R(t) = \lambda e^{-\lambda t}$, giving

$$\int_0^\infty \lambda e^{-\lambda t} \omega \sin \omega t \, dt = \frac{\lambda\omega^2}{\lambda^2 + \omega^2} \geq 0$$

The stability is unconditional for this case.

It can be shown that the stability criterion of the above condition on the integrability of the kernel $R(t)$ is the same for the nonlinear case as it is for the linear situation. It is assumed first that delayed neutrons are absent. This does not detract from the generality of the results which follow, because the delayed neutrons will be shown to always strengthen the stability. Then for $f \equiv 0$, the equation of Q motion is, from Eq. (4-55) with $\dot{K}_c/\Lambda = 0$, which corresponds to neutral stability,

$$\ddot{Q} + \omega^2(e^Q - 1) = \omega^2 \int_0^\infty (e^{Q(t-\tau)} - 1)R(\tau)\,d\tau \tag{4-86}$$

The procedure is to obtain the corresponding lagrangian, from which the hamiltonian is easily obtained. It can be shown that this hamiltonian corresponds to the total energy (total energy excess about equilibrium).† If the hamiltonian is positive definite and bounded, then so is the energy implying conservative motion, and thus unconditional stability is assured.

Here the lagrangian is

$$L(q, Q; \dot{q}, \dot{Q}) = \frac{1}{2}\left(\dot{Q} + \int_0^\infty q(z, t)\,dz\right)^2 - \omega^2(e^Q - Q) + \frac{1}{2}\int_0^\infty \frac{\dot{q}^2(z, t) - z^2 q^2(z, t)}{z\bar{R}(z)}\,dz \tag{4-87}$$

with $\bar{R}(z)$, the Fourier sine transform of the kernel $\bar{R}(t)$, given by

$$\bar{R}(z) = \frac{2}{\pi}\int_0^\infty R(t)\sin zt\,dt \tag{4-88}$$

It is first shown that this lagrangian is suitable. That is, that the Euler–Lagrange equations

$$\frac{\partial L}{\partial q_i} - \frac{d}{dt}\frac{\partial L}{\partial \dot{q}_i} = 0 \qquad i = 1, 2 \tag{4-89}$$

where $q_1 = q$, $q_2 = Q$, and L is the above lagrangian, yield the equation of motion (4-86). The Euler–Lagrange equations in the coordinates Q and q, respectively, yield, by inserting Eq. (4-87) into Eq. (4-89), the two equations

$$\ddot{Q} + \omega^2(e^Q - 1) = -\int_0^\infty \dot{q}(z, t)\,dz \tag{4-90}$$

$$\frac{\ddot{q} + z^2 q}{z\bar{R}(z)} = \dot{Q} + \int_0^\infty q(z, t)\,dz \tag{4-91}$$

If $\dot{Q}$ is eliminated between these two equations by differentiating Eq. (4-69) and inserting it into Eq. (4-90), the result is

$$q + z^2\dot{q} = -z\bar{R}(z)\omega^2(e^Q - 1) \tag{4-92}$$

† NRK I, Sec. 4-1.

Assume that the dependent variable in Eq. (4-92) is $\dot{q}$. The above equation is then integrated with zero initial conditions, that is, $\dot{q}$ should depend *only* on the past history of Q, to give

$$\dot{q} = -\omega^2 \bar{R}(z) \int_0^\infty (e^{Q(t-\tau)} - 1) \sin zt \, d\tau \tag{4-93}$$

With the above $\dot{q}$ inserted in Eq. (4-90), the resulting equation is

$$\ddot{Q} + \omega^2(e^Q - 1) = \omega^2 \int_0^\infty d\tau \int_0^\infty dz \, (e^{Q(t-\tau)} - 1)\bar{R}(z) \sin z\tau \tag{4-94}$$

where the order of integration has been reversed. Equation (4-94) is exactly the Q-motion equation (4-86) since, in the above integrand, the transform mate to $\bar{R}(z)$ is

$$R(\tau) = \int_0^\infty \bar{R}(z) \sin z\tau \, d\tau \tag{4-95}$$

Therefore, the above lagrangian has passed the test. It is not unique, however. From the methods espoused in Sec. 4-5 the hamiltonian or total excess energy is obtained from the lagrangian, Eq. (4-87) to give

$$H = E = \tfrac{1}{2}\dot{Q}^2 + \omega^2(e^Q - Q) + \frac{1}{2}\int_0^\infty \frac{\dot{q}^2 + z^2 q^2}{z\bar{R}(z)} \, dz - \frac{1}{2}\left[\int_0^\infty q \, dz\right]^2 \tag{4-96}$$

Since the first and second terms are positive definite, while the fourth term contributes nothing to the stability determination,[7] the third term is also positive definite if

$$z\bar{R}(z) = \int_0^\infty R(\tau) z \sin z\tau \, d\tau \geq 0 \tag{4-97}$$

which is the same condition for unconditional stability that was obtained for the linear case.†

4-7 STABILITY CRITERIA USING POWER-DEPENDENT TRANSFER FUNCTIONS

For present-day reactor systems, power-dependent transfer functions are a very important means for analyzing the stability at full power prior to actual operation at full power. The following is a description of this method. Equation (1-47) for an equilibrium neutron density (power) level $n(0)$ is

$$n(t) = n(0) + \int_0^t G(t - \tau)K(\tau)n(\tau) \, d\tau \tag{4-98}$$

† For many types of reactors, this is too strong (sufficient but not necessary). For other stability criteria, cf. Akcasu[9] and Gyftopoulos.[10]

As in Sec. 4-1, assume a power reactivity relationship between the neutron density and $K_p(t)$, the reactivity feedback due to the power excursion, as

$$K_p(t) = \int_0^t H(t-\tau)\varepsilon(\tau)\,d\tau \qquad \varepsilon = \frac{n - n(0)}{n(0)} \tag{4-99}$$

As discussed in Sec. 3-4, small excursions from equilibrium allow Eq. (3-9) to be linearized by iterating it once to obtain

$$n(t) = n(0) + n(0)\int_0^t G(t-\tau)K(\tau)\,d\tau \tag{4-100}$$

or

$$\varepsilon(t) = \int_0^t G(t-\tau)K(\tau)\,d\tau \tag{4-101}$$

The Laplace transform of the above is

$$\bar{\varepsilon}(s) = \bar{G}_0(s)\bar{K}(s) \tag{4-102}$$

which defines the zero-power transfer function $\bar{G}_0(s) = \{s[\Lambda + \bar{f}(s)]\}^{-1}$. Again, let the reactivity consist of two additive terms K_c, the controllable reactivity portion and K_p. From Eq. (4-101)

$$\varepsilon(t) = \int_0^t G(t-\tau)(K_c - K_p)n(\tau)\,d\tau \tag{4-103}$$

Using Laplace transforms, Eqs (4-99) and (4-103) combine to yield

$$\bar{\varepsilon}(s) = \frac{\bar{G}_0(s)\bar{K}_c(s)}{1 + \bar{G}_0(s)\bar{H}(s)} \tag{4-104}$$

or

$$\bar{\varepsilon}(s) = \bar{G}(s)\bar{K}(s) \tag{4-105}$$

where

$$\bar{G}(s) = \frac{\bar{G}_0(s)}{1 + \bar{G}_0(s)\bar{H}(s)} \tag{4-106}$$

If both $\bar{G}_0(s)$ and $\bar{H}(s)$ are known, then the stability of the system can be investigated by using the Nyquist criterion as described in Sec. 3-5. This examines whether or not $\bar{G}_0(s)\bar{H}(s)$ encloses the point $(-1, 0)$ in the $\bar{G}_0(s)\bar{H}(s)$ complex plane. That plane, called the Nyquist diagram, is obtained by plotting Im $\{\bar{G}_0(s)\bar{H}(s)\}$ against Re $\{\bar{G}_0(s)\bar{H}(s)\}$. If the $\bar{G}_0(s)\bar{H}(s)$ vector encloses the point $(-1, 0)$, the system is unstable.

If Eq. (4-106) is rewritten to obtain

$$\bar{H}(s) = \frac{1}{\bar{G}(s)} - \frac{1}{\bar{G}_0(s)} \tag{4-107}$$

then $\bar{H}(s)$, and subsequently $H(t)$, can be determined as a function of power level.

This separates the zero-power effects from the operating-power effects so that the latter can be examined through $\bar{H}(s)$. For reasons discussed in Chap. 3, s can be replaced by $i\omega$ to obtain

$$\bar{H}(\omega) = \frac{1}{\bar{G}(\omega)} - \frac{1}{\bar{G}_0(\omega)} \tag{4-108}$$

$\bar{G}(\omega)$ can be experimentally determined by using pile oscillator methods for various power levels, and thereby $\bar{H}(\omega)$ is obtained. Further, as the power is increased, with $\bar{H}(\omega)$ being measured over a range of power levels, it can then be extrapolated to much higher power levels. Then $\bar{G}_0(\omega)\bar{H}(\omega)$ plots, as a function of power level, can be obtained, and the trends toward instability can be noted as a function of increasing power level. This can be done without risking the reactor by being obliged to operate it at the possibly dangerous high power. The power level at which the $\bar{G}_0(s)\bar{H}(s)$ vector just includes the point $(-1, 0)$ on its contour is the upper bound on the stable operating power. In Figs 4-3 and 4-4 are some examples

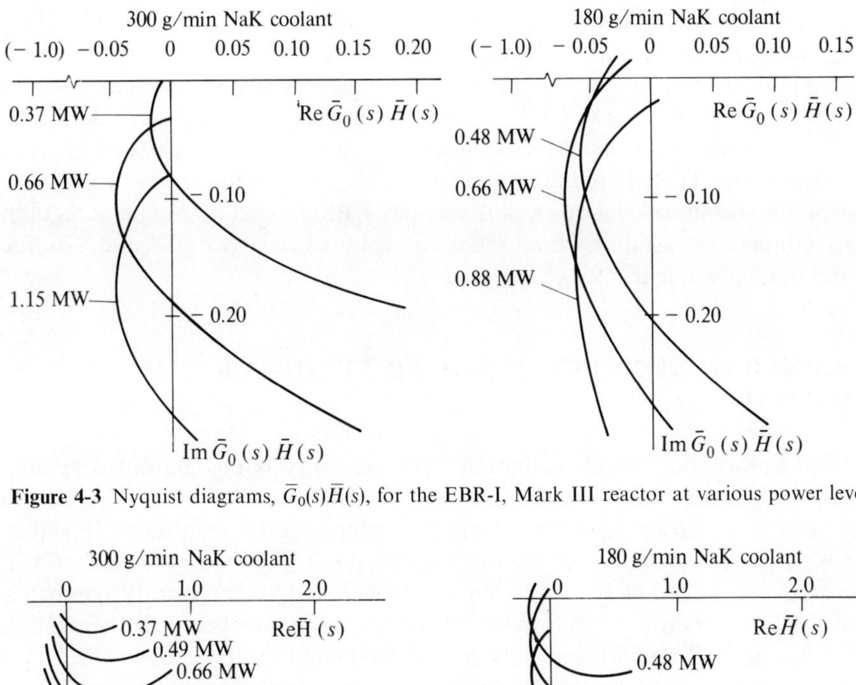

Figure 4-3 Nyquist diagrams, $\bar{G}_0(s)\bar{H}(s)$, for the EBR-I, Mark III reactor at various power levels.[2]

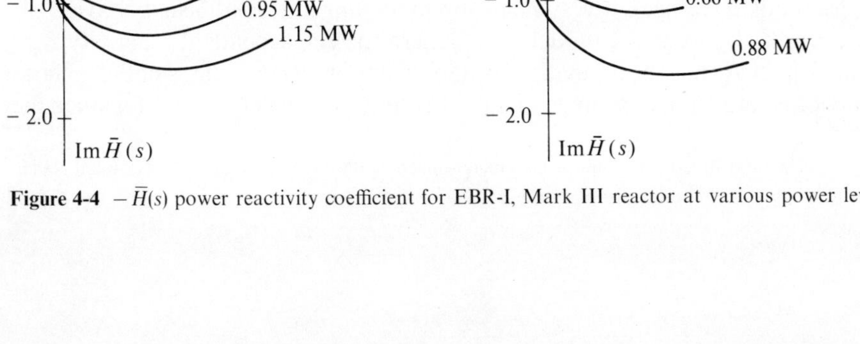

Figure 4-4 $-\bar{H}(s)$ power reactivity coefficient for EBR-I, Mark III reactor at various power levels.[2]

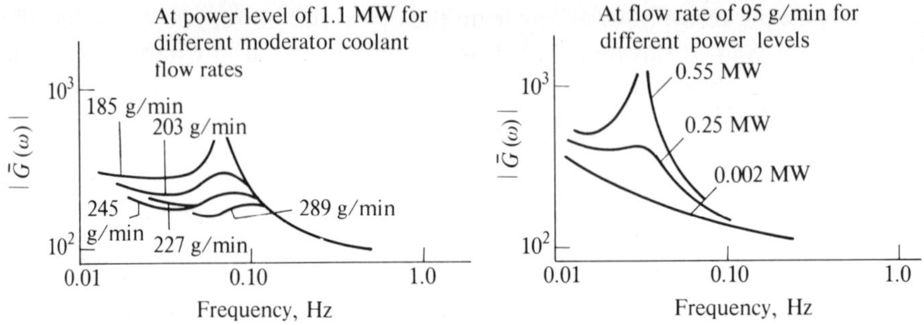

Figure 4-5 Amplitude transfer functions of EBR-I, Mark II.[11]

from the experimental breeder reactor (EBR-I, Mark III).† From Fig. 4-3, it is surmised that this reactor is well within the stable regime at the power levels depicted. Further, it appears from Fig. 4-3 that the reactor is even more stable at higher powers. The apparent linearity of $\bar{H}(s)$ with power is exhibited in Fig. 4-4. That is, over a 3:1 ratio of power, the components of the vector $\bar{G}_0(s)\bar{H}(s)$ in Fig. 4-3 seem to expand the plot by this ratio. There also seems to be a negative pure imaginary component that becomes increasingly evident with increasing power in that the $\bar{G}_0(s)\bar{H}(s)$ curves seem to slip down the graph in Fig. 4-3. This certainly aids the stability of the system as it fends the plots away from the $(-1, 0)$ point. This phase lag of approximately 90° was taken into account in some of the empirical fits to $\bar{K}_p(\omega)$ for the EBR-I reactor.[11] The foregoing extrapolations to the stability limit are discussed further in Secs 8-4 and 8-14 for fast and boiling-water reactors, respectively.

4-8 GENERAL NONLINEAR POWER FEEDBACK REACTIVITY

It is well known that small-oscillation linear stability is no guarantee against unstable, large, nonlinear power excursions. As alluded to in Sec. 4-5 (Example 3), there is some power level for which the system can be conditionally stable. That is, there is a given set of parameters $(\omega^2, \Lambda \ldots)$ for which the reactor will undergo self-sustained oscillatory behavior. Such behavior is not truly symmetric about the time abscissa, in that more power can be pumped into the oscillation on the "up side" of equilibrium than can be dissipated on the "down side." This can cause the power level to rise so that an unstable regime is approached.

Methods have been investigated for predicting the nonlinear behavior of a power reactivity feedback model for a general power-dependent reactivity $K_p(t)$. Some headway has been made for the situation where the small-oscillation behavior is known over a range of equilibrium power levels. These together then

† Experimental Breeder Reactor, a fast, metal-fueled, sodium-cooled reactor at National Reactor Testing Station, Arco, Idaho.

span the nonlinear power excursions resulting in a general $K_p(t)$. One method, due to Brooks,[12] for obtaining such a $K_p(t)$ is now described.

Suppose that Eq. (1-42) is rewritten as, where time is in units of Λ,

$$\dot{n}(t) = (K_c - K_p)n(t) - \int_0^t n'(\tau)f(t-\tau)\,d\tau \tag{4-109}$$

and the reactivity feedback due to the neutron density (power) level is the functional

$$K_p(t) = \int_{-\infty}^{t} z(t, \tau; n(\tau))\,d\tau \tag{4-110}$$

Now a transitivity relationship is assumed, i.e., that Eq. (4-110) can be translated forward or backward in time. This is certainly physically plausible, in that fluctuations from equilibrium due to $K_p(t)$ have the same effect whether they occur yesterday, today, or tomorrow, other things being equal. This means that z does not depend on any given time reference; it merely depends on time differences so that it can be expressed as a displacement kernel. Then let

$$z(t, \tau; n(\tau)) = W(t-\tau; n(\tau)) \tag{4-111}$$

Equation (4-110) now becomes

$$K_p(t) = \int_{-\infty}^{t} W(t-\tau; n(\tau))\,d\tau \tag{4-112}$$

Let the kernel be expressed further as a power series in n; that is

$$W(t, n) = \sum_{r=0}^{\infty} a_r W_r(t) n^r \tag{4-113}$$

where the a_r and $W_r(t)$ are suitable coefficients. For the reactor undergoing small fluctuations about a given neutron density equilibrium level n_0, let

$$n = n_0 + \varepsilon \qquad \varepsilon \ll n_0 \tag{4-114}$$

Inserting this into Eq. (4-113) yields, by expanding $(n_0 + \varepsilon)^r \cong n_0^r + rn_0^{r-1}\varepsilon + \cdots$

$$W(t, n) = \sum_{r=0}^{\infty} a_r W_r(t)(n_0^r + rn_0^{r-1}\varepsilon + \cdots)$$

$$= \sum_{r=0}^{\infty} a_r W_r n_0^r + \sum_{r=0}^{\infty} ra_r W_r n_0^{r-1}\varepsilon + \cdots 0(\varepsilon^2) \tag{4-115}$$

The first two terms are used to represent $W(t, n)$ and are substituted back into Eq. (4-112) to give, where $n(t) \equiv n(0)$ for $t < 0$

$$K_p(t) = K_p(0) + \sum_{r=0}^{\infty} a_r r n_0^{r-1} \int_{-\infty}^{t} W_r(t-\tau)\varepsilon(\tau)\,d\tau \tag{4-116}$$

Equation (4-116), by the definition in Eq. (4-114), can be expressed equivalently as

$$K_p(t) = K_p(0) + \int_{-\infty}^{t} \frac{\partial}{\partial n_0} W(t-\tau; n_0)\varepsilon(\tau)\, d\tau \tag{4-117}$$

From the definitions of $K_p(t)$ made earlier, e.g., Eq. (4-49), the identification can be made that

$$\frac{\partial}{\partial n_0} W(t; n_0) \equiv H(t; n_0) \tag{4-118}$$

where $H(t; n_0)$ is the reactivity power coefficient per unit power, measured over the appropriate range of power levels n_0 mentioned at the outset. Then, from Eq. (4-118)

$$W(t; n_0) = \int_0^{n_0} H(t; n_0')\, dn_0' \tag{4-119}$$

Therefore, if a sequence of $H(t; n_0)$ can be measured over the range of power levels of interest, they can be integrated to obtain the grand kernel $W(t; n_0)$ for a generalized nonlinear K_p. This then yields the nonlinear $K_p(t)$ behavior, assuming the model as given in Eq. (4-110) is valid. Even more general models are discussed by Brooks.[12]

4-9 STABILITY IN THE SENSE OF LIAPUNOV

One of the prominent mathematicians contributing to the stability aspects of differential equations is A. M. Liapunov, whose work in the topological aspects

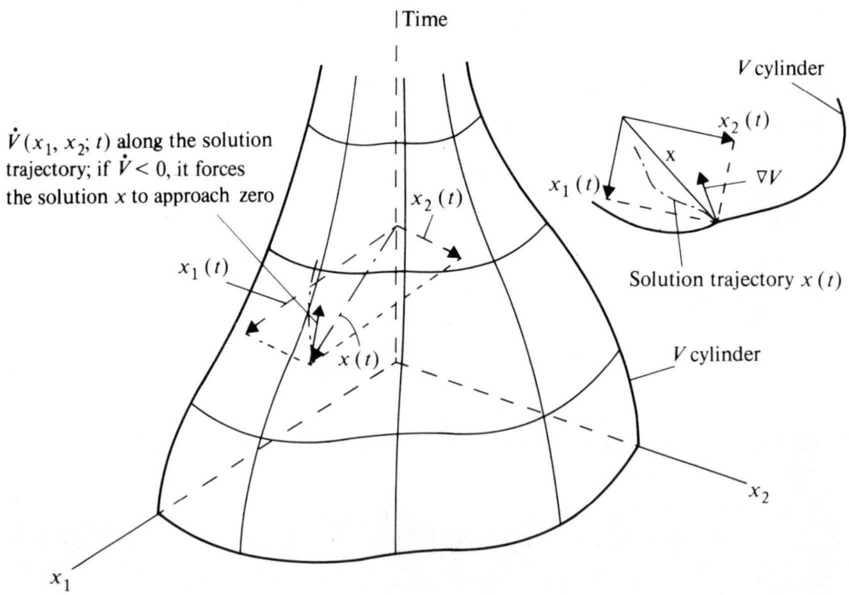

Figure 4-6 Configuration space for depicting Liapunov's theorem for stability in the large.

of stability theory is apropos here.[13] One important theorem of Liapunov is embodied in the fact that the singular points of a nonlinear differential equation can be investigated by using the appropriate linearized version of the differential equation near the singular point. This is termed an example of the investigation of stability "in the small" (local stability). It was used in the investigation of the singular point behavior of the system discussed in Sec. 3-12. The basis of the stability investigations discussed in this section is due to Liapunov's second method, which is construed as a determination of stability in the large (global stability).

Actually, the hamiltonian, or total energy, conceptualization of the stability problem is closely related, and in many instances coincident with, Liapunov's second method. However, the latter is somewhat more general, as will be seen. The essence of the method is to find, somehow, a real, positive definite scalar function of solutions of the differential equation system under consideration. If this function, called the Liapunov function, has a negative definite time derivative along the solution trajectories as well, then the system is stable.

More precisely,[14] let a linear or nonlinear system of differential equations, whose stability is to be investigated, be given as the set of n equations

$$\begin{aligned} \dot{x}_1 &= f_1(x_1, x_2, \ldots, x_n; t) \\ \dot{x}_2 &= f_2(x_1, x_2, \ldots, x_n; t) \\ &\cdots\cdots\cdots\cdots \\ \dot{x}_n &= f_n(x_1, x_2, \ldots, x_n; t) \end{aligned} \tag{4-120}$$

Or, using vector matrix notation, the above can be written compactly as

$$\dot{\mathbf{x}} = A(\mathbf{x}) \qquad A(\mathbf{0}) = 0 \tag{4-121}$$

where $\mathbf{x}$ is a column vector with components $(x_1, x_2, \ldots, x_n)$ satisfying the matrix equation (4-114). $A(\mathbf{0}) = 0$ implies that

$$\mathbf{x} = \mathbf{0} = (0, 0, \ldots, 0)$$

and all of the time derivatives $\{\dot{x}_i\} = 0$ for all i. The Liapunov function $V(\mathbf{x})$ along a solution trajectory satisfies

$$V(0) = 0 \qquad V(\mathbf{x}) \geq 0 \qquad V(\infty) = \infty \qquad \dot{V}(\mathbf{x}) \leq 0 \tag{4-122}$$

If $\dot{V}(\mathbf{x}) = 0$ merely, the stability is called uniform, which means that any excursions are only bounded with increasing time. A stronger condition is $\dot{V}(\mathbf{x}) < 0$, called asymptotic stability, meaning that all of the excursional motion ultimately vanishes. That is, the system reverts to its initial equilibrium state with increasing time. Figure 4-6 depicts the Liapunov cylinder for two dimensions (plus one time dimension). For uniformly stable motion the vector $\mathbf{x} = \{x_1(t), x_2(t)\}$ is merely bounded with time. This implies that the solution begins inside the Liapunov function cylinder and $\dot{V} = 0$ for all time. The theorem addresses itself to what is called the trivial solution, i.e., questions of stability concerning the solution asymptotic to $x_1 = x_2 = \cdots = x_n = 0$ after a sufficiently long time. The solution, $\mathbf{x} \equiv \mathbf{0}$ can always be placed inside a Liapunov function cylinder.

If both x_1 and x_2 in the two-dimensional illustration (Fig. 4-6) are not only bounded but decrease to zero with time, then the trivial solution is said to be uniformly and asymptotically stable in the sense of Liapunov. Any positive definite function with the properties given in Eq. (4-122) can be used as a Liapunov function V to determine whether or not $\mathbf{x}(t)$ is bounded with time. Such a function is a cylinder in the appropriate space enclosing the time axis, as in Fig. 4-6. If the derivative of V along a solution trajectory is negative definite, that is, if

$$\frac{d}{dt}V(x_1(t), x_2(t); t) = \frac{\partial V}{\partial x_1}\dot{x}_1 + \frac{\partial V}{\partial x_2}\dot{x}_2 + \frac{\partial V}{\partial t} \equiv \dot{\mathbf{x}}\cdot\nabla V + \frac{\partial V}{\partial t} < 0 \qquad (4\text{-}123)$$

where $\dot{\mathbf{x}} \equiv (\dot{x}_1, \dot{x}_2)$ is the solution trajectory vector as shown in the inset of Fig. 4-6, then the vector $\mathbf{x}$ is confined by the solution trajectory and the shrinking Liapunov cylinder, and so it must shrink to zero with increasing time as the Liapunov cylinder does so. This renders the $\mathbf{x}$ uniformly and asymptotically stable. An immediate Liapunov function is the hamiltonian function which relates to the physical ideas of energy stability discussed earlier.

Now consider a few examples[15,16] where a Liapunov function can be found. Some of these will be recognized as having been discussed previously. This is again because of the similarity between Liapunov and hamiltonian stability.

Example 1 Consider a homogeneous reactor where the reactivity is a function of temperature only, with no delayed neutrons. At equilibrium, the temperature $T = 0$, and $n = 1$ is the equilibrium neutron density (power) level. Energy is extracted at constant power while equilibrium is maintained. Essentially as in Sec. 3-11

$$\dot{n} = -\frac{\alpha}{\Lambda}Tn \qquad (4\text{-}124)$$

$$C\dot{T} = n - 1 \qquad (4\text{-}125)$$

where α is the temperature reactivity coefficient and C is the heat capacity of the reactor. Letting $n = e^Q$, the above system reduces to

$$\dot{Q} = -\frac{\alpha}{\Lambda}T \qquad \dot{Q}(0) = 0 \qquad (4\text{-}126)$$

$$\dot{T} = \frac{1}{C}(e^Q - 1) \qquad T(0) = 0 \qquad (4\text{-}127)$$

which takes the form of Eq. (4-120). Now "cross multiply" these two equations to obtain

$$\frac{d}{dt}V(Q, T; t) \equiv \frac{d}{dt}\left(e^Q - Q - 1 + \frac{\alpha C}{2\Lambda}T^2\right) = 0 \qquad (4\text{-}128)$$

where the extra term, -1, has been added to make V a Liapunov function, i.e., so that $V(0, 0, t) = 0$. Integrating gives

$$V(Q, T; t) = e^Q - Q - 1 + \frac{\alpha C}{2\Lambda}T^2 = \text{const} \qquad (4\text{-}129)$$

Now, V is positive definite, as can be seen. However, $\dot{V} = 0$, so that asymptotic stability is not ensured. The $V =$ const cylinder does not squeeze the "**x** vector" in the Q, T space to zero with increasing time. Nevertheless, this vector remains bounded by the V cylinder, which implies uniform stability only.

Example 2 Consider the same reactor as in Example 1 except that the amount of power extracted is proportional to the temperature difference between the instantaneous operating temperature T and the equilibrium temperature T_0, that is

$$C\dot{T} = n - \sigma(T - T_0) \tag{4-130}$$

where σ is a constant and the kinetics equation is the same as Eq. (4-126). This type of heat extraction is called Newton's law of cooling. It is an approximation to the situation where energy is radiated to the surroundings from a black-body reactor whose temperature equation is given by

$$C\dot{T} = n - A(T^4 - T_0^4) = n - A(T - T_0)(T + T_0)(T^2 + T_0^2) \tag{4-131}$$

If T is not very different from T_0 so that $T \cong T_0$, then, with $n = e^Q$

$$C\dot{T} \cong e^Q - \sigma(T - T_0) \qquad \sigma = 4AT_0^3 \tag{4-132}$$

Again, multiply Eqs (4-126) and (4-132) together to obtain

$$\frac{d}{dt}\left(e^Q + \sigma T_0 Q + \frac{\alpha C}{2\Lambda} T^2\right) = -\frac{\alpha\sigma}{\Lambda} T^2 \tag{4-133}$$

Define a Liapunov function as

$$V_1(Q, T; t) \equiv e^Q + \sigma T_0 Q - 1 + \frac{\alpha C}{2\Lambda}(T^2 - T_0^2) \tag{4-134}$$

which is positive definite. Substituting into Eq. (4-133) gives

$$\frac{dV_1}{dt} = -\frac{\alpha\sigma}{\Lambda} T^2 \tag{4-135}$$

Then the system is not only uniformly stable but asymptotically stable since V_1 tends to zero with increasing time. Again, the stability question is directed at the trivial solution, implying that if the system is perturbed from its initial equilibrium state, it will ultimately settle back to the state, $Q = 0$, $T = T_0$.

Example 3[15] An obvious generalization is to consider a heterogeneous reactor that can be divided readily into constituent media. Let

$T_i =$ deviation of the temperature from $T_i(0) = 0$ in the ith medium
$-\alpha_i =$ temperature coefficient of reactivity in the ith medium
$\lambda_i =$ fraction of power generated in the ith medium
$C_i =$ heat capacity
$k_{ij} =$ thermal conductivity between the ith and jth medium

Let time be measured in units of Λ and $n = e^Q$; $Q(0) = 0$. Delayed neutrons are neglected. The dynamical equations for a heterogeneous reactor of n media, with heat generated in each medium and with reactivity variations due solely to the temperature, are

$$\dot{Q} = -\sum_{i=1}^{n} \alpha_i T_i \qquad Q(0) = 0$$

$$C_r \dot{T}_r = \lambda_r(e^Q - 1) - \sum_{i=1}^{n} k_{ri}(T_r - T_i) \qquad r = 1, 2, \ldots, n \tag{4-136}$$

$$T_1(0) = T_2(0) = \cdots = T_n(0) = 0$$

The first equation is merely the kinetics equation with no delayed neutrons and reactivity due only to the deviation of temperatures from equilibrium. The second equation is the heat-balance equation in the rth medium. The first term on the right-hand side of this equation is the fraction of the total fission power generated in this medium. The second term is the loss of heat through conduction into the other media. Note that some of the k_{ri} may vanish since all of the other media do not necessarily interface with the rth medium.

Before considering the general case above, first look at the two-medium situation (e.g., coupled-core reactor). The representative equations are, from (4-136)

$$\begin{aligned} \dot{Q} &= -(\alpha_1 T_1 + \alpha_2 T_2) && Q(0) = 0 \\ C_1 \dot{T}_1 &= \lambda_1(e^Q - 1) - k(T_1 - T_2) && T_1(0) = 0 \\ C_2 \dot{T}_2 &= \lambda_2(e^Q - 1) - k(T_2 - T_1) && T_2(0) = 0 \\ & k_{12} = k_{21} = k \end{aligned} \tag{4-137}$$

Transform to a new set of variables given by

$$C_1 T_1 + C_2 T_2 = \Theta \qquad T_1 - T_2 = T \tag{4-138}$$

and define the constants

$$\xi = \frac{\alpha_1 + \alpha_2}{C_1 + C_2} \qquad \eta = \frac{\alpha_1 C_2 - \alpha_2 C_1}{C_1 + C_2} \qquad \zeta = k\frac{C_1 + C_2}{C_1 C_2} \tag{4-139}$$

All of this is substituted into Eq. (4-137) giving an equivalent set of equations

$$\begin{aligned} \dot{Q} &= -(\eta T + \xi\Theta) \\ \dot{\Theta} &= e^Q - 1 \\ \dot{T} &= \left(\frac{\lambda_1}{C_1} - \frac{\lambda_2}{C_2}\right)(e^Q - 1) - \zeta T \end{aligned} \tag{4-140}$$

Now combine the Eqs (4-140) to form

$$\frac{d}{dt}\left[e^Q - Q + \tfrac{1}{2}\xi\Theta^2 + \frac{\tfrac{1}{2}\eta}{\dfrac{\lambda_1}{C_1} - \dfrac{\lambda_2}{C_2}} T^2\right] = -\frac{\zeta\eta}{\dfrac{\lambda_1}{C_1} - \dfrac{\lambda_2}{C_2}} T^2 \tag{4-141}$$

The Liapunov function is

$$V_2(Q, T, \Theta; t) = e^Q - Q - 1 + \tfrac{1}{2}\xi\Theta^2 + \frac{\tfrac{1}{2}\eta}{\lambda_1/C_1 - \lambda_2/C_2} T^2 \tag{4-142}$$

V_2 is a Liapunov function, since it is positive definite and $\dot{V}_2 \leq 0$, if

$$\frac{\lambda_1}{C_1} - \frac{\lambda_2}{C_2} > 0 \qquad \eta > 0 \tag{4-143}$$

which guarantees asymptotic stability. However, if $\eta = 0$, $\dot{V}_2 = 0$, so that the reactor is uniformly stable only with respect to its trivial solution.

Example 4 The general n-medium case is investigated by defining

$$\Theta = \sum_{i=1}^{n} C_i T_i$$

and

$$\dot{Q} = -\sum_{i=1}^{n} \alpha_i T_i = -\sum_{i=1}^{n} \frac{\alpha_i}{C_i}(C_i T_i) \tag{4-144}$$

Now assume that

$$\frac{\alpha_1}{C_1} = \frac{\alpha_2}{C_2} = \cdots = \gamma > 0 \tag{4-145}$$

so that from Eq. (4-144), with γ being a constant

$$\dot{Q} = -\gamma\Theta \tag{4-146}$$

Add together the n equations that involve $\dot{T}_r$ in Eqs (4-136), using the fact that $\sum_{i=1}^{n} \lambda_i = 1$ and the first of Eqs (4-144). This gives

$$\dot{\Theta} = e^Q - 1 \tag{4-147}$$

The condition on the constants in Eq. (4-145) reduces this system of equations to the two generic equations

$$\dot{Q} = -\gamma\Theta \qquad \dot{\Theta} = e^Q - 1 \tag{4-148}$$

which are precisely the equations for the homogeneous reactor considered in the first example. All of the stability results that held for that reactor, i.e., that its trivial solution is bounded and therefore uniformly stable, hold for the n-medium case if

$$\frac{\alpha_1}{C_1} = \frac{\alpha_2}{C_2} = \cdots = \gamma > 0$$

This implies that a given amount of heat has the same effect on the reactivity irrespective of the medium in which it originated, or that the reactivity is determined by the total heat in the reactor independently of the spatial distribution of the latter. Letting $\alpha_i/C_i = \gamma$ for all i effectively homogenizes the reactor.

4-10 POPOV STABILITY

Certain generalizations in the reactivity, among other parameters, are taken up next.[16] Assume that a homogeneous reactor whose power is fluctuating about its equilibrium level has reactivity fluctuations due only to temperature. Let the reactivity be given by

$$K = -f(T) \tag{4-149}$$

in the kinetics equations,† written in units of $\Lambda = \beta = 1$. These are

$$\begin{aligned} \dot{n} &= -f(T)n - n + \sum_{i=1}^{n} \lambda_i c_i \\ \dot{c}_i &= -\lambda_i c_i + a_i n \qquad i = 1, \ldots, 6 \qquad a_i = \frac{\beta_i}{\beta} \end{aligned} \tag{4-150}$$

where the temperature is governed by a general heat balance equation

$$\dot{T} = An - B - E(T - T_0) \tag{4-151}$$

where A, B, and E are constants and T_0 is the ambient temperature. This equation includes the constant-power extraction case ($E = 0$) as well as the Newton law of cooling case ($B = 0$). The trivial solution to the system equations (4-150), (4-151), i.e., the solution at equilibrium, is

$$\begin{aligned} T &= T_1 \\ n &= n_0 = B + \frac{E(T_1 - T_0)}{A} \\ c_i &= c_i(0) = a_i n_0 \qquad \sum_{i=1}^{6} a_i = 1 \end{aligned} \tag{4-152}$$

First, let us examine the stability of the trivial solution for the one delayed neutron group approximation. From above, the corresponding kinetics equations are

$$\begin{aligned} \dot{n} &= -[1 + f(T)]n + \lambda c \\ \dot{c} &= -\lambda c + n \\ \dot{T} &= An - B - E(T - T_0) = A(n - n_0) - E(T - T_1) \end{aligned} \tag{4-153}$$

where the last equation above is obtained from the second of Eqs (4-152). For this system, consider the Liapunov function proposed by Popov[16]

$$V_3(T, n, c) = \int_{T_1}^{T} f(T')\,dT' + A\left[\int_{n_0}^{n} \left(\frac{n' - n_0}{n'}\right) dn' + \int_{c_0}^{c} \left(\frac{c' - c_0}{c'}\right) dc'\right] \geq 0 \tag{4-154}$$

V_3 is positive semidefinite by construction. The derivative of V_3 along a solution trajectory is, from Eqs (4-153) and (4-154)

† $f(T)$ is not the delayed neutron decay function.

$$\dot{V}_3 = \frac{\partial V_3}{\partial T}\dot{T} + \frac{\partial V_3}{\partial c}\dot{n} + \frac{\partial V_3}{\partial c}\dot{c} = -Ef(T)(T - T_1) - \frac{An_0}{\lambda nc}(n - \lambda c)^2 \qquad (4\text{-}155)$$

which is negative definite if $(T - T_1)f(T) > 0$, since all of the other quantities are positive by their physical nature. Therefore, the system is not only uniformly stable but asymptotically stable.

4-11 DELAYED NEUTRONS AND STABILITY

Let us derive the interesting (and intuitive) result that if a reactor is stable without considering delayed neutrons, it is all the more so when delayed neutrons are taken into account. Specifically, it is shown that if the reactor is uniformly stable without delayed neutrons, i.e., if its excursions following a disturbance from equilibrium are merely bounded, then it is asymptotically stable with delayed neutrons. That is, following a disturbance from equilibrium, the excursions are completely damped out with time when delayed neutrons are included.

Consider a generalized reactor system, again in units corresponding to $\Lambda = \beta = 1$.

$$\dot{n} = F(x_1, x_2, \ldots, x_k) - n + \sum_{i=1}^{6} \lambda_i c_i \qquad (4\text{-}156)$$

$$\dot{c}_i = -\lambda_i c_i + a_i n \qquad i = 1, \ldots, 6 \qquad (4\text{-}157)$$

$$\dot{x}_j = H_j(x_1, x_2, \ldots, x_k; n) \qquad j = 1, \ldots, k \qquad (4\text{-}158)$$

Equation (4-156) is the usual neutron density equation with a generalized reactivity function F depending on k other variables of the system such as temperature, pressure, and the heat transfer parameters. Equation (4-157) expresses the concentration of delayed neutron precursors. Equations (4-158) are the k other equations governing the dynamical behavior of the remaining variables. Equilibrium for this system implies the initial conditions

$$\begin{aligned} n &= n_0 \\ x_i &= x_{io} \qquad i = 1, \ldots, k \\ c_i &= c_{io} = \frac{a_i n_0}{\lambda_i} \end{aligned} \qquad (4\text{-}159)$$

The above asserts that the only solutions to the following system of $k + 1$ equations

$$\begin{aligned} F(x_{10}, x_{20}, \ldots, x_{ko}) &= 0 \\ H_j(x_{10}, x_{20}, \ldots, x_{ko}; n_0) &= 0 \qquad j = 1, \ldots, k \end{aligned} \qquad (4\text{-}160)$$

are the obvious set of solutions $\{x_{io}\}$ and $n_0 > 0$. When delayed neutrons are neglected, the analogous set of equations, from Eqs (4-156), (4-157), and (4-158) are

$$\begin{aligned} \dot{n} &= F(x_1, x_2, \ldots, x_k) \\ \dot{x}_j &= H_j(x_1, x_2, \ldots, x_k; n) \qquad j = 1, \ldots, k \end{aligned} \qquad (4\text{-}161)$$

The second term on the right-hand side of Eq. (4-156), $-n$, corresponds to the rate of loss of delayed neutrons to the system, and so it vanishes. The choice of units just happens to disguise this fact therein. The system of equations (4-161), of course, has the same (trivial) solution at equilibrium $\{x_{i0}\}$ and $n > 0$ that the equations that include the delayed neutrons possess. Let the Liapunov function for the system without delayed neutrons, Eq. (4-161), be given by

$$V_4 = L_A(n) + L_B(x_1, x_2, \ldots, x_k) \tag{4-162}$$

where L_A is assumed to be an increasing function of its argument. Now, assume that the no delayed neutron system is uniformly stable. This implies that V_4 is positive definite but that $\dot{V}_4$ is only negative or zero but not negative definite. That is, somewhere, $\dot{V}_4 = 0$ for increasing time. Let a Liapunov function for the delayed neutron system of Eqs (4-156)–(4-158) be

$$V_5 = L_A(n) + L_B(x_1, x_2, \ldots, x_k) + \sum_{i=1}^{6} \left(\frac{a_i}{\lambda_i}\right) L_A\left(\frac{\lambda_i c_i}{a_i}\right) \tag{4-163}$$

$(\lambda_i c_i/a_i)$ is the argument of the function L_A in the third term above. V_5 is positive definite for the initial conditions given. Then

$$\dot{V}_5 = \dot{n}L'_A + \sum_{j=1}^{k} \frac{\partial L_B}{\partial x_j}\dot{x}_j + \sum_{i=1}^{6} L'_A\left(\frac{\lambda_i c_i}{a_i}\right)\dot{c}_i \tag{4-164}$$

where the prime means the derivative with respect to the argument of the function L_A. Being careful in the partial differentiation above to note the difference between the systems of equations with and without delayed neutrons results in

$$\dot{V}_5 = \dot{V}_4 = \sum_{i=1}^{6} a_i\left[n - \frac{\lambda_i c_i}{a_i}\right]\left[L'_A(n) - L'_A\left(\frac{\lambda_i c_i}{a_i}\right)\right] \tag{4-165}$$

If, as mentioned above, $L_A(x)$ is an increasing function of its argument, then

$$(x_i - x_j)[L'_A(x_i) - L'_A(x_j)] > 0 \qquad \text{for } x_i \neq x_j \tag{4-166}$$

In other words, the sets $\{x_i\}$ and $\{L'_A(x_i)\}$ are similarly ordered. Since $L'_A(n) \geq L'_A(\lambda_i c_i/a_i)$ by the delayed neutron precursor equation (4-157), the terms in the square brackets of Eq. (4-165) are positive. Then, if $\dot{V}_4$ is at least negative or zero, $\dot{V}_5$ is negative definite, with V_5 positive definite as discussed. Therefore, if the nondelayed neutron system is uniformly stable, then delayed neutrons guarantee asymptotic stability.

As mentioned, many of the earlier stability arguments in this chapter, employing mechanical analogs stemming from variational principles, are particular versions of Liapunov functions and are thereby special cases of Liapunov's second method. The latter is more general, however, since Liapunov functions are ostensibly available even when the mechanical analog is not apparent. Though this method is powerful, it supplies no means for obtaining the Liapunov functions, which could be difficult to conjure up for a general problem. Further comprehensive work on Liapunov's methods for calculating stability is given in Refs 17–20.

4-12 LURIE–LETOV STABILITY

Certain types of stability determination of nonlinear systems, using specialized Liapunov functions often carry the names of their discoverers. One such method is that of Lurie and Letov. This method proceeds by first rewriting the system of describing differential equations in "canonic" form. This is a form, where the equilibrium is usually at the origin of the state space, so that the initial conditions on all the state variables are zero. Further, the canonic form delineates the nonlinearity in a natural way. Let the state variables be given by $x_i(t)$; $i = 1, 2, 3, \ldots n$ for an n dimensional system. Then the first canonic form is defined by Lurie as

$$\dot{x}_i = \lambda_i x_i + g(e); \qquad i = 1, 2, \ldots n \tag{4-167}$$

where the nonlinearity is $g(e)$, and e is the linear combination of state variables x_i, namely,

$$e = \sum_{i=1}^{n} \alpha_i x_i \tag{4-168}$$

which e also satisfies the differential equation

$$\dot{e} = \sum_{i=1}^{n} \beta_i x_i - rg(e) \tag{4-169}$$

This last is easily obtained by differentiating e in Eq. (4-168) and using Eq. (4-167), and identifying $\beta_i = \alpha_i \lambda_i$ and $r = -\sum_{i=1}^{n} \alpha_i$.

Now, it should be appreciated that the above nonlinear system can be represented by the diagram in Fig. 4-7.

Taking the Laplace transform of Eq. (4-167) yields

$$\bar{x}_i(s) = \frac{\bar{g}(s)}{s - \lambda_i} \tag{4-170}$$

where $\bar{g}(s)$ is the Laplace transform of $g\{e(t)\}$, and is assumed to exist, at least conceptually. Inserting $\bar{x}_i(s)$ into the Laplace transform of Eq. (4-168) gives

$$\bar{e}(s) = \bar{g}(s) \sum_{i=1}^{n} \frac{\alpha_i}{s - \lambda_i} \tag{4-171}$$

Looking at Fig. 4-7, it is seen that the transfer function of the linear part of the differential equation system, $\bar{G}_l(s) = -\bar{e}(s)/\bar{g}(s)$. So that

$$\bar{G}_l(s) = -\sum_{i=1}^{n} \frac{\alpha_i}{s - \lambda_i} \tag{4-172}$$

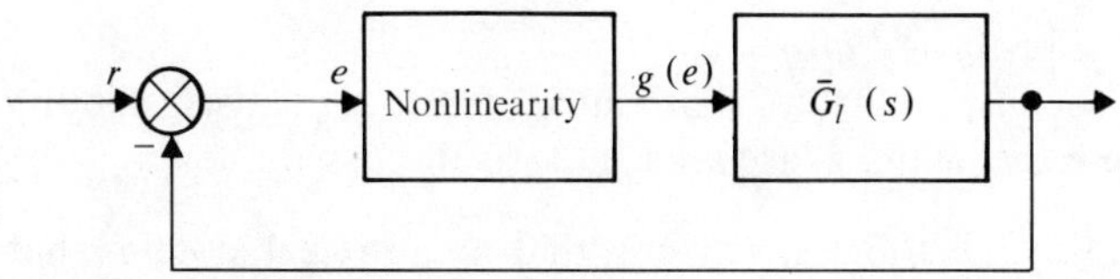

Figure 4-7 Simple separable nonlinear system.

Hence, it is seen that the λ_i are the poles of $\overline{G}_l(s)$, and the α_i are the negative of the residues at these poles.

Lurie has provided a Liapunov function for the above first canonic form, in order to yield stability criteria, which is

$$V = -\int_0^e g(u)\,du + \sum_{i=1}^{n}\sum_{j=1}^{n} \frac{a_i a_j x_i x_j}{\lambda_i + \lambda_j} \tag{4-173}$$

He showed that V is negative semidefinite if (a)

$$\int_0^e g(u)\,du \geq 0, \qquad \text{for } |e| > 0, \text{ and } g(0) = 0 \tag{4-174}$$

and (b) that the a_i that correspond to the real λ_i are themselves real, while the a_i that correspond to complex conjugate pairs of λ_i are themselves complex conjugate pairs of numbers, and (c) that the real parts of λ_i are negative. The time derivative of the above V function is

$$\dot{V} = rg^2(e) + \left(\sum_{i=1}^{n} a_i x_i\right)^2 - g(e)\sum_{i=1}^{n} x_i\left(\beta_i - 2a_i \sum_{j=1}^{n} \frac{a_j}{\lambda_i + \lambda_j}\right) \tag{4-175}$$

which can be made positive-semidefinite by letting the a_i satisfy

$$\beta_i - 2a_i \sum_{j=1}^{n} \frac{a_j}{\lambda_i + \lambda_j} = 0 \tag{4-176}$$

Equation (4-176) is known as a stability equation in this context. Lurie has also shown that $\dot{V}$ can be made positive-definite by adding to the Liapunov function V, a term F given by

$$F = A_1 x_1^2 + A_2 x_2^2 + \cdots A_m x_m^2 + C_1 x_{m+1} x_{m+2} + C_3 x_{m+3} x_{m+4} + \cdots C_{m-n-1} x_{n-1} x_n \tag{4-177}$$

where the A's and the C's are small negative numbers, that are associated with the real and complex conjugate variables x_i in the same way that the above a_i and λ_i correspond.

The Lurie stability criteria is enunciated by his stability theorem. It is that a system described by the first canonic form, Eqs (4-167)–(4-169), is globally asymptotically stable (g.a.s.) if the:

(a) λ_i and α_i are real,
(b) real part of the complex λ_i are negative,
(c) $r \geq 0$,
(d) $\int_0^e g(u)\,du \geq 0$, for all $|e| > 0$, and $g(0) = 0$,
(e) there is at least one set of parameters that can be found to satisfy stability equation (4-176), where the a_i and λ_i are associated as discussed above.

Lurie's theorem, as above, is too restrictive in that it can reject systems that are indeed stable. Aware of this, he relaxed the above requirements, thus including a wider class of systems, by modifying the stability equation (4-176) to

$$2a_i\left(\sum_{j=1}^{n}\frac{a_j}{\lambda_i+\lambda_j}-\sqrt{r}\right)=\beta_i \tag{4-178}$$

which is the result of adding and subtracting $2\sqrt{r}\,g(e)\sum_{i=1}^{n}a_ix_i$ from the above $\dot{V}$. The thus modified $\dot{V}$, is now

$$\dot{V}=\left\{\sqrt{r}\,g(e)+\sum_{i=1}^{n}a_ix_i\right\}^2 \tag{4-179}$$

To illustrate Lurie–Letov stability, let us examine the one dimensional point model kinetic equation for the prompt neutron density with a power-reactivity feedback term. From Sec. 2-2, this neutron density satisfies

$$\Lambda\dot{n}=(K_0-b^2n)n\equiv K_0n-g(e) \tag{4-180}$$

where the nonlinearity is

$$g(e)=e^2=b^2n^2 \tag{4-181}$$

so that $e=bn$, which also satisfies the differential equation, where from Eq. (4-180)

$$\dot{e}=b\dot{n}=\frac{b}{\Lambda}\{K_0n-g(e)\} \tag{4-182}$$

The Laplace transform of the above yields, with the Laplace transform of Eq. (4-180) where $\bar{\varepsilon}(s)$ is the transformed variable $e(t)$.

$$\bar{\varepsilon}(s)=\frac{bn_0}{s-(K_0/\Lambda)}-\frac{b}{\Lambda}\frac{\bar{g}(s)}{s-(K_0/\Lambda)} \tag{4-183}$$

Temporarily ignoring the non-homogeneous first term on the right-hand side of the above, the result is the transfer function of the linear portion of this system

$$\bar{G}_e(s)=-\bar{\varepsilon}(s)/\bar{g}(s)=\frac{b}{\Lambda}\cdot\frac{1}{s-(K_0/\Lambda)} \tag{4-184}$$

The Lurie V function from Eq. (4-173) is, with $\lambda_1=K_0/\Lambda$, $r=b/\Lambda$, and $\beta=bK_0/\Lambda$,

$$V=-\tfrac{1}{3}e^3+\Lambda a_1^2n^2/2K_0 \tag{4-185}$$

or

$$V=-\frac{b^3n^3}{3}+\frac{a_1n^2}{2K_0} \tag{4-186}$$

$\dot{V}$ from Eq. (4-175) becomes

$$\dot{V}=\frac{b}{\Lambda}g^2(e)+a_1^2n^2=n^2\left(\frac{b^5n^2}{\Lambda}+a_1^2\right) \tag{4-187}$$

with the corresponding stability equation

$$a_1^2=bK_0^2/\Lambda^2 \tag{4-188}$$

Substituting for a_1 from above into V and $\dot{V}$, respectively, gives

$$V = -bn^2\left(\frac{b^2 n}{3} - \frac{K_0}{2\Lambda^2}\right) \tag{4-189}$$

and

$$\dot{V} = \frac{bn^2}{\Lambda}\left(b^4 n^2 + \frac{K_0^2}{\Lambda}\right) \tag{4-190}$$

n is always positive, and $\dot{V}$ is always positive-definite for positive or negative controllable reactivity K_0.

For V, there are two cases:

1. K_0 positive; V is negative-definite if $n > 3K_0/2\Lambda^2 b^2$, then the system is globally asymptotically stable (g.a.s.) since V and $\dot{V}$ have "opposite signs."
2. K_0 negative; then the system is always g.a.s. for the same reasons.

Of course, all of the above is assumed within the conditions of Lurie's theorem, particularly

$$\int_0^e g(u)\, du = e^3/3 = b^3 n^3/3 \geq 0, \qquad \text{and } g(0) = 0$$

The solution to Eq. (4-180) is obtained by separating the variables, so that

$$n(t) = n_0\, e^{(K_0/\Lambda)t} \Big/ \left[1 + \frac{b^2 n_0}{K_0}\{e^{(K_0/\Lambda)t} - 1\}\right] \tag{4-191}$$

This is plotted in Fig. 4-8, where it is seen that for negative K, that indeed the system is globally asymptotically stable, as n vanishes for increasing time. For positive K_0, Lurie stability is seen to be sufficient but not necessary, that is, the system is g.a.s. without the Lurie limitation given in (a) above.

This type of stability analysis is, of course, best suited for multidimensional systems, wherein it is difficult, if not impossible, to find closed form solutions to the system equations. Nevertheless, stability information can then be extracted without finding such solutions, which is the *de facto* reason for using this school of stability investigations.

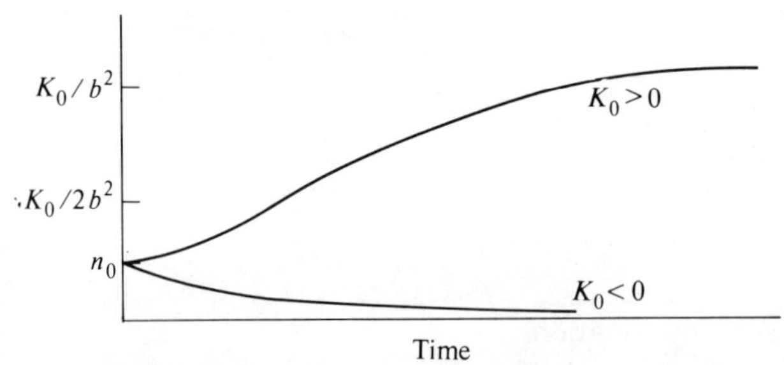

Figure 4-8 Prompt neutron density with power-reactivity feedback.

PROBLEMS

4-1 Show that the limit of the third term in Eq. (4-2) using L'Hopital's rule results in Eq. (4-4).

4-2 With the equation of motion for the nonlinear oscillator given as $\ddot{Q} + \omega^2(e^Q - 1) = \dot{K}_c/\Lambda$ show that by obtaining a first integral, the motion is periodic. Do this by plotting closed curves in the $(Q, \dot{Q})$ plane for the first integral. Show how the curves differ for constant and ramp function K_c.

4-3 Derive Eq. (4-63): $\ddot{Q} + \omega^2(e^Q - 1) = -(\dot{Q}\, e^{-Q})/\Lambda$. Investigate the solution of this equation by letting $\dot{Q} = p$, $\ddot{Q} = p\, dp/dQ$.

4-4 Using the lagrangian

$$L = \frac{1}{2}\left(\dot{Q} + \int_0^\infty q(z,t)\,dz\right)^2 - \omega^2(e^Q - Q) + \frac{1}{2}\int_0^\infty \frac{\dot{q}^2 - z^2q^2}{z\bar{R}(z)}\,dz$$

derive the appropriate equations of motion by inserting L into the Euler–Lagrange equations.

4-5 From the lagrangian given in Prob. 4-4 obtain the hamiltonian as given by Eq. (4-96).

4-6 Derive the hamiltonian time derivative

$$\dot{H}_2 = -kQ\int_0^t (e^Q - 1)\,d\tau \int_{\text{pile}} \alpha(\mathbf{r})\nabla^2\phi(\mathbf{r})\,d^3\mathbf{r}$$

from the heat equation given in the development leading up to Eq. (4-106).

4-7 What is the correspondence between Liapunov uniform stability and neutral stability with regard to a Q motion system which is uniformly stable?

4-8 Obtain the Liapunov function

$$V_2 = e^Q - Q - 1 + \tfrac{1}{2}\xi\Theta^2 + \frac{\frac{1}{2}\eta}{\lambda_1/C_1 - \lambda_2/C_2}T^2$$

from Eq. (4-140). Discuss the physical implication for $\eta = 0$.

4-9 Discuss the heterogeneous reactor system without the homogenizing condition. Attempt to obtain the Liapunov function for this n-medium reactor defined in Eq. (4-144).

4-10 Show that $\dot{V}_3$ along a solution trajectory is negative definite.

4-11 Derive the relationship between the Liapunov functions with and without delayed neutrons as given in Eq. (4-165).

REFERENCES

1. J. G. Truxal: "Control System Synthesis," McGraw-Hill, New York, 1955, chap. 4.
2. F. W. Thalgott et al.: Stability Studies on EBR-I, *Proc. U.N. Intern. Conf. Peaceful Uses At. Energy, 2nd, Geneva*, 1958, **12**:242–266 (1958).
3. Y. Chu: Feedback Control Systems with Dead-Time Lag or Distributed Lag by Root Locus Method, *AIEE Journal*, pp. 291–296 (Nov. 1952).
4. T. A. Welton: Kinetics of Stationary Reactor Systems, *Proc. U.N. Intern. Conf. Peaceful Uses At. Energy, Geneva*, 1955, **5**:377–389 (1955).
5. W. Kaplan: "Ordinary Differential Equations," Addison-Wesley, Reading, Mass., 1959.
6. J. Chernick: The Dependence of Reactor Kinetics on Temperature, *Brookhaven Natl. Labs. Rept. BNL* 173, 1951.
7. W. K. Ergen and A. M. Weinberg: Some Aspects of Non-linear Reactor Dynamics, *Physica*, **20**: 413–426 (1954).
8. D. L. Hetrick: "Dynamics of Nuclear Reactors," Univ. of Chicago Press, 1971, pp. 415–420.
9. A. Z. Akcasu and A. Dalfes: A Study of Non-Linear Dynamics, *Nucl. Sci. Eng.*, **8**(2):89 (1960).

10. E. P. Gyftopoulos and J. Devooght: Boundedness and Stability in Non-Linear Reactor Dynamics, *Nucl. Sci. Eng.*, **7**(6): 533 (1960).
11. F. W. Thalgott et al.: Stability Studies on EBR-I, *Proc. U.N. Intern. Conf. Peaceful Uses At. Energy, 2nd, Geneva*, 1958, **12**: 247 (1958).
12. H. Brooks: Temperature Coefficients and Stability, *Proc. Symp. Appl. Math., Am. Math. Soc.*, **11**: 289–308 (1961).
13. A. M. Liapunov: "Problème Général de la Stabilité du Movement," Princeton University Press, Princeton, N.J., 1947.
14. V. V. Nemytskii and V. V. Stepanov: "Qualitative Theory of Differential Equations," Princeton University Press, Princeton, N.J., 1960, pp. 245–249.
15. W. K. Ergen, H. J. Lipkin, and J. A. Nohel: Application of Liapunov's Second Method in Reactor Dynamics, *J. Math. Phys.*, **36**: 36–48 (1957).
16. V. M. Popov: Notes on the Inherent Stability of Nuclear Reactors, *Proc. U.N. Intern. Conf. Peaceful Uses At. Energy, 2nd, Geneva*, 1958, **11**: 245–250 (1958).
17. W. Hahn: Theorie und Anwendung der Direkte Method von Lyapunov, *Ergeb. Math. Ihrer Grenz* 22, Springer, Berlin, 1949.
18. I. G. Malkin: "Theory of the Stability of Motion," Gostekhizdat, Moscow, 1952. (Available in English translation.)
19. E. A. Barbachin and N. N. Krassovskii: Stability in the Large, *Doklady Akad. Nauk SSSR*, **87**: 453–457 (1952).
20. A. I. Lurie: "Some Non-linear Problems of Automatic Control Theory," Gosenergoizdat, Moscow, 1952.

FURTHER READING

W. K. Ergen: Rearrangement Inequalities and Non-linear Stability Criteria, *Proc. Symp. Appl. Math., Am. Math. Soc.*, **11**: 327–329 (1961).

G. H. Hardy, J. E. Littlewood, and G. Polya: "Inequalities," Cambridge, New York, 1959, p. 43.

Ibid., theorem no. 368, p. 261.

T. A. Welton: A Stability Criterion for Reactor Systems, *Oak Ridge Natl. Labs. Rept. ORNL 1894* (1955).

CHAPTER

FIVE

STATE SPACE KINETICS

5-1 INTRODUCTION

In the earlier chapters, especially in Chap. 3, the modus with respect to modeling a reactor system from the kinetics viewpoint is essentially one of first order perturbation theory in the classical sense. That is, the system is examined about a typical equilibrium state under small variations of the state variables from that state. This results in the transfer function approach and all of its time-frequency domain aspects, as discussed in Chap. 3.

As alluded to in earlier chapters, the kinetics behavior of a nuclear reactor is inherently nonlinear. That is, meaningful results from reactor kinetics analysis should usually construe the reactor nonlinearities as central. This includes especially global variations of the pertinent reactor variables, such as the power level, as well as local variations about an equilibrium state. This means that reactor power changes of, for example, factors of 10^5 from startup to rated full load, as well as planned or accidental shutdowns from normal reactor power scheduling to reactor core meltdown, must be expressed in a credible manner. One of the means for attempting to accomplish this is the reactor state space approach.

State space is the representation of the important reactor system variables as axes in a multidimensional hyperspace from which the formulation of the problem(s) can ensue. This type of geometric representation then allows the use of the power of the theory of mathematical spaces and its application in modern optimal control theory to be brought to bear. Another descriptor for this chapter is that of modern optimal control theory and its application to reactor kinetics and control analysis. Modern optimal control theory is essentially a manifestation of the calculus of variations reinforced with modern day accoutrements, including important developments that have been discovered over the past thirty years. Besides those of the variational calculus, a good share of these developments have been generated in the area of finite mathematics. This is that portion of mathematics which eschews the calculus, examples of which are linear and dynamic programming and certain versions of subdisciplines making up what is nowadays called operations and systems research.

Only in recent years has an apprehensive marriage between aspects of finite mathematics common to the calculus of reasonably continuous functions come to pass. It is felt that the fruition of this union will provide further contributions to modern day nuclear reactor analysis, as it has done over the past two decades. One obvious example is fault-tree theory.

Principally during the past twenty years, the appearance and the development of what is referred to as modern optimal control theory has begun. It can be characterized as the investigation of control using certain types of analysis including differential equations, vector spaces, and the variational calculus. The theory of optimal control has applications to multifarious control problems, such as those that appear in factory process control, space vehicle guidance, military command and control problems, as well as nuclear reactor kinetics and control.

The central problem of modern optimal control in this context is to cause a system, whose state and dynamical behavior are characterized by a state vector whose components are the state variables, and a corresponding set of differential equations in these variables, to operate so as to minimize or maximize a suitably chosen system performance criterion. The sought after control law that effects such behavior is called an optimal control policy. The optimal control policy is a function of the state of the system, while the performance criterion is usually a functional of some or all of the state variables. That is, the state of the system is monitored and fed back to a controller which uses the instantaneous state vector, that is, the set of state variables, as the basis upon which to make an optimal control decision consistent with the performance criterion. Such a feedback process can proceed continuously, or discontinuously by sampling the system state at suitable intervals. An added fillip is adaptive control, where the very parameters that are part of the system characterization are varied as well by virtue of the feedback control decisions.

Of importance are questions of system stability, as discussed in Chap. 4. For example, can the system perform its task discussed above without destructive consequences to itself or to its environment? The emphasis of modern optimal control theory is the investigation and determination of optimal control policies corresponding to various performance criteria, for different classes of systems consistent with stability considerations. This is somewhat distinct from the classical time-frequency domain transfer function analysis, where the determination of stability assumes primary importance, while the performance criterion is usually settled upon as being the minimization of the deviation of the system state from the input reference—usually an equilibrium state. This latter approach to the control problem, when formulated in mathematical terms, limits the application of the classical theory to behavior of systems in the small. That is, one can only consider mathematically linear systems which, in the real world, correspond to processes that deviate only slightly from their equilibrium states.

Modern optimal control theory because of its different emphasis, supplements the local linear control theory by attempting to yield a global description of system behavior. The global description allows a more accurate representation of the system to be controlled through the use of nonlinear dynamical equations, as well as decidedly nonlinear constraints. The classical theory is unable to cope with such nonlinear system descriptions. Again, in this context, most present day optimal control problems can be couched in two principal formulations. One is that of the Optimality Principle of Dynamic Programming of Bellman. The second is the method of the Pontryagin Maximum Principle. These two principles result in complementary descriptions and methodologies for investigating and solving optimal control problems. From the viewpoint of classical mechanics, the Maximum Principle results in a formulation analogous to the canonical equations of Hamilton. The Principle of Optimality of Dynamic Programming, on the other hand, yields a formulation in terms of the solutions of an analogous Hamilton-Jacobi equation. The equivalences of these two formulations will be discussed in later sections of this chapter.

The application of these methods have found their way into reactor kinetics and control, in the form of both point kinetic and spacially dependent kinetic models. These include both the point and spacial aspects of the xenon control problems, as well as analysis of reactor burnup and fuel management. In the space of this chapter, it is not possible to consider the details of all these problems. Sufficient background through the use of simple illustrations is given. This together with a discussion of xenon shutdown analysis, should provide the assiduous reader with the wherewithal to begin to handle substantive optimal control problems.

5-2 SYSTEM REPRESENTATIONS

As already mentioned, early reactor system investigations were concentrated in the time and frequency domains. That is, a given reactor-dependent variable, such as the neutron power level fluctuation was investigated with respect to time, about the reactor equilibrium states in the perturbational sense. Added to this was the corresponding frequency analysis in the Fourier and Laplace transform spaces.

The reactor system has more recently been formulated in a state space for analysis purposes. This, as alluded to earlier, has made available the manifold mathematical apparatus of the various function spaces and their geometries. This is done by determining the important reactor dynamical variables by which the system can be represented, such as power density or level, reactivity, fuel temperature, coolant flow rate, etc., as well as their derivatives. These variables are construed as components of, and axes in a particular kind of vector space—a particular vector in a vector state space. The peregrinations of the reactor system are now examined by investigating the nature of transformations of these variables in this state space. The transformations correspond to the system dynamics, while time rides along as an ever available parameter. Hence, all of the theorems of these vector spaces can be brought to bear, in the form of transformations of the state vector, upon the problem at hand.

Matrices are the usual representation of the linear transformations for a particular system linear dynamical behavior. Matrix operators are ever present throughout these investigations. As an example, consider an n-dimensional linear system given by the following differential equation

$$\dot{\overline{X}}(t) = \overline{A}\overline{X}(t) + \overline{U}(t) \tag{5-1}$$

where $\overline{X}(t)$ is the state (column) vector with components $(x_1, x_2, x_3, \ldots x_n)$. That is, x_1 could be the power level, x_2 through x_7 the delayed neutron precursor concentrations, x_8 the fuel temperature, $x_9 \ldots$, etc. $\overline{A}$ is a constant matrix expressing the right-hand sides of all the system equations. Since the system is assumed linear, the matrix elements are all constant, and do not involve the components, that is, the x_i's themselves. $\overline{U}$ is a constant vector with components $(u_1, u_2, \ldots u_n)$. In Eq. (5-1), the control is spoken of as non-homogeneous or "right-hand side" forcing, as opposed to coefficient forcing, which is discussed later. An example of the latter is the reactivity forcing in the point model reactor kinetics equations discussed in the earlier chapters. The well-known solution to Eq. (5-1) is

$$\overline{X}(t) = e^{\overline{A}t} X(0) + \int_0^t e^{\overline{A}(t-\tau)} U(\tau)\, d\tau \tag{5-2}$$

where $e^{\overline{A}t}$ is understood as the matrix operator

$$e^{\overline{A}t} = \sum_{l=0}^{\infty} t^l \overline{A}^l / l! \tag{5-3}$$

and $\overline{A}^l$ means the l-fold application of the matrix operator $\overline{A}$ on the function to its immediate right.

If $\overline{A}$ is a variable matrix, that is, the matrix elements of $\overline{A}$ are functions of time, then the corresponding solution to Eq. (5-1) is

$$X(t) = \exp\left(\int_0^t \overline{A}\, dt'\right) X(0) + \int_0^t \exp\left(\int_\tau^t \overline{A}\, dx\right) U(\tau)\, d\tau \tag{5-4}$$

where $\exp\left(\int_0^t \overline{A}\, dt'\right)$ is the solution to the homogeneous part of Eq. (5-1) with unity initial conditions, that is to

$$\dot{\overline{X}}^{(0)}(t) = \overline{A}(t)\overline{X}^{(0)}(t); \qquad \overline{X}(0) = \overline{I}$$

where $\overline{I}$ is the identity matrix. Such solutions, as well as those like Eqs (5-2) and (5-4) are usually obtained by numerical integration, using computing machinery.

Sylvester's theorem[1] also can be used to obtain expressions for the power series development of a matrix, for example as in Eq. (5-3).

Generally, when the dynamical behavior of a system is described by an nth order differential equation, successive substitutions are made to reduce it to n first order differential equations in the state variables. The latter make up the state vector, and an equivalent vector-matrix differential equation in this state vector can be written. This then constitutes the mathematical expression of the system state variable representation. As a simple example, consider the nth order homogeneous differential equation

$$\sum_{p=0}^{n} a_p x^{(p)} = 0 \tag{5-5}$$

where (p) is the pth derivative of the dependent variable $x(t)$, and the coefficients can be a function of time. If the simple substitution $x_1 = x'$, $x_2 = x_1'$, $x_3 = x_2'$, $\ldots x_{n-1} = x_{n-2}'$ is made, then the above reduces to n first order differential equations in the n state variables $(x, x_1, \ldots x_{n-1})$.

A second example of reduction to a first order system is a nonlinear illustration from a reactivity pulsed Godiva-like reactor which is modeled using the Hansen–Fuchs model discussed in Chap. 9. Here, one group of delayed neutrons, with an average decay constant, λ, is included in the H–F model as well, to give the requisite kinetics equation as

$$\Lambda \dot{n} = \left(K_0 - b \int_0^t n\, d\tau\right) n - \int_0^t n'(\tau)\, e^{-\lambda(t-\tau)}\, d\tau \tag{5-6}$$

The integral terms are eliminated in Eq. (5-6) by two successive differentiations of the latter to yield, in terms of $E = \int_0^t n\, d\tau$, the energy generated, where $\Lambda\lambda \ll 1$ and is so neglected,

$$\Lambda \dddot{E} = (K_0 - bE - 1)\ddot{E} - b\dot{E}^2 + \lambda(K_0 - bE)\dot{E} \tag{5-7}$$

Now, $\dot{E} = n$, and let $\dot{n} = y$, so that this third order system reduces to three first order differential equations in the state variables n, E, and y, namely

$$\dot{y} = (K_0 - 1 - bE)y - bn^2 + \lambda(K_0 - bE)n; \qquad y(0) = K_0 n_0/\Lambda \tag{5-8}$$

$$\dot{n} = y; \qquad n(0) = n_0 \tag{5-9}$$

$$\dot{E} = n; \qquad E(0) = 0 \tag{5-10}$$

Parenthetically, this third order system can be reduced to one first order differential equation in different state variables, namely, n and K as discussed in Chap. 9.

Again, only to illustrate a linear system, assume b is very small so that the nonlinear terms can be neglected in Eq. (5-8). This results in the linear vector differential equation $\dot{\overline{V}} = \overline{A}\overline{V}$, where the vector $\overline{V}$ has state variable components (y, n, E), that is,

$$\frac{d}{dt}\begin{bmatrix} y \\ n \\ E \end{bmatrix} = \begin{bmatrix} K_0 - 1 & \lambda K_0 & 0 \\ 1 & 1 & 0 \\ 0 & 1 & 0 \end{bmatrix}\begin{bmatrix} y \\ n \\ E \end{bmatrix} \tag{5-11}$$

and the matrix $\bar{A}$ is explicitly identified above. Equation (5-11) can be solved by well-known methods. To do so, determine the eigenvalues of the matrix $\bar{A}$, that is, solve, $\bar{A}\bar{V} = \bar{\alpha}\bar{V}$ where $\bar{\alpha}$ is the matrix of eigenvalues

$$\bar{\alpha} = \begin{pmatrix} \alpha & 0 & 0 \\ 0 & \alpha & 0 \\ 0 & 0 & \alpha \end{pmatrix} \tag{5-12}$$

The eigenvalues are then given by the roots of the determinant

$$\begin{vmatrix} K_0 - 1 - \alpha & \lambda K_0 & 0 \\ 1 & -\alpha & 0 \\ 0 & 1 & -\alpha \end{vmatrix} = 0 \tag{5-13}$$

Equation (5-13) results in a cubic equation in the eigenvalues α, with three roots α_1, α_2, and α_3. The solutions are then written down as

$$\begin{aligned} y &= A_1\, e^{\alpha_1 t} + A_2\, e^{\alpha_2 t} + A_3\, e^{\alpha_3 t} \\ n &= B_1\, e^{\alpha_1 t} + B_2\, e^{\alpha_2 t} + B_3\, e^{\alpha_3 t} \\ E &= C_1\, e^{\alpha_1 t} + C_2\, e^{\alpha_2 t} + C_3\, e^{\alpha_3 t} \end{aligned} \tag{5-14}$$

where the constants $A_1, \ldots C_3$, are determined from the initial conditions given in Eqs (5-8)–(5-10).

What has been done was to transform the matrix $\bar{A}$ to its diagonal representation, so that

$$\dot{\bar{V}} = \bar{A}\bar{V} = \bar{\alpha}\bar{V} \tag{5-15}$$

Then in the transformed space

$$\frac{d}{dt}\begin{bmatrix} y \\ n \\ E \end{bmatrix} = \begin{bmatrix} \alpha & 0 & 0 \\ 0 & \alpha & 0 \\ 0 & 0 & \alpha \end{bmatrix}\begin{bmatrix} y \\ n \\ E \end{bmatrix} \tag{5-16}$$

or $\dot{y} = \alpha y$, $\dot{n} = \alpha n$, $\dot{E} = \alpha E$. Then linear combinations of these primitives, which are also solutions to Eq. (5-11), resulted in Eq. (5-14), or the state vector form of the solution is

$$\begin{bmatrix} y \\ n \\ E \end{bmatrix} = \begin{bmatrix} A_1 & A_2 & A_3 \\ B_1 & B_2 & B_3 \\ C_1 & C_2 & C_3 \end{bmatrix}\begin{bmatrix} e^{\alpha_1 t} \\ e^{\alpha_2 t} \\ e^{\alpha_3 t} \end{bmatrix} \tag{5-17}$$

For more details on the above, especially how it relates to the transfer function, see Weaver.[1]

5-3 OPTIMAL CONTROL THEORY AND THE CALCULUS OF VARIATIONS

Modern optimal control theory is viewed by its many practitioners as part of what is termed systems theory or systems research. Most of the former is subsumed by optimal control, and statistical prediction and estimation analysis. Modern day calculus of variations provides a unifying framework for the solutions of problems springing from such analysis. For especially the deterministic aspects, a good portion of optimal control theory can be considered here a rephrasal of the problem of Lagrange in the calculus of variations. The Lagrange problem is the minimization of, usually, an integral, subject to constraints or side conditions which can include differential equations in the state variables. Notice the emphasis—the calculus of variations is concerned with minimizing some functional, usually an integral, with ancillary state conditions, which can be differential equations. Optimal control theory, on the other hand, reverses the emphasis. That is, the differential equations of motion are the dynamical embodiment of the system behavior, for which state variable solutions are sought, but with the side condition on the solutions that they simultaneously achieve a minimum or maximum of a performance criterion, usually an integral or some other functional, containing the state variable solutions.

To be more explicit, assume a dynamical system to be represented by the state vector differential equation

$$\dot{\overline{X}}(t) = F(\overline{X}(t), \overline{U}(t), t) \tag{5-18}$$

where the above quantities were defined earlier. Now consider the performance criterion functional

$$V(\overline{X}, \overline{U}, t) = H(\overline{X}, t) + \int_{t_0}^{t_f} G(\overline{X}, \overline{U}, t)\, dt \tag{5-19}$$

where the beginning and end of the control phase is reckoned by t_0 and t_f, respectively. The optimal feedback control problem is to find the optimal control policy $\overline{U}^*$, such that for the corresponding state dynamical behavior expressed by the motion

$$\dot{\overline{X}}(t) = F(\overline{X}, \overline{U}^*, t) \tag{5-20}$$

the functional V attains the desired extremum (minimum or maximum). It is well now to enumerate a number of performance criterion types.[2]

1. *Minimum time control:* It is desired that the system state reaches a given final state $\overline{X}_f$, from its initial state $\overline{X}_0$ in minimum time. Then, in Eq. (5-19) $H = 0$ and $G = 1$. It can be shown that this problem has a solution only if $\overline{U}$ is composed of a closed set. Here, this merely implies that $\overline{U}$ be bounded for $t_0 \le t \le t_f$.

2. *Terminal control:* It is desired that the system reaches a final state $\bar{X}_f$ as closely as possible at the final time t_f. That is, the system is to achieve a minimum miss distance from the final (target) state $\bar{X}_f$. Then in Eq. (5-19), $H = \bar{X}_f$, the distance from the target, and $G = 0$.

3. *Optimal regulator or controller:* It is desired that the system return to an equilibrium state, from some not necessarily equilibrium initial state, with a dynamical behavior, that is, a motion that minimizes a performance criterion. Then in Eq. (5-19), G and H are positive.

4. *Optimal pursuit:* A moving target with instantaneous state is given. It is desired to make the system state $\bar{X}(t)$ coincide with $\bar{Y}(t)$ as quickly as possible. This is a generalization of problem (*a*).

5. *Servo control:* It is desired to keep the system state $\bar{X}(t)$ as close to coincidence with $\bar{Z}(t)$, the desired state in the following sense, over a given time interval $t_f - t_0$. The sense is that in Eq. (5-19) G is some function of $|\bar{Z} - \bar{X}|$, the distance (deviation) of the system from the desired state. This is a generalization of problem (*c*).

6. *Minimum energy control:* It is desired to move the system from an initial state $\bar{X}_0$ to a final state $\bar{X}_f$, utilizing a minimum amount of control energy. In Eq. (5-19) this then means that $G = G(\bar{U})$ only, and $G > 0$.

The problem of Lagrange in the calculus of variations is to minimize

$$\int_{t_0}^{t_f} L(\bar{X}, \dot{\bar{X}}, t)\, dt \tag{5-21}$$

by employing a smooth extremal path $\bar{X}(t)$ in the state space between $\bar{X}_0$ and $\bar{X}_f$, that satisfies the constraints given by a constraint relation

$$g(\bar{X}, \dot{\bar{X}}, t) = 0 \tag{5-22}$$

It is seen that the problem of Lagrange and the optimal control problem differ in two non-essential ways. First, the function G in Eq. (5-19) depends on the control policy vector $\bar{U}$, and not $\dot{\bar{X}}$, and second that the optimal control constraints are of a mixed variety in the sense of the variational calculus, that is

$$g \equiv \dot{X} - F(\bar{X}, \bar{U}, t) = 0 \tag{5-23}$$

As discussed, there is no essential difference between the two problems, as it can be shown that a suitable though complicated transformation involving $\dot{\bar{X}}$ and $\bar{U}$ is available to go back and forth between the two.

5-4 DYNAMIC PROGRAMMING

As mentioned in the introduction, there are two complementary methods for couching optimal control problems from the viewpoint of the modern day calculus of variations. One is dynamic programming and the Hamilton–Jacobi equation. The other is the maximum principle and the Hamilton equations. Both of these will be discussed, with examples, in the next two sections. Their equivalence will be established in a later section. First is an introductory exposition of dynamic programming.

The method of dynamic programming, invented by the mathematician Bellman, might be described as a variant of multistage decision theory.[3] The latter lies in the area of complicated control problems in which it is desired to optimize the control so that a performance criterion, or cost functional is extremized. By compartmentalizing the particular control problem under study into a series of coupled subproblems, dynamic programming renders solutions to quite complex optimal control processes. The mathematical aspects of the coupling are implied in a novel functional equation. This equation is a recurrence relation in the discrete representation of the motion of the state variables. This recurrence relation or algorithm is employed to solve for the optimal control stepwise or stagewise, also obtaining a stagewise picture of the state motion. In the limiting case of an infinite number of stages, each of infinitesimally short duration, the sequence of optimal control decisions becomes the continuous control law or policy, while the recurrence relation becomes a novel partial differential equation called Bellman's equation. As will be discussed, Bellman's equation is essentially the Hamilton–Jacobi partial differential equation description of the control process.

The essence of the method of dynamic programming is contained in its principle of optimality. First the control process is compartmentalized into stages in a natural way depending on the particular process. With the application of the principle of optimality, a functional equation unique to the dynamic programming method is obtained. The principle of optimality is: "An optimal policy has the property that whatever the initial state and initial decisions are, the remaining decisions must constitute an optimal policy with regard to the state resulting from the first decision." This is a statement of principle about optimal policies, which are to be understood as those policies that extremize a performance criterion unique to the control process being investigated. Although the phrase "optimal policy" appears in its own statement of principle, it will be realized from the illustrations to follow that where an optimality is defined, the optimality principle is not a mere tautology.

Five introductory examples follow, chosen to illustrate the operative features of the optimality principle as well as the appearance of the Bellman equation as a Hamilton–Jacobi-type partial differential equation.

Example 1. A breeder reactor.[4] Consider a hypothetical enriched uranium breeder reactor, which breeds ^{239}Pu from the fertile ^{238}U. For a given fuel charge of x kilograms, assume that normal operation produces a total of rx kilograms per fuel cycle ($r > 1$). At the end of each fuel cycle, certain fuel elements are rearranged or removed from the core, with new elements added according to a particular fuel management schedule. Here, the fuel is considered homogeneous in the sense that at the end of each fuel cycle, any given amount up to $(r-1)x$ kilograms can be removed, since x kilograms must be retained in the core as the critical mass. It is assumed that the reactor control system, as well as the reactivity coefficients inherent in this reactor, can handle the additional fuel produced over the x kilograms.

The reactor will be operated for N fuel cycles before it is dismantled for overhaul. At the end of each cycle, excess fuel is removed from the core and utilized according to a known utility function $G(y_k)$, where y_k is the amount of fuel removed at the end of the k-1st cycle, that is, at the beginning of the kth cycle. The optimal control problem is to determine the optimal fuel removal policy, which is the one that maximizes the overall utility for N cycles of operation.

Let $f_N(x)$ be the optimal, that is, the maximum overall utility obtained from the fuel removed from the reactor after N cycles of operation using an optimal fuel removal policy. First, consider a one cycle operation, which will be embedded in a two cycle operation, which, in turn, is embedded in a three cycle operation, etc. Assuming that $G(y)$ is an increasing function, then for the trivial one cycle operation, one obtains for the one cycle optimal utility

$$f_1(x) = \max_{0 \le y_1 \le rx} G(y_1) = G(rx); \qquad y_1 = rx \tag{5-24}$$

Obviously, after a one cycle operation, the optimal policy is to remove all the fuel from the core, obtaining the maximum one cycle utility $G(rx)$. For a two cycle operation, at the beginning of the second and last cycle, the x reactor fuel charge x has increased to rx. At this time, if an arbitrary amount of fuel y_2, where $0 \le y_2 \le (r-1)x$, is removed from the reactor and a corresponding utility $G(y_2)$ realized, the two cycle utility is $G(y_2) + f_1(rx - y_2)$. That is, $G(y_2)$ is now the return from the first cycle of a two cycle process, while the remainder of the fuel, $rx - y_2$, is used in an optimal manner, yielding the one cycle utility $f_1(rx - y_2)$. Now for an optimal, instead of arbitrary choice of y_2 for an optimal two stage process, from the definition of f_N, and from the optimality principle, one has

$$f_2(x) = \max_{0 \le y_2 \le (r-1)x} \{G(y_2) + f_1(rx - y_2)\} \tag{5-25}$$

Generalizing to a k cycle operation, one obtains the functional equation of dynamic programming for this process, as

$$f_k(x) = \max_{0 \le y_k \le (r-1)x} \{G(y_k) + f_{k-1}(rx - y_k)\}; \qquad k = 2, 3, \ldots N \tag{5-26}$$

That is, the optimal utility or return from a k cycle process is obtained by maximizing the return $G(y_k)$ plus that from a $k-1$ cycle optimal process starting with an amount $rx - y_k$ of fuel. The maximization is over $0 \le y_k \le (r-1)x$. As already mentioned, $0 \le y_k \le (r-1)x$ holds, except for $k = 1$, because x kilograms must remain to maintain criticality except for the single cycle case, where all fuel is removed before the reactor is dismantled.

Letting $y_k = (r-1)x$, for all k, which implies removing as much fuel as possible after each cycle does not necessarily make for an optimal policy. This is because the utility function G dictates whether or not it is better to further irradiate more fuel than that required by the critical equation. Of course, the optimal policies depend sensitively on the utility function.

Reiterating, the system of functional equations to be solved are the recurrence relations

$$f_1(x) = G(rx); \qquad y_1 = rx \tag{5-27}$$

$$f_k(x) = \max_{0 \le y_k \le (r-1)x} \{G(y_k) + f_{k-1}(rx - y_k)\}; \qquad k = 2, 3, \ldots N \tag{5-28}$$

The solution consists of the optimal fuel removal policy, which is the sequence $\{y_k\}$, and the sequence of optimal returns $\{f_k\}$ of which f_N is the return for the N cycle process. Consider two types of utility functions.

1. $G(y) = y$. Then from the above

$$f_1(x) = rx; \qquad y_1 = rx \tag{5-29}$$

$$f_k(x) = \max_{0 \le y_k \le (r-1)x} \{y_k + f_{k-1}(rx - y_k)\}; \qquad k = 2, 3, \ldots N \tag{5-30}$$

Beginning the solution, using the above as a recurrence relation, the one cycle process is trivial. For the two cycle process

$$f_2(x) = \max_{0 \le y_2 \le (r-1)x} \{y_2 + r(rx - y_2)\} = \max_{0 \le y_2 \le (r-1)x} \{r^2x - (r-1)y_2\} \tag{5-31}$$

As the bracket on the right-hand side of Eq. (5-31) is linear in the control variable y_2, the bracket maximum must lie at one or the other of the extremities in the y_2 range. That is, either at $y_2 = 0$ or $y_2 = (r-1)x$. Here then $y_{2\,\max} = 0$ giving $f_2(x) = r^2x$. Continuing in a straightforward manner, the optimal control policy is $y_1 = rx$, $y_2 = y_3 = \cdots = y_n = 0$, and $f_N = r^N x$. The index N is the number of cycles remaining, so that an N cycle process telescopes into an $N-1$ cycle process, etc. until only a one cycle operation remains. In other words, the optimal policy for this case is to remove no fuel until the very end, and then remove all. This makes physical sense with this type of increasing utility function.

2. $G(y) = y^{1/2}$. Then

$$f_1(x) = (rx)^{1/2} \tag{5-32}$$

$$f_k(x) = \max_{0 \le y_k \le (r-1)x} \{y_k^{1/2} + f_{k-1}(rx - y_k)\} \qquad (k = 2, 3, \ldots, N) \tag{5-33}$$

Then

$$f_2(x) = \max_{0 \le y_2 \le (r-1)x} [y_2^{1/2} + \{r(rx - y_2)\}^{1/2}] \tag{5-34}$$

Now the bracket is seen to be concave in y_2, so that its maximum lies in the interior of the y_2 range. Therefore, equating the derivative of the bracket with respect to y_2 to zero gives $y_{2\,\max} = rx/(1 + r)$. Inserting this value of y_2 back into the bracket above gives $f_2 = \{(1 + r)rx\}^{1/2}$. By induction, it is realized that the return function f_k is proportional to $(rx)^{1/2}$. Then let $f_k = c_k(rx)^{1/2}$, where the coefficient c_k is to be determined. Inserting this f_k into Eq. (5-34) and carrying out the maximization operation yields the recurrent relations

$$y_k = rx/(1 + rc_{k-1}^2) \qquad c_1 = 1 \qquad c_k = (1 + rc_{k-1}^2)^{1/2} \qquad (k = 2, 3, \ldots, N) \tag{5-35}$$

so that

$$f_k = \left(\frac{r^k - 1}{r - 1} rx\right)^{1/2} \qquad y_k = \frac{r(r-1)x}{r^k - 1} \qquad (k = 1, 2, \ldots, N) \tag{5-36}$$

For this type of utility, the optimal policy is seen to decrease quickly for small k, less rapidly for large k. Since k is to be construed as the number of cycles remaining, the optimal policy asserts that relatively small amounts of fuel are removed at the beginning of the N-cyle operation, with increasingly large amounts removed as the irradiation progresses, as depicted in Fig. 5-1.

Example 2. A simple one-dimensional reactor.[5] This example, which is closer to the context of optimal control processes in the sense of the variational calculus, is the consideration of a simple one-dimensional reactor whose kinetic differential equation of state is, in time units such that $\Lambda = 1$,

$$\dot{n} = Kn \tag{5-37}$$

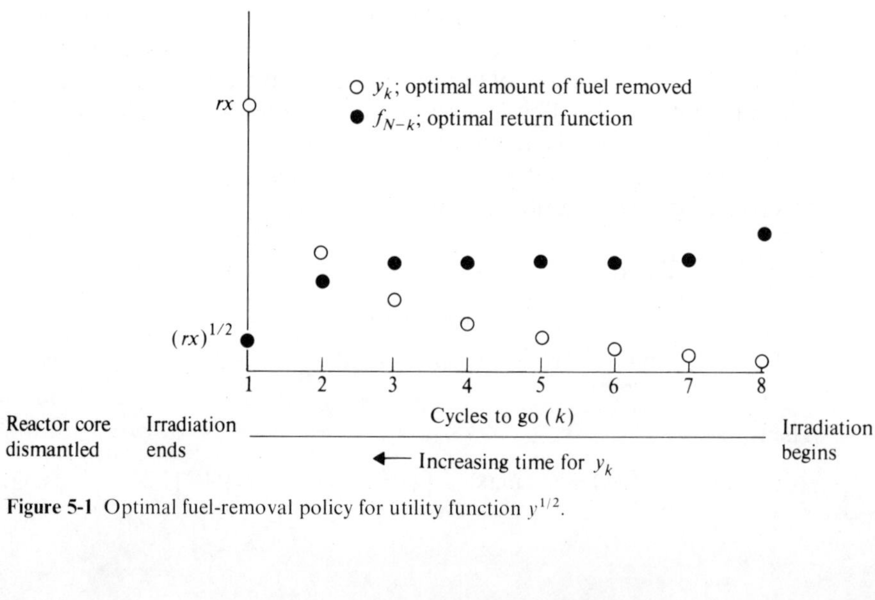

Figure 5-1 Optimal fuel-removal policy for utility function $y^{1/2}$.

where $n(t)$ is the state variable and K is the control variable. The object is to find the optimal control function u that minimizes the following criterion, or "cost," functional

$$J = \alpha \log^2 n(T) + \beta \int_0^T K^2\,dt \tag{5-38}$$

when the system proceeds from its present non-equilibrium state to its final equilibrium state c, accomplishing this task in time T. In the functional J, the first term is the cost of control at the end of the duration of control T. The second term is a "mean-squared" cost of control as it is exerted during the given control duration T. First define the functional

$$F(c, T) = \min_K \beta \int_0^T K^2\,dt \qquad F(c, 0) = \alpha \log^2 c \tag{5-39}$$

$F(c, T)$, even though it is a functional of the control variable u, for a unique optimal control K, is a function only of the state c and the control duration T. Now $F(c, T)$ is rewritten

$$F(c, T) = \min_K \beta \left[\int_0^a K^2(t)\,dt + \int_a^T K^2(t)\,dt \right] \tag{5-40}$$

where a is an intermediate time between 0 and T. Using the optimality principle, this can be rewritten

$$F(c, T) = \min_K \left[\beta \int_0^a K^2(t)\,dt + F\left\{ c + \int_0^a \dot{n}(t)\,dt, T - a \right\} \right] \tag{5-41}$$

That is, the optimal cost $F(c, T)$ is given by the minimum, over K, of the cost from 0 to a seconds, plus the optimal cost F of a new control process (embedded in the original control process) with the system in the state $c + \int_0^a \dot{n}(t)\,dt$, but with a control duration of $T - a$ seconds. The cost of the "initial condition" $F(c, 0)$, which is the terminal cost, since the time remaining to control is zero, is $F(c, 0) = \alpha \log^2 c$.

The process can be thought of as being compartmentalized into two stages. The first stage lasts from 0 to a seconds, during which the optimal control is exerted. This control thereby advances the system state to $c + \int_0^a \dot{n}(t)\,dt$, where now only $T - a$ seconds of control duration remain in the control phase. From a seconds on, the system is controlled optimally, incurring the cost $F(c + \int_0^a \dot{n}(t)\,dt, T - a)$. Then the optimal cost $F(c, T)$ over the whole control duration T is obtained by choosing K to minimize the right side of Eq. (5-41).

Now let a become a small time interval Δ so that, after F is expanded about (c, T) in the right side of Eq. (5-41), the result is

$$F(c, T) = \min_K \left[\beta K^2 \Delta + F(c, T) + \frac{\partial F}{\partial c} \dot{n}(\Delta)\Delta - \frac{\partial F}{\partial T}\Delta + O(\Delta^2) + \cdots \right] \tag{5-42}$$

Canceling F on both sides of Eq. (5-42), dividing by Δ, then letting Δ approach zero and substituting from the equation of state Eq. (5-34) yields Bellman's equation,

$$\frac{\partial F}{\partial T} = \min_{K} \left[\beta K^2 + Kc\frac{\partial F}{\partial c} \right] \qquad F(c,0) = \alpha \log^2 c \tag{5-43}$$

As mentioned before, the above is essentially the Hamilton–Jacobi equation for the control process. In general, such equations are difficult to solve analytically, so that resort must be made to numerical integration on large-scale digital machinery.

For this example, however, the optimal K^* is found by first equating the derivative of the right side of Eq. (5-43), with respect to K, to zero. This is because the right side is quadratic in K, and not linear as it will be for the problems to be considered later. The vanishing of the derivative with respect to K gives

$$K^* = -\frac{c}{2\beta}\frac{\partial F}{\partial c} \tag{5-44}$$

When this is substituted back in Eq. (5-43), the result is

$$\frac{\partial F}{\partial T} = -\frac{c^2}{4\beta}\left[\frac{\partial F}{\partial c}\right]^2 \qquad F(c,0) = \alpha \log^2 c \qquad (c > 0) \tag{5-45}$$

With the solution, the optimal cost, given by

$$F(c, T) = \frac{\alpha \log^2 c}{1 + (\alpha/\beta)T} \tag{5-46}$$

Knowing $F(c, T)$, the optimal control is obtained by inserting Eq. (5-46) in Eq. (5-44), to give

$$K^* = -\frac{\alpha \log c}{\beta + \alpha T} \tag{5-47}$$

It is seen that the optimal cost decreases with the control duration T. This simply reflects the type of cost criterion selected. The same holds for the optimal control K^*, which is seen to be negative if the system (reactor) is supercritical, that is, $c > 1$, and positive if subcritical $c < 1$. This assumes that the critical or equilibrium state is $c = 1$. If the system is in the desired $c = 1$ state, $F = K^* = 0$, so that no control is exerted and thereby no control cost is incurred.

Example 3. Minimax criteria. To lead into control problems involving minimax criteria it is well to first investigate one employing an absolute magnitude criterion. For example, what is the optimal control policy that yields the minimum of the absolute value of the final state? That is

$$|n_N| = \text{minimum} \tag{5-48}$$

This is a terminal control problem, since the final value of the state variable is involved in the criterion functional. Assume a discrete version of an equation of state for a simple reactor with the delayed neutrons ever at equilibrium concentration. From Eq. (2-3), in time units such that $\Lambda = 1$

$$n_{r+1} = K_r n_r + 1; \qquad n_0 = c \tag{5-49}$$

The reactivity control is constrained so that $-M \leq K_r \leq M$. $|M| \leq 1$. Let

$$f_N(c) = \min_{-M \leq K \leq M} |n_N| \tag{5-50}$$

Then, working backwards in time, at the final state, trivially

$$f_0(c) = |n_0| = c \tag{5-51}$$

since zero stages remain to be controlled. For one stage remaining, the state is displaced to $Kc + 1$ so that

$$f_1(c) = \min_{-M \leq K \leq M} f_0(Kc + 1) = \min(-Mc + 1, Mc + 1) = -Mc + 1; \qquad K_0 = -M \tag{5-52}$$

That is, since f_0 is linear in M, its minimum occurs at $K_0 = -M$. So in general

$$f_j(c) = \min_{-M \leq K_j \leq M} f_{j-1}(K_j c + 1) \tag{5-53}$$

The optimal control law is found as a problem at the end of this chapter.

Now consider a minimax criterion problem: to find the reactivity control sequence $\{K_j\}$ that yields, where $-M \leq K_j \leq M$

$$\min_{\{K_j\}} \max |n_j|; \qquad j = 1, 2, \ldots N \tag{5-54}$$

That is, to find the optimal policy K that minimizes the maximum $|n_j|$ in whatever interval j it happens to occur, using the above state equation. Let

$$f_N = \min_{K_j} \max_{j=1,2,\ldots N} |n_j| \tag{5-55}$$

Again, trivially

$$f_0(c) = |c| \tag{5-56}$$

since there are no further control decisions to make. Now

$$f_1(c) = \min_{K_j} \max_{j=0,1} |n_j| \tag{5-57}$$

or

$$f_1(c) = \min_{K_1} \max(|c|, |Kc + 1|) \tag{5-58}$$

Since $|Kc + 1| = f_0(Kc + 1)$, substituting above gives

$$f_1(c) = \min_{K_1} \max\{|c|, f_0(Kc + 1)\} \tag{5-59}$$

so that with j stages remaining to control, the recurrence relation becomes

$$f_j(c) = \min_{-M \le K_j \le M} \max \{|c|, f_{j-1}(K_j c + 1)\} \tag{5-60}$$

Notice that since the minimum operator depends on the index j and that min and max commute in this case and as explained below, Eq. (5-60) can also be written as

$$f_j(c) = \max \left\{ |c|, \min_{-M \le K_j \le M} f_{j-1}(K_j c + 1) \right\} \tag{5-61}$$

Example 4 As a fourth example, examine the following two-dimensional system,[6] which achieves a similar solution, but in a more deductive manner. It is desired to find the optimal control sequence $\{u_j\}$ that yields

$$\min_{u_j} \max_{j=1,2,\ldots,N} |x_j|, \tag{5-62}$$

where the system state is governed by the pair of difference equations

$$\begin{aligned} x_{r+1} &= x_r + y_r \Delta \qquad & x_0 &= c_1 \\ y_{r+1} &= y_r + g(x_r, y_r)\Delta \qquad & y_0 &= c_2 \end{aligned} \tag{5-63}$$

It is important to appreciate the following seemingly simple idea. For a sequence of k numbers $\{N_k\}$, any chosen number N_i obeys the partition relationship

$$\max(N_1, N_2, \ldots, N_k) = \max\{N_i, \max(N_1, N_2, \ldots, N_h, N_j, N_k)\} \tag{5-64}$$

Now let

$$f_N(c_1, c_2) = \min_{u_j} \max_{j=1,2,\ldots,N} |x_j| \tag{5-65}$$

Using the partition relationship above, this can be rewritten

$$f_N(c_1, c_2) = \min_{u_j} \max \left(|x_N|, \max_{j=1,2,\ldots,N-1} |x_j| \right) \tag{5-66}$$

As the minimum is over the set u_j, i.e., the minimum operation is on the index j only, it can be placed inside the parentheses to give

$$f_N(c_1, c_2) = \max \left(|x_N|, \min_{u_j} \max_{j=1,2,\ldots,N-1} |x_j| \right) \tag{5-67}$$

or

$$f_N(c_1, c_2) = \max(|x_N|, f_{N-1}\{c_1 + c_2\Delta, c_2 + g(c_1, c_2)\Delta\} \tag{5-68}$$

since with one less j choice to make for u_j, the system now finds itself confronted with a similar minimax problem but now advanced from the old state (c_1, c_2) to the new state $\{c_1 + c_2\Delta, c_2 + g(c_1, c_2)\Delta\}$ as given by the difference equations (5-63). Note that this method, if applied to the previous one-dimensional problem, will yield Eq. (5-61) as one of its solutions.

Example 5 A fifth example is that of a simplified minimax criterion functional. Consider the old chestnut of finding a bogus coin from a seemingly identical pile of coins, using an ungraduated balance, where it is known that the bogus coin is heavier (or lighter) than the others.[6] That is, find the minimum number of weighings required to guarantee finding the bogus coin from among N coins.

Let $f(N)$ be the minimum number of weighings required, for N coins, to guarantee finding the bogus coin using an optimal weighing policy. The principle of optimality asserts that

$$f(N) = 1 + \min_k \max \{f(k), f(N - 2k)\} \tag{5-69}$$

It is obvious that the N coins should be divided into three equal stacks, as nearly as possible. Then two stacks each contain k coins, while the third contains $N - 2k$ coins, yielding $k = [N/3]$; $[n]$ is read as the nearest integer contained in n. The two equal stacks containing k coins each are weighed. This accounts for the first term (unity) on the right side of Eq. (5-69). If the scale becomes unbalanced, the search is immediately narrowed to k coins. That case is fortuitous, because if the scale balances, the search narrows to the possibly larger stack of $N - 2k$ coins, since $N - 2k \geq N/3$ is possible. Then, at worst, the search narrows to the stack that has one more coin than the other two. Thence, $k = [N/3]$, and

$$f(N) = 1 + \min_{k=[N/3]} \max \{f(k), f(N - 2k)\} \tag{5-70}$$

Since the minimum operator commutes with the maximum operator in the above equation, it becomes

$$f(N) = 1 + \max \{f([N/3), f(N - 2[N/3])\} \tag{5-71}$$

This implies that the minimum number of weighings, $f(N)$, equals the first weighing plus the subsequent minimum number of weighings, $f([N/3])$, or $f(N - 2[N/3])$, whichever is the larger, to guarantee finding the bogus coin. To solve Eq. (5-71), it is convenient first to let $N = 3m$ to obtain

$$f(3m) = 1 + \max \{f(m), f(m)\} = 1 + f(m) \tag{5-72}$$

Second, let $N = 3m + 1$ in Eq. (5-71), to obtain

$$f(3m + 1) = 1 + \max \{f(m), f(m + 1)\} = 1 + f(m + 1) \tag{5-73}$$

Third, let $N = 3m + 2$ in Eq. (5-71), to obtain

$$f(3m + 2) = 1 + \max \{f(m + 1), f(m)\} = 1 + f(m + 1) \tag{5-74}$$

Using these recurrence relations, with $m = 1, 2, \ldots$, the nontrivial solutions are easily obtained; they are,

$$f(N) = M, \qquad \text{where } 3^{M-1} < N \leq 3^M \tag{5-75}$$

For example, six weighings are sufficient to guarantee finding a bogus coin (known to be lighter or heavier) from among collections ranging from 243 to 729 coins.

5-5 THE MAXIMUM PRINCIPLE

As mentioned in the earlier discussion, the modern theory of optimal control has seen the emergence in the last decade of two complementary methodologies for treating control processes. These are Bellman's optimality principle of dynamic programming and Pontryagin's maximum principle.[7] Both of these possess advantages and disadvantages with respect to the formulation and solution of particular control problems. The central problem herein, optimal xenon shutdown control, will be mathematically formulated later, using both principles. The difficulties will be pointed out, and it will be seen that for this class of problems, especially from the computational viewpoint, dynamic programming is the more straightforward and hence the more efficacious method.

The aim of this section is to introduce the maximum principle. It will be stated but not proved, and illustrative examples and ramifications will be presented. Proof of the maximum principle can be found in many places in the control-theory literature of today.[7]

Consider a system described by the set of ordinary differential equations written here in expanded form

$$\begin{aligned} \dot{x}_1 &= f_1(x_1, x_2, \ldots, x_n; u_1, u_2, \ldots, u_n; t) \qquad & x_1(0) &= x_{10}, \\ \dot{x}_2 &= f_2(x_1, x_2, \ldots, x_n; u_1, u_2, \ldots, u_n; t) \qquad & x_2(0) &= x_{20}, \\ &\vdots \\ \dot{x}_n &= f_n(x_1, x_2, \ldots, x_n; u_1, u_2, \ldots, u_n; t) \qquad & x_n(0) &= x_{n0}, \end{aligned} \tag{5-76}$$

where the state of the system is described by the vector $(x_1, x_2, \ldots, x_n)$, and the control vector is $(u_1, u_2, \ldots, u_n)$. The number of components in the state vector $\bar{x}$ and the control vector $\bar{u}$ need not be the same.

Assume that it is desired to find the optimal control policy, that is, the vector $\bar{u} = (u_1, u_2, \ldots, u_n)$ which will transform the system described by Eqs (5-76) from its initial state $(x_{10}, x_{20}, \ldots, x_{n0})$ to a final state in duration T such that the criterion or cost functional

$$J[\bar{u}] = \int_0^T f_{n+1}(x_1, x_2, \ldots, x_n; u_1, u_2, \ldots, u_n; t)\, dt \tag{5-77}$$

is minimized. If it turns out that a particular criterion is to be maximized, one merely minimizes the negative in this context. Some typical examples of cost functionals are:

1. minimum time control, $f_{n+1} = 1$, so that trivially $J = T$;
2. minimum mean-squared control, $f_{n+1} = |\bar{u}|^2$; then $J = \int_0^T |\bar{u}|^2\, dt$;

3. minimum mean-squared deviation, $f_{n+1} = |\bar{x}|^2$; then $J = \int_0^T |\bar{x}|^2 \, dt$;

or various combinations of these.

Now define a new additional state variable

$$x_{n+1} = \int_0^t f_{n+1}(x_1, x_2, \ldots, x_n; u_1, u_2, \ldots, u_n; z) \, dz \tag{5-78}$$

with its differential equation

$$\dot{x}_{n+1} = f_{n+1}(x_1, x_2, \ldots, x_n; u_1, u_2, \ldots, u_n; t) \tag{5-79}$$

and initial condition $x_{n+1}(0) = 0$ and final condition $x_{n+1}(T) = J$. x_{n+1} is an additional state variable which should be considered as added to the system described by Eqs (5-76). As can be appreciated, the above general optimal control process is written in the language of the calculus of variations. With the integral constraint on the motion of the system, Eq. (5-77), such a formulation is called a Lagrange problem. Defining an additional variable x_{n+1} as above, so that now the motion of the system, augmented by the additional variable, is controlled, in order that $x_{n+1}(T) = \min$ is obtained, converts it to a Mayer problem. That is, the augmented system is transformed to a final state where one particular variable, x_{n+1}, has its extremum, while the others acquire certain fixed values. This is done because the maximum principle is couched in terms of a Mayer formulation.

Now define a hamiltonian H, by letting

$$H = p_1 f_1 + p_2 f_2 + \cdots + p_{n+1} f_{n+1} \tag{5-80}$$

where a set of adjoint (dual, auxiliary) variables $p_1, p_2, \ldots, p_{n+1}$ are defined in that they satisfy the following system of equations, adjoint to the system given by Eqs (5-76):

$$\begin{aligned}
-\dot{p}_1 &= \frac{\partial f_1}{\partial x_1} p_1 + \frac{\partial f_2}{\partial x_1} p_2 + \cdots + \frac{\partial f_{n+1}}{\partial x_1} p_{n+1} && p_1(T) = 0 \\
-\dot{p}_2 &= \frac{\partial f_1}{\partial x_2} p_1 + \frac{\partial f_2}{\partial x_2} p_2 + \cdots + \frac{\partial f_{n+1}}{\partial x_2} p_{n+1} && p_2(T) = 0 \\
&\vdots \\
-\dot{p}_n &= \frac{\partial f_1}{\partial x_n} p_1 + \frac{\partial f_2}{\partial x_n} p_2 + \cdots + \frac{\partial f_{n+1}}{\partial x_n} p_{n+1} && p_n(T) = 0 \\
-\dot{p}_{n+1} &= \frac{\partial f_1}{\partial x_{n+1}} p_1 + \frac{\partial f_2}{\partial x_{n+1}} p_2 + \cdots \frac{\partial f_{n+1}}{\partial x_{n+1}} p_{n+1} && p_{n+1}(T) = -1
\end{aligned} \tag{5-81}$$

From the definition of the hamiltonian, it is seen that the state vector $\bar{x}$, and its dual vector $\bar{p}$ (generalized momentum of classical mechanics), satisfy the following canonical equations of Hamilton; that is, from Eqs (5-76) and (5-81)

$$\dot{x}_i = \frac{\partial H}{\partial p_i}, \qquad \dot{p}_i = -\frac{\partial H}{\partial x_i}, \qquad (i = 1, 2, \ldots, n+1) \tag{5-82}$$

x_{n+1} and p_{n+1} are seen to be "special" variables, since the definition of x_{n+1} from Eq. (5-78) implies

$$\dot{p}_{n+1} = -\frac{\partial H}{\partial x_{n+1}} = 0 \tag{5-83}$$

because the hamiltonian does not depend on x_{n+1} explicitly. In classical mechanics, it is said that the hamiltonian is cyclic in the coordinate x_{n+1}, so that the corresponding generalized momentum p_{n+1} is a constant of the motion. Then, for mathematical convenience, $p_{n+1} = -1$ is chosen, which is the reason for the final condition on p_{n+1} in Eqs (5-81).

With regard to the control vector $\bar{u}$, the maximum principle requires that it belongs to a closed set of functions. As mentioned, the maximum principle is an extension of the Weierstrass necessary condition for the existence of extremal, in the variational calculus, to closed sets of control functions. In our terms, this merely implies that the control vector is constrained, so that for each of its components, $a_i \le u_i \le b_i, i = 1, \ldots, n$. The proof of the maximum principle requires closed sets of $\bar{u}$ vectors; otherwise the preceding statement of the control problem is meaningless. This often has a physical analog as well, since an unconstrained control vector implies the availability of infinite amounts of energy, momentum, thrust, etc., which is physically impossible.

The maximum principle, briefly stated, is that the sought-for optimal control $\bar{u}^*$, which is the control vector that guides the system satisfying Eqs (5-76) so that J in Eq. (5-77) is minimized, is also that control vector which maximizes the hamiltonian defined in Eq. (5-80). That is

$$\max_{\bar{u}} H(\bar{p}(t), \bar{x}(t), \bar{u}(t)) = H(\bar{p}(t), \bar{x}(t), \bar{u}^*(t)) = \text{constant} \tag{5-84}$$

where the adjoint vector $\bar{p}(t)$ satisfies Eqs (5-81) and $p_{n+1} < 0$. In other words, considering the hamiltonian as a functional of $\bar{u}$, the optimal control $\bar{u}^*$ is the vector that maximizes H, which maximum is a constant of the motion. $H(\bar{p}, \bar{x}, \bar{u}^*)$ is a constant throughout the control duration T, and further $H(\bar{p}, \bar{x}, \bar{u}^*) = 0$, if T is not fixed a priori.

The maximum principle is a necessary, but not sufficient, condition. However, for most control processes, physical intuition will provide sufficiency arguments as well for the calculated optimal $\bar{u}^*$.

With respect to the $\bar{u}^*$ that gives the sought-for $H(\bar{p}, \bar{x}, \bar{u}^*) = \max$, the optimal control behavior falls into two general categories. The first is that if the system equations (5-76) (usually linear) are such that the resulting hamiltonian is linear in the u_i, then with the constraints $a_i \le u_i \le b_i$, $i = 1, 2, \ldots, n$, the optimal control lies on the boundary of its closed set. That is, all the u_i will equal either a_i or b_i over all or part of the control duration T. If the preceding is true only over portions of the control duration, then the u_i will switch back and forth between a_i and b_i during the control phase, so that $H(\bar{p}, \bar{x}, \bar{u}^*) = \max$ will be maintained throughout the control duration. This type of control is called bang-bang, for obvious reasons. The second category is the case where $H(\bar{p}, \bar{x}, \bar{u})$ is nonlinear in

$\bar{u}$. Then the u_i are obtained from $\partial H/\partial u_i = 0$, which generally results in a continuous function of time for the optimal control. Both of these types will be illustrated in the following examples.

Five Examples

Example 1 The first example is one-dimensional (first-order) and is quite similar to the one in Sec. 5-4 (which was solved by dynamic programming). The equation of state is

$$\dot{n} = un \qquad n(0) = 1 \tag{5-85}$$

The cost functional is $J = \beta \int_0^T u^2\,dt$. Following the scheme of the maximum principle, define the additional variable

$$y = \beta \int_0^t u^2\,dz \tag{5-86}$$

with the corresponding additional equation of state

$$\dot{y} = \beta u^2 \qquad y(0) = 0 \tag{5-87}$$

thereby converting the Lagrange problem of minimizing $\beta \int_0^T u^2\,dt$ to the Mayer problem of seeking $y(T) = \text{min}$. The problem now is one of finding the optimal control u that transfers the system from its initial state $n(0) = 1$ to a prescribed final state n_T, in control duration T, such that $y(T)$ is a minimum.

The hamiltonian is simply, from Eq. (5-80),

$$H = p_1 un + p_2 \beta u^2 \tag{5-88}$$

and the adjoint variables satisfy

$$\dot{p}_1 = -\frac{\partial H}{\partial n} = -p_1 u \qquad \dot{p}_2 = -\frac{\partial H}{\partial y} = 0 \qquad p_2 = -1 \tag{5-89}$$

Since the hamiltonian is nonlinear in u, $\partial H/\partial u = 0$, together with $p_2 = -1$, yields

$$u^* = np_1/2\beta \tag{5-90}$$

Then to obtain u^* explicitly, it is necessary to solve the following two-point boundary-value problem:

$$\dot{n} = un \qquad n(0) = 1 \tag{5-91a}$$

$$\dot{y} = \beta u^2 \qquad y(0) = 0 \tag{5-91b}$$

$$\dot{p}_1 = -up_1 \qquad p_1(T) = 0 \tag{5-91c}$$

$$\dot{p}_2 = 0 \qquad p_2(T) = -1 \tag{5-91d}$$

That is, both n and p_1 must be obtained from solutions of Eqs (5-91). In this case it is easy, since multiplication of Eq. (5-91c) by n yields

$$n\dot{p}_1 + p_1 un = 0 \tag{5-92}$$

Using Eq. (5-91a), this is equivalent to

$$\frac{d}{dt}(np_1) = 0 \qquad \text{or} \qquad np_1 = c_1 \text{ (constant)} \tag{5-93}$$

This implies that $u^* = c_1/2\beta$, a constant of the motion. Integrating Eq. (5-91a) yields

$$n = \exp(c_1 t/2\beta) \tag{5-94}$$

In terms of the final state n_T, $c_1 = (2\beta/T)\ln n_T$, which follows from Eq. (5-94).

In turn, the optimal control is

$$u^* = \frac{1}{T}\ln n_T \tag{5-95}$$

and the corresponding state behavior is

$$n = (n_T)^{t/T} \tag{5-96}$$

If the final state is greater than the initial state, i.e., $n_T > 1$, then u^* is positive, so that the system increases exponentially to its final state, reaching there in time T. On the other hand, if $n_T < 1$, then $u^* < 0$, and the system decreases exponentially to its final state in time T. If $n_T = 1$, $u^* = 0$, as the system is already in its desired state, since the initial and final states coincide. From the foregoing, it is easily shown that

$$\max H = \beta u^{*2} = \frac{\beta}{T^2}\ln^2 n_T \tag{5-97}$$

with

$$\min J = \beta T u^{*2} = \frac{\beta}{T}\ln^2 n_T \tag{5-98}$$

or that

$$\max H = \beta\frac{1}{T}\int_0^T u^{*2}\,dt \tag{5-99}$$

so that the optimal control u^* transforms the system motion so that the hamiltonian is proportional to the minimum mean-squared control cost. Though u^* remained constant throughout the control duration, this is not true bang-bang control, as u^* does not switch between its bounds at a certain time, or times, during the control duration T. A bang-bang control example will be presented later.

Example 2 As the next example, change only the control criterion to $J_1 = \int_0^T (u^2 + Q^2)\, dt$, where the equation of state is the same as before, but the state variable is redefined as $Q = \log n$. Then the augmented equation of state is, from Eq. (5-85)

$$\dot{Q} = u \tag{5-100}$$

As before, define a second variable $Z = \int_0^t (u^2 + Q^2)\, dz$ with

$$\dot{Z} = u^2 + Q^2 \tag{5-101}$$

The hamiltonian is

$$H = p_1 u + p_2(u^2 + Q^2) \tag{5-102}$$

and

$$\dot{p}_1 = -\frac{\partial H}{\partial Q} = -2p_2 Q \qquad \dot{p}_2 = -\frac{\partial H}{\partial Z} = 0 \qquad \text{(as usual, } p_2 = -1) \tag{5-103}$$

From

$$\partial H/\partial u = 0 \tag{5-104a}$$

$$u^* = p_1/2 \tag{5-104b}$$

Using Eqs (5-100), (5-103), and (5-104), it is seen that the optimal control satisfies

$$\ddot{u}^* - u^* = 0 \tag{5-105}$$

Since now $\dot{Q} = u^*$, the initial conditions on u^* needed to solve Eq. (5-105) can be inferred from those on $\dot{Q}$. Let $\dot{Q}(0) = u_0^*$, so that the system is initially disturbed from equilibrium, and $\dot{Q}_T = u_T^* = 0$, for equilibrium to be achieved at the end of the control duration. With these conditions, the optimal control is obtained from Eq. (5-105) as

$$u^* = \frac{u_0^* \sinh (T - t)}{\sinh T} \tag{5-106}$$

which is seen to be an example of a continuous optimal control function. The state-variable motion then follows simply from $\dot{Q} = u^*$ and $Q(0) = 0$.

Example 3. Bang-Bang Control[9] Consider the linear second-order system described by

$$\dot{x}_1 = -x_1 + x_2 \qquad x_1(0) = 0 \tag{5-107a}$$

$$\dot{x}_2 = \tfrac{5}{4}x_1 - 3x_2 + u \qquad x_2(0) = 0 \tag{5-107b}$$

It is desired to transform this system from its initial equilibrium state (0, 0) to a final state (2, 2) in minimum time. There are constraints on the magnitude of u given by $0 \leq u \leq 7$. For minimum time control, the cost functional is evidently $J = T$. The additional variable is $x_3 = t$, so $\dot{x}_3 = 1$. The hamiltonian is simply

$$H = p_1(-x_1 + x_2) + p_2(\tfrac{5}{4}x_1 - 3x_2 + u) - 1 \tag{5-108}$$

The adjoint variables satisfy

$$\dot{p}_1 = -\frac{\partial H}{\partial x_1} = p_1 - \tfrac{5}{4}p_2, \quad \dot{p}_2 = -\frac{\partial H}{\partial x_2} = -p_1 + 3p_2, \quad \dot{p}_3 = -\frac{\partial H}{\partial x_3} = 0 \tag{5-109}$$

The fact that the control is bang-bang can be ascertained immediately, since H can be rewritten

$$H = p_2 u + \text{terms not containing } u \tag{5-110}$$

Thence, to maximize H

$$u^* = \begin{cases} 7 & p_2 > 0 \\ 0 & p_2 < 0 \end{cases} \tag{5-111}$$

Then, over the initial part of the control epoch, $u^* = 7$, switching at a time $t_s < T$ to $u^* = 0$ until the control duration is ended. If $p_2 = 0$ over part of the control duration, u^* is arbitrary, which behavior is called *singular control.* This occurs here only at isolated points,† so that no difficulties of this type arise. Singular control has only recently begun to come under scrutiny by investigators in modern optimal control theory.

The solution of Eqs (5-107) are, within arbitrary constants, given by

1. $0 \le t \le t_s$:

$$\begin{aligned} x_1 &= A\,e^{-(7/2)t} + B\,e^{-(1/2)t} + \tfrac{4}{7}u \\ x_2 &= -\tfrac{5}{4}A\,e^{-(7/2)t} + \tfrac{1}{2}B\,e^{-(1/2)t} + \tfrac{4}{7}u \end{aligned} \tag{5-112}$$

2. $t_s \le t \le T$:

$$x_1 = F\,e^{-(7/2)t} + G\,e^{-(1/2)t} \qquad x_2 = -\tfrac{5}{2}F\,e^{-(7/2)t} + \tfrac{1}{2}G\,e^{-(1/2)t} \tag{5-113}$$

and the solutions of Eqs (5-109) are

$$p_1 = C\,e^{(7/2)t} + D\,e^{(1/2)t} \tag{5-114}$$

$$p_2 = -2C\,e^{(7/2)t} + \tfrac{2}{5}D\,e^{(1/2)t} \tag{5-115}$$

$$p_3 = -1 \tag{5-116}$$

Using the initial and final states, the two-point boundary-value solution to (x_1, x_2) is as follows:

† On a set of measure zero (cf. reference 8).

1. $0 \le t \le t_s, u = 7$:

$$x_1 = \tfrac{2}{3}e^{-(7/2)t} - \tfrac{14}{3}e^{-(1/2)t} + 4 \tag{5-117}$$

$$x_2 = -\tfrac{5}{3}e^{-(7/2)t} - \tfrac{7}{3}e^{-(1/2)t} + 4 \tag{5-118}$$

2. $t_s \le t \le T, u = 0$:

$$x_1 = -\tfrac{1}{3}e^{(7/2)(T-t)} + \tfrac{7}{3}e^{(1/2)(T-t)} \tag{5-119}$$

$$x_2 = \tfrac{5}{6}e^{(7/2)(T-t)} + \tfrac{7}{6}e^{(1/2)(T-t)} \tag{5-120}$$

Since x_1 and x_2 are continuous (but not their derivatives) at the switching time, t_s, the switching time itself, can be determined by matching the above solutions at $t = t_s$. This results in two transcendental equations involving the control duration T and the switching time t_s. These can be solved using numerical approximation methods. It turns out that $t_s = 1.605$ and $T = 1.802$. The constants C and D are obtained from $H\{\bar{p}(0), \bar{x}(0), \bar{u}^*(0)\} = 0$, since the time T is unspecified† and $p_2(t_s) = 0$ in order that $H\{\bar{p}(t_s), \bar{x}(t_s), \bar{u}^*(t_s)\} \equiv 0$ holds.

Example 4. Continuous and Bang-Bang Control Let the linear one-dimensional equation of state be given by

$$\dot{x} = -x + u \qquad x(0) = 1 \tag{5-121}$$

It is desired to control the system so that it reaches equilibrium at an unspecified time T, that is, $x(T) = 0$, while minimizing the cost functional[9]

$$J = \int_0^T (x^2 + u^2)\,dt \tag{5-122}$$

There are two cases: (1) no constraint on u, and (2) the control constraint $|u| \le \frac{1}{4}$.

Case (1). The additional variable y satisfies

$$\dot{y} = x^2 + u^2 \tag{5-123}$$

with the hamiltonian given by

$$H = p_1(-x + u) - (x^2 + u^2) \tag{5-124}$$

As H is nonlinear in u, $\partial H/\partial u = 0$ yields the optimal control as

$$u^* = p_1/2 \tag{5-125}$$

and

$$\dot{p}_1 = -\frac{\partial H}{\partial x} = p_1 + 2x \qquad \dot{p}_2 = -\frac{\partial H}{\partial y} = 0 \qquad p_2 = -1 \tag{5-126}$$

† Since the time is unspecific a priori, $H\{\bar{p}(t), \bar{x}(t), \bar{u}^*(t)\} \equiv 0$ during the entire control duration, as stated in the maximum principle.

Eliminating p_1 from Eqs (5-121), (5-125), and (5-126) results in

$$\ddot{x} - 2x = 0 \qquad \ddot{u}^* - 2u^* = 0 \tag{5-127}$$

with solutions

$$x = A\,e^{2^{1/2}t} + B\,e^{-2^{1/2}t} \tag{5-128}$$

$$u^* = (1 + 2^{1/2})A\,e^{2^{1/2}t} + (1 - 2^{1/2})B\,e^{-2^{1/2}t} \tag{5-129}$$

A and B are found by using $x(0) = 1$ and $H(\bar{p}, \bar{x}, u^*) = 0$. Two pairs occur, namely, $A = 0$, $B = 1$ and $A = 1$, $B = 0$. The first pair is chosen because the desired final (equilibrium) state could not be reached otherwise. Then the optimal path, control, and duration are given, respectively, by

$$x = e^{-2^{1/2}t} \qquad u^* = (1 - 2^{1/2})\,e^{-2^{1/2}t} \qquad T = \infty \tag{5-130}$$

Case (2). This is the same problem as in case (1), but with the additional control constraint $|u| \le \frac{1}{4}$. Hence the equations of state, the hamiltonian, and the adjoint equations are the same as in case (1). u^* is depicted in Fig. 5-2.

It is first appropriate to find the switching time, t_s, at which the optimal control u^* changes from $u^* = -\frac{1}{4}$ to a continuous function. Since $u^* = p_1/2$ holds where u^* is continuous, which is the region for the validity of obtaining u^* from $\partial H/\partial u = 0$, then t_s is found from

$$-\tfrac{1}{4} = u^*(t_s) = p_1(t_s)/2 \tag{5-131}$$

Therefore p_1 must be found in the region $0 \le t \le t_s$, i.e., where $u^* = -\frac{1}{4}$. This is done by first solving Eq. (5-121) with $u = -\frac{1}{4}$. Its solution is

$$x = \tfrac{5}{4}\,e^{-t} - \tfrac{1}{4} \tag{5-132}$$

With this x inserted in Eq. (5-126), its solution gives

$$p_1 = \{p_1(0) + \tfrac{3}{4}\}\,e^{t} + \tfrac{1}{2} - \tfrac{5}{4}\,e^{-t} \tag{5-133}$$

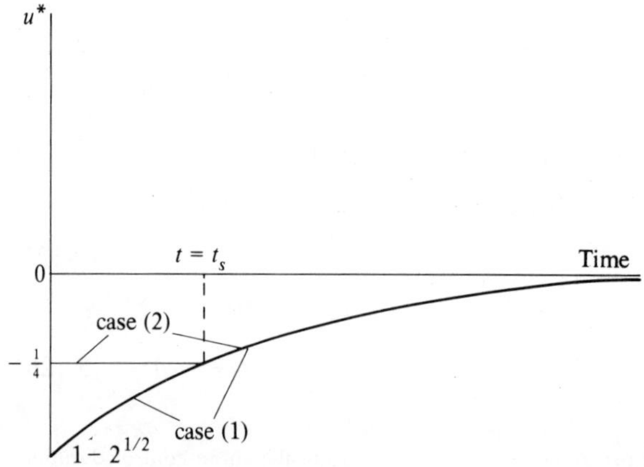

Figure 5-2 Optimal control u^* for continuous and bang-bang control.

$p_1(0)$ is obtained from $H\{\bar{p}_1(0), \bar{x}(0), -\frac{1}{4}\} = 0$ in Eq. (5-124). This yields

$$p_1 = -\tfrac{1}{10}e^t + \tfrac{1}{2} - \tfrac{5}{4}e^{-t} \qquad 0 \le t \le t_s \tag{5-134}$$

Hence t_s is obtained from Eqs (5-131) and (5-134) evaluated at t_s. The behavior of u^* for times following the switching time t_s is obtained in the same way as in case (1).

Example 5. Optimal Orbital-Rendezvous Control It is desired to transform a system (space vehicle) from an initial state to a final state that is not completely fixed but lies on a given curve.[9] This can be construed as orbital-rendezvous control, where a space vehicle is to be launched and controlled to minimize, for example, fuel consumption and to rendezvous with an orbiting space station within a given time T. The additional orbital rendezvous condition that the optimal path must satisfy comes from the calculus of variations, where it is referred to as a transversality condition. This derived condition is a generalized orthogonality property, as can be realized by consulting references in the variational calculus.[6]

Consider the second-order system whose equations of state are

$$\dot{x}_1 = x_2 \qquad x_1(0) = 2 \tag{5-135}$$

$$\dot{x}_2 = u \qquad x_2(0) = 0 \tag{5-136}$$

It is desired to find the optimal control u^* that transforms the state of the system, initially at $(2, 0)$, to the orbit (circle) $x_1^2(T) + x_2^2(T) = 1$ in the control duration time T. The control is constrained so that $|u| \le 1$, and the cost functional to be minimized is

$$J = \int_0^T (|u| + \tfrac{1}{2})\, dt \tag{5-137}$$

$\int_0^T |u|\, dt$ can be construed as a fuel-consumption cost, while $T/2$ is the rendezvous cost. The additional variable $y = \int_0^t (|u| + \frac{1}{2})\, dz$ satisfies the differential equation

$$\dot{y} = |u| + \tfrac{1}{2} \tag{5-138}$$

The hamiltonian is simply

$$H = p_1 x_2 + p_2 u - |u| - \tfrac{1}{2} \tag{5-139}$$

The adjoint equations are given by

$$\dot{p}_1 = 0 \qquad \dot{p}_2 = -p_1 \qquad \dot{p}_3 = 0 \qquad p_3 = -1 \tag{5-140}$$

The control is seen to be of the "bang-off-bang" type. That is, to maximize H,

$$u = \begin{cases} -1 & p_2 < -1 \\ 0 & -1 < p_2 < 1 \\ 1 & p_2 > 1 \end{cases} \tag{5-141}$$

The solutions of Eqs (5-140) are, within arbitrary constants

$$p_1 = C \qquad p_2 = B - Ct \tag{5-142}$$

To determine the constant C, assume that the control regime begins with $u^* = -1$, which is prior to the switching time t_s. That is, let $u^* = -1$ for $0 \le t \le t_s$, so that $H(\bar{x}(0), \bar{p}(0), -1) = 0$ can be used as one condition to determine B and C in Eqs (5-142).

The other condition is obtained from Eq. (5-141), i.e., $p_2(t_s) = -1$ if $u^* = -1$ marks the first phase of control. These two conditions result in $B = -\frac{3}{2}$ and $C = -\frac{1}{2}t_s$, so that

$$p_1 = -\frac{1}{2t_s} \qquad p_2 = -\frac{3}{2} + \frac{1}{2}\frac{t}{t_s} \tag{5-143}$$

Hence, from the case where the first phase of the optimal control is $u^* = -1$, the second-phase control is $u^* = 0$. Now suppose that the rendezvous occurs during this phase of the control duration. Then the transversality condition discussed earlier yields

$$p_1(T)/x_1(T) = p_2(T)/x_2(T) \tag{5-144}$$

Using Eqs (5-135), (5-136), (5-143), and (5-144) at the end of the control duration T, together with the orbit $x_1^2(T) + x_2^2(T) = 1$, will determine the switching time t_s and the control duration T. To provide a measure of confidence in the solution, it is readily shown that $H(\bar{x}, \bar{p}, u^*) \equiv 0$ over the control duration T.

5-6 EQUIVALENCE BETWEEN THE OPTIMALITY PRINCIPLE AND THE MAXIMUM PRINCIPLE

There is an equivalence between the optimality principle of dynamic programming and the maximum principle, which is readily apparent when both principles are applied to find the solution to a particular optimal control process.[8]

This equivalence will be demonstrated by first using the principle of optimality to derive the functional equation of dynamic programming for a general class of optimal control processes. Using this functional equation, the adjoint equations of the maximum principle will follow, as well as the statement of the maximum principle itself.

Consider the general control process of finding the optimal control policy that minimizes

$$J \equiv \int_0^T g(x_1, x_2, \ldots, x_N; u_1, u_2, \ldots, u_m)\, dt \qquad (m \le N) \tag{5-145}$$

where T is the allowable duration of control, $\bar{x} = (x_1, x_2, \ldots, x_N)$ is the state vector, and $\bar{u} = (u_1, u_2, \ldots, u_m)$ is the control vector. The components x_i of the state vector satisfy the following equations of state (equations of motion):

$$\begin{aligned}
\dot{x}_1 &= f_1(x_1, x_2, \ldots, x_N; u_1, u_2, \ldots, u_m) \qquad x_1(0) = c_1 \\
\dot{x}_2 &= f_2(x_1, x_2, \ldots, x_N; u_1, u_2, \ldots, u_m) \qquad x_2(0) = c_2 \\
&\vdots \\
\dot{x}_N &= f_N(x_1, x_2, \ldots, x_N; u_1, u_2, \ldots, u_m) \qquad x_N(0) = c_N
\end{aligned} \tag{5-146}$$

That is, it is desired to find the optimal control policy $\bar{u}^* = (u_1^*, u_2^*, \ldots, u_m^*)$ that yields

$$J^* \equiv \int_0^T g(x_1, x_2, \ldots, x_N; u_1^*, u_2^*, \ldots, u_m^*)\, dt = \min \tag{5-147}$$

where the $\{x_i\}$ satisfy Eqs (5-146) with initial conditions $\bar{x}(0) = (c_1, c_2, \ldots, c_N)$ such that the system proceeds to some given final state, $\bar{x}_T$, at the end of the given control duration, T.

Now, let a "cost," or criterion, function $S(c_1, c_2, \ldots, c_N, T)$ be defined as

$$S(c_1, c_2, \ldots, c_N, T) = \int_0^T g(x_1, x_2, \ldots, x_N; u_1, u_2, \ldots, u_m)\, dt \tag{5-148}$$

Using the principle of optimality of dynamic programming allows one, as explained in previous chapters, to write the following functional equation for S:

$$S(c_1, c_2, \ldots, c_N, T) = \min_{\bar{u}} [g(c_1, c_2, \ldots, c_N; u_1, u_2, \ldots, u_m)\Delta + S(c_1 + \dot{x}_1\Delta, \ldots, c_N + \dot{x}_N\Delta, T - \Delta)] \tag{5-149}$$

This equation enunciates the optimality principle in that the cost functional S is given by the minimum, over the allowable $\{u_i\}$, of the sum of two terms on the right side of Eq. (5-149). The first term is the cost of control for a duration of Δ units of time, and the second is the cost of a "new" optimal control process $S(c_1 + \dot{x}_1\Delta, c_2 + \dot{x}_2\Delta, \ldots, c_N + \dot{x}_N\Delta, T - \Delta)$, beginning in the state $(c_1 + \dot{x}_1\Delta, c_2 + \dot{x}_2\Delta, \ldots, c_N + \dot{x}_N\Delta)$ but of duration $T - \Delta$ units of time.

In the usual manner, expansion of the second term on the right side of Eq. (5-149) about the initial state $(c_1, c_2, \ldots, c_N)$ yields

$$S(c_1, c_2, \ldots, c_N, T) = \min_{\bar{u}} \left[g(c_1, c_2, \ldots, c_N; u_1, u_2, \ldots, u_m)\Delta + S + \sum_{i=1}^{N} \frac{\partial S}{\partial c_i} \dot{x}_i \Delta - \frac{\partial S}{\partial T}\Delta + O(\Delta^2) + \cdots \right] \tag{5-150}$$

Canceling S on both sides of Eq. (5-150), dividing the result by Δ, taking the limit of zero Δ, and substituting from the equations of state (5-146) results in Bellman's equation, namely,

$$\frac{\partial S}{\partial T} = \min_{\bar{u}} \left[g(c_1, c_2, \ldots, c_N; u_1, u_2, \ldots, u_m) + \sum_{i=1}^{N} \frac{\partial S}{\partial c_i} f_i(c_1, c_2, \ldots, c_N; u_1, u_2, \ldots, u_m) \right] \tag{5-151}$$

This is a partial differential equation from which the optimal control vector $\bar{u}^*$ can be obtained in principle, together with the minimum cost, S. As already discussed a number of times, this equation is too difficult to solve in closed form, except for tutorial examples, so that one must resort to numerical computations, using its discrete analog, Eq. (5-149).

Now define, for convenience, an additional component, x_{N+1}, of the state vector $\bar{x}$ by letting x_{N+1} satisfy

$$\begin{aligned}\dot{x}_{N+1} &= g(x_1, x_2, \dots, x_N; u_1, u_2, \dots, u_m) \\ &\equiv f_{N+1}(x_1, x_2, \dots, x_N; u_1, u_2, \dots, u_m) \qquad x_{N+1}(0) = 0\end{aligned} \tag{5-152}$$

That is

$$x_{N+1}(T) = J = \int_0^T g(x_1, x_2, \dots, x_N; u_1, u_2, \dots, u_m)\, dt \tag{5-153}$$

It should be noted that the right side of Eq. (5-152) does not depend explicitly on x_{N+1}. This fact plays a role in the formulation of the solution using the maximum principle. Here x_{N+1} is introduced so that the resemblance between the two principles will become more apparent; Bellman's equation (5-151) can be rewritten

$$\frac{\partial S}{\partial T}(c_1, c_2, \dots, c_N; T) = \min_{\bar{u}} \sum_{i=1}^{N+1} \frac{\partial S}{\partial c_i} f_i(c_1, c_2, \dots, c_N; u_1, u_2, \dots, u_m) \tag{5-154}$$

To obtain the correspondence between this equation and the maximum principle, define a set of adjoint variables as

$$p_i(t) = \frac{\partial S}{\partial c_i} \qquad (i = 1, \dots, N), p_{N+1} \equiv 1 \tag{5-155}$$

From the definition of $p_i(t)$ in Eq. (5-155), its total derivative is

$$\dot{p}_i(t) = \sum_{j=1}^{N} \frac{\partial^2 S}{\partial c_i\, \partial c_j} f_j \tag{5-156}$$

Furthermore, along an extremal in the phase space, Eq. (5-154) asserts that

$$\frac{\partial S}{\partial T} = \sum_{i=1}^{N+1} \frac{\partial S}{\partial c_i} f_i^* = \min = \text{const} \tag{5-157}$$

where the optimal control vector $\bar{u}^*$ has been inserted in the f_i, giving $f_i^* \equiv f_i(c_1, c_2, \dots, c_N; u_1^*, u_2^*, \dots, u_m^*)$.

Differentiation of Eq. (5-157) with respect to, e.g., u_m^*, assumed continuous, gives

$$\sum_{i=1}^{N+1} \frac{\partial S}{\partial c_i} \frac{\partial f_i^*}{\partial u_m^*} = 0 \tag{5-158}$$

If Eq. (5-157) is also differentiated with respect to c_m, the result is

$$\sum_{i=1}^{N+1} \left\{ \frac{\partial^2 S}{\partial c_i \, \partial c_m} f_i^* + \frac{\partial S}{\partial c_i} \left[\frac{\partial f_i^*}{\partial c_m} + \frac{\partial f_i^*}{\partial u_m^*} \frac{\partial u_m^*}{\partial c_m} \right] \right\} = 0 \tag{5-159}$$

From Eq. (5-158) it is seen that the third term of Eq. (5-159) vanishes. Comparing Eq. (5-159) with Eqs (5-155) and (5-156) supplies the equation that the adjoint variables p_i must satisfy, namely

$$\dot{p}_i(t) = -\sum_{j=1}^{N+1} p_j \frac{\partial f_j}{\partial c_i} \tag{5-160}$$

Identifying a *hamiltonian*, after Pontryagin, as

$$H = -\sum_{i=1}^{N+1} p_i f_i \tag{5-161}$$

then along an extremal, from Eq. (5-157), H must satisfy

$$H^* = -\sum_{i=1}^{N+1} p_i f_i^* = \max = \text{const} \tag{5-162}$$

That is, the maximum principle asserts that for a system satisfying Eqs (5-146), the optimal control that minimizes J, also maximizes H, where $-H = \sum_{i=1}^{N+1} p_i f_i$ and where the p_i are defined by Eq. (5-160).

As can now be seen from Eq. (5-154), Bellman's equation is the Hamilton–Jacobi equation, that is, along an extremal

$$H\left(\frac{\partial S}{\partial c_1}, \ldots, \frac{\partial S}{\partial c_N}; u_1^*, \ldots, u_m^* \right) + \frac{\partial S}{\partial t} = 0 \tag{5-163}$$

and the adjoint equations are the canonical equations of Hamilton. In other words, from Eq. (5-160), the definition of H, and identifying c_i with x_i yields

$$\dot{p}_i(t) = \frac{\partial H}{\partial x_i} \qquad \dot{x}_i = -\frac{\partial H}{\partial p_i} \qquad (i = 1, \ldots, N+1) \tag{5-164}$$

5-7 OPTIMALLY CONTROLLED REACTORS

An increasing interest in the optimal control of nuclear reactors has been manifest in the multifarious control problems evident in the varied families of reactors studied and built over the past three decades. Besides the control diversities of nuclear reactors, such as the nuclear rocket, circulating fuel, cavity, aircraft, space borne, pulsed, and microreactors, etc., there has been a definite trend toward substantive real world control with an emphasis on reactor safety, albeit with less exotic reactors such as pressurized water, boiling water, and fast breeder reactors. That is, the recent emphasis seems to be not on the diversity or novel aspect of esoteric reactors, but for straightforward incisive, correct control analysis of systems like PWR reactors. Good control schemata will include a panoply of techniques involving computer programs using well thought out algorithms, buttressed by previous hand calculations, cut and try designs, systems tests and simulations. In addition to the fact that safety is the prime consideration in reactor design, it must be tempered by performance, especially in the exacerbatingly competitive commercial power industry replete with other potential energy sources such as fusion, geothermal, coal, oil, solar, etc.

The seemingly inherent rift between the abstractionists and the journeyman reactor designers and builders should be narrowed by a closer cooperation between the two. Perhaps the more practical ramifications of optimal control theory can aid this. The following rather detailed version of the optimal shutdown xenon control process is a case in point.

5-8 XENON SHUTDOWN CONTROL

As is well known, large, high power, thermal energy, nuclear reactors produce various fission products which are unusually large neutron absorbers. The principal fission product poisons in terms of their thermal neutron absorption cross sections are samarium-149 and xenon-135, with corresponding cross sections of 50,000 and 3.5×10^6 barns, respectively. This xenon cross section is the largest known neutron absorption cross section of any nucleus on the Segrè chart. During World War II one of the Hanford plutonium producing reactors was slowly shutting itself off for a hitherto unknown reason. Because of the exigencies of those times, a top priority investigation was immediately launched into an attempt to ferret out the mechanism causing the dwindling neutron population of those reactors. Within two days E. Fermi found that ^{135}Xe was the terribly large neutron absorber. The Hanford reactors were then loaded with the additional fuel required to accommodate the xenon poison. That is, enough additional fissionable material was added to the core to compensate for the negative reactivity contributed by the accumulated xenon.

Another, perhaps not too hypothetical, situation would be one in which the commander of a Polaris submarine orders the engines immediately shut down to "lie on the bottom" for repairs, or other reasons. With the neutron density in the reactors now far below that at criticality at operating power, the ^{135}I fission product fragments keep decaying to, and therefore build up the amount of, ^{135}Xe. With relatively few neutrons available to burn up the ^{135}Xe by transmuting it to ^{136}Xe, a harmless isotope from the neutron standpoint, the ^{135}Xe now accumulates in inordinately large amounts. If the amount of ^{135}Xe were too large (too much negative reactivity) for the reactors, the commander could not restart them if the repairs took more than a half hour or so, and he would be obliged to wait for a period of 30 to 60 hr until the xenon decayed naturally, with a mean life of about 2.5 days. This would make the submarine extremely vulnerable. Of course, to accommodate the modern submarines, the fuel requirements are greatly increased, at considerable additional expense, for burnup over the military life of the reactor, so that the xenon reactivity is always capable of being overridden.

All of the above is based upon a practical immediate power shutdown, or as nearly immediate as practicality allows. If the reactor is, a priori, not shutdown, then a xenon peak never occurs. Somewhere between these two extremes lies a power shutdown program that should minimize the xenon maximum concentration whenever in the post shutdown phase it might occur. The remainder of this chapter is given over to all aspects of that question, as well as the investigation of other associated criteria, such as optimal shutdown in minimum time, observing a xenon concentration constraint, which latter is dictated by the amount of extra reactor fuel allowed for this xenon negative reactivity.

There is another class of xenon reactor problems. These are caused by spacial oscillations of the power throughout a large thermal reactor, due primarily to the space–time kinetic behavior of ^{135}Xe. Such flux tilt oscillations are discussed in Chap. 7 on space-dependent kinetics.

As already alluded to, constraints play a major role in the behavior of optimal control processes in general and xenon shutdown policies in particular. There are three principal constraints associated with reactor shutdown behavior when xenon is used: those on (1) flux magnitude, (2) inverse period sign and magnitude, and (3) xenon concentration magnitude.

1. The flux magnitude $u = \phi/\phi_0$.

$$0 \le u \le M \tag{5-165}$$

which implies that the lower bound on the flux is taken to be zero. However, in an operating reactor, there are many and diverse nuclear isotopic reactions taking place. A number of these result in end products that emit neutrons for a long time compared to the useful life of the reactor. Hence when the reactor is shut down after normal operation, these neutron emitters provide sufficient neutrons during any conceivably long shutdown period that the flux will not decrease to zero in the reactor. Also, in many systems, the original radioactive-isotope neutron source used to start the nascent reactor is simply left inside from that time forward. In sum, the asymptotic reactor flux never drops to zero following shutdown, but to a low value which is usually about 10^{-6} to 10^{-9} times its equilibrium value.

The upper bound M is usually taken as rated operating power, so that in terms of the normalized flux, $M = 1$. However, it is often desired to make $M > 1$ if only for reasonably short periods of time. This property will be seen to be of advantage later, in the sense that over a portion of the optimal shutdown program it may be desired that $M > 1$. Such an M behavior may be preferred, because it reduces the number of shutdowns and restarts, if the cost of an elaborate shutdown control procedure involving much alternate reactor shutdown and startup is felt to be too high.

2. The "α" or inverse period $\dot{\phi}/\phi$.

$$-\alpha_L \le \dot{\phi}/\phi \le \alpha_H \tag{5-166}$$

The lower (negative) bound, $-\alpha_L$, is automatically present by virtue of the reactor population decrease following normal shutdown. This is due to the mechanical inertia of the control-rod mechanisms and the phenomenon of delayed neutrons. Delayed neutrons are a concomitant of fission, whereby certain fission-product isotopes emit neutrons in seconds to minutes following their creation from fission. Such production of delayed neutrons is exploited in the control of reactors, since they can be made to comprise the additional neutron source by which the system is kept in exact equilibrium (critical). Since delayed neutrons act as a source with a time lag of approximately 10 seconds, their amount can be manipulated so that the reactor state changes in reasonable time with respect to present "state of the art" control systems.[4] If delayed neutrons were nonexistent, the time behavior of the reactor neutron population would be capricious indeed, since the corresponding system time constant would then be measured in fractions of a millisecond.[4] The reactor difficulties would become horrendous, owing to the atom-bomb-like quality of the consequent reactor temporal behavior.

The positive upper bound, α_H, is limited by reactor standard operating safety procedures, since for very large α_H the neutron population will increase on a too-rapid exponential behavior. If this occurs for a supercritical ($k > 1$) reactor, the burgeoning neutron population would cause the reactor to far exceed its rated power and an accident could result. Even if the shutdown alarm (reactor scram) system operated in a satisfactory manner, thus avoiding an accident, it is economically very costly to restart a reactor following an unscheduled shutdown.

However, both $|\alpha_L|$ and $|\alpha_H|$ are measured in seconds, while the time behavior of the xenon and iodine concentration is measured in hours. Hence the approximation will be made in the ensuing solutions for optimal xenon shutdown policies that the reactor can be shut down and restarted instantaneously. To actually include the inverse period constraints would result in a modified problem formulation. It is just this type of derivative constraint that would pose difficulties if optimal xenon shutdown were to be investigated using the classical calculus of variations, even using the Valentine artifice.[10]

3. The xenon override constraint x_c.

$$x \leq x_c \tag{5-167}$$

This is an inequality constraint on one of the state variables. When using the classical approach, the constraint x_c is handled by the Valentine artifice, which results in an additional variational constraint equation. In the maximum principle formulation, inequality constraints on the state variables give rise to additional difficulties, since, for one thing, the hamiltonian must be modified to include a Lagrange multiplier, thus introducing an additional variable in the problem. Further, at the bounds of the region of phase space in which $x = x_c$, the corresponding adjoint variable p_1 experiences a jump discontinuity.[7] As can be appreciated, the constraint $x \leq x_c$ is most important, since if x_c were very large, the xenon concentration could increase almost without bound, so that no optimal shutdown programs would be necessary. This is the situation in which there is sufficient fuel in the reactor, corresponding to enough positi... reactivity, to override the negative xenon reactivity in the shutdown phase at

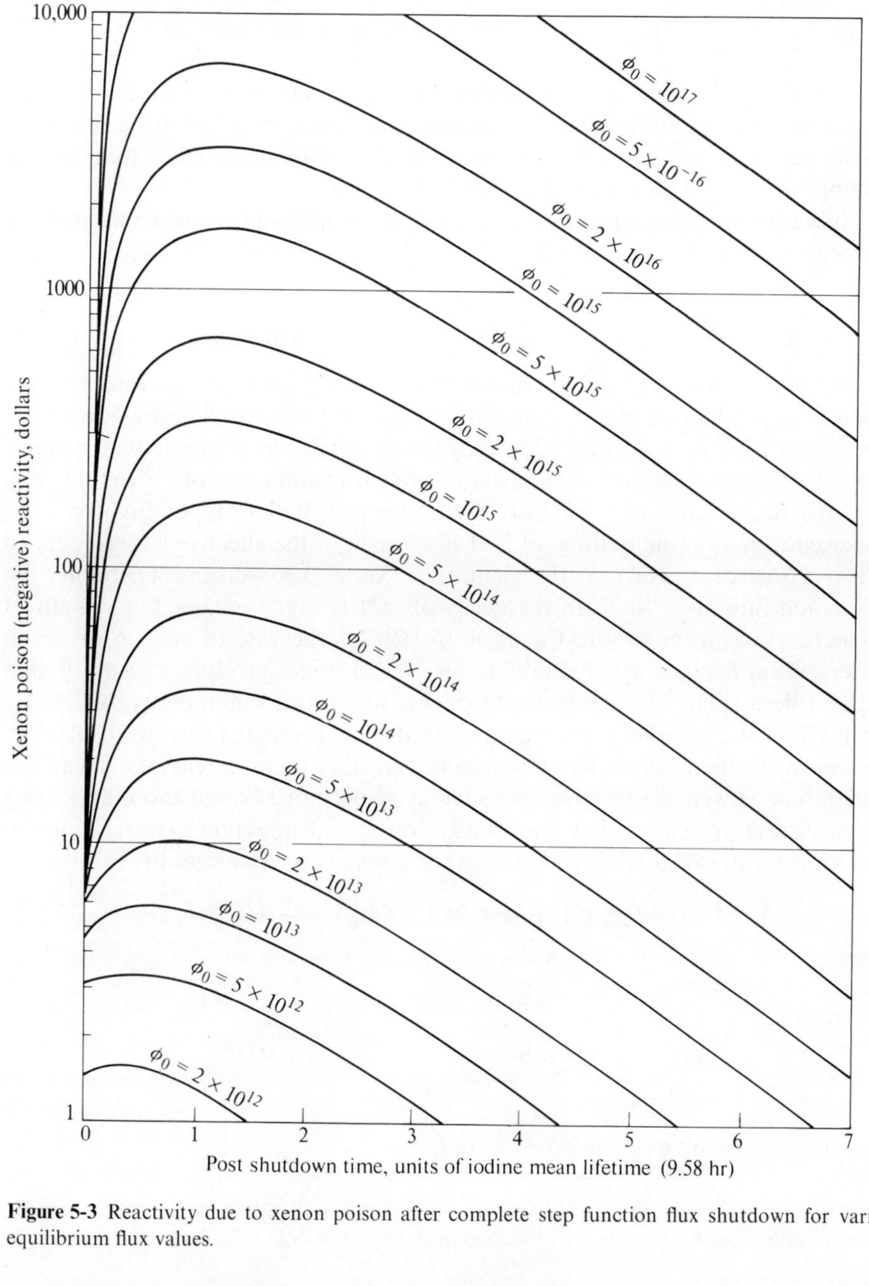

Figure 5-3 Reactivity due to xenon poison after complete step function flux shutdown for various equilibrium flux values.

all times. Then the reactor can be restarted at will following shutdown, irrespective of the amount of xenon in the system. As mentioned before, this amount of fuel inventory is prohibitively large, and therefore extremely costly, in the conventional thermal-reactor power plant. Therefore, the normal situation is that the system possesses sufficient fuel to override xenon only partially following shutdown. This situation can be ameliorated by using optimal shutdown programs, as will be developed subsequently.

The statement of the problem under discussion follows the introduction that the xenon concentration following abrupt shutdown to $u = 0$, from its prior equilibrium power level, can rise to orders of magnitude over that at the equilibrium value. This is seen in Fig. 5-3.

By way of introduction, it is well to recall the xenon and iodine concentration balance equations†

$$\dot{I} = \gamma_1 \Sigma_f \phi - \lambda_1 I \qquad \dot{I}(0-) = 0 \tag{5-168}$$

$$\dot{X} = \gamma_2 \Sigma_f \phi + \lambda_1 I - (\lambda_2 + \sigma\phi)X \qquad \dot{X}(0-) = 0 \tag{5-169}$$

The first equation asserts that the rate of change of the iodine (^{135}I) concentration is due to its production rate as a fission product, less its natural decay into ^{135}Xe. The fission product tellurium 135 decays with a half-life of about two minutes into ^{135}I. This is construed as an instantaneous transmutation of ^{135}Te into ^{135}I, since the decay rates of ^{135}I and ^{135}Xe, the principal isotopes involved, are measured in hours (the half-life of ^{135}I is 6.7 hr). γ_1, the effective fission yield of ^{135}I, is 5.6 percent, while γ_2, the yield of ^{135}Xe, is 0.30 percent. Therefore, the production rate of ^{135}Xe from the decay of ^{135}I is much greater than its direct production rate from fission. Equation (5-169) for the rate of change of xenon concentration is made up of source terms due to direct production from fission and ^{135}I decay. The loss terms are those due to natural xenon decay, which has a half-life of 9.2 hr, plus a prodigious loss due to thermal neutron absorption, the reason for the concern about xenon in high-flux thermal reactors. There are other terms, as well as other poisons such as samarium 149, but those appearing above are the principals that are usually considered in xenon investigations. If the iodine terms are eliminated between the above two equations, the result is

$$\ddot{X} + (\lambda + \sigma\phi)\dot{X} + [\lambda_1(\lambda_2 + \sigma\phi) + \sigma\dot{\phi}]X = \Sigma_f(\lambda_1 \gamma \phi + \gamma_2 \dot{\phi}) \tag{5-170}$$

where

$$\lambda = \lambda_1 + \lambda_2 \qquad \lambda_1 = 2.9 \times 10^{-5}\ \text{sec}^{-1} \qquad \lambda_2 = 2.1 \times 10^{-5}\ \text{sec}^{-1}$$

$$\gamma = \gamma_1 + \gamma_2 \qquad \gamma_1 = 0.056 \qquad \gamma_2 = 0.003$$

$$\sigma = 3.5 \times 10^6\ \text{barns}$$

Σ_f is the macroscopic fission cross section.

† The symbology and numerical constants are obtained from S. Glasstone and M. C. Edlund: "Elements of Nuclear Reactor Theory," Van Nostrand, Princeton, N.J., 1952, chap. 11.

An important parameter is K_x, the negative reactivity due to the xenon present in the reactor. This can be shown to be given to good approximation† by

$$K_x = \frac{\sigma X}{\Sigma_{\text{fuel}}} \tag{5-171}$$

where Σ_{fuel} is the total macroscopic absorption cross section of the fuel. At equilibrium the initial conditions in Eqs (5-168) and (5-169) assert that $X(0-) = X_0 = \gamma\Sigma_f\phi_0/(\lambda_2 + \sigma\phi_0)$, where ϕ_0 is the equilibrium flux level. Then the equilibrium xenon reactivity is

$$|K_x(0-)| = \frac{\gamma\sigma\Sigma_f\phi_0}{(\lambda_2 + \sigma\phi_0)\Sigma_{\text{fuel}}} \tag{5-172}$$

For low-power thermal reactors, $\sigma\phi_0 \ll \lambda_2$ corresponding to $\phi_0 \lesssim 10^{11}$. Also $\Sigma_f/\Sigma_{\text{fuel}} = 0.84$ for ^{235}U. The low-power equilibrium xenon reactivity is then given by

$$|K_x(0-)| \cong \frac{\gamma\sigma\Sigma_f\phi_0}{\lambda_2\Sigma_{\text{fuel}}} \cong 8 \times 10^{-13}\phi_0 \tag{5-173}$$

For $\phi_0 = 10^{11}$ the low-power equilibrium xenon reactivity is about 10 cents, which is considered negligible. When the flux is very high, as seen from Eq. (5-172) the equilibrium reactivity is $\sim\gamma\Sigma_f/\Sigma_{\text{fuel}} \cong \$8.00$.

As mentioned earlier, the xenon buildup following a reactor shutdown poses difficulties with respect to the amount of negative reactivity represented by the xenon. Ordinarily provision must be made by including sizable amounts of additional fuel over that called for by the criticality equation to "override" the xenon when it is desired to restart the reactor. To illustrate, consider the xenon buildup following a negative step function *flux* shutdown to essentially zero flux. This of course, is far more severe than the manner by which an actual reactor can be shut down. To mention one factor, delayed neutrons, as discussed in Chap. 2, provide a constraint on the maximum asymptotic shutdown flux rate following a negative step reactivity insertion. Their existence implies a lower bound on the shutdown period through the decay constant of the longest lived delayed emitter. Its corresponding mean life is 80 sec for large negative reactivity steps, so that asymptotically $|\dot{\phi}/\phi| \leq \frac{1}{80}$. For the above abrupt shutdown $\phi(0-) = \phi_0$, $\phi(0+) = 0$. Then Eqs (5-168) and (5-169) combine to give

$$\ddot{X} + \lambda\dot{X} + \lambda_1\lambda_2 X = 0 \tag{5-174}$$

It should be noted from Eqs (5-168) and (5-169) for abrupt shutdown that $\dot{X}(0)$ is discontinuous. The second initial condition is then

$$\dot{X}(0+) = \lambda_1 I_0 - \lambda_2 X_0 = \Sigma_f\phi_0\left(\gamma_1 - \frac{\gamma}{1 + \sigma\phi_0/\lambda_2}\right) \tag{5-175}$$

† Glasstone and Edlund, *ibid.*

giving the solution

$$X(t) = \frac{X_0}{1-\zeta}\left[\left(1-\zeta+\frac{I_0}{X_0}\right)e^{-\lambda_2 t} - \frac{I_0}{X_0}e^{-\lambda_1 t}\right]$$

$$\zeta = \frac{\lambda_2}{\lambda_1} = 0.725 \qquad \frac{I_0}{X_0} = \frac{\gamma_1}{\gamma}\left(\zeta + \frac{\sigma\phi_0}{\lambda_1}\right) \tag{5-176}$$

The negative reactivity due to the xenon poison is obtained by multiplying by $\sigma_x/\Sigma_{\text{fuel}}$ to obtain

$$|K_x| = \frac{|K_x(0)|}{1-\zeta}\left[\left(1-\zeta+\frac{I_0}{X_0}\right)e^{-\lambda_2 t} - \frac{I_0}{X_0}e^{-\lambda_1 t}\right] \tag{5-177}$$

For very large equilibrium flux, $\phi_0 \geq 10^{14}$, it is easy to show that the maximum xenon reactivity, using Eq. (5-176), following a negative step shutdown to zero flux is proportional to ϕ_0 to good approximation as seen from Eq. (5-175). Then, for example, $\phi_0 \cong 10^{15}$ gives a $K_{x,\max} \cong -\$350$.† From the foregoing, it is seen that a xenon peak or maximum occurs about 10 to 12 hr after shutdown, after an initially rapid buildup. Depending on the steady-state neutron density level (reactor power), the xenon maximum for a high-flux thermal reactor amounts to many dollars of reactivity. Further, it is almost an order of magnitude larger than the xenon-induced reactivity corresponding to the equilibrium xenon concentration.

Perhaps by an ingenious shutdown program of the flux over time, the xenon maximum, whenever it occurs, can be minimized. More explicitly, suppose that it is required to shut down the reactor to a negligible neutron density, or flux, in a given time t_0. After time t_0 no further control is exerted, and at a time later than t_0 the xenon concentration has built up to its maximum. What is the shutdown program during the control phase to minimize the maximum xenon concentration when it occurs? This problem is analogous to the control of a ballistic missile during its powered phase, a small time fraction of its orbital flight, with control terminating at engine cutoff so that the missile, now following a ballistic trajectory, minimizes the target circular error probability (CEP). Mathematically, these problems are of a nonclassical implicit variational type. Dynamic programming can handle such problems in a natural way by using the optimality principle to derive a functional equation in the criterion functions with an optimal control policy obtained to extremize the criterion. Unfortunately, perhaps, an analytic solution for the optimal policy is not usually obtainable, since the criterion cannot be couched in terms of a simple analytic functional.

In addition to the flux magnitude and inverse period constraints, there is the very important constraint that sufficient positive reactivity must be available to override the xenon at the onset of the first flux control pulse. Such reactivity can be provided in part by a large pulse of flux. In the following, it is assumed that this reactivity is available to realize the optimal shutdown.

† For large reactivities, $K_{x,\max}$ is replaced by $K_{ex(x,\max)} = (k-1)/\beta$ since $\lim_{k\to\infty} K = 1/\beta$ dollars.

The solution described herein provides an optimum search algorithm for the shutdown program to satisfy a "minimax" xenon concentration criterion. The algorithm demands the services of large digital electronic processing machinery (EDPM), as will be discussed. For serious inquiry into the vagaries of this problem, the algorithm provides an invaluable parameter sensitivity analysis. Optimal shutdown programs can be provided for a whole series of control phase durations, inverse periods, flux equilibrium levels, etc. The sensitivity analysis is, of course, more important than merely one optimal solution for one set of parameters, especially from the design standpoint. Dynamic programming provides this sensitivity analysis in a completely natural way.

Now, in the xenon-iodine state space, Eqs (5-168) and (5-169) are rewritten in dimensionless form, with $\gamma_1 = \gamma_i/(\gamma_x + \gamma_i)$, $\gamma_2 = \gamma_x/(\gamma_x + \gamma_i)$, $\gamma_1 + \gamma_2 = 1$, as

$$\dot{x} = -(w + r_0 u)x + \gamma_1(w + r_0)y + \gamma_2(w + r_0)u \tag{5-178}$$

$$\dot{y} = u - y \tag{5-179}$$

where the time scale is now in units of λ_1^{-1} (9.58 hr). The dimensionless form of the fission yields used here are now such that $\gamma_1 = 56/59$, $\gamma_2 = 3/59$, and that $\gamma_1 + \gamma_2 = 1$, with $x = X/X_0$, $r_0 = \sigma\phi_0/\lambda_1 \simeq 1.2 \times 10^{-13}\phi_0$, $y = I/I_0$, $w = \lambda_2/\lambda_1 \simeq 21/29$, and $u = \phi/\phi_0$. In the xenon-iodine state space, Eq. (5-178) is divided by Eq. (5-179) to yield the corresponding differential equation

$$\frac{dx}{dy} = \frac{\gamma_2(w + r_0)u + \gamma_1(w + r_0)y - (w + r_0 u)x}{u - y} \tag{5-180}$$

The immediate flux shutdown trajectories, that is, those equivalent to the solutions depicted in Fig. 5-3 with time as abscissa, are obtained by first letting $u = 0$ in Eq. (5-180) to give

$$\frac{dx}{dy} = \frac{wx - \gamma_1(w + \gamma_0)y}{y}; \qquad x(1) = 1 \tag{5-181}$$

Then the above is easily integrated to yield

$$x = \left[1 + \frac{\gamma_1(w + r_0)}{1 - w}\right]y^w - \frac{\gamma_1(w + r_0)}{1 - w}y \tag{5-182}$$

These trajectories are shown in Fig. 5-4.

Similarly, from Eq. (5-180), for constant $u = M \neq 0$, the solutions are given by

$$x = \frac{(w + r_0)M}{w + r_0 M} + \frac{\gamma_1(w + r_0)(M - y)}{1 - (w + r_0 M)} + \frac{c(M - y)^{-(w + r_0 M)}}{w + r_0 M} \tag{5-183}$$

where c can be evaluated from the equilibrium conditions $x = x_0$, $y = y_0$. These trajectories are shown as dashed lines in Fig. 5-4.

First dynamic programming will be used to study the shutdown. Then the maximum principle will be used, and comparisons made. The problem statement is: to find the optimal control policy u^* that yields the minimum of the xenon

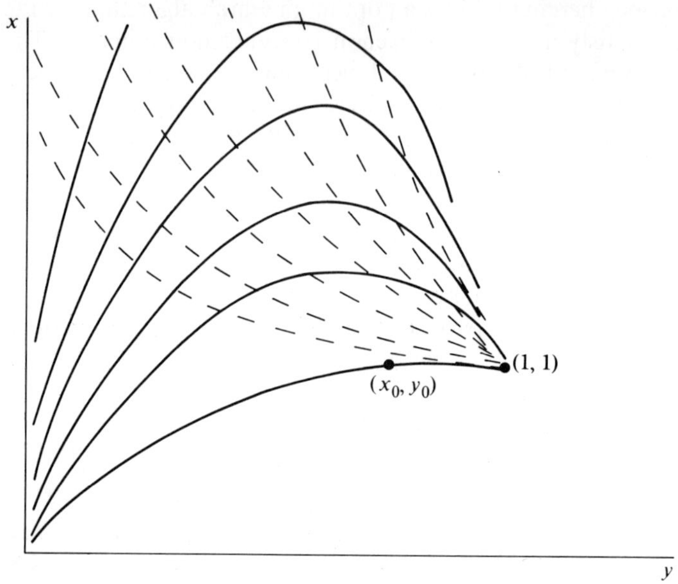

Figure 5-4 Xenon-iodine trajectories in dimensionless form.

maximum whenever it occurs in the post shutdown phase. The constraints are: maximum xenon override $x \leq x_c$ where x_c is the negative reactivity represented by the xenon concentration for which the reactor can just override. If $x > x_c$ then one must wait until enough xenon decays until $x = x_c$ again so that the control system can take over once more. A second constraint is simply $0 \leq u \leq M$.

Now the Bellman equation for the control process will be derived. Let the allowable shutdown control duration T be divided into N intervals, each of length $\Delta = T/N$. As this class of problems is of the terminal control type discussed earlier, define a xenon criterion or cost functional as

$$F_N\{x(0), y(0)\} = \min_{0 \leq u \leq M} \Phi(x_T, y_T) \tag{5-184}$$

which implies that N control decisions on the set $0 \leq u \leq M$ must be made to minimize the right side of Eq. (5-184). Even though F_N is a functional of u, it is assumed that the optimal flux shutdown sequence $\{u_N^*\}$ is unique, so that F_N must be a function of the initial state $\{x(0), y(0)\}$ only.

To derive the functional equation, which is simultaneously a recurrence relation algorithm from which $F_k, k = 0, 1, 2, \ldots N$, and $\{u_N^*\}$ are obtained, first consider the trivial case beginning at the termination of the control duration, where the system is in one of the set of final states (x_T, y_T). In this state, the reactor is shut down, so that $u^* \equiv 0$ from there on, and no further control decisions are made. Since the subscript on F implies the number of control decisions remaining in the shutdown phase, the corresponding relationship is that

$$F_0(x_T, y_T) = \Phi(x_T, y_T) \tag{5-185}$$

Now proceeding backward along an extremal toward the initial state to one state removed from the termination of control (a one-stage process), the principle of optimality of dynamic programming asserts that

$$F_1(x_1, y_1) = \min_{0 \le u_1 \le M} F_0\{x_0 + \dot{x}_0(u_1)\Delta, y_0 + \dot{y}_0(u_1)\Delta\} \tag{5-186}$$

since $(x_0, y_0) \equiv (x_T, y_T)$. That is, the optimal cost for a one-stage process $F_1(x_1, y_1)$ is given by the minimum over u of the terminal cost F_0 but one state removed from the terminal state, namely, $(x_1, y_1) = \{x_0 + \dot{x}_0(u_1^*)\Delta, y_0 + \dot{y}_0(u_1^*)\Delta\}$. Equation (5-186) can, of course, be rewritten

$$F_1(x_1, y_1) = \min_{0 \le u_1 \le M} \Phi\{x_0 + \dot{x}_0(u_1)\Delta, y_0 + \dot{y}_0(u_1)\Delta\} \tag{5-187}$$

$\dot{x}_0(u_1)$ and $\dot{y}_0(u_1)$ are obtained from the xenon and iodine state equations.

By induction then, for k stages *remaining* in the shutdown phase of duration T, one can write the following functional equation:

$$F_k(x_k, y_k) = \min_{0 \le u_k \le M} F_{k-1}\{x_{k-1} + \dot{x}_{k-1}(u_k)\Delta, y_{k-1} + \dot{y}_{k-1}(u_k)\Delta\} \qquad (k = 1, 2, \ldots, N) \tag{5-188}$$

That is, in the kth stage, F_k is obtained by choosing $0 \le u_k \le M$ so as to minimize F_{k-1} one state removed from (x_{k-1}, y_{k-1}), thus obtaining an increment of the extremal arc lying between (x_k, y_k) and (x_{k-1}, y_{k-1}). By repeating this procedure (i.e., $k = 1, 2, \ldots$), one proceeds along an extremal in the (x, y) phase space to obtain ultimately the F_N and, more important, the optimal flux shutdown sequence $\{u_N^*\}$.

The xenon override constraint $x \le x_c$ is implied in the above and does not detract from the generality of the derivational argument thus far. More will be said about the xenon constraint later, where it will be seen to constrain a portion of the extremal arc, thereby rendering a modified xenon shutdown program.

The important point should be made that along an extremal in the phase space, $F_N = F_{N-1} = \cdots = F_0$, so that $F_k(u_k^*)$, $k = 0, 1, 2, \ldots, N$, is a constant of the motion. It is easily shown that along an extremal in the phase space, $F = -H$, the hamiltonian of the maximum principle, which is one manifestation of the equivalence between the optimality principle of dynamic programming and the maximum principle.

5-9 DERIVATION OF BELLMAN'S EQUATION

The continuous analog of the functional recurrence relation Eq. (5-188) is obtained by first elevating the time-like stage index k into the argument of the criterion functional F_k, so that for a time increment Δ,

$$F(x, y, T) = \min_{0 \le u \le M} F\{x + \dot{x}(u)\Delta, y + \dot{y}(u)\Delta, T - \Delta\} \tag{5-189}$$

with terminal control, or boundary, surfaces $F(x, y, 0) = \Phi(x, y)$. Again Eq. (5-189) enunciates the optimality principle of dynamic programming in that the criterion functional $F(x, y, T)$, with allowable control duration T and system state (x, y), is given by the minimum over the possible control set $0 \leq u \leq M$ of the criterion functionals $F\{x + \dot{x}(u)\Delta, y + \dot{y}(u)\Delta, T - \Delta\}$ embedded in a set of control processes beginning in neighboring states $\{x + \dot{x}(u)\Delta, y + \dot{y}(u)\Delta\}$ but of allowable control duration $T - \Delta$. $\dot{x}(u)$ and $\dot{y}(u)$ are, of course, obtained from Eqs (5-178) and (5-179).

Now expand F on the right side of Eq. (5-189) in a Taylor series about (x, y, T) through linear terms explicitly, so that

$$F(x, y, T) = \min_{0 \leq u \leq M} \left[F(x, y, T) + \dot{x}\Delta \frac{\partial F}{\partial x} + \dot{y}\Delta \frac{\partial F}{\partial y} - \frac{\partial F}{\partial T}\Delta + O(\Delta^2) + \cdots \right] \tag{5-190}$$

$F(x, y, T)$ is canceled on both sides, since it does not depend on u explicitly. Then the result is divided by Δ, and the limit of zero Δ is taken, plus substituting from Eqs (5-178) and (5-179), which yields finally

$$\frac{\partial F}{\partial T} = \min_{0 \leq u \leq M} \left\{ [-(w + r_0 u)x + \gamma_1(w + r_0)y + \gamma_2(w + r_0)u] \frac{\partial F}{\partial x} + [u - y] \frac{\partial F}{\partial y} \right\} \tag{5-191}$$

which is Bellman's differential equation. The boundary curve is

$$F(x, y, 0) = \Phi(x, y) \tag{5-192}$$

The novel aspect of Bellman's equation is due to the minimum operator, which, if removed, would result in Eq. (5-191) becoming the garden variety of first-order partial differential equation. Such an equation would then be meaningless in our context. If the optimal flux shutdown program u^* is known and substituted in Eq. (5-191), the result is the Hamilton–Jacobi equation for this optimal control process.

Bellman's equation, as is the case for most substantive partial differential equations, is in general difficult to solve analytically except for illustrative examples, so that resort must be had to large-scale high-speed digital computers to obtain solutions by numerical integration. In principle, the right side of Eq. (5-191) must first be minimized with respect to u, and then solutions to the resulting equation obtained.[8] For numerical computations the discrete analog Eq. (5-188) is used instead, where the minimization and the solutions are generated together in a stepwise manner, as will be discussed later.

As in the maximum principle formulation of the xenon shutdown problem to be given later, the discontinuous, or bang-bang, nature of the optimal shutdown control u^* is immediately evident. This is seen by merely rewriting Eq. (5-191) as

$$\frac{\partial F}{\partial T} = [\gamma_1(w + r_0)y - wx] \frac{\partial F}{\partial x} - y \frac{\partial F}{\partial y} + \min_{0 \leq u \leq M} u\Gamma \tag{5-193}$$

where

$$\Gamma = [\gamma_2(w + r_0) - r_0 x] \frac{\partial F}{\partial x} + \frac{\partial F}{\partial y} \tag{5-194}$$

It is seen that the minimization in Eq. (5-193) calls for an optimal u^* given by

$$u^* = \begin{cases} M & \Gamma < 0 \\ 0 & \Gamma > 0 \end{cases} \tag{5-195}$$

That is, if Γ is positive, the optimal flux u^* immediately drops to zero. For Γ negative, u^* immediately jumps to its maximum value M. Therefore, the optimal flux shutdown program consists of pulses, switching back and forth between 0 and M, where Γ changes sign. If $\Gamma = 0$ for a finite interval in the allowable control duration T, then u^* is arbitrary. However, $\Gamma = 0$ above only at isolated points. Arbitrary u^* is termed singular control, which is largely an unexplored area in modern optimal control theory. Where are the points at which $\Gamma = 0$? They are the points at which the control is switched from zero to M, or vice versa, and are found by solving Bellman's equation to yield an expression for Γ. As mentioned, the discrete analog Eq. (5-188) is used for numerical solution.

5-10 BANG-BANG CONTROL DILEMMA

For bang-bang control processes in general, Bellman's equation poses an interesting dilemma. Referring to Eqs (5-193) to (5-195), Γ must be known since the times of its zero crossings from one sign to another, determine the switching of u to obtain u^*. However, Γ cannot be known until Bellman's equation is solved, which in turn cannot be accomplished until the correct u^* is inserted therein. As mentioned, to break out of this difficulty additional information about the nature of the optimal process must be known and applied to "start the solution off." Again, if the discrete analog is used, as it must be in a numerical computation, this difficulty is not encountered since the optimal control u^* and the optimal

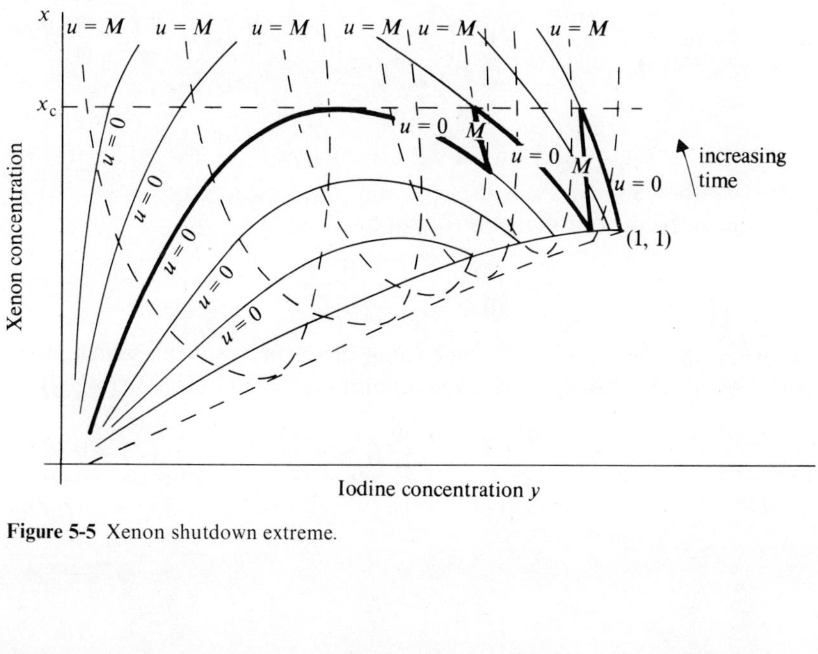

Figure 5-5 Xenon shutdown extreme.

cost functional F are determined stepwise together. No such difficulty occurs when u appears in a nonlinear fashion in Bellman's equation, since u^* can ordinarily be determined by differentiation and then eliminated by substitution back into the latter equation.

In sum then, using the discrete analog of Bellman's equation (5-188) with the initial condition $F_0(x_0, y_0) = \Phi(x_0, y_0)$ with constraints $0 \le u \le M$, and $x \le x_c$, solutions are obtained which are depicted in Fig. 5-5. It is seen that the optimal u^*, switching back and forth between 0 and M, yields a discontinuous extremal as shown in Fig. 5-5. The extremal begins at $(1, 1)$ and "scallops" its way across to the left, always just under the $x = x_c$ constraint, until it coincides with a $u = 0$ trajectory which is tangent to the $x = x_c$ constraint line. The final u^* line switch is to $u^* = 0$ onto the latter $u = 0$ trajectory on which the system "coasts" until the xenon (and iodine) vanish asymptotically at the origin.

5-11 MAXIMUM PRINCIPLE FORMULATION

To investigate this problem using the maximum principle, the equations of state are now

$$\dot{x} = -(w + r_0 u)x + \gamma_1(w + r_0)y + \gamma_2(w + r_0)u; \qquad u(0) = x(0) = 1 \tag{5-196}$$

$$\dot{y} = u - y; \qquad y(0) = 1 \tag{5-197}$$

$$\dot{z} = \frac{\partial \Phi}{\partial x}\dot{x} + \frac{\partial \Phi}{\partial y}\dot{y}; \qquad z(0) = 0 \tag{5-198}$$

The state space is now three-dimensional (x, y, z), so that the adjoint vector $\bar{p} = (p_1, p_2, p_3)$ and $p_3 = -1$ is assumed as usual. The hamiltonian follows from the schema outlined earlier as

$$H = u\Gamma + y\left[\gamma_1(w + r_0)\left(p_1 - \frac{\partial \Phi}{\partial x}\right) - \left(p_2 - \frac{\partial \Phi}{\partial y}\right)\right] - wx\left(p_1 - \frac{\partial \Phi}{\partial x}\right) \tag{5-199}$$

where

$$\Gamma = \left(p_1 - \frac{\partial \Phi}{\partial x}\right)(\gamma_2(w + r_0) - r_0 x) + p_2 - \frac{\partial \Phi}{\partial y} \tag{5-200}$$

so that to maximize H, the optimal control policy u^* is

$$u^* = \begin{cases} M; & \Gamma > 0 \\ 0; & \Gamma < 0 \end{cases} \tag{5-201}$$

Γ is known when the two point boundary value problem is solved, which comprises Eqs (5-196)–(5-198) together with the adjoint system Eqs (5-202)–(5-204)

$$\dot{p}_1 = -\left(p_1 - \frac{\partial \Phi}{\partial x}\right)\frac{\partial f_1}{\partial x} - \left(p_2 - \frac{\partial \Phi}{\partial y}\right)\frac{\partial f_2}{\partial x} + \frac{\partial^2 \Phi}{\partial x^2} f_1 + \frac{\partial^2 \Phi}{\partial x\, \partial y} f_2; \qquad p_1(T) = 0 \tag{5-202}$$

$$\dot{p}_2 = -\left(p_1 - \frac{\partial \Phi}{\partial x}\right)\frac{\partial f_1}{\partial y} - \left(p_2 - \frac{\partial \Phi}{\partial y}\right)\frac{\partial f_2}{\partial y} + \frac{\partial^2 \Phi}{\partial y^2} f_2 + \frac{\partial^2 \Phi}{\partial x\,\partial y} f_1; \qquad p_2(T) = 0 \tag{5-203}$$

$$\dot{p}_3 = 0; \qquad p_3(T) \tag{5-204}$$

where f_1 and f_2 are the right-hand sides of Eqs (5-196) and (5-197), respectively. In either case, using the optimality or the maximum principle, this system is sufficiently complicated so that the use of digital computation is in order to numerically compute the solutions.

5-12 MINIMUM TIME SHUTDOWN

The preceding xenon minimax solution is found by numerical methods. The dynamic programming algorithms lend themselves to digital computation in a natural way, as discussed earlier. Here, the minimum time shutdown will be examined for at least two reasons: (*a*) much of the solution can be obtained analytically mainly because the cost functional is much simpler than the minimax cost functional, and (*b*) there is an equivalence between minimum time shutdown and minimax shutdown, which will be elucidated.

From the xenon-iodine equations (5-178) and (5-179), it is seen that at reactor power equilibrium $\dot{x} = \dot{y} = 0$, the latter implying that $u = y$ for equilibrium so that the locus of equilibrium states in the xenon-iodine state space is simply then

$$x = (w + r_0)y/(w + r_0 y) \tag{5-205}$$

It is also seen from Fig. 5-5 that there are two kinds of trajectory curves important in this context. The first are the abrupt shutdown ($u = 0$) curves given from Eqs (5-178) and (5-179) for $u = 0$ that satisfy

$$\frac{dx}{dy} = \{w_x - \gamma_1(w + r_0)y\}/y \tag{5-206}$$

with solutions

$$x = cy^w - \frac{\gamma_1(w + r_0)}{1 - w}\,y \tag{5-207}$$

where c is evaluated at the appropriate point on the above locus of equilibrium states corresponding to the particular reactor equilibrium power level. The second kind of trajectory curves are the full flux ($u = 1$) curves, which happen to be tangent to, and open out from, the locus of their minima which is the straight line

$$x = \gamma_1 y + \gamma_2 \tag{5-208}$$

obtained by letting $\dot{x} = 0$, and $u = 1$ in Eq. (5-178).

The minimum time cost functional is simply $J = \int_0^T dt = T$. Then the third variable in the maximum principle treatment is $z = t$, with the hamiltonian given as

$$H = p_1\dot{x} + p_2\dot{y} + p_3\dot{z} = p_1[-(w + r_0u)x + \gamma_1(w + r_0)y + r_2(w + r_0)u] + p_2[u - y] + p_3 \tag{5-209}$$

For non-singular extremal trajectories, the bang-bang nature of the control is immediately apparent since the hamiltonian is linear in u and so can be rewritten as

$$H = u[p_1(\gamma_2(w + r_0) - r_0x) + p_2] + \text{terms not containing } u \tag{5-210}$$

so that the optimal control, that which minimizes H, is simply

$$u = \begin{cases} 1; & \Gamma > 0 \\ 0; & \Gamma < 0 \end{cases} \tag{5-211}$$

where

$$\Gamma = p_1(\gamma_2(w + r_0) - \Gamma_0 x) + p_2 \tag{5-212}$$

The adjoint equations are given by

$$\dot{p}_1 = -\frac{\partial H}{\partial x} = (w + r_0u)p_1; \qquad \dot{p}_2 = -\frac{\partial H}{\partial y} = p_2; \qquad \dot{p}_3 = -\frac{\partial H}{\partial z} = 0$$

$$\text{that is, } p_3 = 1 \tag{5-213}$$

5-13 TRANSVERSALITY CONDITION

Following the control phase, the reactor state is on a final shutdown ($u = 0$) curve "coasting" to its ultimate shutdown state of $x = y = 0$, if allowed. Immediately prior to the "coast" phase, the reactor optimal trajectory must have intersected this final shutdown curve, called the target curve. In the calculus of variations, when an extremal curve is not only required to extremize a cost functional, and satisfy side conditions which include the intersection of a given curve at the end of its control phase, another condition must be satisfied. It is called the transversality condition—a kind of generalized orthogonality condition, here at the point where the extremal and the adjoint curve intersect the target curve. For our case the transversality condition is given by

$$p_1/p_2 = -\dot{y}(t, 0)/\dot{x}(t, 0) \tag{5-214}$$

where $\dot{y}(t, u)$, $\dot{x}(t, u)$, $p_1(t, u)$, and $p_2(t, u)$ are evaluated at the intersection of the extremal with the particular target ($u = 0$) curve.

In addition, the maximum principle asserts that when the control duration is not fixed a priori, as in this case, that $\max_u H \equiv H^* \equiv 0$. That is, the maximum H is just zero, and it stays that way all along the extremal curve.

5-14 OPTIMAL CONTROL CURVE—THE EXTREMAL SOUGHT

Integrating the adjoint equations (5-213) along (*a*) the $u = 0$ shutdown curves gives

$$p_1 = A_1\, e^{wt}, \qquad p_2 = A_2\, e^{t}, \qquad p_3 = -1 \tag{5-215}$$

and (*b*), along the full flux ($u = 1$) curves gives

$$p_1 = A_1\, e^{(w+r_0)t}, \qquad p_2 = A_2\, e^{t}, \qquad p_3 = -1 \tag{5-216}$$

As mentioned, for non-singular regions, the control u "bang-bangs" between 0 and 1 whenever Γ changes sign, that is, whenever Γ goes through one of its zeros. In the singular segments of the extremal where $\Gamma = 0$, then the optimal $u = u^*$ can vary with time. On such an extremal $\Gamma = \dot{\Gamma} = H = 0$, will hold. Besides $H = 0$ from Eq. (5-209), the vanishing of Γ and $\dot{\Gamma}$ give, respectively, from Eq. (5-212) and its derivative, using Eq. (5-213).

$$0 = p_2[1 - (w + r_0)] + r_0 p_1[(w + r_0 u)x - \gamma_1(w + r_0)y - \gamma_2(w + r_0)u] \tag{5-217}$$

Here, arguments that are given in Roberts and Smith[12] show that the xenon constraint line can comprise a segment of the optimal shutdown extremal. On this xenon constraint line $\dot{x} = 0$ and $x = x_c$. From Eqs (5-196) and (5-197) this yields the corresponding optimal shutdown flux behavior as

$$u = \frac{wx_c}{r_0(x_c - \gamma_1)}\left\{\left[1 + \frac{r_0(x_c - \gamma_1)}{wx_c}\right]\exp\left[\left(1 - \frac{\gamma_1}{x_c}\right)(t_2 - t)\right] - 1\right\}; \quad t_1 \le t \le t_2 \tag{5-218}$$

where without loss of generality a high power reactor is assumed, implying $r_0 \gg w$ and $x_c \gg \gamma_2$. The time t_1 is the time at which the initial unconstrained shutdown $u = 0$ curve intersects the xenon constraint locus, and t_2 is the time at which the constrained extremal leaves the xenon constraint line. Again, argumentation given in Ref. 12 is used to chart the minimum time optimal shutdown curve depicted therein. Also, unconstrained optimal shutdown curves can be derived, as well as extremals for other cost functionals such as minimum energy $\int u\, dt$, or minimum power deviation, $\int (u - 1)\, dt$, etc.

5-15 MINIMUM TIME AND MINIMAX XENON EQUIVALENCE

For a given xenon override constraint x_c and allowable shutdown time T, the optimal minimax shutdown control and the minimum time shutdown control are equivalent, in a sense to be discussed. Consider the time t_p at which the xenon peak occurs, which is obtained by solving Eqs (5-196) and (5-197) with $u = \dot{x} = 0$, in terms of the shutdown duration T. This results in the following relationship between these parameters, which is given by

$$\exp\left[-(1-w)(t_p - T)\right] = w\left[1 + \frac{1-w}{\gamma_1(w+r_0)} \cdot \frac{x_T}{y_T}\right] \tag{5-219}$$

where x_T and y_T are the state point coordinates at the final state, that is, at the onset of the coasting phase. The corresponding magnitude of the xenon peak is

$$x_p = \gamma_1(w+r_0)w^{w/(1-w)}y_T\left[1 + \frac{1-w}{\gamma_1(w+r_0)} \cdot \frac{x_T}{y_T}\right]^{1/(1-w)} \tag{5-220}$$

Combining these two equations gives the xenon peak explicitly in terms of the shutdown duration time T and t_p as

$$x_p = \gamma_1\left(1 + \frac{r_0}{w}\right)y_T \exp - (t_p - T) \tag{5-221}$$

Ignoring the constant $\gamma_1[1 + (r_0/w)]$, the minimax problem for a fixed shutdown control phase duration time T is to find the optimal $u = u^*$ so that, from Eq. (5-221)

$$\min_{0 \le u \le 1} x_p = e^T \min_{0 \le u \le 1} y_T \exp - t_p \tag{5-222}$$

On the other hand, for fixed x_p, the minimal time problem is, also from Eq. (5-221)

$$\min_{0 \le u \le 1} e^T = x_p \max_{u^*} y_T \exp - t_p \tag{5-223}$$

u^* is the optimal shutdown policy that minimizes the shutdown duration time T required to transform the reactor, with given xenon constraint x_c, to achieve a fixed x_p. It is in this sense that this is equivalent to an optimal shutdown minimax policy that yields a minimum x_p for a fixed shutdown time T. A geometrical proof by contradiction is given in Ref. 12. Other equivalences, especially with reference to optimal routing problems are given in Ref. 8.

To recapitulate, it is shown in Ref. 12 that the time optimal extremal must satisfy (*a*) the maximum principle, (*b*) a transversality condition where the extremal intersects the "coasting phase" $u = 0$ shutdown curve, and (*c*) jump conditions where the initial $u = 0$ shutdown extremal segment first intersects the xenon constraint line x_c, and where it finally leaves x_c on a $u = 1$ full flux extremal segment, which in turn intersects the "coasting phase" $u = 0$ shutdown curve. The control phase is over at that point, that is, $t = T$ just where the latter intersection occurs. It is easily shown that this extremal satisfies the minimax cost functional as well. That is, the minimum time extremal also is the extremal that satisfies the xenon minimax criterion discussed above, but not conversely.

PROBLEMS

5-1 It is desired to choose the N numbers $x_1, x_2, \ldots, x_N$ so as to maximize their product $x_1 \cdot x_2 \cdots x_N$, but with two side conditions (*a*) that all the numbers are positive definite integers, that is $x_i \ge 0$, and (*b*) that the numbers sum to a given constant c, that is $x_1 + x_2 + \cdots x_N = c$. Letting $f_N(c) = \max \prod_{i=1}^{N} x_i$, derive the functional equation

$$f_j(c) = \max_{0 \le x_1 \le c} x_j f_{j-1}(c - x_j), \qquad j = 2, \ldots N$$

and

$$f_1(c) = \max_{0 \le x_1 \le c} x_1 = c$$

and show that its solution is $f_N(c) = (c/N)^N$, with optimal policy $y_N = c/N$.

5-2 For "bang-bang" control processes in general, Bellman's equation poses an interesting dilemma. For example, if x is the rate of profit, and u is the rate of profit reinvested (reinvestment policy), it is desired to find the reinvestment policy that maximizes the total investment produced at the end of the period T years. Assume a growth industry so that $\dot{x} = au$; $x(0) = c$; $(a > 0)$. Define the criterional functional $f(c, T)$ as the optimal return on the investment in that

$$f(c, T) = \max_{0 \le u \le c} \int_0^T (x - u)\, dt$$

Derive the corresponding Bellman's equation

$$\frac{\partial f}{\partial T} = c + \max_{0 \le u \le c} u \left(a \frac{\partial f}{\partial c} - 1 \right); \qquad f(c, 0) = 0$$

The dilemma raises its head at this point, since the sign of $\{a(\partial f/\partial c) - 1\}$ is not known, without knowing the solution f. The solution f cannot be found until the optimal control policy $u = u^*$ is known. u^* cannot be known (except that it is obviously "bang-bang") until $\{a(\partial f/\partial c) - 1\}$ is known, etc. In this case more information must be brought to bear to break out of this dilemma. Consider the situation for a newly launched industry, that is, small T. Since $f(c, 0) = 0$, $\partial f/\partial c$ is then also small, since the return on a new industry is usually small. Then $a(\partial f/\partial c) \ll 1$, hence $a(\partial f/\partial c) - 1 < 0$, so that u should be made as small as possible to maximize the above Bellman's equation. Since u^* is "bang-bang," make $u^* = 0$ to maximize. Then $\partial f/\partial T = c$, so $f = cT$ until $T = 1/a$. Continue this development to find that

$$u^* = \begin{cases} c; & T > \dfrac{1}{a} \\ 0; & T \le \dfrac{1}{a} \end{cases}, \qquad \text{and } f = \begin{cases} \dfrac{c}{a} e^{aT-1}; & T \ge \dfrac{1}{a} \\ cT; & T < \dfrac{1}{a} \end{cases}$$

5-3 Assume that the neutron density is governed by the difference equation $n_{r+1} = (1 + K_r)n_r$; $n_0 = c$. Let the cost criterion function be of the "squared deviation" type, that is

$$C = \sum_{j=0}^{N-1} (n_j - 1)^2 + (n_N - 1)^2 + \sum_{j=0}^{N-1} K_j^2$$

The N decision control phase is assumed short compared to the delayed neutron periods, so that the delayed neutrons are neglected. Derive the functional equations

$$F_0 = \alpha(c - 1)^2 \qquad F_r = \min_K \left[(c - 1)^2 + \gamma K^2 + F_{r-1}\{c(1 + K)\}\right] \qquad r = 1, 2, \ldots, N$$

where $F_N(c)$ is the minimum of the criterion C over the reactivity K. Find the optimal reactivity policy.

5-4 Let the equation of state be $\dot{n} = (K - 1)n + c$; $n(0) = c$, where the delayed neutron emitters remain at their equilibrium concentration because of the short duration of the control phase. The cost function is

$$C_2 = \int_0^T (n-1)^2\,dt + \alpha\{n^2(T) - 1\} + \beta \int_0^T K^2\,dt$$

Derive the equation for $F(c, T) = \min_K C_2$ and show that

$$\frac{\partial F}{\partial T} = \min_K \left[(c-1)^2 + \beta K^2 + Kc \frac{\partial F}{\partial c} \right] \qquad F(c, 0) = \alpha(c^2 - 1)$$

5-5 Using the xenon and iodine equations of state following a flux shutdown program with zero flux in the post-shutdown period, find the time at which the xenon maximum occurs.

5-6 Show that, for $\phi_0 > 10^{14}$, the maximum xenon reactivity following a negative step shutdown to essentially zero flux is proportional to ϕ_0. Plot this maximum as a function of ϕ_0. Also, derive a simple expression for the increase in fuel concentration, over that called for in the critical equation, needed to override the above maximum xenon reactivity, as a function of the maximum xenon reactivity. Plot this for $10^{13} \le \phi_0 \le 10^{15}$.

5-7 Consider the diffusion equation for an unreflected slab reactor as

$$\phi_0'' + K_0^2 \phi_0 = 0; \qquad K_0 = \frac{k_0 - 1}{M_0^2}$$

where M_0^2 is the migration area. If the reactor is poisoned by xenon, then the diffusion equation becomes

$$\phi_0'' + K^2 \phi_0 = 0; \qquad K = \frac{k-1}{M^2}$$

where $k \cong k_0/(1+P)$, $P = (\sigma\gamma\Sigma_f\phi)/\{(\lambda_2 + \sigma\phi)\Sigma_a\}$ the xenon poison factor, and $M^2 = M_0^2/(1+P)$. Letting $m\phi = u$ reduces the latter diffusion equation to

$$u'' + K^2 u \left(\frac{1 + \alpha u}{1 + u} \right) = 0$$

where

$$\alpha = 1 - \frac{\gamma}{k_0 - 1} \cdot \frac{\Sigma_f}{\Sigma_0}, \qquad m = \frac{\sigma}{\lambda_2}$$

The problem is to calculate the shift in eigenvalue due to the poison. Let $f(a, c)$ be the minimization of the associated variational integral, that is, $f(a, c) = \min_u \int_0^1 (u'^2 - K^2 G(u))\,dx$ and $G(u)$ is given below, $u(a) = c$, $u(1) = 0$. Obtain the corresponding Bellman's equation (Hamilton–Jacobi equation) namely $\partial f/\partial a = \frac{1}{4}(\partial f/\partial c)^2 + K^2 c^2$. Solve the equation using the fact that $f(a, c) = c^2 g(a)$, and show that the minimizing u satisfies $u' = -K\{G(1) - G(u)\}^{1/2}$, for $G(u) = \alpha u^2 + 2(1-\alpha)(u - \log(1+u))$. Finally show that the eigenvalue has shifted from K_0 to $K = \int_0^1 dx/\{G(1) - G(x)\}^{1/2}$.

REFERENCES

1. L. E. Weaver: "Reactor Dynamics and Control," Amer. Elsevier Publ. NY, p. 29, 1968.
2. G. Leitman (ed.): "Optimization Techniques with Application to Aerospace Systems," Academic Press, New York, 1962.
3. R. Bellman: "Dynamic Programming," Princeton Univ. Press, 1957.
4. M. Ash: "Nuclear Reactor Kinetics," 1st ed., McGraw-Hill, 1965, chap. 6.
5. *loc. cit.*

6. S. M. Roberts: "Dynamic Programming in Chemical Engineering and Process Control," Academic Press, N.Y., 1964, chap. 7.
7. L. S. Pontryagin, et al.: "The Modern Theory of Optimal Control Processes," Wiley, N.Y. (Interscience), 1962, chap. 2.
8. R. Bellman: "Adaptive Control Processes—A Guided Tour," Princeton Press, 1961.
9. Y. Takahashi: The Maximum Principle and its Applications, *Amer. Soc. Mech. Engrs, Trans.* Winter Mtg ASME Paper 63-WA-333, 8–10, Phila., Pa., 1963.
10. F. A. Valentine: "The Problem of Lagrange with Differential Inequalities as Added Side Conditions," Contr. Var. Calc. Univ. Chicago, pp. 403–407, 1937.
11. L. I. Rosonoer: Pontryagin's Maximum Principle in Optimal Control, *Autom. Telemekhan.*, **20**: 10–12, 1959.
12. J. J. Roberts, H. P. Smith JR.: Time Optimal Solution to the Reactivity-Xenon Shutdown Problem, *Nucl. Sci., Eng.*, **22**: 470–478, 1965.

CHAPTER

SIX

IMPORTANCE AND KINETICS REVISITED

6-1 INTRODUCTION

In this chapter, one of the most elegant concepts associated with reactor theory and analysis—that of importance will be investigated. Both the importance, and the importance function will be treated. This will lead into a rederivation of the point model kinetics equations from a much more fundamental viewpoint than the derivation given in the first chapter. Further, this will allow a reinterpretation of certain of the kinetics parameters in their full generality, thus facilitating the gleaning of added insight into the hierarchy of kinetics descriptions of an operating reactor. Also, a treatment of the generalized kinetics equations as derived from different variational principles will be compared and contrasted, with regard to their efficacy of kinetic description, and with an eye to the practicality of implementing kinetics experimentation. Some aspects of this treatment are amenable to experimental verification, and corresponding experiments will be discussed. However, some definitions first must be supplied from which to build the importance edifice, and its associated mathematical apparatus. Certain of these follow briefly in the next section. The others will be given as their need arises.

6-2 ADJOINT FUNCTIONS AND OPERATORS

The following is a brief explanation of the idea of adjoint function and adjoint operators, oriented toward their use herein. The mathematical rigor supplied is only enough to aid in their clear exposition.

Consider two functions $f(\bar{r})$, and $g(\bar{r})$, and their integral, called an inner product, namely

$$(f, g) = \int f^*(\bar{r})g(\bar{r})\, d^3r \tag{6-1}$$

The integral is over the configuration space of interest, $\bar{r}$ is a position vector, and $d^3\bar{r}$ is an infinitesimal volume element in that space. f^* is the complex conjugate of f, and quantities to the left of the comma in (f, g) above are always meant as the complex conjugate. Some obvious properties of Eq. (6-1) then follow

(a) $(f, g) = (g, f)^*$

(b) $(f + h, g) = (f, g) + (h, g)$

(c) $(f, ag) = a(f, g)$ where a is a scalar

(d) $(af, g) = a^*(f, g)$

(e) $(f, f) \geq 0$, with equality holding iff (if and only if) $f(\bar{r}) \equiv 0$.

Now, consider a multidimensional vector space that is represented by a set of unit vectors $\{\phi_i\}$, one unit vector for each axis of the space. In ordinary xyz space, the usual unit vector basis is the triplet $(\bar{i}, \bar{j}, \bar{k})$. This basis set $\{\phi_i\}$ is orthonormal, that is, its inner product $(\phi_i, \phi_j) = \delta_{ij}$, where δ_{ij} is the well-known kronecker delta ($\delta_{ij} = 0$ unless $i = j$, where $\delta_{ii} = \delta_{jj} = 1$). The basis vectors of the above three space is orthonormal, that is, $(\bar{i}, \bar{j}) = (\bar{j}, \bar{k}) = (\bar{k}, \bar{i}) = 0$, and $(\bar{i}, \bar{i}) = (\bar{j}, \bar{j}) = (\bar{k}, \bar{k}) = 1$.

If it is desired to express the known vector (function) $f(\bar{r})$ in the space of the $\{\phi_i\}$, then the expansion of $f(\bar{r})$, where $\bar{r}$ is the multidimensional version of the usual three space $\bar{r}$, is given by

$$f(\bar{r}) = \sum_i c_i \phi_i \tag{6-2}$$

where the coefficients c_i are given using the orthonormal properties of the basis vectors, such as $c_l = (\phi_l, f)$. To see that this does indeed give them, note that

$$c_l = (\phi_l, f) = (\phi_l, \sum_i c_i \phi_i) = \sum_i c_i (\phi_l, \phi_i) = \sum_i c_i\, \delta_{li} \tag{6-3}$$

Suppose now that an operator A in this space is represented by a matrix A (same symbol). When this operator operates on the basis vectors to yield the kth component, f_k, of the vector $f(\bar{r})$, it is meant in the sense that

$$f_k = A\phi_k = \sum_j \phi_j a_{jk} \tag{6-4}$$

that is, f_k is reproduced as a linear combination of the basis vectors. Again, because of the orthogonality of the $\{\phi_i\}$, the jkth element of the matrix A is given by the inner product

$$a_{jk} = (\phi_j, A\phi_k) \tag{6-5}$$

As in the case of the operator A being represented by the matrix A, then the adjoint operator $A^\dagger$ is represented by an adjoint matrix $A^\dagger$. The defining relationship for the adjoint operator is, that for any two vectors $f(\bar{r})$ and $g(\bar{r})$

$$(f, Ag) = (A^\dagger f, g) \tag{6-6}$$

From this definition of adjoint operator, the relationship between the A and $A^\dagger$ matrices is found simply as, using Eqs (6-1) and (6-6)

$$a_{ij} = (\phi_i, A\phi_i) = (A^\dagger \phi_i, \phi_j) = (\phi_j, A^\dagger \phi_i)^* = (a^\dagger)^*_{ji} \tag{6-7}$$

That is, the ijth element of the matrix A is the complex conjugate of the jith element of the adjoint matrix $A^\dagger$. Or, vice versa, that is

$$a^\dagger_{ij} = a^*_{ji} \tag{6-8}$$

Therefore, the adjoint matrix $A^\dagger$ is obtained from the transpose of A, that is, $A^\dagger$, and taking the complex conjugate of each element. The transpose of A can be obtained by reflecting the elements of A about its main diagonal. If $A = A^\dagger$, then A is called a self-adjoint operator.

For f and g considered as vectors, in the expression

$$(f, g) \tag{6-9}$$

the quantity to the left of the comma is a row vector, while the quantity to the right of the comma is a column vector. Equation (6-9) is considered to be a scalar or inner product of the two vectors $f \cdot g$ and so is a scalar. Hence in

$$(f, Ag) \tag{6-10}$$

the matrix A operating on the vector g gives Ag, a column vector, or if

$$(A^\dagger f, g) \tag{6-11}$$

then $A^\dagger f$ is a row vector.

The solution, $\phi^\dagger$, of an adjoint transport equation, say

$$L^\dagger \phi^\dagger = 0 \tag{6-12}$$

is called the adjoint density. Whether the adjoint dagger is on an operator or on a function ($\phi^\dagger$) will be apparent from the context.

The foregoing precis on adjoint functions and operators will be used later in the full development of the importance idea. Now we shall start from certain simple ideas regarding importance and use these as a basis for building the importance concept from that point.

6-3 IMPORTANCE FUNCTIONS AND GENERALIZED KINETICS EQUATIONS

If a neutron burst from a pulsed source of neutrons is introduced into a critical reactor, the flux will rapidly increase. The time increase itself, not involving any spatial dependence, is discussed in Sec. 2-7. If the burst is not too strong, so that the reactor does not run away on a prompt critical excursion, the flux will finally settle back to a new higher equilibrium level. Initially, let the flux excursion be construed as a perturbation, because of the small "size" of the burst, so that well-known results can be used. A thermal reactor is considered, but delayed neutrons are neglected.[1]

Let a thermal neutron burst of unit intensity be introduced into a bare thermal reactor at $t = 0$ at the coordinates $\mathbf{r}'$. Initially then, the thermal flux is increased by this single thermal neutron at $\mathbf{r}'$ so that

$$\phi(\mathbf{r}, 0) = \phi_0(\mathbf{r}) + v\delta(\mathbf{r} - \mathbf{r}') \tag{6-13}$$

where v is the speed of the thermal neutrons and $\phi_0(\mathbf{r})$ satisfies the appropriate steady-state diffusion equation, which is

$$\nabla \cdot D\nabla\phi_0(\mathbf{r}) + (\nu\Sigma_f - \Sigma_a)\phi_0(\mathbf{r}) = 0 \tag{6-14}$$

In Eq. (6-14) D, the diffusion constant, is a function of position in general and Σ_f and Σ_a are the macroscopic fission and total macroscopic absorption cross sections. The time-dependent thermal flux satisfies a similar equation

$$\nabla \cdot D\nabla\phi(\mathbf{r}, t) + (\nu\Sigma_f - \Sigma_a)\phi(\mathbf{r}, t) = \frac{1}{v}\frac{\partial}{\partial t}\phi(\mathbf{r}, t) \tag{6-15}$$

Again, assume that $\phi(\mathbf{r}, t)$ can be expanded in eigenfunctions $\psi_m(\mathbf{r})$:

$$\phi(\mathbf{r}, t) = \sum_{m=0}^{\infty} a_m\psi_m(\mathbf{r})\, e^{\lambda_m t} \tag{6-16}$$

where the eigenfunctions satisfy the space-dependent equation obtained by substituting this expansion into Eq. (6-15)

$$\nabla \cdot D\nabla\psi_m(\mathbf{r}) + \left(\nu\Sigma_f - \Sigma_a - \frac{\lambda_m}{v}\right)\psi_m(\mathbf{r}) = 0 \tag{6-17}$$

and the $\psi_m(\mathbf{r})$ vanish on the extrapolated boundary. The $\psi_m(\mathbf{r})$ are assumed to form a complete orthonormal set, that is

$$\int_{\text{pile}} \psi_n^*(\mathbf{r})\psi_m(\mathbf{r})\, d^3\mathbf{r} \equiv (\psi_n, \psi_m) = \delta_{nm}$$

To determine the a_m, the initial conditions given in Eq. (6-13) are used, as the orthogonal property of the $\psi_m(\mathbf{r})$ is exploited, to get

$$a_m = \delta_{m0} + v\psi_m(\mathbf{r}') \tag{6-18}$$

so that

$$\phi(\mathbf{r}, t) = [1 + v\psi_0(\mathbf{r}')]\psi_0(\mathbf{r}) + \sum_{m=1}^{\infty} v\psi_m(\mathbf{r}')\psi_m(\mathbf{r})\, e^{\lambda_m t} \tag{6-19}$$

The λ_m are found by solving the eigenvalue problem—Eq. (6-17) for the particular reactor involved. It turns out that with $\lambda_0 = 0$, $\lambda_m < 0$ for $m > 0$. That is, the higher-flux modes are ultimately damped to zero, leaving

$$\phi(\mathbf{r}, \infty) = [1 + v\psi_0(\mathbf{r}')]\psi_0(\mathbf{r}) = [1 + v\phi_0(\mathbf{r}')]\phi_0(\mathbf{r}) \tag{6-20}$$

Therefore, following the burst, the flux has settled back to essentially its equilibrium shape ϕ_0, but with an additional amplitude of $v\phi_0(\mathbf{r}')$. This increase is due to the neutron burst at $\mathbf{r}'$. It is called the importance of the neutron introduced at $\mathbf{r}'$, and it is denoted as $\phi_0^\dagger(\mathbf{r}')$. For a thermal reactor only, $\phi_0^\dagger = \phi_0$, so that the importance at $\mathbf{r}'$ is just the thermal flux at $\mathbf{r}'$. It is sometimes said that the importance is proportional to $\phi_0(\mathbf{r})\phi_0^\dagger(\mathbf{r}') \approx \phi_0^2$, the "square" of the flux. This is a measure of how "important" a neutron is with respect to creating progeny neutrons at a given position in a reactor.

A more general formulation of the concept of importance, and one which brings home the idea of adjointness, is the two energy group diffusion equations. As in the preceding discussion, a critical reactor with one thermal neutron introduced initially and no delayed neutrons is considered. Here, there is a slow (thermal) group of neutrons and a fast group of neutrons. The slow group, only, provides fissions. Then for this group, with the diffusion constants now assumed constant with respect to position, the diffusion equation is

$$D_s\nabla^2\phi_s - \Sigma_s\phi_s + \Sigma_f\phi_f = \frac{1}{v_s}\frac{\partial}{\partial t}\phi_s \tag{6-21}$$

where the subscripts s and f stand for slow (thermal) and fast flux, respectively. Σ_s is the total macroscopic absorption cross section for the thermal group of neutrons and Σ_f is the two-group cross section for fast neutrons transferring (by moderation) into the thermal group. The fast flux satisfies

$$D_f\nabla^2\phi_f + k\Sigma_s\phi_s - \Sigma_f\phi_f = \frac{1}{v_f}\frac{\partial}{\partial t}\phi_f, \qquad k = \frac{\nu\Sigma_{\text{fuel}}}{\Sigma_s} \tag{6-22}$$

These two equations are written together conveniently in vector matrix form as

$$\begin{bmatrix} D_s\nabla^2 - \Sigma_s & \Sigma_f \\ k\Sigma_s & D_f\nabla^2 - \Sigma_f \end{bmatrix}\begin{bmatrix}\phi_s \\ \phi_f\end{bmatrix} = \frac{\partial}{\partial t}\begin{bmatrix}\dfrac{1}{v_s}\phi_s \\ \dfrac{1}{v_f}\phi_f\end{bmatrix} = \frac{\partial}{\partial t}\begin{bmatrix}\dfrac{1}{v_s} & 0 \\ 0 & \dfrac{1}{v_f}\end{bmatrix}\begin{bmatrix}\phi_s \\ \phi_f\end{bmatrix} \tag{6-23}$$

With the neutron burst introduced into the thermal group, the initial conditions are

$$\phi_s(\mathbf{r}, 0) = \phi_{s0}(\mathbf{r}) + v_s\delta(\mathbf{r} - \mathbf{r}') \qquad \phi_f(\mathbf{r}, 0) = \phi_{f0}(\mathbf{r}) \tag{6-24}$$

where the equilibrium thermal and fast flux, ϕ_{s0} and ϕ_{f0}, respectively, satisfy the steady-state counterpart of Eq. (6-23), namely

$$\begin{bmatrix} D_s\nabla^2 - \Sigma_s & \Sigma_f \\ k\Sigma_s & D_f\nabla^2 - \Sigma_f \end{bmatrix}\begin{bmatrix}\phi_{s0} \\ \phi_{f0}\end{bmatrix} = 0 \tag{6-25}$$

Just as in the one-group case, assume that the slow and fast time-dependent fluxes can be expanded in eigenfunctions as

$$\phi_s = \sum_m a_m \psi_{sm}\, e^{\lambda_m t} \qquad \phi_f = \sum_m a_m \psi_{fm}\, e^{\lambda_m t} \tag{6-26}$$

respectively. As before, Eqs (6-26) are substituted into their respective diffusion equations to obtain

$$\begin{bmatrix} D_s\nabla^2 - \Sigma_s - \dfrac{\lambda_m}{v_s} & \Sigma_f \\ k\Sigma_s & D_f\nabla^2 - \Sigma_f - \dfrac{\lambda_m}{v_f} \end{bmatrix} \begin{bmatrix} \psi_{sm} \\ \\ \psi_{fm} \end{bmatrix} \equiv Q\psi = 0 \qquad m = 1, 2, \ldots \tag{6-27}$$

where Q is the above matrix and ψ, the above column vector. Now define the adjoint system of equations, in the adjoint fluxes $\psi_{sm}^\dagger$ and $\psi_{fm}^\dagger$. These are so called because they satisfy

$$\begin{bmatrix} D_s\nabla^2 - \Sigma_s - \dfrac{\lambda_j}{v_s} & k\Sigma_s \\ \Sigma_f & D_f\nabla^2 - \Sigma_f - \dfrac{\lambda_j}{v_f} \end{bmatrix} \begin{bmatrix} \psi_{sj}^\dagger \\ \\ \psi_{fj}^\dagger \end{bmatrix} \equiv Q^\dagger\psi^\dagger = 0 \qquad j = 1, 2, \ldots \tag{6-28}$$

which is like Eq. (6-27) except that the elements of the Q matrix have been reflected about the main diagonal, which is the definition of the adjoint, or $Q^\dagger$, matrix. In general, if Q_{ij} is the element of the ith row and jth column of the Q matrix, then $Q_{ij}^\dagger$, the corresponding element in the adjoint matrix, satisfies $Q_{ij}^\dagger = Q_{ji}^*$, where the asterisk means the complex conjugate. Here, all of the elements are real, so that the asterisk is superfluous.

One interesting interpretation of the adjoint fluxes can be made from Eq. (6-28), in keeping with the classical concept of the "adjoint." It is that the neutron cycle seems to be time-reversed such that fast adjoint phantom neutrons cause fissions which yield k slow adjoint neutrons per fast neutron absorbed. These neutrons are moderated upward until they become fast; then they cause more fissions yielding more slow adjoint phantom neutrons, etc. Hurwitz[2] and Ussachoff[3] suggest another interpretation of the adjoint flux; it is given later in connection with the derivation of a transport equation for this quantity.

Consider the matrix equations (6-27) and (6-28) in terms of their individual equations and fluxes. Then multiply Eqs (6-27) by $\psi_{sj}^\dagger$ and $\psi_{fj}^\dagger$ and Eqs (6-28) by ψ_{sm} and ψ_{fm}, respectively. Subtract the corresponding equations to get

$$(\psi_{sj}^\dagger D_s\nabla^2\psi_{sm} - \psi_{sm}D_s\nabla^2\psi_{sj}^\dagger) + (\psi_{fj}^\dagger D_f\nabla^2\psi_{fm} - \psi_{fm}D_f\nabla^2\psi_{fj}^\dagger)$$
$$= (\lambda_m - \lambda_j)\left(\frac{\psi_{sj}^\dagger\psi_{sm}}{v_s} + \frac{\psi_{fj}^\dagger\psi_{fm}}{v_f}\right) \tag{6-29}$$

Integrate over the reactor volume and use Green's theorem.† This reveals that the left-hand side vanishes, since all fluxes vanish on the extrapolated boundaries. Then

$$0 = (\lambda_m - \lambda_j) \int_{\text{pile}} \left(\frac{\psi_{sj}^\dagger \psi_{sm}}{v_s} + \frac{\psi_{fj}^\dagger \psi_{fm}}{v_f} \right) d^3\mathbf{r} \tag{6-30}$$

Assume that $\lambda_m \neq \lambda_j$; then the following must hold:

$$\int_{\text{pile}} \left[\frac{\psi_{sj}^\dagger \psi_{sm}}{v_s} + \frac{\psi_{fj}^\dagger \psi_{fm}}{v_f} \right] d^3\mathbf{r} = \delta_{mj} \tag{6-31}$$

which provides a binormalization of the eigenfunctions (fluxes) and their adjoints. With the initial conditions given in Eq. (6-24), the expansion coefficients are obtained as before:

$$a_m = \delta_{m0} + \psi_{sm}^\dagger(\mathbf{r}') \tag{6-32}$$

As earlier then, the slow and fast fluxes are

$$\begin{aligned} \phi_s(\mathbf{r}, t) &= [1 + \psi_{s0}^\dagger(\mathbf{r}')]\psi_{s0}(\mathbf{r}) + \sum_{m=1}^{\infty} \psi_{sm}^\dagger(\mathbf{r}')\psi_{sm}(\mathbf{r})\, e^{\lambda_m t} \\ \phi_f(\mathbf{r}, t) &= [1 + \psi_{s0}^\dagger(\mathbf{r}')]\psi_{f0}(\mathbf{r}) + \sum_{m=1}^{\infty} \psi_{sm}^\dagger(\mathbf{r}')\psi_{fm}(\mathbf{r})\, e^{\lambda_m t} \end{aligned} \tag{6-33}$$

Again $\lambda_0 = 0$ and $\lambda_m < 0$ for $m > 0$, so that all higher harmonics are damped, giving

$$\begin{aligned} \phi_s(\mathbf{r}, \infty) &= [1 + \psi_{s0}^\dagger(\mathbf{r}')]\psi_{s0}(\mathbf{r}) \\ \phi_f(\mathbf{r}, \infty) &= [1 + \psi_{s0}^\dagger(\mathbf{r}')]\psi_{f0}(\mathbf{r}) \end{aligned} \tag{6-34}$$

Therefore the slow importance function is $\psi_{s0}^\dagger(\mathbf{r}')$, the slow adjoint flux at $\mathbf{r} = \mathbf{r}'$. This is a measure of the increase in all neutrons due to the insertion of a slow neutron initially at $\mathbf{r}'$. A fast importance function can be defined and obtained in an analogous manner by introducing a single fast neutron initially at $\mathbf{r}'$.

In the following, the quintessence of the idea of importance will be used to derive an equilibrium transport equation in the importance function: the adjoint flux. This will be used in the derivation of a very general formulation of the reactor kinetics equations obtained from the transport equation. In Chap. 1, the reactor kinetics equations were derived, with an uncomfortable number of

† Green's theorem transforms the resulting volume integrals to surface integrals over the closed surfaces that envelope the respective volumes. That is

$$\int_v (\psi_{sj}^\dagger \nabla^2 \psi_{sm} - \psi_{sm} \nabla^2 \psi_{sj}^\dagger)\, d^3\mathbf{r} = \int_s (\psi_{sj}^\dagger \nabla \psi_{sm} - \psi_{sm} \nabla \psi_{sj}^\dagger)\, d^2\mathbf{r}$$

This surface integral vanishes because the fluxes in the integrand are assumed to vanish on the extrapolated bounding surfaces.

approximations, via the diffusion equation approximation to the one-velocity transport equation. Here, no such approximations will be made. The tenor of this derivation closely approximates the work of Henry[4] and Ussachoff.[3] As in Chap. 1 consider an angular flux extended now to $\phi(\mathbf{r}, \mathbf{\Omega}, u, t)$, which is the number of neutrons per square centimeter per second at $\mathbf{r}$ traveling in direction $\mathbf{\Omega}$ with speed $v(u)$, where u is the lethargy ($u = \ln E_0/E$; E_0 is the energy of the neutron at fission) at time t. As an aside, the ordinary "flux"

$$\phi_0(\mathbf{r}, t) = \int_0^\infty \int_\Omega \phi(\mathbf{r}, \mathbf{\Omega}, u, t)\, d\mathbf{\Omega}\, du$$

The transport equation for the angular flux is

$$\begin{aligned}\frac{1}{v(u)}\frac{\partial \phi}{\partial t}(\mathbf{r}, \mathbf{\Omega}, u, t) &+ \mathbf{\Omega} \cdot \nabla\phi(\mathbf{r}, \mathbf{\Omega}, u, t) + \Sigma(\mathbf{r}, u, t)\phi(\mathbf{r}, \mathbf{\Omega}, u, t) \\ &= \int_0^\infty \int_{\Omega'} \Sigma_s(\mathbf{r}; u, u'; \mathbf{\Omega}, \mathbf{\Omega}'; t)\phi(\mathbf{r}, \mathbf{\Omega}', u', t)\, d\mathbf{\Omega}'\, du' \\ &\quad + (1-\beta)\int_0^\infty \int_\Omega \nu(u') f(u)\Sigma_f(\mathbf{r}, u', t)\phi(\mathbf{r}, \mathbf{\Omega}, u', t)\, d\mathbf{\Omega}\, du' \\ &\quad + \sum_i \lambda_i C_i(\mathbf{r}, t) f_i(u) + S(\mathbf{r}, \mathbf{\Omega}, u, t)\end{aligned} \tag{6-35}$$

This equation is quite similar to its one-velocity counterpart except that, as mentioned, an additional variable, the lethargy, is included in the description of the angular flux to yield the general transport equation. $\Sigma(\mathbf{r}, u, t)$ is the total macroscoptic capture cross section at $(\mathbf{r}, u, t)$ for all processes (including fission). The terms on the left-hand side of Eq. (6-35) are essentially the same as those in the transport equation in Sec. 1-6. The first term on the right-hand side of Eq. (6-35) gives the contribution of neutrons into $\mathbf{r}$, u, and $\mathbf{\Omega}$ from scattering collisions that have taken place at $\mathbf{r}$, u', and $\mathbf{\Omega}'$. The second term on the right-hand side is the prompt fission neutron contribution from all solid angle and lethargies u' to u. $f(u)$ is the prompt neutron fission spectrum; $f_i(u)$ is that for the ith group of delayed neutrons. The last two terms on the right-hand side are the delayed neutron emitter concentrations and the contribution from a generalized external source. Similarly, for the concentration of delayed neutron emitters,

$$\begin{aligned}\frac{\partial C_i}{\partial t}(\mathbf{r}, t) &= -\lambda_i C_i(\mathbf{r}, t) + \int_0^\infty \int_\Omega \beta_i \nu(u)\Sigma_f(\mathbf{r}, u, t)\phi(\mathbf{r}, \mathbf{\Omega}, u, t)\, d\mathbf{\Omega}\, du \\ \frac{\partial C_i}{\partial t}(\mathbf{r}, 0) &= 0 \qquad i = 1, 2, \ldots\end{aligned} \tag{6-36}$$

As implied earlier, simultaneously with the neutron cycle, there is the cycle of importance or adjoint flux fictitious neutrons. As mentioned, the importance is a measure of how "important" a particular neutron is with respect to generating progeny neutrons in a reactor.† With this idea in mind, the transport equation in the adjoint flux for a critical reactor can be written down by analogy to the above transport equation. It will be needed in the subsequent development of the kinetics equations. It is

$$-\mathbf{\Omega}\cdot\nabla\phi_0^\dagger(\mathbf{r},\mathbf{\Omega},u)+\Sigma^0(\mathbf{r},u)\phi_0^\dagger(\mathbf{r},\mathbf{\Omega},u)$$
$$=\int_0^\infty\int_{\Omega'}\Sigma_s^0(\mathbf{r};u',u;\mathbf{\Omega}',\mathbf{\Omega})\phi_0^\dagger(\mathbf{r},\mathbf{\Omega},u')\,d\mathbf{\Omega}'\,du' \tag{6-37}$$
$$+\int_0^\infty\int_{\Omega'}f(u')\nu(u)\Sigma_f^0(\mathbf{r},u)\phi_0^\dagger(\mathbf{r},\mathbf{\Omega}',u')\,d\mathbf{\Omega}'\,du'$$

For a critical reactor, the flux (and adjoint flux) is assumed to completely dwarf the source term, so that it is neglected. There is, of course, no time derivative term because the reactor is in equilibrium. The leakage or diffusional term above has a negative sign to indicate the net efflux of importance in a sense opposite to that of the neutron flux. Also, the first and second terms on the right-hand side of Eq. (6-37) have u, u' and $\mathbf{\Omega}$, $\mathbf{\Omega}'$ reversed. Again, this corresponds to the notion that as the angular flux is being scattered (moderated) from u' to u and changed in direction from $\mathbf{\Omega}'$ to $\mathbf{\Omega}$ the importance (adjoint flux) is being up-scattered and is traveling in the opposite direction as well since its time, and so neutron cycle is reversed. The fissions are inverse in the sense of the adjoint neutron cycle discussed earlier, which is reflected in the second term on the right-hand side of Eq. (6-37).

Now let the angular flux be made up of two factors, Henry,[4]

$$\phi(\mathbf{r},\mathbf{\Omega},u,t)=\psi(\mathbf{r},\mathbf{\Omega},u,t)n(t) \tag{6-38}$$

where $\psi(\mathbf{r},\mathbf{\Omega},u,t)$ is a time-dependent shape function and $n(t)$ is the time function. Note that this is more general than merely separating $\phi(\mathbf{r},\mathbf{\Omega},u,t)$ into a space and time product. Its generality will be discussed later. First, we define the fluxes

$$\phi_0^\dagger(\mathbf{r},u)=\int_\Omega\phi_0^\dagger(\mathbf{r},\mathbf{\Omega},u)\,d\mathbf{\Omega} \tag{6-39}$$

$$\phi(\mathbf{r},u,t)=\int_\Omega\psi(\mathbf{r},\mathbf{\Omega},u,t)\,d\mathbf{\Omega} \tag{6-40}$$

Equation (6-35) is now integrated over all solid angle and multiplied by $\phi_0^\dagger(\mathbf{r},u)$, and the result is integrated over all lethargy and the reactor as well. Equation (6-37) is also integrated over all solid angle and then multiplied by $n(t)\phi(\mathbf{r},u,t)$, and the result is also integrated over all lethargy and the reactor. The latter is then subtracted from the former, with

† Ussachoff,[3] Hurwitz,[2] and Lewins[5] give essentially the same physical interpretation to the importance or adjoint flux. It is that of the asymptotic production rate of neutrons in the reactor resulting from the initial introduction of a test neutron at $(\mathbf{r},\mathbf{v})$ in the phase space of a critical neutronless reactor.

$$\int_{\text{pile}} \int_0^\infty \frac{\phi_0^\dagger(\mathbf{r}, u)\phi(\mathbf{r}, u, t)}{v(u)}\, d^3\mathbf{r}\, du \tag{6-41}$$

assumed to be an invariant of the reactor, to yield the formal kinetics equations, where the parameters are redefined in a generalized manner, that is

$$\dot{n}(t) = \frac{(K-1)n(t)}{\Lambda} + \sum_i \lambda_i c_i(t) + R(t)$$

$$\dot{c}_i(t) = -\lambda_i c_i(t) + \frac{\bar{a}_i n(t)}{\Lambda} \qquad \bar{a}_i \equiv \frac{\beta_{i,\text{eff}}}{\beta_{\text{eff}}} \equiv \frac{\overline{\beta}_i}{\overline{\beta}} \qquad i = 1, 2, \ldots \tag{6-42}$$

where

$$I(t) = \int_{\text{pile}} \int_0^\infty \int_0^\infty f(u)v(u')\Sigma_f(\mathbf{r}, u', t)\phi_0^\dagger(\mathbf{r}, u)\phi(\mathbf{r}, u', t)\, d^3\mathbf{r}\, du\, du' \tag{6-43}$$

is called the importance of fission neutrons, after Ussachoff.[3] The reactivity K, in dollars (units of $\overline{\beta}$), is identified in the derivation as†

$$\begin{aligned} K = (\overline{\beta} I)^{-1} \int_{\text{pile}} d^3\mathbf{r} \int_0^\infty du \Bigg[\int_0^\infty du' \Bigg(\Delta\Sigma_{\varepsilon 0} + \Bigg[(1-\beta)f(u) \\ + \sum_i f_i(u)\beta_i \Bigg] v(u')\, \Delta\Sigma_f(\mathbf{r}, u', t) \Bigg) \phi_0^\dagger(\mathbf{r}, u)\phi(\mathbf{r}, u, t) \\ - \Delta\Sigma(\mathbf{r}, u, t)\phi_0^\dagger(\mathbf{r}, u)\phi(\mathbf{r}, u, t) - \phi_0^\dagger(\mathbf{r}, u)\nabla \cdot \vec{j}(\mathbf{r}, u, t) \\ - \phi(\mathbf{r}, u, t)\nabla \cdot \vec{j}_0^\dagger(\mathbf{r}, u) \Bigg] \end{aligned} \tag{6-44}$$

$\Sigma_{s0}(\mathbf{r}; u, u'; t) = \int_\Omega \Sigma_s(\mathbf{r}; u, u'; \mathbf{\Omega}, \mathbf{\Omega}', t)\, d\mathbf{\Omega}$ and $\Delta\Sigma_{s0}$, $\Delta\Sigma_f$, and $\Delta\Sigma$ are merely the differences between the time-dependent and equilibrium values of the corresponding cross sections, where

$$\vec{j}(\mathbf{r}, u, t) = \int_\Omega \psi(\mathbf{r}, \mathbf{\Omega}, u, t)\mathbf{\Omega}\, d\mathbf{\Omega}$$

$$\vec{j}_0^\dagger(\mathbf{r}, u) = \int_\Omega \phi_0^\dagger(\mathbf{r}, \mathbf{\Omega}, u)\mathbf{\Omega}\, d\mathbf{\Omega}.$$

The prompt neutron lifetime is defined as

$$\Lambda = (\overline{\beta} I)^{-1} \int_{\text{pile}} \int_0^\infty \frac{\phi_0^\dagger(\mathbf{r}, u)\phi(\mathbf{r}, u, t)}{v(u)}\, d^3\mathbf{r}\, du \tag{6-45}$$

† By premultiplying the fission cross sections above by an adjustable eigenvalue, the importance $I(t)$ can be redefined, so that the reactivity K becomes instead equal to $k_{\text{ex}} = k_{\text{eff}} - 1$. The other parameters will change slightly, but the form of the kinetics equations remains invariant (cf. Henry[4]). Therefore, the kinetics parameters do not have the precise physical meanings usually associated with them. However, when the flux can be factored into a space and time product to good approximation, then the parameters do delineate to their commonly understood meanings.

The delayed neutron emitter concentrations are generalized to

$$c_i(t) = (\Lambda I)^{-1} \int_{\text{pile}} \int_0^\infty c_i(\mathbf{r}, t) f_i(u) \phi_0^\dagger(\mathbf{r}, u)\, d^3\mathbf{r}\, du \tag{6-46}$$

and the generalized source is

$$R(t) = (\Lambda I)^{-1} \int_{\text{pile}} \int_0^\infty S(\mathbf{r}, \mathbf{\Omega}, u, t) \phi_0^\dagger(\mathbf{r}, u)\, d^3\mathbf{r}\, du \tag{6-47}$$

More importantly, an effective $\overline{\beta}_i$ (and thereby an effective $\overline{\beta}$) weighed by the importance in the reactor of the position of the particular delayed neutron and its spectrum $f_i(u)$ is defined as, with $\overline{\beta} = \sum_i \overline{\beta}_i$

$$\overline{\beta}_i = I^{-1} \int_{\text{pile}} \int_0^\infty \int_0^\infty \nu(u') f_i(u) \beta_i \Sigma_f(\mathbf{r}, u', t) \phi_0^\dagger(\mathbf{r}, u) \phi(\mathbf{r}, u', t)\, d^3\mathbf{r}\, du\, du' \tag{6-48}$$

Note that in the case where the angular flux $\phi(\mathbf{r}, \mathbf{\Omega}, u, t)$ can be made separable, that is, $\phi(\mathbf{r}, \mathbf{\Omega}, u, t) = \psi(\mathbf{r}, \mathbf{\Omega}, u) n(t)$, such as a long time after a reactivity disturbance, the spatial factor of the expression in Eq. (6-41) becomes exactly invariant. Thereby importance and the prompt neutron lifetime become secular invariants. That is, these quantities vary only through secular changes in the fission cross section with burnup, as discussed in Sec. 1-5.

The above derivation of the point kinetics equations stems from the fundamental equations of forward and adjoint transport theory, as opposed to writing them out from a semi-empirical balance equation approach using the usual definitions of reproduction factor, neutron generation time, and other kinetics parameters. This should be likened to the derivation of Ohm's Law from Maxwell's equations of electromagnetic theory, instead of the semi-empirical approach of the well-known hydraulic analogy using the definition of voltage, current, and impedance.

The above method transformed the general time-dependent transport equation into the reactor kinetics equations, where appropriate identification is made with the pertinent kinetics parameters such as K, $\overline{\beta}$, and Λ. This method depended on the partitioning of the angular flux, that is, $\phi(\mathbf{r}, \mathbf{\Omega}, u, t) = \psi(\mathbf{r}, \mathbf{\Omega}, u, t) n(t)$, where $\psi(\mathbf{r}, \mathbf{\Omega}, u, t)$ is the shape function and $n(t)$ is the time-varying amplitude. The shape of the neutron flux depends mainly on the prompt neutron source terms, since prompt neutrons comprise better than 99 percent of the neutrons. Following a reactivity change, large or small, the shape $\psi(\mathbf{r}, \mathbf{\Omega}, u, t)$ thereby quickly adjusts to the new reactor state in a few prompt lifetimes. That is, the asymptotic form of the shape $\psi_{\text{asymp}}(\mathbf{r}, \mathbf{\Omega}, u, \infty)$ usually ensues very rapidly. Therefore ψ_{asymp} can be found to good approximation by solving a series of reactor configurations which were attained during the reactivity excursion by assuming the configurations represent reactor equilibrium states.[6] For example, if a control rod is quickly inserted into a reactor, a sequence of ψ_{asymp} can be calculated for a corresponding set of rod positions assumed during the insertion. These then multiply the $n(t)$ obtained from the kinetics equations to solve the problem.

6-4 OPERATIONAL AND VARIATIONAL IMPORTANCE

The preceding section included a description of the connection between importance and the point kinetics equations as derived from the fundamental basis of importance. The following takes up the further development of the idea of importance both from the physical-operational viewpoint, as well as the relatively abstract viewpoint of the variational principle. The latter aspects are paraphased from the work of Lewins,[7] and more recently the work of Becker.[8] The initial portion of the following is a paraphrasal of Lewin's[5] work in the interpretation of importance for reactor theory and kinetics, with emphasis on both the physical and variational aspects. The dynamics of non-conservative systems such as nuclear reactors can be couched in a variational principle through the use of an adjoint system. Classically, the adjoint system can be interpreted in terms of the reversal of cause and effect, which usually implies time reversal, so that the concept of causality is defied. Since in our universe causality seems to make sense at least classically, retaining this principle implies that adjoint quantities do not usually possess a physical reality, hence the earlier reference to phantom adjoint neutrons. The forward and adjoint systems become the two major constituents in a synthesis that yields a conservative stationary system, in the sense of the variational calculus.

The adjointness idea was introduced into reactor physics by Wigner in a World War II document (CP-3048, 1945) in the application of the Rayleigh–Schroedinger perturbation theory to reactor theory. Normally, the adjoint equation used in this context is the time-independent (steady state) equation. However, it is important to realize the efficacy of a time-dependent adjoint equation, corresponding to a hypothetical non-critical reactor in which the behavior can be so construed that causality is not necessarily obeyed.

6-5 OPERATIONAL DEFINITION OF IMPORTANCE

A basic assumption embodying the physical or operational concept of importance is that the reactor behavior description is not complete, even when the solutions to the corresponding most sophisticated reactor equations (e.g. time- and energy-dependent transport equation) are available (Lewins[7]). The description must include the response of the reactor to a given distribution of reactor detectors at a given time T. The detectors are considered at least conceptually to be wired together in such a way that their combined output is read on a single meter to give a meaningful result. This observable or operational response is called the meter reading by Lewins. The spacial distribution of detector readings over the reactor, per reactor neutron at time T is defined to be the importance function. This definition will be seen to conform to the previous definitions already encountered. This detector distribution can be construed as the contribution of a single neutron at $\bar{r}$ and T to the meter reading at time T. This distribution function depends on variables that distinguish the manner by which the neutron can be detected in practice. The importance at time T will also depend on ancestor neutrons through their progeny. That is, the importance of an earlier neutron depends on how it generates progeny, and how important that progeny is at time T. The forward, or ordinary, neutron density function depends on conditions in the past, that is, initial conditions, while the importance function depends on conditions in the future, in other words, how important each neutron is to the future reading of the reactor behavior (meter reading).

There is no absolute operational detector meter reading, in that, for example, all equilibrium reactor power levels enjoy the same importance status. Therefore, there will always be an arbitrary normalization of the importance, as well as for the scalar product of the importance with the forward neutron density, as will be seen. So that, as the initial conditions on the neutron density are arbitrary, so are the final conditions on the importance. A further implication is that the forward neutron density and the importance are independent of each other. So that different kinds of importances can be defined, e.g. the iterated fission density, the ultimate progeny, etc. With the above concepts of importance, the corresponding neutron importance density equations and their boundary conditions will be seen to be identical to the equations in the adjoint neutron density. For example, it can be shown that for the multigroup diffusion equation, if

$$L\psi = \lambda V^{-1}\psi \tag{6-49}$$

where ψ is the flux or angular density vector, L is the matrix representation of the multigroup diffusion operator, and V^{-1} is a diagonal matrix of reciprocal neutron speeds for the corresponding multigroup energies, and λ are the eigenvalues, then

$$L^T\psi^\dagger = -\lambda_T V^{-1}\psi^\dagger \tag{6-50}$$

where $\psi^\dagger$ is the adjoint density, L^T is the transpose of L, and λ_T are the corresponding eigenvalues, and that $|\lambda| = |\lambda_T|$.

Since the adjoint equations are homogeneous, then normalization is arbitrary as mentioned already. So the normalization can be chosen to coincide with a detector calibration fiducial. For example, if the final condition on the importance is taken to be the uniform detection of all neutrons, for example, an infinite homogeneous reactor with its fuel acting as the detector distribution, then the importance is construed as the normal importance, or ultimate progeny definition. On the other hand, if the final condition is the power level at time T, then the importance is construed as the iterated fission rate probability. Other normalizations are possible (Selengut in Lewins[7]) depending on what other kinds of detectors are available and chosen. For example, if eigenfunction solutions of a source free reactor are applicable, the importance contribution may be compartimentalized into that from each eigenfunction. That is, each eigenfunction mode contributes a fractional importance such that the following normalization ensues, namely

$$\int_E dE' \int \psi_i^\dagger V^{-1}\psi_j \, d^3\bar{r} = \delta_{ij} \tag{6-51}$$

where the integration is over all energies and the reactor volume V. Note that the normalization is independent of T, and that all modes are orthogonal to each other.

Lewins[7] puts forth an axiom: one neutron is as important as its progeny. This means that the only way a neutron can affect the meter reading at time T, due to the reactor detector distribution, is through its progeny at T. This axiom is used to derive the equation which the importance satisfies, that is, the equation for the adjoint neutron density. First, the forward or ordinary neutron density satisfies

$$V^{-1}\frac{\partial \psi}{\partial t} = L\psi + S \tag{6-52}$$

In the source free case, and if the time dependence of ψ is exponential (in the eigenvalue regime) then $\partial\psi/\partial t = \lambda\psi$ and the previously discussed Eq. (6-49) results. Now consider the adjoint to Eq. (6-52) with no source, namely

$$-V^{-1}(\partial\psi^{\dagger}/\partial t) = L^{\dagger}\psi^{\dagger} \tag{6-53}$$

The fact that there is no source, is connected with the idea of arbitrary normalization of importance discussed earlier. That is, arbitrary normalization implies a homogeneous adjoint equation, so that any $S^{\dagger} = 0$. Also the time dependence of the adjoint is reversed with respect to the forward equation (6-52). This is also in consonance with the idea of importance. Because, a reactor density *decreasing* with time, but fed by a source to keep the power level constant, corresponds to an increasing importance situation. Simply, as the neutron progeny are decreasing with time, but the power level is constant, the importance per neutron progeny is increasing.

The second axiom of Lewins[7] is that: for given initial conditions on neutrons and precursors, and with the source function assumed as given up until the meter reading time T, then the meter reading, $R(T)$, is

$$R(T) = \int \phi^{\dagger}(\bar{r}, t; T)V^{-1}\psi(\bar{r}, t)\, d^3\bar{r} + \int_t^T \int \phi^{\dagger}(\bar{r}, t', T)S(\bar{r}, t')\, d^3\bar{r}\, dt' \tag{6-54}$$

where $\bar{r}$ is a generic variable standing for $r, \Omega, u, \ldots$ with the exception of time. Differentiating Eq. (6-54) with respect to time yields

$$0 = \frac{\partial}{\partial t}\int \phi^{\dagger} V^{-1}\psi\, d^3\bar{r} - \int \phi^{\dagger} S\, d^3\bar{r} \tag{6-55}$$

or

$$\int \phi^{\dagger} V^{-1}\frac{\partial \psi}{\partial t}\, d^3\bar{r} + \int \psi V^{-1}\frac{\partial \phi^{\dagger}}{\partial t}\, d^3\bar{r} = \int \phi^{\dagger} S\, d^3\bar{r} \tag{6-56}$$

If Eqs (6-52) and (6-53) are inserted into Eq. (6-56) the result is

$$\int \phi^{\dagger} L\psi\, d^3\bar{r} = \int L^{\dagger}\phi^{\dagger}\psi\, d^3\bar{r} \tag{6-57}$$

the well-known definition for the adjoint property, that is, $(\phi^{\dagger}, L\psi) = (L^{\dagger}\phi^{\dagger}, \psi)$.

A variational principle can be derived from the second axiom, which is embodied in Eq. (6-56). Substitute for ψ from its equation $V^{-1}(\partial\psi/\partial t) = L\psi + S$, into Eq. (6-56) to give, using $(\phi^\dagger, L\psi) = (L^\dagger\phi^\dagger, \psi)$

$$\int \psi\left[V^{-1}\frac{\partial\phi^\dagger}{\partial t} + L^\dagger\phi^\dagger\right] d^3\bar{r} = 0 \tag{6-58}$$

Taking the variation $\delta\psi$ about a persisting distribution such as an equilibrium or period equilibrium state, gives for all arbitrary $\delta\psi$,

$$V^{-1}\frac{\partial\phi^\dagger}{\partial t} + L^\dagger\phi^\dagger = 0 \tag{6-59}$$

which is seen to be the second "field" equation (6-53).

Now alternatively, if a substitution for $\phi^\dagger$ from its equation $-V^{-1}(\partial\phi^\dagger/\partial t) = L^\dagger\phi^\dagger$ is made into the second axiom, Eq. (6-55), then the result is

$$\int \phi^\dagger\left[L\psi - V^{-1}\frac{\partial\psi}{\partial t} + S\right] d^3\bar{r} = 0 \tag{6-60}$$

Then for arbitrary variations $\delta\phi^\dagger$, the above yields simply

$$V^{-1}\frac{\partial\psi}{\partial t} - L\psi = S \tag{6-61}$$

which is the first "field" equation.

Hence the second axiom is a variational principle, in that with it as a physical model plus the field equation in the forward flux, the variation operation yields the field equation in the adjoint flux, or vice versa.

To examine the consequences of an optimum distribution of detectors, suppose that somehow an estimate ψ' is obtained which deviates from the actual time-dependent flux in the reactor. Now the (importance weighted) error in the meter reading, E, due to ψ' is, for a given detector distribution $\phi^\dagger$

$$E = \int \phi^\dagger \, dE(\bar{r}, t)\, d^3\bar{r} \tag{6-62}$$

The infinitesimal error dE is construed as that due to ψ', in that

$$dE(\bar{r}, t) = V^{-1}\frac{\partial\psi'}{\partial t} - L\psi' - S \tag{6-63}$$

So that if $dE = 0$ then $\psi' = \psi$, the correct flux satisfying Eq. (6-63). Hence

$$E = \int \phi^\dagger\left(V^{-1}\frac{\partial\psi'}{\partial t} - L\psi' - S\right) d^3\bar{r} \tag{6-64}$$

Now to the question of what is the distribution of detectors $\phi^\dagger$ that minimizes the above error. Again, similar to the above development, using the boundary conditions and the second axiom

$$E = \int \psi' \left[V^{-1} \frac{\partial \phi^\dagger}{\partial t} + L\phi^\dagger \right] d^3\bar{r} \tag{6-65}$$

It is seen that the error E will vanish if the $\phi^\dagger$ satisfies the adjoint flux or importance function equation. Also from Eq. (6-65) it is seen that if ψ' satisfies the first field equation, that is, $\psi' = \psi$, then the error also vanishes.

In the source free case, the critical reactor satisfies an eigenvalue relationship as is well known. Here it is expressed as

$$\frac{\partial \psi_l}{\partial t} = \alpha_l(t)\psi_l: \quad \frac{\partial \phi_l^\dagger}{\partial t} = \alpha_l^\dagger(t)\phi_l^\dagger: \quad l = 1, 2, \ldots \tag{6-66}$$

where α_l and $\alpha_l^\dagger$ are the eigenvalues for the flux and the adjoint (importance) flux, respectively. Now consider any $l \neq 0$ mode, which is thereby a decaying mode. As the mode decays, its neutron population of course, fractionally decreases as

$$\frac{\partial \psi_l}{\partial t} \Big/ \psi_l = -\alpha_l$$

Now, the recurring neutrons in that mode each have an importance that has increased thereby, since those neutrons are contributing to the conceptual detector of this mode. This fractional increase is

$$\frac{\partial \phi_l^\dagger}{\partial t} \Big/ \phi_l^\dagger = \alpha_l^\dagger$$

Since, by the definition of importance

$$\frac{\partial \psi_l}{\partial t} \Big/ \psi_l = \frac{\partial \phi_l^\dagger}{\partial t} \Big/ \phi_l^\dagger$$

then $\alpha_l^\dagger = -\alpha_l$. This result is very well known and is mathematically rigorous, at least for the neutron diffusion equation. Also it can be shown that

$$(\phi_l^\dagger, V^{-1}\psi_m) = \delta_{ml} \tag{6-67}$$

One proof of this is to substitute the source-free eigenfunction assumption into the second axiom equation to obtain $(\alpha_l^\dagger + \alpha_m)(\phi_l^\dagger, V^{-1}\psi_m) = 0$, from which $(\phi_l^\dagger, V^{-1}\psi_m) = 0$ follows.

Becker[8] proposes another variational principle for the derivation of the point kinetics equations from fundamental principles, as in the Henry derivation. Recall that in the earlier kinetics equation development, that if the normalization constraint

$$\frac{d}{dt}\left\{ n(t) \int\int \phi_0^\dagger(\bar{r}, \bar{\Omega}, u) V^{-1} \psi(\bar{r}, \bar{\Omega}, u, t)\, d^3r\, du\, d\bar{\Omega} \right\} = 0 \tag{6-68}$$

held, then the point model kinetics equations (6-42) were the result. However, unless the shape function $\psi(\bar{r}, \bar{\Omega}, u, t)$ is separable into time and space factors, then $n(t)$ cannot be construed as proportional to the reactor power level and still be consistent with equation (6-68). Besides separability, other stipulations are that the reactor is in equilibrium or period equilibrium, that is, on a long inhour type period. If Eq. (6-68) is not satisfied, then Gyftopolous[9] derives a corresponding formal point model kinetics system with the additional factor $\Lambda(d/dt)\ln(\Lambda F)$, that can be incorporated into the precursor decay constant and the reactivity definitions. F is an arbitrary normalizing factor, which is similar to the reactor invariant of Eq. (6-41).

Here, let Eq. (6-68) be construed as a time-dependent normalization. Therefore, it is desirable to extend this reactor kinetics reformulation which permits an importance shape function, as well as the forward flux shape to vary with time. This is done, paraphrasing Becker,[8] using a variational principle. His functional is symbolically, where Q is an external source,

$$
\begin{aligned}
L[\cdot] = \iiint d^3\bar{r}\, d\bar{\Omega}\, du \int_0^T dt \Big[\{ \phi^\dagger [\text{time-dep. transport eq. with source}] \\
+ Q^\dagger \phi \} + \Big\{ \sum_{i=1}^{6} c_i^\dagger \chi_i [i\text{th percursor eq.}] \Big\} \Big] \\
+ \iiint d^3\bar{r}\, d\bar{\Omega}\, du \Big\{ \phi_F^\dagger V^{-1} \phi(\bar{r}, \bar{\Omega}, u, T) \\
+ \phi^\dagger(\bar{r}, \bar{\Omega}, u, 0) V^{-1} [\phi_0(\bar{r}, \bar{\Omega}, u) - \phi(\bar{r}, \bar{\Omega}, u, 0)] \\
+ \sum_{i=1}^{6} c_{iF}^\dagger \chi_i c_i(\bar{r}, \bar{\Omega}, u, T) \\
+ c_i^\dagger(\bar{r}, \bar{\Omega}, u, 0) \chi_i [c_{i0}(\bar{r}, \bar{\Omega}, u) - c_i(\bar{r}, \bar{\Omega}, u, 0)] \Big\}
\end{aligned}
\qquad (6\text{-}69)
$$

where the latter five terms are boundary, initial, and final conditions. Substitutions of the following sort are made in the above functional.

$$
\begin{aligned}
&\text{(a)} \quad \phi^\dagger = \psi^\dagger \eta^\dagger(t) \\
&\text{(b)} \quad c_i = \theta_i \xi_i(t) \\
&\text{(c)} \quad c_i^\dagger = \theta_i^\dagger \xi_i^\dagger(t)
\end{aligned}
\qquad (6\text{-}70)
$$

Inserting Eqs. (6-70) into Eq. (6-69) and setting the variation with respect to $\eta^\dagger$ equal to zero, and then the same for $\xi_i^\dagger$, one finally also obtains corresponding kinetics equations

$$
\dot{n} = \left(\frac{K-1}{\Lambda} \right) n + \lambda_i b_i c_i + S - W_2 n \qquad (6\text{-}71)
$$

and

$$
\dot{c}_i = -\lambda_i c_i + g_i(t) n/\Lambda + W_{3i} c_i \qquad (6\text{-}72)
$$

where

$$
W_2 = \frac{1}{\Lambda F} \iiint d^3\bar{r}\, d\bar{\Omega}\, du \psi^\dagger(\bar{r}, \bar{\Omega}, u, t) V^{-1} \frac{\partial \psi}{\partial t}(\bar{r}, \bar{\Omega}, u, t) \qquad (6\text{-}73)
$$

and

$$W_{3i} = \iiint d^3\bar{r}\, d\bar{\Omega}\, du \frac{\partial}{\partial t} (\theta_i^\dagger/\Lambda F)\chi_i\theta_i \bigg/ \frac{1}{\Lambda F} \iiint d^3\bar{r}\, d\bar{\Omega}\, du\, \theta_i^\dagger \chi_i \theta_i \tag{6-74}$$

and

$$g_i(t) = \iiint d^3\bar{r}\, d\bar{\Omega}\, du\, \theta_i^\dagger \chi_i L\psi \bigg/ \iiint d^3\bar{r}\, d\bar{\Omega}\, du\, \psi^\dagger \chi_i L\psi \tag{6-75}$$

Both W_2 and W_{3i} can be made zero by imposing differential equation constraints that are satisfied by the auxiliary shape function, cf. Becker.[8]

6-6 IMPORTANCE EXPERIMENTS

There have been very few experiments to measure or at least infer the adjoint flux, and its behavior within a reactor. A recent series of experiments along these lines was conducted at the Advanced Epithermal Thorium Reactor (AETR) facility of Rockwell-Atomics International Division, Los Angeles.[10] The purpose of the AETR facility is to study critical assemblies containing ^{233}U for feasibility investigations of thorium breeder reactors. The corresponding critical assembly is made up of two halves of a honeycomb type of physical matrix containing reactor materials. One half of the honeycomb is stationary, while the other is moveable. The latter can be retracted mechanically to and from the stationary half. Safety control rods as well as other types can be included in such an

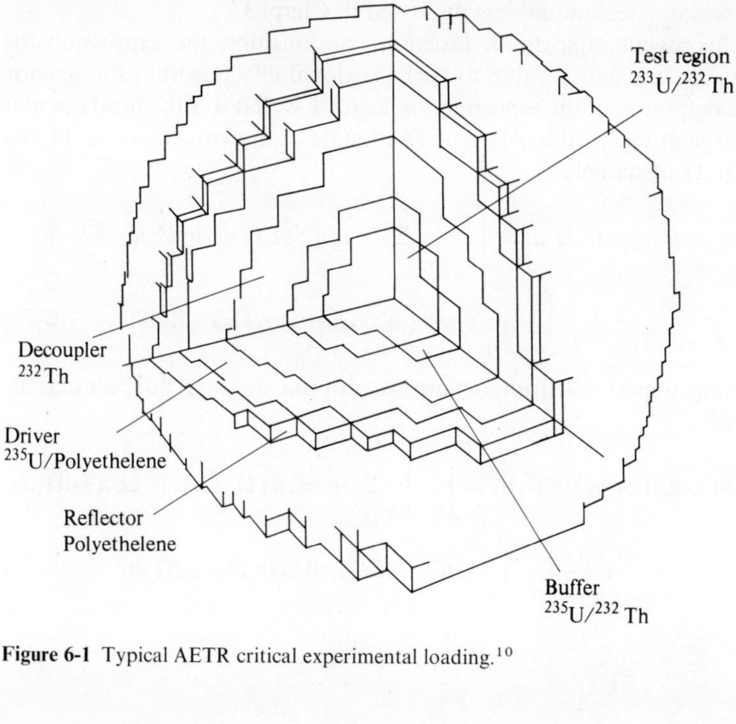

Figure 6-1 Typical AETR critical experimental loading.[10]

assembly. An oscillator for mechanically reciprocating samples or sources within the assembly core is also available.

The importance experiments were done on essentially spherical assemblies that simulate fast or intermediate thorium breeders, as shown in Fig. 6-1. They are constructed by loading drawers, which traverse each half of the matrix, with various materials. Each spherical assembly consists of concentric shells of material, each of which bears a descriptor label. They include the innermost test (core) region, consisting mainly of ^{233}U fuel and ^{232}Th. Surrounding this is a concentric buffer region shell mostly consisting of ^{235}U and ^{232}Th. The buffer region is identical to the test region except for the substitution of ^{235}U for ^{233}U, simply because there was not enough ^{233}U available at the time to constitute an all ^{233}U core. Following the buffer region is a concentric decoupler region made essentially of ^{232}Th. The decoupler region is for the purpose of acting mainly as an absorber for incoming thermal neutrons that might enter the test region-buffer critical core from the (outer) driver region. The latter is an enriched uranium and polyethylene concentric shell. The driver region materials are composed so as to minimize the reactivity–temperature coefficient. The outermost shell is a polyethylene reflector, which completes the critical assembly. Safety and regulating rods are composed of driver and reflector region materials threaded on solid rods attached to their respective drives.

Nine different critical assembly cores were investigated where the individual core median neutron energy ranged from 8.5 eV to 0.422 MeV. The detailed description of the composition of the nine cores are given in Table I of Tuttle.[11] Besides importance investigations, included were those on the doppler-reactivity coefficient, and reactivity coefficients of many different material compositions using the pile oscillator technique[12] as discussed in Chap. 3.

To derive the cogent importance function relationships, the expression for the resultant change in reactivity due to perturbational effects within the reactor is required. One definition of the reproduction factor k is that it is the fundamental (algebraically largest) eigenvalue of the steady state transport equation in the angular flux $\phi_0(\bar{r}, \bar{\Omega}, u)$, namely,

$$\bar{\Omega} \cdot \nabla\phi_0(\bar{r}, \bar{\Omega}, u) + \Sigma^0(\bar{r}, u)\phi_0(\bar{r}, \bar{\Omega}, u) = \int_0^\infty \int_{\bar{\Omega}'} \Sigma_s^0(\bar{r}, u, u'; \bar{\Omega}, \bar{\Omega}')\phi_0(\bar{r}, \bar{\Omega}, u')\, d\bar{\Omega}'\, du'$$

$$+ \frac{1}{k} \int_0^\infty \int_{\bar{\Omega}'} \chi(u')v(u)\Sigma_f^0(\bar{r}, u)\phi_0(\bar{r}, \bar{\Omega}, u')\, d\bar{\Omega}'\, du' \qquad (6\text{-}76)$$

The corresponding unperturbed importance or adjoint angular flux, similar to Eq. (6-37), satisfies

$$-\bar{\Omega} \cdot \nabla\phi_0^\dagger(\bar{r}, \bar{\Omega}, u) + \Sigma^0(\bar{r}, u)\phi_0^\dagger(\bar{r}, \bar{\Omega}, u) = \int_0^\infty \int_{\bar{\Omega}'} \Sigma_s^0(\bar{r}; u', u; \bar{\Omega}', \bar{\Omega})\phi_0(\bar{r}, \bar{\Omega}, u')\, d\bar{\Omega}'\, du'$$

$$+ \frac{1}{k^\dagger} \int_0^\infty \int_{\bar{\Omega}'} \chi(u')v(u)\Sigma_f^0(\bar{r}, u)\phi_0^\dagger(\bar{r}, \bar{\Omega}, u')\, d\bar{\Omega}\, du' \qquad (6\text{-}77)$$

If the reactor is exactly critical $k = k^\dagger = 1$.

For a small perturbation the reproduction factor is perturbed to k^* (not complex conjugate) since certain cross sections take on perturbed values. So that the perturbed forward angular flux $\phi_0^*(\bar{r}, \bar{\Omega}, u)$ satisfies

$$\bar{\Omega} \cdot \nabla \phi_0^*(\bar{r}, \bar{\Omega}, u) + \Sigma^*(\bar{r}, u)\phi_0^*(\bar{r}, \bar{\Omega}, u) = \int_0^\infty \int_{\bar{\Omega}'} \Sigma_s^*(r; u, u'; \bar{\Omega}, \bar{\Omega}')\phi_0(\bar{r}, \bar{\Omega}, u)\, d\bar{\Omega}'\, du'$$

$$+ \frac{1}{k^*}\int_0^\infty \int_{\bar{\Omega}'} \chi(u')\nu(u)\Sigma_f^*(r, u')\phi(\bar{r}, \bar{\Omega}, u')\, d\bar{\Omega}\, du' \qquad \text{(6-78)}$$

In the usual perturbation theory manner, multiplication of Eq. (6-77) by ϕ_0^*, and Eq. (6-78) by $\phi_0^\dagger$, with the resultant difference integrated over all directions, energies, and the reactor volume, the well-known result is, to first order, with $\phi_0^* \simeq \phi_0$,

$$\frac{\Delta k}{k} \simeq = \frac{\displaystyle\iiint \Delta\Sigma(\bar{r}, u)\phi_0^\dagger(\bar{r}, \bar{\Omega}, u)\phi_0(\bar{r}, \bar{\Omega}, u)\, d\bar{\Omega}\, du\, d^3\bar{r} + \int 5\text{ fold} \int \Delta\Sigma_s^0(\bar{r}; u, u'; \bar{\Omega}, \bar{\Omega}')\phi_0^\dagger(\bar{r}, \bar{\Omega}, u)\phi_0(\bar{r}, \bar{\Omega}, u)\, d\bar{\Omega}\, d\bar{\Omega}\, du\, du'\, d^3\bar{r}}{\displaystyle\int 5\text{ fold} \int \chi(u)\nu(u)\Sigma_f^0(\bar{r}, u, \phi_0^\dagger(\bar{r}, \bar{\Omega}, u)\phi_0(\bar{r}, \bar{\Omega}, u)\, d\bar{\Omega}\, d\bar{\Omega}'\, du\, du'\, d^3\bar{r}} \qquad \text{(6-79)}$$

Assuming that neutron scattering is unaffected by the perturbation, the second term of Eq. (6-79) vanishes, so that one can write for the so particularized first order perturbed reactivity, to within a constant

$$k_p \simeq \langle \phi_0^\dagger \,|\Delta P|\, \phi_0 \rangle / \langle \phi_0^\dagger \,|\chi\nu\Sigma_f|\, \phi_0 \rangle \qquad \text{(6-80)}$$

where Dirac notation is used to abbreviate the above ponderous integrations, and $\Delta\Sigma = \Delta P$ is the operator that corresponds to the remaining perturbation.

Suppose that instead of a control rod motion or similar type of perturbation, a perturbing point source per unit neutron flux is introduced into the critical assembly at $\bar{r} = \bar{r}_0$, so the perturbation operator is given by

$$\Delta P(\bar{r}, u, u') = AS(u)\delta(\bar{r} - \bar{r}_0)/\phi_0(\bar{r}, u)n(t) \qquad \text{(6-81)}$$

where A is the number of source neutrons emitted per second, and $S(u)$ is its lethargy spectrum. If this is substituted into Eq. (6-80), with the average fission importance defined by

$$\overline{\phi_f^\dagger} = \langle \phi_0^\dagger \,|\chi\nu\Sigma_f|\, \phi_0 \rangle / (\phi_0, \bar{\nu}\Sigma_f) \qquad \text{(6-82)}$$

where

$$(\phi_0, \bar{\nu}\Sigma_f) = \int \bar{\nu}(\bar{r}, u)\Sigma_f(\bar{r}, u)\phi_0(\bar{r}, u)\, du\, d^3\bar{r} \tag{6-83}$$

then the corresponding perturbed reactivity $K_p = l\Delta\phi_s^\dagger(r_0)/\bar{\phi}_f^\dagger n$, or the importance ratio is

$$\phi_s^\dagger(\bar{r}_0)/\bar{\phi}_f^\dagger = \frac{n}{l} \cdot \frac{K_p}{A} \tag{6-84}$$

where the source importance is $\phi_s^*(\bar{r}) = \int \phi^\dagger(\bar{r}, u)S(u)\, du$, and l is the neutron generation time.

Once the perturbational reactivity is related to the importance as above, it is necessary in this experimental series to obtain a relationship between the power level, n, and the reactivity K. The latter is straightforward from Sec. 2-9 where the reactivity and neutron density (power level) are connected by the relation

$$K = \left\{\Lambda\dot{n} + \int_0^t n'(\tau)f(t-\tau)\, d\tau - \Lambda S\right\} \Big/ n \tag{6-85}$$

Or, the point kinetics integral equation containing a neutron source rate S; that is, Eq. (1-45) can be rewritten slightly as

$$n(t) = n_0 + \int_0^t G(t-\tau)\left(K + \frac{\Lambda S}{n}\right) n\, d\tau \tag{6-86}$$

From the above equation it is seen that there is a total effective reactivity K_T given by

$$K_T = K + \frac{\Lambda S}{n} \tag{6-87}$$

where K is the usual reactivity, and $\Lambda S/n$ is the equivalent reactivity contribution per unit neutron due to the source rate S.

From the previous importance function reactivity relations

$$K_T = K + \frac{\Lambda}{n} S + \frac{\Lambda}{n}\left(\frac{A\phi_s^\dagger(\bar{r}(t))}{\bar{\phi}_f^\dagger}\right) \tag{6-88}$$

where the second term is the equivalent reactivity due to the inherent sources in the critical assembly. The third term is that due to the effective importance weighted source strength, from the inserted source that traverses within the assembly which is the reason that $\bar{r} = \bar{r}(t)$. Now the series of experiments for each core are conducted at least at two power levels corresponding to two values of n. This provides data for two simultaneous equations, which are taken from Eq. (6-88). From these the radial importance $\phi_s^\dagger(\bar{r})/\bar{\phi}_f^\dagger$ can be obtained. The source position as a function of time is known as part of the experiments, and so can be easily related to radial position within the cores. Corrections for the first two terms in Eqs (6-88) are determinate, and are incorporated into the data reduction.

PROBLEMS

6-1 Calculate the coefficients of the eigenfunction expansion, Eq. (6-16), to obtain $a_m = \delta_{m0} + v\psi_m(\mathbf{r}')$. Also obtain $a_m = \delta_{m0} + \psi_{sm}^{\dagger}(\mathbf{r}')$ for the two-group equations using the biorthogonality relation.

6-2 Show that both in the one-group and two-group cases, for a multiplying medium, $\lambda_0 = 0$ and that the higher modes are damped, that is, $\lambda_m < 0$ for $m > 0$.

6-3 Sketch the derivation of the determination of the fast neutron importance function.

6-4 Trace the derivation of the generalized kinetics equations from the time-dependent transport equation in detail, as outlined in Sec. 7-6. Show how the separation of the angular flux into space and time factors delineates the time-invariant property of the principal kinetics parameters, such as Λ and β.

6-5 What physical attributes can be ascribed to the adjoint flux?

6-6 Does a neutronless reactor that is critical possess an adjoint flux?

6-7 If a reactor becomes supercritical and maintains its supercritical state until it consumes all of its fuel, discuss the time history of its adjoint flux.

6-8 Is the importance equal to the adjoint flux, or is it equal to $\phi\phi^{\dagger}$ where ϕ is the forward flux?

6-9 Does the importance vary with the physical size of the reactor in an absolute manner?

6-10 Suppose that there exists an extraterrestrial spherical reactor whose critical radius coincides with the radius approaching that of the Earth. How would its importance vary with the reactor geometry, and with time to include the fact that the neutron itself is unstable decaying with a half-life of about 12 min?

6-11 If a "hyperon" nuclear reactor could be built that would cause hyperon fissions which in turn produced more hyperons, would the hyperon importance differ from that of neutron importance in a neutron reactor?

REFERENCES

1. A. Weinberg and E. Wigner: "Physical Theory of Neutron Chain Reactors," University of Chicago Press, Chicago, 1958, chap. 16.
2. H. Hurwitz Jr.: "Note on the Theory of Danger Coefficients," General Electric KAPL 98, 1948; "On the Derivation and Integration of the Pile Kinetics Equations," AECD 2438, 1948.
3. L. N. Ussachoff: "Equation for the Importance of Neutrons," Reactor Kinetics and Perturbation Theory, *Proc. U.N. Intern. Conf. Peaceful Uses At. Energy, Geneva I*, 1955, **5**: 503–570 (1955).
4. A. F. Henry: Application of Reactor Kinetics to the Analysis of Experiments, *Nucl. Sci. Eng.* **3**: 52–70, 1958.
5. J. Lewins: The Time Dependent Importance of Neutrons and Precursors, *Nucl. Sci. Eng.* **1**:268, 1960.
6. A. F. Henry, N. J. Curlee: Verification of a Method for Treating Neutron Space-Time Problems, *Nucl. Sci. Eng.* **4**:727–744, 1958.
7. J. Lewins: "Importance—The Adjoint Function," Pergamon Press, 1965.
8. M. Becker: A Generalized Formulation of Point Nuclear Reactor Kinetics Equations, *Nucl. Sci. Eng.* **31**:458–464, 1968.
9. E. P. Gyftopoulos: "The Technology of Nuclear Reactor Safety," Vol. 1, pp. 175–204, T. J. Thompson, J. G. Beckerley (eds) MIT Press, Cambridge, Mass., 1964.
10. H. A. Morewitz and L. A. Mountford: The Advanced Epithermal Thorium Reactor (AETR) Critical Experiments, *Nucl. Sci. Eng.* **21**:421–428, 1965.
11. R. J. Tuttle: Neutron Importance Measurements in the Advanced Epithermal Thorium Reactor (AETR) Critical Experiments. *Nucl. Sci. Eng.* **21**:451–462, 1965.
12. S. G. Carpenter: Pile Oscillator Techniques in the Advanced Epithermal Thorium Reactor (AETR), *Nucl. Sci. Eng.* **21**:429–450, 1965.

CHAPTER

SEVEN

SPACE-DEPENDENT KINETICS

7-1 INTRODUCTION

In Chap. 1 the kinetics equations were derived amid a welter of approximations. One of the principal approximations was that the neutron density and the delayed neutron precursor concentrations $c_i(r, t)$ are, respectively, separable into two factors. One is a function of the space variables; the other depends on time only. This implies that the neutron level in the reactor waxes and wanes as a whole with time, similarly to standing waves on a vibrating string, as the flux shape remains proportionately constant. Such an assumption holds only under certain conditions which include (1) a bare homogeneous reactor, simplifying the type of boundary condition for space–time product eigen function expansions,[1] and/or (2) small perturbations of the reactor parameters, such as the diffusion constant, age, and cross sections, from their equilibrium values. This implies that the flux undergoes only small deviations from equilibrium as a result of these perturbations, thereby tending to maintain its shape proportionately constant. For the reasons above, these same parameters should only undergo time changes whose frequencies are low compared to the delayed neutron periods.[2]

The investigations below will be directed toward situations where separability is not valid. Since there are pressing practical reactor system situations where the separability assumption simply does not hold, such as heavily reflected reactors operating at high power densities under all sorts of operational vagaries, it is important that these systems be examined.

7-2 NEUTRON WAVES[3]

A milestone in the derivation of the kinetics equations from the monoenergetic transport equation is the telegrapher's equation (1-31) in the thermal flux. It is

$$\frac{3D}{v^2}\frac{\partial^2}{\partial t^2}\phi_0(\mathbf{r}, t) + \left(\frac{1 + 3D\Sigma_a}{v}\right)\frac{\partial}{\partial t}\phi_0(\mathbf{r}, t) = D\nabla^2\phi_0(\mathbf{r}, t) - \Sigma_a\phi_0(\mathbf{r}, t) + S_0(\mathbf{r}, t) \qquad (7\text{-}1)$$

The general solution of this equation exhibits the phenomenon of retardation. This means that the general solution possesses a retardation, $t - |\mathbf{r}|/v$, term in its argument. This is the wavefront property, in that a source-generated disturbance takes $|\mathbf{r}|/v$ units of time to travel a distance $|\mathbf{r}|$ from the source. In addition to the wavefront property, these solutions leave a residual disturbance (a "wake"). These two aspects will be discussed in problems at the end of the chapter.

The zero mean free path approximation to Eq. (7-1) is now made and results in the diffusion equation. That is, the limits of zero transport mean free path and infinite velocity, such that the product Dv remains finite, yield the time-dependent diffusion equation

$$\frac{1}{v}\frac{\partial}{\partial t}\phi_0(\mathbf{r}, t) = D\nabla^2\phi_0(\mathbf{r}, t) - \Sigma_a\phi_0(\mathbf{r}, t) + S_0(\mathbf{r}, t) \qquad (7\text{-}2)$$

Since the propagation velocity is assumed very large, all parts of the reactor immediately experience a neutron-level disturbance originating in any part thereof. The wavefront is therefore absent. Equation (7-2) can be considered the temporal asymptotic equation corresponding to Eq. (7-1). Equation (7-2) is a good approximation to what occurs at any point $\mathbf{r}$ in the reactor after the disturbance front has actually passed through $\mathbf{r}$. Then for sufficiently large time, it holds everywhere in the reactor. Spherical wave solutions will be obtained for Eq. (7-2) by realizing that they are the asymptotic forms of solutions of Eq. (7-1) for times long after the initiation of the disturbance. Let the source have a periodic character, so that

$$\frac{1}{v}\frac{\partial}{\partial t}\phi_0(\mathbf{r}, t) = D\nabla^2\phi_0(\mathbf{r}, t) - \Sigma_a\phi_0(\mathbf{r}, t) + S_0(\mathbf{r})\, e^{i\omega t} \qquad (7\text{-}3)$$

with the mental reservation that only the real or imaginary portions of the solution are of interest. This type of source can be written in this manner, instead of merely $\sin \omega t$ or $\cos \omega t$, since these equations are linear as opposed to the general kinetics equations, which are usually nonlinear. Such a periodic source is easily obtained experimentally. One example is that of a sinusoidally modulated source of γ rays impinging on a beryllium target submerged in a swimming pool reactor to produce photoneutrons with a sinusoidal intensity. Now let the source be localized, that is, $S_0(\mathbf{r}, t) = \delta(\mathbf{r})\, e^{i\omega t}$, and let $\phi_0(\mathbf{r}, t) = \psi(\mathbf{r})\, e^{i\omega t}$ in Eq. (7-3) be written in the *source free* region (which is almost everywhere) of the reactor to yield

$$\nabla^2\psi(\mathbf{r}) - \left(\frac{1}{L^2} + \frac{i\omega}{Dv}\right)\psi(\mathbf{r}) = 0 \qquad L^2 = \frac{D}{\Sigma_a} \qquad (7\text{-}4)$$

In spherical coordinates this becomes, with spherical symmetry

$$\psi''(r) + \frac{2}{r}\psi'(r) - \left(\frac{1}{L^2} + \frac{i\omega}{Dv}\right)\psi(r) = 0 \tag{7-5}$$

For $\psi(r) = e^{\gamma r}/r$, inserted in Eq. (7-5), the result gives the propagation constant γ as

$$\gamma = \left[\frac{1}{L^2} + \left(\frac{i\omega}{Dv}\right)\right]^{1/2} = \left(\frac{1}{2L^2} + \frac{\xi^2}{2}\right)^{1/2} + i\left(\frac{\xi^2}{2} - \frac{1}{2L^2}\right)^{1/2}$$
$$\xi^2 = \left[\left(\frac{1}{L^2}\right)^2 + \left(\frac{\omega}{Dv}\right)^2\right]^{1/2} \tag{7-6}$$

The spherical wave solutions are then

$$\phi_0(\mathbf{r}, t) \approx \frac{1}{r} e^{\pm r/\mu} e^{i(\omega t - 2\pi r/\lambda)} \tag{7-7}$$

with an attenuation length μ (where the amplitude falls to e^{-1} of its maximum value)

$$\mu = \left(\frac{2L^2}{L^2\xi^2 + 1}\right)^{1/2} \tag{7-8}$$

a wavelength

$$\lambda = 2\pi\left[\frac{2L^2}{L^2\xi^2 - 1}\right]^{1/2} \tag{7-9}$$

and a velocity

$$V = \omega\left[\frac{2L^2}{L^2\xi^2 - 1}\right]^{1/2} \tag{7-10}$$

As is evident, the reactor medium is dispersive, since the velocity depends on the frequency of the excitation. Two cases are considered: (1) $\omega \gg v\Sigma_a$, then from above $V \cong (2Dv\omega)^{1/2}$, and (2) $\omega \ll v\Sigma_a$, giving correspondingly $V \cong 2Dv/L$. These two cases are tabulated for a graphite-moderated reactor, see Table 7-1. Since

Table 7-1 Wave parameters in graphite†
($D = 0.9$ cm, $L = 50$ cm)

	Source excitation frequency, Hz	
Parameter	1.6	160
Attenuation length, m	0.5	0.19
Wavelength, m	50	1.32
Velocity, m/sec‡	79	210

† After Weinberg and Wigner.[3]
‡ Compare with thermal neutron speed of 2200 m/sec.

the above neutron wave parameters are usually easy to measure, these measurements can be used to determine particular reactor constants such as the diffusion constant or thermal diffusion length. This method is directly analogous to Angstrom's method for measuring thermal constants of various heat-conducting materials. The method exploits the well-known fact that the thermal conductivity of a metal can be measured by heating a metal rod at one end, periodically, to produce temperature waves within the rod. Measurement of the temperature wave parameters by using thermocouples becomes an easy task. In the area of reactor technology, the method has been applied to measure the transport mean free path in heavy water.[4]

7-3 SPATIAL FLUX OSCILLATIONS

One method for investigating neutron wave phenomena in a reactor is to introduce an oscillating absorber into the reactor in the same manner as pile transfer functions are obtained. The actual oscillator can be a physically reciprocating length of an absorber within the reactor, or it can be a butterfly arrangement of interleaving absorbers exposing more or less absorber area in a time-periodic fashion inside the reactor. As discussed in Chap. 3, this method was used to some extent in lieu of the danger coefficient method to determine, more accurately, the neutron capture properties of weak absorbers. Since the absorber depresses the flux locally, its displacement oscillations will propagate these flux depressions in the same way that sound waves in a medium are propagated. The following will show that for oscillation frequencies which are low compared to the delayed neutron periods, the flux response essentially maintains its shape as it rises and falls periodically as a whole. At frequencies high compared to the delayed neutron periods, the neutron traveling wave character becomes evident as the absorber propagates damped neutron waves whose properties are a function of the excitation frequency.

Paraphrasing the excellent pedagogical rendition by Weinberg and Schweinler,[2] consider a bare homogeneous reactor, with fluxes of all energies vanishing at the same extrapolation distance beyond the reactor boundaries. The diffusion equation for the thermal flux is

$$\frac{1}{v}\frac{\partial}{\partial t}\phi_0(\mathbf{r},t) = D\nabla^2\phi_0(\mathbf{r},t) - \Sigma_a\phi_0(\mathbf{r},t) + (1-\beta)\frac{k_\infty}{p}\Sigma_a\int_{\text{pile}}\phi_0(\mathbf{r}',t)P(\mathbf{r},\mathbf{r}';E_{\text{th}})\,d^3\mathbf{r}'$$

$$+\sum_{i=1}^{6}\lambda_i\int_{\text{pile}}C_i(\mathbf{r}',t)P(\mathbf{r},\mathbf{r}';E_{\text{th}})\,d^3\mathbf{r}' \qquad (7\text{-}11)$$

where the symbols are as defined earlier, and the slowing-down kernel[6] $P(\mathbf{r},\mathbf{r}';E_{\text{th}})$ is assumed identical for the delayed and prompt neutrons. $P(\mathbf{r},\mathbf{r}';E_{\text{th}})$ is the probability density that a neutron born of a fission at $\mathbf{r}'$, in a finite reactor, will be moderated to thermal energies with final position vector $\mathbf{r}$. For such a reactor, assume that the flux $\phi_0(\mathbf{r},t)$ can be expanded in eigenfunctions of the Helmholtz equation

$$\nabla^2\psi_m(\mathbf{r}) + B_m^2\psi_m(\mathbf{r}) = 0 \qquad m = 0, 1, 2, \ldots \tag{7-12}$$

with the $\psi_m(\mathbf{r})$ vanishing on the extrapolated boundary. B_m^2 are the corresponding eigenvalues. Then let

$$\phi_0(\mathbf{r}, t) = \sum_{m=0}^{\infty} A_m\psi_m(\mathbf{r})T_m(t) \tag{7-13}$$

Assume further that the delayed neutron precursors can be similarly expanded,

$$C_i(\mathbf{r}, t) = \sum_{m=0}^{\infty} C_{im}\psi_m(\mathbf{r})R_{im}(t) \qquad i = 1, \ldots, 6 \tag{7-14}$$

where $C_i(\mathbf{r}, t)$ satisfies

$$\frac{\partial}{\partial t}C_i(\mathbf{r}, t) = -\lambda_i C_i(\mathbf{r}, t) + \frac{\beta_i k_\infty}{p}\Sigma_a\phi_0(\mathbf{r}, t) \tag{7-15}$$

Now, writing down only the two thermal neutron source terms (the third and fourth) of the right-hand side of Eq. (7-11) with the expansions above results in

$$(1-\beta)\frac{k_\infty}{p}\Sigma_a \sum_{m=0}^{\infty} A_m T_m(t)\int_{\text{pile}} \psi_m(r')P(r, r'; E_{\text{th}})\, d^3\mathbf{r}'$$
$$+ \sum_{i=1}^{6}\lambda_i \sum_{m=0}^{\infty} C_{im}R_{im}(t)\int_{\text{pile}} \psi_m(\mathbf{r}')P(\mathbf{r}, \mathbf{r}'; E_{\text{th}})\, d^3\mathbf{r}' \tag{7-16}$$

Consider for a moment that the above reactor is critical. Then the left-hand sides of Eqs (7-11) and (7-15) vanish, so that (assuming $T_m \equiv 1$ for criticality) combining the remainder of these two equations yields

$$D\nabla^2\phi_0(\mathbf{r}) - \Sigma_a\phi_0(\mathbf{r}) + \frac{k_\infty}{p}\Sigma_a\int_{\text{pile}} \phi_0(\mathbf{r}')P(\mathbf{r}, \mathbf{r}'; E_{\text{th}})\, d^3\mathbf{r}' = 0 \tag{7-17}$$

where $\phi_0(\mathbf{r})$ is the time-independent portion of $\phi_0(\mathbf{r}, t)$, that is, from Eq. (7-13)

$$\phi_0(\mathbf{r}) = \sum_{m=0}^{\infty} A_m\psi_m(\mathbf{r}) \tag{7-18}$$

Then, similarly, from Eq. (7-17) the thermal neutron source term is

$$\frac{k_\infty}{p}\Sigma_a \sum_{m=0}^{\infty} A_m \int_{\text{pile}} \psi_m(\mathbf{r}')P(\mathbf{r}, \mathbf{r}'; E_{\text{th}})\, d^3\mathbf{r}' \tag{7-19}$$

The flux of a finite reactor can be calculated by assuming that the reactor is infinite and calculating the asymptotic flux for the infinite reactor.[6] This is because the asymptotic flux is a superposition of spatial flux modes that add up to the correct flux for the finite reactor and that all vanish on the extrapolated boundary. The situation is analogous to the superposition of plane waves to form a wave packet in quantum mechanics or a superposition of image charges (image reactors) to form an equivalent electric field. This allows one to relax the limits on the above integrals over the reactor to integrals over all space. Then the slowing-down probability density can be written more simply as $P_\infty(|\mathbf{r}-\mathbf{r}'|; E_{th})$. It is written as a displacement kernel, since there is no coordinate datum in an infinite reactor; only differences of coordinates (distances) make sense (cf. Sec. 7-3). With this type of slowing-down probability density, the integral in Eq. (7-19) can be simplified by the second fundamental pile theorem,[6] which is that if $\phi_0(\mathbf{r})$ vanishes on the extrapolated boundary and is a solution of Eq. (7-17), then it also satisfies the Helmholtz equation $\nabla^2\phi_0(\mathbf{r}) + B_0^2\phi_0(\mathbf{r}) = 0$ if the critical equation for the lowest eigenvalue B_0^2

$$\frac{k_\infty}{p}\bar{P}_\infty(B_0^2) - L^2B_0^2 - 1 = 0 \tag{7-20}$$

holds. $\bar{P}_\infty(B_0^2)$ is the three-dimensional Fourier transform of $P_\infty(|\mathbf{r}-\mathbf{r}'|; E_{th})$. Since B_0^2 also depends on the geometry of the pile, through the solutions of the Helmholtz equation, then, say, for a slab reactor of thickness l, $B_0^2 = (\pi/l)^2$; so that Eq. (7-20) relates the critical size to the neutronics parameters of the reactor. This is what is meant by a critical equation.

The proof is straightforward; for let $\phi_0(\mathbf{r})$ be represented by its inverse Fourier transform

$$\phi_0(\mathbf{r}) = \frac{1}{(2\pi)^3}\int Z_0(\boldsymbol{\omega})\, e^{-i\boldsymbol{\omega}\cdot\mathbf{r}}\, d^3\omega \tag{7-21}$$

where $\boldsymbol{\omega} = (\omega_x, \omega_y, \omega_z)$ is the (vector) Fourier transform variable. Note that in spherical coordinates in $\boldsymbol{\omega}$ space, Eq. (7-21) becomes

$$\phi_0(\mathbf{r}) = \frac{1}{(2\pi)^3}\int_0^\infty\int_{-1}^{+1}\int_0^{2\pi} Z_0(\omega)\, e^{-i\omega\mu r}\,\omega^2\, d\mu\, d\phi\, d\omega \tag{7-22}$$

with $\mu = \cos\theta$.

Since $\phi_0(\mathbf{r})$ satisfies the Helmholtz equation

$$\nabla^2\phi_0(\mathbf{r}) + B_0^2\phi_0(\mathbf{r}) = 0 \tag{7-23}$$

its Fourier transform $Z_0(\boldsymbol{\omega})$, obtained from taking the Fourier transform of the above equation, satisfies

$$(B_0^2 - \omega^2)Z_0(\boldsymbol{\omega}) = 0 \tag{7-24}$$

which implies that $Z_0(\boldsymbol{\omega})$ is proportional to a three-dimensional delta function, i.e.,

$$Z_0(\boldsymbol{\omega}) = \delta(|\boldsymbol{\omega}| - B_0) \tag{7-25}$$

Insert $\phi_0(\mathbf{r})$, as defined by its inverse transform, Eq. (7-21), into the source term of Eq. (7-17) to obtain, by using Eq. (7-23),

$$-DB_0^2\phi_0(\mathbf{r}) - \Sigma_a\phi_0(\mathbf{r}) - \frac{1}{(2\pi)^3}\frac{k_\infty}{p}\Sigma_a\int_{\text{pile}} d^3\mathbf{r}'\, P_\infty(|\mathbf{r}-\mathbf{r}'|; E_{\text{th}})$$

$$\cdot\int_\infty d^3\omega Z_0(\boldsymbol{\omega})\, e^{-i\boldsymbol{\omega}\cdot\mathbf{r}'} = 0 \qquad (7\text{-}26)$$

Substituting the delta function behavior of $Z_0(\boldsymbol{\omega})$ into this equation gives, after some manipulation, the critical equation (7-20), where

$$\bar{P}_\infty(B_0^2) = 4\pi\int P_\infty(r, E_{\text{th}})\frac{\sin B_0 r}{B_0 r}r^2\, dr \qquad r = |\mathbf{r}| \qquad (7\text{-}27)$$

is the three-dimensional Fourier transform of the slowing-down probability density. This completes the proof.

Then the two terms constituting Eq. (7-16) can be written as

$$(1-\beta)\frac{k_\infty}{p}\Sigma_a\sum_{m=0}^{\infty} A_m T_m(t)\bar{P}_\infty(B_m^2)\psi_m(\mathbf{r}) + \sum_{i=1}^{6}\lambda_i\sum_{m=0}^{\infty} C_{im}R_{im}(t)\bar{P}_\infty(B_m^2)\psi_m(\mathbf{r}) \quad (7\text{-}28)$$

For clarity, and to reduce the amount of algebraic manipulation involved, we shall consider the one-dimensional version of the oscillating absorber in the remainder of this section. Assume that an absorber of absorption cross section $\Sigma(x, t)$ is placed in a semi-infinite slab reactor at position x and time t. The reactor must now be readjusted to a new k'_∞ to maintain criticality. During this transient period, the one-dimensional thermal flux satisfies

$$D\phi''(x,t) - [\Sigma_a + \Sigma(x,t)]\phi(x,t) + \frac{k'_\infty}{p}(1-\beta)\Sigma_a\int_{-\infty}^{+\infty}\phi(x',t)P_\infty(|x-x'|)\, dx'$$

$$+\sum_{i=1}^{6}\lambda_i\int_{-\infty}^{+\infty}C_i(x',t)P_\infty(|x-x'|)\, dx' = \frac{1}{v}\frac{\partial}{\partial t}\phi(x,t) \qquad (7\text{-}29)$$

When the reactor was critical, from Eqs (7-15) and (7-17), prior to the introduction of the absorber, the flux ϕ_0 satisfied

$$D\phi_0''(x) - \Sigma_a\phi_0(x) + \frac{k_\infty}{p}\Sigma_a\int_{-\infty}^{+\infty}\phi_0(x')P_\infty(|x-x'|)\, dx' = 0 \qquad (7\text{-}30)$$

The delayed neutron precursors satisfy

$$\frac{\partial}{\partial t}C_i(x,t) = -\lambda_i C_i(x,t) + \beta_i\frac{k'_\infty}{p}\Sigma_a\phi(x,t) \qquad i = 1, \ldots, 6 \qquad (7\text{-}31)$$

Let $\phi(x, t)$ be expanded in eigenfunctions $\psi_m(x)$

$$\phi(x,t) = \sum_{m=0}^{\infty} A_m T_m(t)\psi_m(x) \qquad (7\text{-}32)$$

where the $\psi_m(x)$ satisfy a one-dimensional Helmholtz equation

$$\psi_m''(x) + B_m^2 \psi_m(x) = 0 \tag{7-33}$$

with $\psi_m(x)$ vanishing on the extrapolated boundary of the slab. These eigenfunctions are, of course,

$$\psi_m(x) = e^{\pm iB_m x} \qquad B_m = \frac{m\pi}{l} \qquad m = 0, 1, 2, \ldots \tag{7-34}$$

Now multiply Eq. (7-29) by $\phi_0(x)$, after realizing that the source terms therein can be written as

$$\cdots + \frac{k'_\infty}{p}(1-\beta)\bar{P}_\infty(B_0^2)\phi(x,t) + \sum_{i=1}^{6} \lambda_i C_i(x,t)\bar{P}_\infty(B_0^2) \tag{7-35}$$

Also, multiply Eq. (7-30) by $\phi(x,t)$. Subtract the two results and integrate over the reactor whose thickness is l to get

$$\begin{aligned} D\int_0^l [\phi(x,t)\phi_0''(x) - \phi''(x,t)\phi_0(x)]\,dx \\ + \int_0^l \Sigma(x,t)\phi_0(x)\phi(x,t)\,dx + \frac{K\bar{P}_\infty}{pv\Lambda}\int_0^l \phi(x,t)\phi_0(x)\,dx \\ - \sum_{i=1}^{6} \lambda_i \bar{P}_\infty(B_0^2)\int_0^l [C_i(x,t) - C_i(x,0)]\phi_0(x)\,dx \\ = \frac{1}{v}\int_0^l \phi_0(x)\frac{\partial}{\partial t}\phi(x,t)\,dx \end{aligned} \tag{7-36}$$

where K is the reactivity, in dollars. Integrating the first term by parts reveals that it vanishes, because all fluxes vanish at the boundary. This is simply the one-dimensional equivalent of Green's theorem. Then take the time average of the remainder of the above equation. That is, operate on each term with

$$\langle\cdot\rangle \equiv \lim_{T\to\infty} \frac{1}{T}\int_0^T \cdot\, dt$$

to obtain

$$\begin{aligned} \int_0^l \phi_0(x)\langle\Sigma(x,t)\,\phi(x,t)\rangle\,dx + \frac{K\bar{P}_\infty}{pv\Lambda}\int_0^l \phi_0(x)\langle\phi(x,t)\rangle\,dx \\ = \lim_{T\to\infty}\left[\sum_{i=1}^{6}\lambda_i\bar{P}_\infty\int_0^l \langle C_i(x,t) - C_i(x,0)\rangle\phi_0\,dx + \int_0^l \phi_0\left\{\frac{\phi(x,T)-\phi(x,0)}{v}\right\}dx\right] \end{aligned} \tag{7-37}$$

In the limit of large T, the right-hand side vanishes, since the reactor is assumed to settle back to equilibrium after a long time. In other words, the reactor is stable. Then, in terms of the reactivity K introduced by the absorber, the above

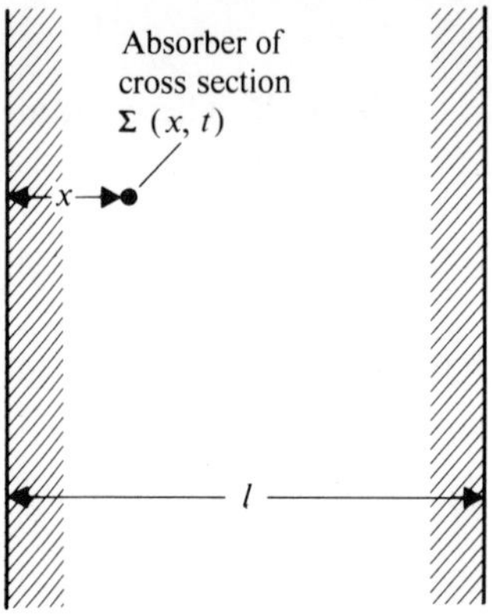

Figure 7-1 Semi-infinite slab reactor with oscillating absorber.

reduces to

$$K = \frac{pv\Lambda}{\bar{P}_\infty(B_0^2)} \frac{\displaystyle\int_0^l \phi_0(x)\langle\Sigma(x,t)\phi(x,t)\rangle\,dx}{\displaystyle\int_0^l \phi_0(x)\langle\phi(x,t)\rangle\,dx} \tag{7-38}$$

which is a well-known result obtained in perturbation theory.

Now consider an absorber of length x_a and cross section Σ_1 oscillating to and fro about $x = x_0$ with amplitude α and frequency ω, cf Fig. 7-1. Let the cross section $\Sigma(x, t)$ be represented by the delta function, that is

$$\Sigma(x,t) = x_a\Sigma_1\delta(x - x_0 + \alpha\,e^{i\omega t}) \tag{7-39}$$

Further, assume that $\alpha \ll (x - x_0)$ so that the first two terms in its Taylor series expansion about α suffice to good approximation. Or

$$\Sigma(x,t) \cong x_a\Sigma_1[\delta(x - x_0) + \alpha\,e^{i\omega t}\,\delta'(x - x_0)] \tag{7-40}$$

Since the amplitude α is small, let the flux and the delayed neutron precursors be expressed as the sum of an equilibrium term and an additional time-dependent oscillatory term, that is

$$\phi(x,t) = \bar{\phi}(x) + \varepsilon(x)\,e^{i\omega t} \tag{7-41a}$$

$$C_i(x,t) = \bar{C}_i(x) + \eta_i(x)\,e^{i\omega t} \qquad i = 1,\ldots,6 \tag{7-41b}$$

with the reservation that only the real (or imaginary) parts of any complex quantity will be retained.

With Eqs (7-40) and (7-41), Eq. (7-29) yields, after retaining only the time-dependent portions and only those through first order, equations for $\varepsilon(x)$ and $\eta_i(x)$; $i = 1,\ldots,6$.

$$D\varepsilon''(x) - \Sigma_a\varepsilon(x) + \frac{k'_\infty}{P}(1-\beta)\Sigma_a\int_{-\infty}^{+\infty}\varepsilon(x')\bar{P}_\infty(|x - x'|; E_{\text{th}})\,dx'$$

$$+\sum_{i=1}^{6}\lambda_i\int_{-\infty}^{+\infty}\eta_i(x')\overline{P}_\infty(|x-x'|;E_{\text{th}})\,dx' - \frac{i\omega}{v}\varepsilon(x) = \phi_0 x_a \Sigma_1 \delta(x-x_0) \tag{7-42}$$

where ϕ_0 is inserted for $\overline{\phi}$, since the absorber is assumed weak. Also

$$(i\omega+\lambda_i)\eta_i(x) = \beta_i\frac{k'_\infty}{p}\Sigma_a\varepsilon(x) \qquad i=1,\ldots,6 \tag{7-43}$$

Expand $\phi_0(x)$ around x_0 in Eq. (7-42) to obtain

$$D\varepsilon''(x) - \Sigma_a\varepsilon(x) + \frac{k'_\infty}{p}(1-\beta)\Sigma_a\int_{-\infty}^{+\infty}\varepsilon(x')\overline{P}_\infty(|x-x'|;E_{\text{th}})\,dx'$$
$$+\sum_{i=1}^{6}\lambda_i\int_{-\infty}^{+\infty}\eta_i(x')\overline{P}_\infty(|x-x'|;E_{\text{th}})\,dx' - \frac{i\omega}{v}\varepsilon(x) = x_a\Sigma_1\phi_0(x_0)\delta(x-x_0) \tag{7-44}$$

Now eliminate η_i by inserting its value from Eq. (7-43) into Eq. (7-44). Then expand both $\varepsilon(x)$ and $\delta(x-x_0)$ in the complete orthonormal-set $\{\psi_m(x)\}$, or

$$\varepsilon(x) = \sum_{m=0}^{\infty}\varepsilon_m\psi_m(x) \tag{7-45}$$

$$\delta(x-x_0) = \sum_{m=0}^{\infty}\psi_m(x)\psi_m(x_0) \tag{7-46}$$

Equation (7-46) is called[7] the closure relation for the set $\{\psi_m(x)\}$. Remembering that

$$\int_{-\infty}^{+\infty}\psi_m(x')\overline{P}_\infty(|x-x'|;E_{\text{th}})\,dx' = \overline{P}_\infty(B_m^2)\psi_m(x) \tag{7-47}$$

and therefore

$$\int_{-\infty}^{+\infty}\varepsilon(x')\overline{P}_\infty(|x-x'|;E_{\text{th}})\,dx' = \sum_{m=0}^{\infty}\varepsilon_m\overline{P}_\infty(B_m^2)\psi_m(x) \tag{7-48}$$

where $\overline{P}_\infty(|x-x'|;E_{\text{th}})$ and $\overline{P}_\infty(B_m^2)$ are the one-dimensional analogs of these quantities defined earlier. All of this inserted into Eq. (7-44) yields finally

$$\varepsilon(x)\,e^{i\omega t} = \frac{x_a\Sigma_1\phi_0}{\Sigma_a}\sum_{m=0}^{\infty}\frac{\phi_0(x_0)\psi_m(x)\psi_m(x_0)\,e^{i\omega t}}{(k'_\infty/p)[\overline{P}_\infty(B_m^2) - i\omega\beta p\{\Lambda + [\overline{P}_\infty(B_m^2)/p]\overline{f}(\omega)\}] - L^2B_m^2 - 1} \tag{7-49}$$

where $\overline{f}(\omega)$ is the Fourier transform of the delayed neutron decay function $f(t)$ discussed in Chap. 1. The real part of $\varepsilon(x)\,e^{i\omega t}$ is the oscillatory fluctuation of the flux about its equilibrium value for a weak absorber.

Now consider (1) low-frequency and (2) high-frequency oscillations. For low-frequency oscillations, $\omega/\lambda_i \ll 1$ for all i, and vice versa for high-frequency oscillations. Then, for low frequencies, Eq. (7-49) gives, where

$$\lim_{\omega \to 0} \bar{f}(\omega) = \sum_{i=1}^{6} \frac{\alpha_i}{\lambda_i} = \bar{\tau}$$

the weighted mean life of the delayed neutrons

$$\varepsilon(x)\, e^{i\omega t} \cong \frac{x_a \Sigma_1 \phi_0}{\Sigma_a} \sum_{m=0}^{\infty} \frac{\phi_0(x_0)\psi_m(x)\psi_m(x_0)\, e^{i\omega t}}{(k'_\infty/p)\bar{P}_\infty(B_m^2) - L^2 B_m^2 - 1 - i\omega\beta k'_\infty[\Lambda + \bar{P}_\infty(B_m^2)\bar{\tau}/p]} \tag{7-50}$$

Since the reactor is critical, the lowest eigenvalue B_0^2 is a root of the critical equation

$$\frac{k}{p}\bar{P}_\infty(B_0^2) - L^2 B_0^2 - 1 = 0 \tag{7-51}$$

Then, the ratio r of the fundamental ($m = 0$) to the mth harmonic is, from Eqs (7-50) and (7-51)

$$r = \frac{\psi_0(x_0)\psi_0(x)}{\psi_m(x_0)\psi_m(x)} \cdot \frac{(k'_\infty/p)\bar{P}_\infty(B_m^2) - L^2 B_m^2 - 1 - i\omega\beta k'_\infty\{\Lambda + [\bar{P}_\infty(B_m^2)/p]\bar{\tau}\}}{-i\omega\beta\{\Lambda + [\bar{P}_\infty(B_0^2)/p]\bar{\tau}\}} \tag{7-52}$$

As ω approaches the zero-frequency limit, r approaches ∞. This implies that the fundamental predominates as the excitation frequency becomes lower and lower, so that the flux level fluctuates as a whole, with the flux in all parts of the reactor in constant phase.

For high-frequency oscillations, $\omega/\lambda_i \gg 1$ for all i. Then Eq. (7-49) becomes, since $\psi_m(x) = e^{\pm im\pi x/l}$; $m = 0, 1, 2, \ldots,$

$$\varepsilon(x)\, e^{i\omega t} \approx \frac{x_a \Sigma_1 \phi_0}{\Sigma_a} e^{i\omega t} \cdot \sum_{m=-\infty}^{m=+\infty} \frac{e^{im\pi(x - x_0)/l} - e^{im\pi(x + x_0)/l}}{(k'_\infty/p)\{\bar{P}_\infty(B_m^2) - i\omega\beta p[\Lambda + [\bar{P}_\infty(B_m^2)/p]\bar{f}(\omega)]\} - L^2 B_m^2 - 1} \tag{7-53}$$

Using the Poisson summation formula[8]

$$\sum_{n=-\infty}^{+\infty} f(n) = \sum_{k=-\infty}^{+\infty} \int_{-\infty}^{+\infty} f(\xi)\, e^{-2\pi i k\xi}\, d\xi \tag{7-54}$$

in the above, transforms it to

$$\varepsilon(x)\, e^{i\omega t} \approx \frac{l x_a \Sigma_1 \phi_0}{\pi \Sigma_a} e^{i\omega t} \cdot \sum_{\mu=-\infty}^{+\infty} \int_{-\infty}^{+\infty} \frac{e^{2\pi i\xi[(x - x_0)/l - \mu]} - e^{-2\pi i\xi[(x + x_0)/l - \mu]}}{(k'_\infty/p)\{\bar{P}_\infty(\xi^2) - i\omega\beta p[\Lambda + [\bar{P}_\infty(\xi^2)/p]\bar{f}(\xi)]\} - L^2\xi^2 - 1} \tag{7-55}$$

The integrals in Eq. (7-55) can be evaluated by contour integration over a semicircular contour of infinite radius in the upper half of the complex ξ plane and including the real axis. Let B_s be the roots (assumed distinct) of the denominator in the integrand. They are assumed to possess a positive imaginary part so that the poles of the integrand lie in the upper half ξ plane and are enclosed by the contour. The integral then simply evaluates to the sum of the residues at the poles.

Let the denominator of the integrand be abbreviated by $k'M(\xi^2) - L^2\xi^2 - 1$. Then $k'M(\xi^2) - L^2\xi^2 - 1$ can be expressed as a sum of partial fractions, in the usual manner, as

$$[k'M(\xi^2) - L^2\xi^2 - 1]^{-1} = \sum_s \left\{\left[k'\frac{dM}{d\xi^2}(B_s) - L^2\right](\xi^2 - B_s)\right\}^{-1} \tag{7-56}$$

With the integral thus evaluated

$$\varepsilon(x)\, e^{i\omega t} \approx \frac{l}{\pi}\frac{x_a\Sigma_1\phi_0}{\Sigma_a} e^{i\omega t} \sum_\mu \sum_s [k'M'(B_s) - L^2]^{-1}$$

$$\cdot \left\{\exp\left[2\pi i B_s\left(\frac{x - x_0}{l} - \mu\right)\right] - \exp\left[-2\pi i B_s\left(\frac{x + x_0}{l} - \mu\right)\right]\right\} \tag{7-57}$$

This can be interpreted as an infinite sequence of positive and negative neutron image sources at the coordinates $x = \pm x_0 + \mu l$; $\mu = 0, 1, 2, \ldots$ which superimpose to the time portion of the flux and which vanish at the extrapolated reactor boundary.

Equation (7-57), then, is a superposition of traveling-wave disturbances with phases given by $2\pi B_s(x \pm x_0)/l - \mu + i\omega t$ traveling in opposite directions with complex wavelength B_s^{-1}. The complex wavelength provides the propagation characteristics of the traveling waves.

From Eq. (7-40) it is seen that for small-amplitude oscillations, the excitation consists of a zero-order (monopole) plus a dipole traveling wave. Far from the oscillator, only the zero-order wave persists. Actually, all multipoles are excited. The source generates a wave for each B_s which determines its phase, amplitude, and attenuation characteristics. The large s waves are important close to the oscillating absorber. As s gets large, the wave attenuates more quickly with distance, since the real parts of iB_s (the attenuation factor) become large with increasing s. In the reactor at positions remote from the oscillator, only the $s = \mu = 0$ wave persists. From Eq. (7-57) this is

$$\varepsilon_{00}(x)\, e^{i\omega t} \approx \frac{l}{\pi}\frac{x_a\Sigma_1\phi_0}{\Sigma_a}\frac{e^{-i(2\pi B_0 x_0/l - \omega t)}}{k'M'(B_0^2) - L^2}\sin\frac{2\pi B_0 x}{l} \tag{7-58}$$

which is a sinusoidal wave in the reactor traveling from left to right. The wavelength, velocity, and attenuation factor for this wave are obtained, of course, from an examination of B_0^{-1}. This is given as a problem at the end of the chapter. Results analogous to those in Sec. 7-2 are obtained.

Figure 7-2 depicts the thermal flux transients from equilibrium in a hypothetical three-region slab reactor consisting of an inner control region, a graphite-moderated ^{235}U core, and a graphite reflector. This resulted from computations by Garabedian et al.[13] made by using large-scale EDPM machinery.

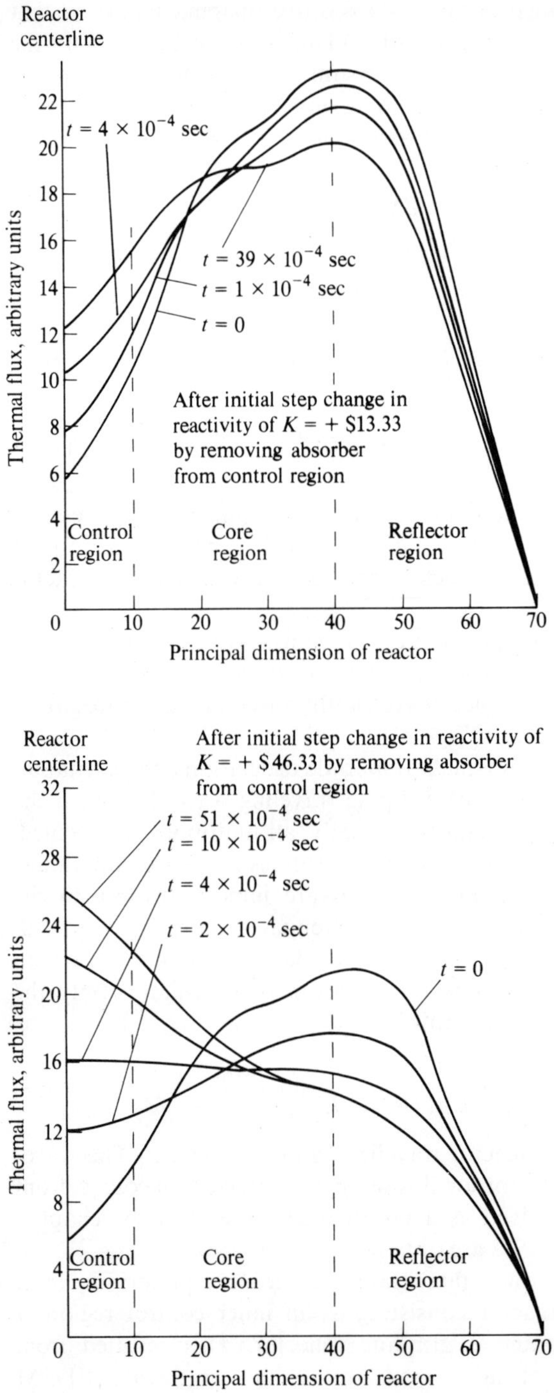

Figure 7-2 Transient thermal flux for ^{235}U, three-region slab reactor with graphite moderator, and reflector.[13]

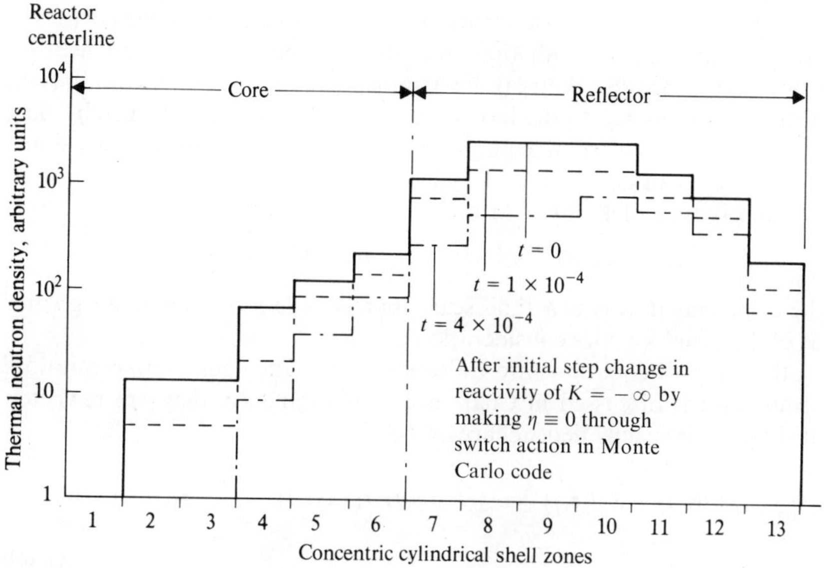

Figure 7-3 Monte Carlo results for transient thermal flux for two-region cylindrical reactor with ^{235}U graphite core and heavy-water reflector.†

7-4 XENON SPACIAL OSCILLATIONS

Besides the effect on the reactor neutron kinetics due to delayed neutrons, temperature, moderator coolant flow, etc., there is an additional major factor in the presence of ^{135}Xe. As discussed in Chap. 6, the ^{135}Xe fission product constitutes a poison to be reckoned with because of its enormous cross section for thermal neutrons, which is in the neighborhood of 3.5 million barns. Since the xenon concentration depends on the fission rate and neutron density in a complicated manner, it is of interest to investigate its temporal and spacial effect on large, high-flux, thermal reactors. It is in these reactors that xenon becomes an important consideration, as will be described in detail.

Because xenon oscillations occur in large thermal reactors having relatively flat flux, usually with low frequencies of oscillation on the order of one cycle per day, the thermal flux can be treated to good approximation by one group diffusion theory. Then the flux satisfies

$$\nabla^2\phi_0(\mathbf{r}, t) + B^2(\mathbf{r}, t)\phi_0(\mathbf{r}, t) \cong 0 \tag{7-59}$$

The flux time dependence, proportional to $\partial\phi_0/\partial t$, usually on the right-hand side is assumed negligible, again since the flux time changes are slow and the reactor

† This is the reactor that is discussed in Chap. 5 of M. Ash, "Nuclear Reactor Kinetics," first edn, 1965, and for which the Monte Carlo code therein was written.

is assumed to be not far from criticality. The material buckling $B^2(\mathbf{r}, t) = (k-1)/M^2$ is a function of both time and space in general. Delayed neutron effects are neglected because they can be assumed to remain at their equilibrium concentration, again owing to the low frequency of the xenon, and thereby flux, oscillations. The main source of xenon is the decay of the fission product iodine isotope ^{135}I. The mean decay time of ^{135}I is about 10 hr, while the xenon itself has a mean decay time of about 14 hr

$$(\lambda_I = 2.87 \times 10^{-5}\ \text{sec}^{-1}, \lambda_X = 2.09 \times 10^{-5}\ \text{sec}^{-1})$$

As can be appreciated, this is a time scale that is orders of magnitude greater than that of the usual kinetics considerations.

The other pair of equations are those for xenon and iodine concentration. Their formulation is discussed in Chap. 6. For convenience, they are rewritten here to include their spacial dependence as well.†

$$\frac{\partial X}{\partial t}(\mathbf{r}, t) = \gamma_2 \Sigma_f \phi_0(\mathbf{r}, t) + \lambda_1 I(\mathbf{r}, t) - (\lambda_2 + \sigma\phi_0(\mathbf{r}, t))X(\mathbf{r}, t) \qquad \frac{\partial X}{\partial t}(\mathbf{r}, 0-) = 0 \tag{7-60}$$

$$\frac{\partial I}{\partial t}(\mathbf{r}, t) = \gamma_1 \Sigma_f \phi_0(\mathbf{r}, t) - \lambda_1 I(\mathbf{r}, t) \qquad \frac{\partial I}{\partial t}(\mathbf{r}, 0-) = 0 \tag{7-61}$$

Initially,[9] the type of oscillation to be considered is that excited when only a small change in reactivity occurs, so that the system is not far removed from equilibrium. Then the buckling, thermal flux, xenon, and iodine concentration can be represented by small excursions about equilibrium values, i.e., after Randall et al.[9]

$$\begin{aligned} \phi_0 &= \phi_0^0(\mathbf{r}) + \varepsilon(\mathbf{r}, t) \qquad & X(\mathbf{r}, t) &= X_0^0(\mathbf{r}) + x(\mathbf{r}, t) \\ I(\mathbf{r}, t) &= I_0^0(\mathbf{r}) + i(\mathbf{r}, t) \qquad & B^2(\mathbf{r}, t) &= B_0^{02} + a_T\varepsilon(\mathbf{r}, t) + a_x x(\mathbf{r}, t) \end{aligned} \tag{7-62}$$

$\varepsilon(\mathbf{r}, t)$, $x(\mathbf{r}, t)$, and $i(\mathbf{r}, t)$ are the small perturbations about equilibrium and a_T and a_x are proportional to the power and xenon reactivity coefficients, respectively. Inserting the above into Eqs (7-59) through (7-61) and subtracting the original equations gives

$$\nabla^2\varepsilon(\mathbf{r}, t) + B_0^{02}\varepsilon(\mathbf{r}, t) + [a_T\varepsilon(\mathbf{r}, t) + a_x x(\mathbf{r}, t)]\phi_0^0(\mathbf{r}) = 0 \tag{7-63}$$

$$\frac{\partial x}{\partial t}(\mathbf{r}, t) = \gamma_2\Sigma_f\varepsilon(\mathbf{r}, t) + \lambda_1 i(\mathbf{r}, t) - \sigma X_0^0(\mathbf{r})\varepsilon(\mathbf{r}, t) - \sigma\phi_0^0(\mathbf{r})x(\mathbf{r}, t) \tag{7-64}$$

$$\frac{\partial i}{\partial t}(\mathbf{r}, t) = \gamma_1\Sigma_f\varepsilon(\mathbf{r}, t) - \lambda_1 i(\mathbf{r}, t) \tag{7-65}$$

† The first term on the right-hand side of Eq. (7-60) is the xenon production rate directly from fission. Because of the relatively small fission-product xenon yield (0.3 per cent), it is usually much smaller than the xenon produced because of the decay of iodine. This is the second term on the right-hand side. The smaller of the two terms is then often neglected.

where products of the perturbational quantities are assumed to be of second order and are therefore neglected in this approximation.

Now the (unperturbed) equilibrium sequence of orthonormal fluxes satisfies the boundary value problem

$$\nabla^2\phi_n(\mathbf{r}) + (B_0^{02} + k_n^2)\phi_n(\mathbf{r}) = 0 \qquad n = 0, 1, 2, \ldots \qquad k_0^2 = 0 \tag{7-66}$$

where these fluxes vanish on the extrapolated reactor boundary. ϕ_0^0 is proportional to $\phi_0(\mathbf{r})$. Following the usual scheme of perturbation theory, we expand the perturbational quantities in terms of the unperturbed flux harmonics, and thus

$$\varepsilon(\mathbf{r}, t) = \sum_n^{\infty} E_n\phi_n(\mathbf{r})\, e^{\omega_n t} \qquad x(\mathbf{r}, t) = \sum_n^{\infty} X_n\phi_n(\mathbf{r})\, e^{\omega_n t} \qquad i(\mathbf{r}, t) = \sum_n^{\infty} I_n\phi_n(\mathbf{r})\, e^{\omega_n t} \tag{7-67}$$

Eliminating the iodine terms between Eqs (7-64) and (7-65), inserting the expansions, and using Eq. (7-66) yields

$$-\sum_n^{\infty} E_n k_n^2\phi_n(\mathbf{r}) + a_T\sum_n^{\infty} E_n\phi_0^0(\mathbf{r})\phi_n(\mathbf{r}) + a_x\sum_n^{\infty} X_n\phi_0^0(\mathbf{r})\phi_n(r) = 0 \tag{7-68}$$

$$\sum_n^{\infty} [\omega_n + \lambda_2 + \sigma\phi_0^0(\mathbf{r})]X_n\phi_n(\mathbf{r}) - \sum_n^{\infty}\left[\gamma_2\Sigma_f + \frac{\lambda_1\gamma_1\Sigma_f}{\omega_n + \lambda_1} - \sigma X_0^0(\mathbf{r})\right]E_n\phi_n(\mathbf{r}) = 0 \tag{7-69}$$

Multiply both equations by $\phi_m(\mathbf{r})$ and integrate over the reactor volume, giving, respectively

$$[a_T\langle m|\phi_0^0|m\rangle - k_m^2]E_m + a_x\langle m|\phi_0^0|m\rangle X_m + \sum_{n\neq m}^{\infty}[\langle n|\phi_0^0|m\rangle(a_T E_n + a_x X_n)] = 0 \tag{7-70}$$

$$\left[\sigma\langle m|X_0^0|m\rangle - \gamma_2\Sigma_f - \frac{\lambda_1\gamma_1\Sigma_f}{\omega_m + \lambda_1}\right]E_m + \left[\omega_m + \lambda_2 + \sigma\langle m|\phi_0^0|m\rangle\right]X_m + \sigma\sum_{n\neq m}^{\infty}\left[\langle n|\phi_0^0|m\rangle X_n + \langle n|X_0^0|m\rangle E_n\right] = 0 \tag{7-71}$$

where

$$\langle n|\phi_0^0|m\rangle = \int_{\text{pile}} \phi_n(\mathbf{r})\phi_0^0(\mathbf{r})\phi_m(\mathbf{r})\, d^3\mathbf{r}$$

$$\langle n|X_0^0|m\rangle = \int_{\text{pile}} \phi_n(\mathbf{r})X_0^0(\mathbf{r})\phi_m(\mathbf{r})\, d^3\mathbf{r}$$

Since the unperturbed flux and thereby the unperturbed xenon concentration are considered to be approximately spacially flat, together with the fact that the $\{\phi_n\}$ are orthonormal, it is assumed that $\langle n|\phi_0^0|m\rangle$ and $\langle n|X_0^0|m\rangle$ are vanishingly small for $n \neq m$ as compared to the $n = m$ terms.† Then, omitting the former results in the following for $m = 0, 1, 2, \ldots,$

$$[a_T\langle m|\phi_0^0|m\rangle - k_m^2]E_m + a_x\langle m|\phi_0^0|m\rangle X_m = 0 \tag{7-72}$$

$$\left[\sigma\langle m|X_0^0|m\rangle - \gamma_2\Sigma_f - \frac{\lambda_1\gamma_1\Sigma_f}{\omega_m + \omega_1}\right]E_m + \left[\omega_m + \lambda_2 + \sigma\langle m|\phi_0^0|m\rangle\right]X_m = 0 \tag{7-73}$$

These equations constitute a set of homogeneous linear algebraic equations in the coefficients E_m and X_m. For nontrivial solutions, the determinant of the coefficients must vanish. With the further abbreviations $\langle m|\phi_0^0|m\rangle \equiv \phi_{mm}^0$, $\langle m|X_0^0|m\rangle \equiv X_{mm}^0$; then for every m

$$\begin{vmatrix} (a_T\phi_{mm}^0 - k_m^2) & (a_x\phi_{mm}^0) \\ \left(\sigma X_{mm}^0 - \gamma_2\Sigma_f - \dfrac{\lambda_1\gamma_1\Sigma_f}{\omega_m + \lambda_1}\right) & (\sigma\phi_{mm}^0 + \lambda_2 + \omega_m) \end{vmatrix} = 0 \tag{7-74}$$

which yields a quadratic equation in ω_m, the allowed exponents in the exponential factor of the perturbational quantities. That is, the determinant reduces to

$$\omega_m^2 + b_m\omega_m + c_m = 0 \tag{7-75}$$

where

$$b_m = \lambda + \sigma\phi_{mm}^0 - \frac{a_x\phi_{mn}^0(\sigma X_{mm}^0 - \gamma_2\Sigma_f)}{(a_T\phi_{mm}^0 - k_m^2)} \qquad \lambda = \lambda_1 + \lambda_2 \qquad \gamma = \gamma_1 + \gamma_2$$

$$c_m = \lambda_1\lambda_2 + \sigma\phi_{mm}^0 - \left(\frac{a_x\phi_{mm}^0}{a_T\phi_{mm}^0 - k_m^2}\right)(\sigma X_{mm}^0 - \gamma\Sigma_f) \tag{7-76}$$

It is easily shown that $b_m^2 - 4c_m < 0$, so that ω_m has complex roots, that is,

$$\omega_m = -\frac{(b_m/2) \pm i(4c_m - b_m^2)^{1/2}}{2} \tag{7-77}$$

then for sustained oscillations, the damping factor b_m must vanish. Equation (7-76) for $b_m = 0$ yields the following equation in $k_m^2 (m > 0)$, the bucklings of the flux modes,

$$k_m^2 = a_T\phi_{mm}^0 - a_x\phi_{mm}^0\left(\frac{\sigma X_{mm}^0 - \gamma_2\Sigma_f}{\lambda + \sigma\phi_{mm}^0}\right) \equiv a_T\phi_{mm}^0 - a_xF(\phi_{mm}^0) \tag{7-78}$$

where

$$F(\phi_{mm}^0) \equiv \frac{\sigma X_{mm}^0 - \gamma_2\Sigma_f}{\lambda + \sigma\phi_{mm}^0}$$

† This is not always the case, since for very large reactors under certain conditions, modal interaction can lessen the oscillation stability threshold. Cf. Gyorey[10] and Lellouche.[11] Large thermal reactors are assumed above.

Let $k_m^2/B_g^2 = \lambda_m$, $a_x = a_x/M^2$, and $a_T = \alpha_T/M^2$. M^2 is the migration area, B_g^2 is the geometrical buckling, α_x is the reactivity coefficient of the xenon, and α_T is the power coefficient of reactivity. For $\alpha_x < 0$, which is the usual case, the above can be written as

$$F(\phi_{mm}^0) = -\frac{\alpha_T}{|\alpha_x|}\phi_{mm}^0 + \frac{M^2 B_g^2}{|\alpha_x|}\lambda_m \tag{7-79}$$

Approximate $F(\phi_{mm}^0)$ by

$$F(\phi) \cong \frac{\dfrac{\gamma\Sigma_f}{\sigma}\left[\dfrac{1}{1+\lambda_2/\phi\sigma} - \dfrac{\gamma_2}{\gamma}\right]}{1+\lambda/\phi\sigma} \tag{7-80}$$

where it is assumed that $X_{mm}^0 = \gamma\Sigma_f\phi_{mm}^0/(\lambda_2 + \sigma\phi_{mm}^0)$. That is, X_{mm}^0 is approximated by the equilibrium xenon concentration, where the equilibrium flux is replaced by ϕ_{mm}^0. This is a good approximation, since for flat flux $\phi_{mm}^0 \cong \phi_0^0$. For a given reactor, Eq. (7-79) determines the flux threshold (lower bound) for sustained oscillations to occur. This is shown in Fig. 7-4.

$F(\phi)$ vanishes at a flux, $\phi \cong 3.5 \times 10^{11}$ neutrons per square centimeter per second. Below this equilibrium flux level, the amount of negative reactivity represented by the xenon concentration is negligible, as discussed in Chap. 6. Inspection of Eq. (7-78) and Fig. 7-4 for positive α_x shows that for an increasing power reactivity coefficient α_T, the flux threshold for sustained oscillations is lowered. If α_T is large enough, implying closely coupled power reactivity feedback, $F(\phi)$ approaches zero. This is tantamount to assuming that the xenon is being "burned out" as fast as it is formed ($\sigma X_{mm}^0 = \gamma_2\Sigma_f$) so that sustained oscillations could occur without the effective presence of xenon. On the other hand, if $\alpha_T < 0$ and $|\alpha_T|$ is sufficiently large, then the flux threshold is raised to very high flux levels.

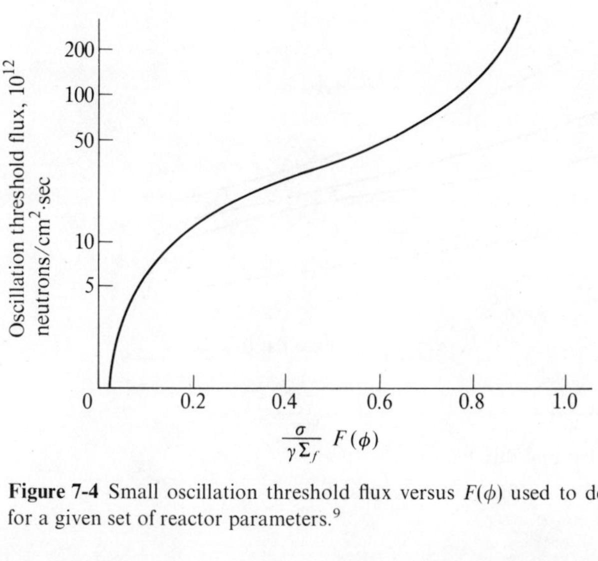

Figure 7-4 Small oscillation threshold flux versus $F(\phi)$ used to determine flux oscillation threshold for a given set of reactor parameters.[9]

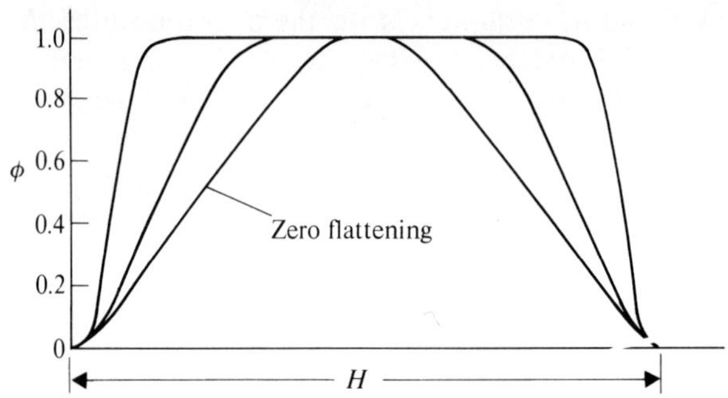

Figure 7-5 Flux shapes for flattened one-dimensional piles.

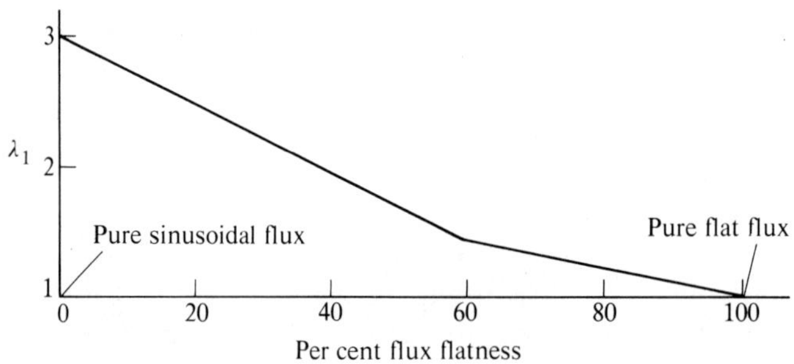

Figure 7-6 Lowest mode buckling λ_1 versus one-dimensional flux flatness.[11]

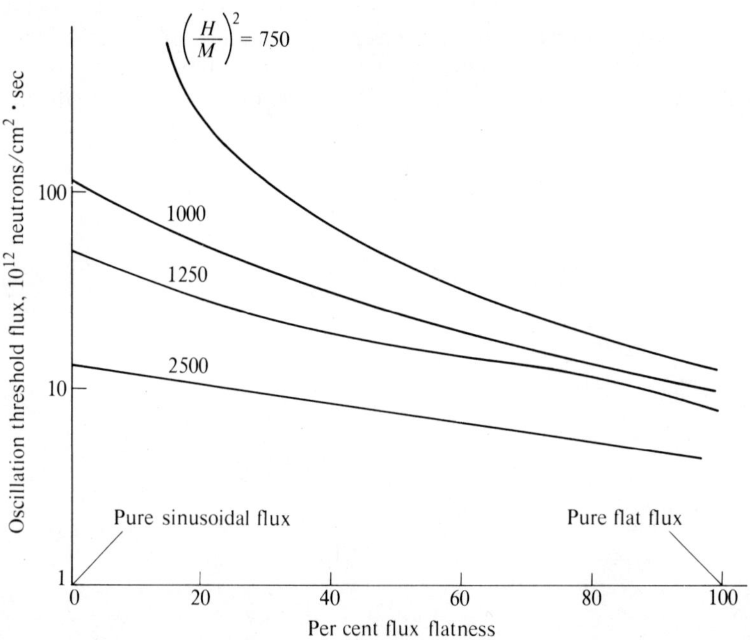

Figure 7-7 Oscillation threshold flux as a function of flatness for various one-dimensional reactor heights.[11]

Equations (7-78) and (7-80) yield a cubic equation in ϕ^0_{mm}, which is essentially the equilibrium flux level whose leading term is $\alpha_T(\phi^0_{mm})^3$. For $|\alpha_x|$ and $|\alpha_T|$ both small, $|\alpha_T/\alpha_x| \cong 1$, and $\alpha_T < 0$, this cubic term is considered negligible, resulting in a quadratic equation in ϕ^0_{mm}. The two roots represent the upper and lower flux thresholds that bound the region of sustained flux oscillations. The upper bound corresponds to the fact that the equilibrium xenon approaches its asymptotic value with increasing flux which is in the neighborhood of 10^{15} neutrons per square centimeter per second. This can be seen by investigating Eq. (7-60) for $\partial X/\partial t\,(\mathbf{r}, 0) = 0$ in the limit of very high flux.

The lower bound is due, as mentioned before, to the negligible effect of the xenon at low flux levels. The rather large flux upper bound implies that the stability of a high-flux reactor can be maintained, in so far as xenon is concerned, by a small negative reactivity power coefficient.

Consider a one-dimensional pile of height H, depicted in Fig. 7-5, for various degrees of flux flatness. The lowest eigenvalue, λ_1, is plotted as a function of the degree of one-dimensional flux flattening in Fig. 7-6.

Figure 7-7 depicts the flux level threshold required for sustained oscillations as a function of the pile size measure $(H/M)^2 = (\pi/MB_g)^2$. Since $b_m = 0$ in Eq. (7-75) implies sustained oscillations, the period of oscillation is given by $T_m = 2\pi/(c_m)^{1/2}$. Equation (7-74) yields for the period of the lowest mode, T_1, the relation

$$T_1 = 2\pi\left[\frac{\gamma_1\lambda_1\Sigma_f(\lambda + \sigma\phi^0_{11})}{(\sigma X^0_{11} - \gamma_2\Sigma_f)} - \lambda_1^2\right]^{-1/2} \tag{7-81}$$

which is plotted in Fig. 7-8.

From Figs 7-7 and 7-8 it can be appreciated that "flux tilt" oscillations are prevalent in large, high-power, thermal reactors that have a high percentage of

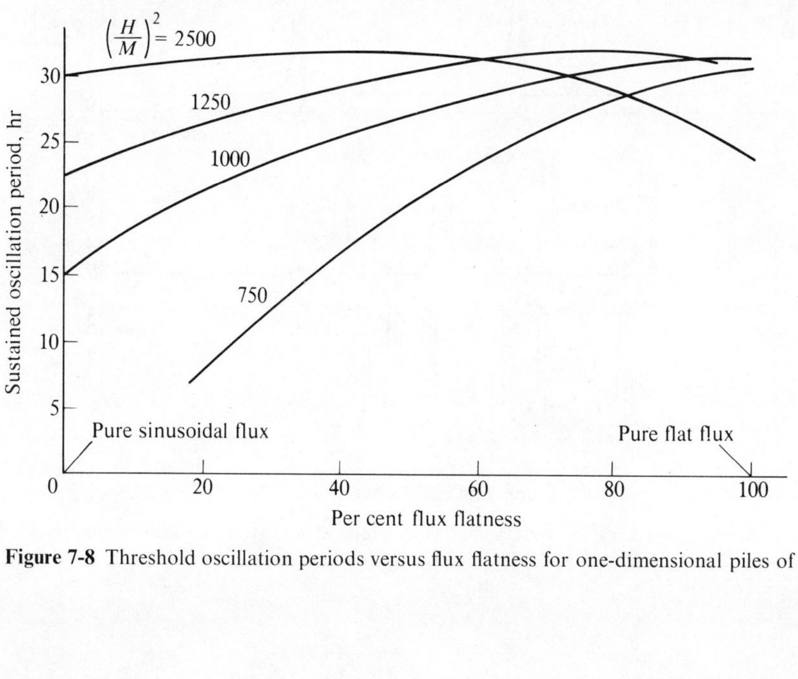

Figure 7-8 Threshold oscillation periods versus flux flatness for one-dimensional piles of height H.[11]

flux flatness. It is seen that the oscillation period increases with flux flattening. What happens in such reactors is that a positive perturbation in flux on one side of the reactor increases the local xenon burnout rate immediately. The decrease in negative reactivity due to the burnout enhances this effect as the flux continues to increase. However, the increase in flux increases the fission product ^{135}I production rate, which subsequently increases the xenon concentration. The appearance of this new xenon begins to shift the positive reactivity to the other side of the reactor where comparatively less xenon is being formed. This completes the first half-cycle of oscillation, which now begins the second half-cycle in the same way on the other side of the reactor. If $b_m < 0$ in Eq. (7-75), this corresponds to divergent oscillations in this approximation. Actually, this pushes the oscillation amplitude into the large-amplitude nonlinear regime where the foregoing results are no longer valid. Chernick[12] has investigated this regime and has shown that nonlinear oscillations of the relaxation type ensue. He begins with the *space-independent* counterparts of Eqs (7-60) and (7-61) for xenon and iodine and eliminates the iodine terms between the two equations, resulting in a second-order differential equation in the xenon concentration given here for completeness (relaxation oscillations are not flux tilt oscillations).

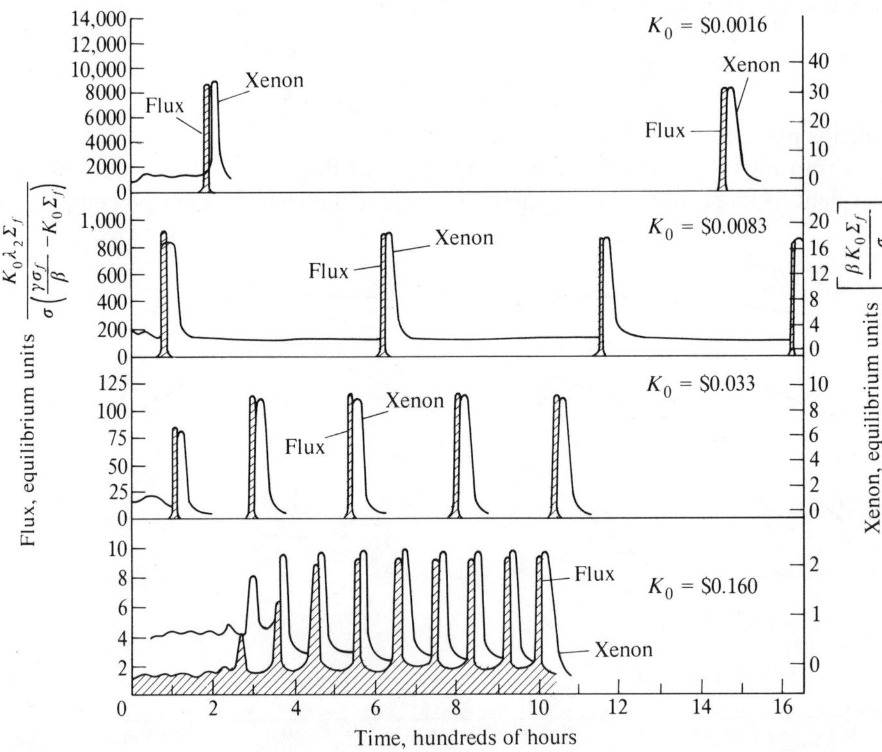

Figure 7-9 Xenon and flux relaxation oscillations. (Free-running multivibrator-like oscillations, of very long period, where K_0 is the trigger pulse.) For various initial step reactivities K_0.[12]

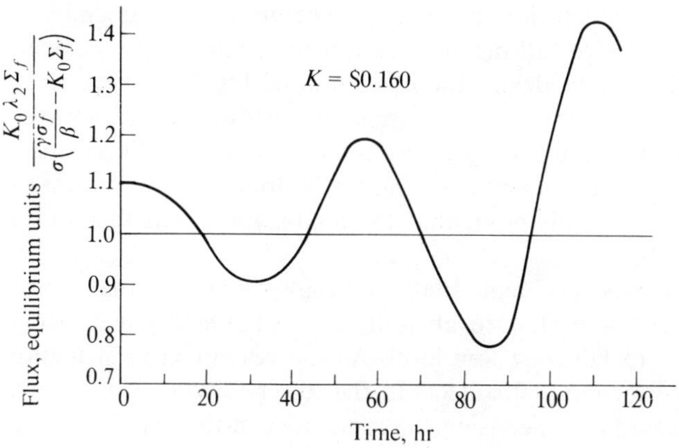

Figure 7-10 Small-amplitude xenon oscillations calculated from linearized equations.[12]

$$\ddot{X}(t) + \dot{X}(t)(\lambda + \sigma\phi(t)) + X(t)(\lambda_1(\lambda_2 + \sigma\phi(t)) + \dot{\phi}(t)) = \Sigma_f(\gamma_2\dot{\phi}(t) + \lambda_1\gamma\phi(t)) \tag{7-82}$$

As discussed earlier, delayed neutron periods are short compared to xenon oscillation periods. Therefore, to account for the asymptotic effects of the delayed neutrons, simply neglect the delayed neutron emitter terms in the kinetics equations, but extend the prompt neutron lifetime, as discussed in Chap. 3, to correspond to $\bar{\tau}$, the mean life of the delayed neutrons, so that the flux is assumed to satisfy

$$\bar{\tau}\dot{\phi}(t) = \left(K_0 - \frac{\sigma K}{\Sigma_f}\right)\phi(t) \qquad \phi(t) = nv_{\text{th}} \tag{7-83}$$

The additional reactivity due to the xenon is contained in the second term on the right-hand side of Eq. (7-83). $\sigma X/\Sigma_f$ is ratio of thermal neutrons absorbed by xenon to those absorbed by the fuel. This is easily shown to be essentially the reactivity due to the xenon, assuming a homogeneous system, as mentioned in Chap. 6. When this reactivity equals K_0, the reactivity due to all other causes, the reactor is critical.

Then Eqs (7-82) and (7-83) are combined and linearized to provide a threshold for sustained oscillations. For the divergent oscillation regime, second-order approximations[12] reveal the existence of relaxation oscillations in the flux and xenon as shown in Figs 7-9 and 7-10.

7-5 PROMPT NEUTRONS AND EXPLODING CORES IN HYPOTHETICAL ACCIDENTS

Another interesting case of the coupling of space and time kinetics is that of "exploding" critical assemblies, studied with respect to safety hazard and accident analysis. At this time it is rather apparent that this kind of analysis and certain aspects of nuclear weapons considerations have much in common. A disassembling nuclear device is similar to a bare "all metal"-fueled reactor having an accident. It is well known that in a nuclear device the conventional high-explosive (HE) periphery implodes an assemblage of fissionable material surrounded by a tamper which provides mass inertia at the instant of maximum implosion. This supercritical assembly now begins to disassemble owing to the tremendous amount of fission energy generated. It is at this point that the analogy between this and a reactor accident occurs.

The *improbable worst type* accident analysis usually assumes that some external agent, e.g., a coolant pump failure, abruptly causes the heat power being extracted from the reactor to fall to a low level. As the reactor core structure begins to melt thereby, the melted fuel collects in the lower regions of the core, thereby weakening and subsequently causing the remainder of the core structure to collapse on top of the already melted fuel. This is assumed to produce a supercritical configuration at a rapid (gravity-accelerated) reactivity insertion rate. Owing to the tremendously large pressures also generated at supercriticality, the structure begins to disassemble with the reactivity then being reduced at a prodigious rate.

The gigantic power densities attained because of prompt supercritical multiplication allow the neglect of delayed neutrons, since they simply do not appear in time to affect the neutron kinetics. In terms of reactivity measured in dollars over prompt critical

$$\dot{n}(t) = K_p(t)n(t) \qquad n(0) = 1 \qquad K_p(0+) = K_0 \tag{7-84}$$

where $K_p(t)$ is the reactivity due to prompt neutrons only, with respect to reactivities above prompt critical. A nuclear accident corresponds to

$$K_p(t) = K_0 - K_{\text{ex}}(t) \tag{7-85}$$

K_0 is the reactivity inserted by the accidental implosion or core collapse, reckoned in dollars over prompt critical, and it will be approximated by a positive step function.[14] $-K_{\text{ex}}(t)$ is the reactivity decrease due to disassembly of the supercritical mass because of the large amounts of energy and tremendous pressures that are generated when K_0 is inserted. Usually $K_{\text{ex}}(0-) = 0$ and $K_0(0-) = 1$, which is to say that, in this context, the reactor is prompt critical initially with the neutron population multiplying at a very rapid exponential rate.

A number of reactivity withdrawal models will be discussed;[14,15] they will describe how such reactivity depends on the energy generated during the implosion. This implies various dependencies between the pressure and energy generated, i.e., various equations of state. The first is that, for small times following maximum implosion, the mass inertia of the disassembly is such that its outward momentum lags the heat transfer, which causes thermal expansion of the constituents. This expansion of the assembly is proportional to the heat energy generated, $Q = \int_0^t n(\tau)\,d\tau$, so that

$$-K_{\text{ex}} = -\alpha Q \tag{7-86}$$

Then, simply,

$$\dot{n} = \left[K_0 - \alpha \int_0^t n(\tau)\,d\tau\right] n(t) \qquad n(0+) = 1 \qquad K_p(0+) = K_0 > 1 \tag{7-87}$$

A characteristic equation that typifies the kinetic behavior of these models is written in terms of the reactivity K_p as a function of the fission energy generated. For this case, from Eqs (7-85) and (7-86), such an equation is

$$K_p(t) = K_0 - \alpha Q \qquad \text{or} \qquad \dot{K}_p(t) = -\alpha n(t) \tag{7-88}$$

To obtain an expression for the neutron density under these circumstances, Eq. (7-87) is differentiated to obtain

$$\ddot{n}(t) = \frac{\dot{n}^2}{n} - \alpha n^2 \qquad n(0) = 1 \qquad \dot{n}(0+) = K_0 \tag{7-89}$$

This is easily solved to give

$$n(t) = \frac{K_0^2}{2\alpha}\operatorname{sech}^2\frac{K_0}{2}(t + c) \qquad c = \frac{4}{\alpha K_0}\ln K_0 \qquad \alpha \ll K_0^2 \tag{7-90}$$

This equation is quite similar to that obtained in the analysis of pulsed reactors discussed in Sec. 9-9. An equation in the prompt reactivity can be obtained by eliminating the neutron density between Eqs (7-86) to (7-88) to give

$$\ddot{K}_p = K_p \dot{K}_p \qquad K_p(0) = K_0 \qquad \dot{K}_p(0) = -\alpha \tag{7-91}$$

The maximum neutron density n_m is reached when $\dot{n}$ vanishes. Equation (7-81) with $\dot{n} = 0$ implies $K_p(t_m) = 0$, where t_m is the time at which the maximum occurs. A first integral of Eq. (7-91) is

$$\dot{K}_p + \alpha = \tfrac{1}{2}(K_p^2 - K_0^2) \tag{7-92}$$

If the neutron density from Eq. (7-88) is inserted above, the result is

$$n(t) = 1 - \frac{1}{2\alpha}(K_p^2 - K_0^2) \tag{7-93}$$

so that, when $n(t)$ reaches its maximum, $K_p(t_m) = 0$ and

$$n_m = 1 + \frac{K_0^2}{2\alpha} \qquad Q_m = \frac{K_0}{\alpha} \tag{7-94}$$

using Eq. (7-88). The solution of Eq. (7-92) is

$$K_p(t) = (K_0^2 + 2\alpha)^{1/2}\tanh\frac{(K_0^2 + 2\alpha)^{1/2}}{2}(t_m - t) \tag{7-95}$$

t_m can be found above by using $K_p(0) = K_0$. It is

$$t_m = \frac{1}{K_0}\log\frac{2K_0^2}{\alpha} \qquad \text{for} \qquad \alpha \ll K_0 \tag{7-96}$$

Instead of the disassembly reactivity $-K_{ex}$ being proportional to the energy generated, assume that the negative reactivity *rate* is proportional to the energy generated. This models the situation where the fission energy expels a constant-velocity pressure shock wave which disassembles the supercritical mass at a constant rate, resulting in the following expression for $\dot{K}_{ex}$:

$$\dot{K}_{ex} = -\alpha\omega Q \tag{7-97}$$

where the parameter ω has dimensions of inverse time. Then Eq. (7-84) becomes

$$\dot{n} = \left[K_0 - \alpha\omega \int_0^t Q(\tau)\, d\tau \right] n \tag{7-98}$$

In terms of the prompt reactivity, eliminating the neutron density as before results in

$$\dddot{K}_p = K_p \ddot{K}_p \qquad K_p(0) = K_0 \qquad \dot{K}_p(0) = 0 \qquad \ddot{K}_p(0) = -\alpha\omega \tag{7-99}$$

For very large amounts of inserted reactivity, the energy generated by the implosion can be of such magnitude as to accelerate strongly the above compressive shock wave. This acceleration is transmitted to the assembly, which now comes apart corresponding to a disassembly reactivity

$$-\ddot{K}_{ex} = -\alpha\omega^2 Q \qquad \text{that is} \qquad K_{ex} = \alpha\omega^2 \int_0^t \int_0^{t'} Q(\tau)\, d\tau\, dt' \tag{7-100}$$

and

$$\dot{n} = \left[K_0 - \alpha\omega^2 \int_0^t d\tau \int_0^\tau dx\, Q(x) \right] n \tag{7-101}$$

With Eqs (7-100) and $Q = \int_0^t n\, d\tau$, the reactivity differential equation for this case is

$$K_p^{(\mathrm{iv})} = K_p \dddot{K}_p \quad K_p(0) = K_0 \quad \dot{K}_p(0) = 0 \quad \ddot{K}_p(0) = 0 \quad \dddot{K}_p(0) = -\alpha\omega^2 \tag{7-102}$$

An important type of approximation is termed a threshold model. It is simply that a minimum amount of fission energy Q^* must be generated before the onset of disassembly. This approximation has been used effectively by Bethe and Tait[16] in reactor accident calculations. Later we shall discuss it in detail and apply it to a hypothetical fast reactor accident. The approximation, combined with the assumption that the disassembly reactivity is proportional to the excess energy generated over threshold, infers

$$K_{ex} = \begin{cases} 0 & Q < Q^* \\ \alpha(Q - Q^*) & Q \geq Q^* \end{cases} \tag{7-103}$$

where α is a suitable reactivity coefficient. For times $t < t^*$, that is, prior to the onset of the disassembly reactivity

$$n = e^{K_0 t} \qquad Q = \frac{e^{K_0 t} - 1}{K_0} \tag{7-104}$$

Then $Q = (n - 1)/K_0$ and $Q^* = (n^* - 1)/K_0$, so that

$$t^* = \frac{1}{K_0} \ln(1 + K_0 Q^*) \tag{7-105}$$

For $t \geq t^*$ any of the above models can be invoked. As mentioned, they are each obtained by assuming an appropriate empirical equation of state for the supercritical mass through the dependence of pressure on some power of the energy generated during the implosion. The models can also be manifest as particular cases of the generic reactivity expression

$$K_p^{(r)} = -\omega^r(\alpha Q)^s \tag{7-106}$$

where (r) signifies the rth derivative of K_p with respect to time. By redefining the time as $\tau = K_0 t/\Lambda$ and the neutron density as

$$N(\tau) = \frac{\alpha\omega^{r/s}}{K_0^{1/s}}\left(\frac{\Lambda}{K_0}\right)^{1+r/s} n(\tau)$$

the above is rewritten in dimensionless form as

$$\left(\frac{K_p}{K_0}\right)^{(r)} = -\left[\int_0^\tau N(x)\,dx\right]^s \tag{7-107}$$

with initial conditions $K_p(0)/K_0 = 1$ and $(K_p(0)/K_0)^{(r)} = 0$ for all r derivatives except for the $(r + s)$th derivative, which does not vanish, but is given by

$$\left(\frac{K_p}{K_0}\right)^{(r+s)} = -s!\,N^s(0) \tag{7-108}$$

The total energy generated in the accident is $Q_\infty = \int_0^\infty n\,d\tau$, or

$$Q_\infty = \frac{K_0^{1/s}}{\alpha\omega^{r/s}}\left(\frac{K_0}{\Lambda}\right)^{1+r/s} N_\infty \qquad N_\infty = \int_0^\infty N(x)\,dx \tag{7-109}$$

These expressions are plotted in Fig. 7-11, which shows how the neutron density (power) depends on the prompt reactivity from the beginning of reactivity insertion to the later stages of reactivity withdrawal.

The threshold model implies that following an accidental assembly, the heat and pressure generated are insignificant until a threshold heat energy Q^* is attained. Prior to disassembly, the reactivity insertion rate is sufficiently high that the total reactivity inserted is large enough to generate fission heat energy at an exceptionally large rate. The energy is being generated at such a stupendous rate that the inertia of the structure causes the disassembly to lag the heat generated until the threshold energy Q^* is reached.

The total positive reactivity inserted during an accidental assembly is severely limited, however. Even though a large amount of positive reactivity is available, it is not realized because disassembly begins "too soon," since assembly is usually due to gravity forces only.

This is opposed to a nuclear device assembly, where the constituents must be

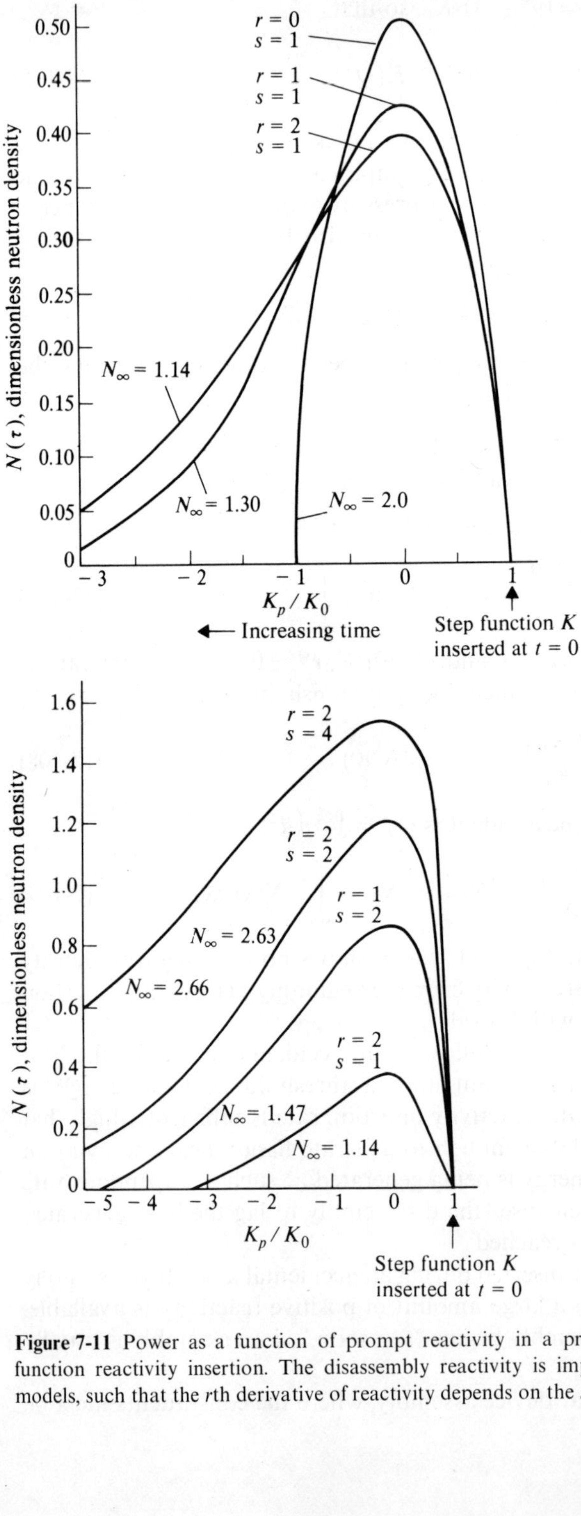

Figure 7-11 Power as a function of prompt reactivity in a prompt excursion initiated by a step function reactivity insertion. The disassembly reactivity is implied by various equations of state models, such that the rth derivative of reactivity depends on the sth power of energy generated.[14]

imploded together with HE, which is no mean feat. The reactivity reduction rate during reactor disassembly is much faster than the reactivity increase during the accidental assembly phase. If the reactivity insertion velocity is known, assuming linear reactivity insertion and given the reactor operating power prior to the accident, knowledge of the threshold power level (which generates the threshold pressure) is sufficient to determine the total amount of reactivity inserted before disassembly.

Let the reactivity insertion rate, in dollars per second, be termed $\dot{K}$. Then relatively long after prompt critical, but prior to disassembly, the neutron density (power) satisfies

$$\Lambda \dot{n}(t) \approx \dot{K} t n(t) \tag{7-110}$$

Assume that $\dot{K}$ is essentially constant during the time interval involved. Then the integral is

$$n \approx n(0)\, e^{1/2 \dot{K} t^2/\Lambda} \tag{7-111}$$

At $n = n(0)$, the reactor is assumed to be already at prompt critical and on the way to the maximum inserted reactivity, whatever that turns out to be. The amount of heat generated is (within some dimensionality constants),

$$Q(t) = \int_0^t n(\tau)\, d\tau = n(0) \int_0^t e^{1/2 \dot{K} \tau^2/\Lambda}\, d\tau \tag{7-112}$$

Since the times of interest in units of Λ are large, the desired asymptotic expansion is obtained by successively integrating Eq. (7-112) by parts to get, for a two-term expansion,

$$Q(t) \sim \frac{\Lambda n(0)}{\dot{K} t} e^{1/2 K t^2/\Lambda} \left(1 + \frac{\Lambda}{\dot{K} t^2}\right) \tag{7-113}$$

This can be solved for the time t^* when the threshold energy Q^* is reached. This time is given by successively iterating the first term above. It is

$$t^* \cong \left[\frac{\Lambda}{\dot{K}}\left(\ln \frac{Q^{*2} \dot{K}}{\Lambda n^2(0)} + \ln \ln \frac{Q^{*2} \dot{K}}{\Lambda n^2(0)} + \cdots\right)\right]^{1/2} \tag{7-114}$$

or the accident has caused the core to coalesce into a supercritical mass with a maximum inserted reactivity determined from Eq. (7-114) as

$$K^* = \dot{K} t^* \cong \left[\Lambda \dot{K}\left(\ln \frac{Q^{*2} \dot{K}}{\Lambda n^2(0)} + \ln \ln \frac{Q^{*2} \dot{K}}{\Lambda n^2(0)} + \cdots\right)\right]^{1/2} \tag{7-115}$$

The total reactivity inserted, K^*, is seen to depend only slowly on Q^*.

According to the model, when the threshold energy is attained ($Q = Q^*$), the core starts to disassemble owing to the tremendous hydrostatic pressure generated. The reactivity begins to decrease at a very rapid rate. Further, it can be shown that the reactivity is such that first-order perturbation theory is applicable, since the pressure shock wave has not had time to be expelled. The reactivity is then limited only by the mass inertia of the assembly. Also, the one-group diffusion equation is assumed to hold in the reactor prior to disassembly. That is

$$\nabla \cdot D\nabla\phi + [(\nu - 1)\Sigma_f - \Sigma_a]\phi = 0 \tag{7-116}$$

where Σ_f and Σ_a are the macroscopic fission and total absorption cross sections, respectively. From first-order perturbation theory,[17] assuming a spherical assembly, the resultant reactivity reduction is

$$\Delta K = K - K^* = -4\int_{\text{core}} r_0 D\phi'^2(r) r\, dr \Big/ \int_{\text{core}} \nu\Sigma_f \phi^2(r) r^2\, dr \tag{7-117}$$

where r_0 is the outward displacement of the exploding system. Let the "fission importance" in the core be given by (r_1 is the core plus tamper boundary and b is the core boundary, $0 < b < r_1$)

$$\zeta = \frac{\displaystyle\int_0^b \Sigma_f \phi^2 r^2\, dr}{\displaystyle\int_0^{r_1} \Sigma_f \phi^2 r^2\, dr} \tag{7-118}$$

with $\phi(r_1) = 0$, and with a parabolic approximation to the flux, that is, $\phi = 1 - (q^2/b^2)r^2$. Σ_{tr}/ρ_c (ρ_c is the core density) is assumed to be the same in the core and tamper. These quantities, as well as Σ_f, are sensibly constant, so that they may be factored from integrals where necessary. Inserting all of this into Eq. (7-117) yields

$$\Delta K = -\frac{16q^4\zeta \displaystyle\int_0^{r_1} \rho r_0 r^3\, dr}{\nu\Sigma_{tr}\Sigma_f \rho_c b^7(1 - \frac{6}{5}q^2 + \frac{3}{7}q^4)} \tag{7-119}$$

Now, the displacement r_0 satisfies the wave equation[18] on a per unit mass basis

$$\frac{\partial^2}{\partial t^2} r_0(r, t) = -\frac{1}{\rho}\frac{\partial p}{\partial r} + \frac{c^2}{r^2}\left[\frac{\partial}{\partial r}\left(r^2 \frac{\partial r_0}{\partial r}\right) - 2r_0\right] \tag{7-120}$$

where c is the velocity of the outward displacement wave and p is the pressure. As assumed previously, the outward pressure shock wave is usually much slower than the thermal shock wave. Then the diffusional terms are neglected as compared to the force term resulting from the thermal shock, so that the simplified pressure wave equation is

$$\frac{\partial^2}{\partial t^2} r_0 \cong -\frac{1}{\rho}\frac{\partial p}{\partial r} \tag{7-121}$$

Equation (7-119) is differentiated with respect to time twice, so as to eliminate r_0 by using Eq. (7-121), which yields after an integration by parts,

$$\ddot{K} = -\frac{48q^4 \int_0^{r_1} pr^2\,dr}{\nu\Sigma_{\text{tr}}\Sigma_f \rho_c b^7(1 - \frac{6}{5}q^2 + \frac{3}{7}q^4)} \tag{7-122}$$

Considering a particular spherical shell of volume $4\pi r^2\,dr$ of the exploding reactor, it is seen that $\ddot{K}$ is proportional to the total outward force. This expression can be used to calculate the reactivity reduction, as the disassembly proceeds, provided the pressure is known. The pressure is obtained from an equation of state, if it is known. Equations of state are usually associated with gases, e.g., for one mole of a perfect gas, $PV = RT$. However, equations of state for solids or semiliquids are usually manufactured empirically as needed. For example, for the Godiva† reactor accident,[19] the equation of state used in the calculation was

$$p(\mathbf{r}, t) = -0.545649 + 0.278460\theta(\mathbf{r}, t) + 0.02873\rho(\mathbf{r}, t) \tag{7-123}$$

where p = pressure, megabars (essentially mega-atmospheres)
θ = temperature, eV
ρ = density, g/cm^3

Another interesting set of equations of state for an exploding core, for $0.75 \le \rho/\rho_0 \le 1.0$, where ρ_0 is the density of the fuel at STP, is[19]

$$p = -0.54569 + 0.278460\theta(\mathbf{r}, t) + 0.02873\rho(\mathbf{r}, t) \qquad \text{solid} \tag{7-124a}$$

$$p = -2.99148 + 0.2470870\theta(\mathbf{r}, t) \qquad \text{transition} \tag{7-124b}$$

$$p = 0.56441 + 0.278460\theta(\mathbf{r}, t) - 0.538688(\rho_0/\rho) \qquad \text{liquid} \tag{7-124c}$$

These types of equations of state are usually used in reactor explosion codes written for digital computers. For "hand" calculations, much simplified equations of state are used for the sake of practicality. Two such are discussed now. The first is (1) if the neutron generation time (the time for the neutron population to increase e-fold) is much longer than the time for neutron acoustical waves to traverse the core, then a post-expansion semiequilibrium state is assumed attained. This corresponds to $\alpha^* \equiv K^*/\Lambda \ll c/b$, where c is the wave velocity and b is the core radius. The pressures are now negligible, so that the reactivity can be computed by using Eq. (7-119) with an assumed known displacement. The second, and more interesting case, mentioned earlier and discussed below, is (2) during the pre-expansion phase, where the heat is being generated so fast ($\alpha^* \gg c/b$) that the expansion lags, owing to mass inertia, to the point of being negligible. Then the pressure can be computed on a constant-volume basis, as in Eqs. (7-124), and the corresponding reactivity is available from Eq. (7-122). Consider the parabolic flux shape given earlier, and the energy generated within the core being separable as

$$E(\mathbf{r}, t) = Q(t)\phi(\mathbf{r})$$

† Godiva is a bare sphere of ^{235}U fabricated as a sandwich, with the two outer slices being retractable, while the center filling is separately supported.

and assumed zero outside the core. Then Eq. (7-122) can be rewritten as

$$\ddot{K} = \frac{-24q\zeta \int_{1-q^2}^{1} p(E)[1 - \phi(E)]^{1/2}\, d\phi(E)}{v\Sigma_{\text{tr}}\Sigma_f \rho_c b^4[1 - \frac{6}{5}q^2 + \frac{3}{7}q^4]} \tag{7-125}$$

where the integration is taken over the core only.

Invoking the threshold idea, in that the pressure is assumed zero until the threshold energy density Q^* is reached and then is linear with energy, gives a simplified empirical equation of state as[16]

$$p(E) = \begin{cases} 0 & E \leq Q^* \\ (\gamma - 1)\rho_c(E - Q^*) & E > Q^* \end{cases} \tag{7-126}$$

where γ is the usual ratio of specific heats. Inserting this pressure–energy relationship into the integrand of the reactivity equation gives, with Eq. (7-84)

$$\ddot{K} = 0, \qquad Q < Q^* \text{ threshold energy exists in no part of assembly} \tag{7-127a}$$

$$\ddot{K} = -\frac{16q^4\zeta(\gamma - 1)[(1 - 0.6q^2)Q - Q^*]}{v\Sigma_{\text{tr}}\Sigma_f b^4[1 - \frac{6}{5}q^2 + \frac{3}{7}q^4]},$$

$$Q^* < Q < \frac{Q^*}{1 - q^2} \text{ threshold energy exists in core only} \tag{7-127b}$$

$$\ddot{K} = -\frac{16\zeta(\gamma - 1)(\frac{2}{5}Q - Q^*)}{v\Sigma_{\text{tr}}\Sigma_f b^4[1 - \frac{6}{5}q^2 + \frac{3}{7}q^4]},$$

$$Q > \frac{Q^*}{1 - q^2} \text{ threshold energy exists in core plus tamper (blanket)} \tag{7-127c}$$

The above, together with the definition of $Q = \int_0^t n(\tau)\, d\tau$ in terms of the neutronics above prompt critical so that, from $n \sim \exp\left[\int_0^t K(\tau)\, d\tau/\Lambda\right]$

$$K = \Lambda \frac{d}{dt} \log \frac{dQ}{dt} \tag{7-128}$$

constitute the kinetics equations for a disassembling reactor.

Assume that the energy is being generated so rapidly that the the preponderant contribution to it is derived from the last few neutron generations above prompt critical for $Q > Q^*$. Then the asymptotic form of the energy density is given by

$$Q = Q^* e^{K^* t/\Lambda} \tag{7-129}$$

until the reactivity, as given by the appropriate equation (7-127), drops to zero. The total energy yield developed would simply be

$$Y(t_0) = 4\pi \int_{\text{core+tamper}} Q(t_0)\phi(r)r^2\, dr = \tfrac{8}{15}\pi b^3 Q(t_0) \tag{7-130}$$

where the parabolic form of $\phi(r)$ is used and the integration extends to the zero

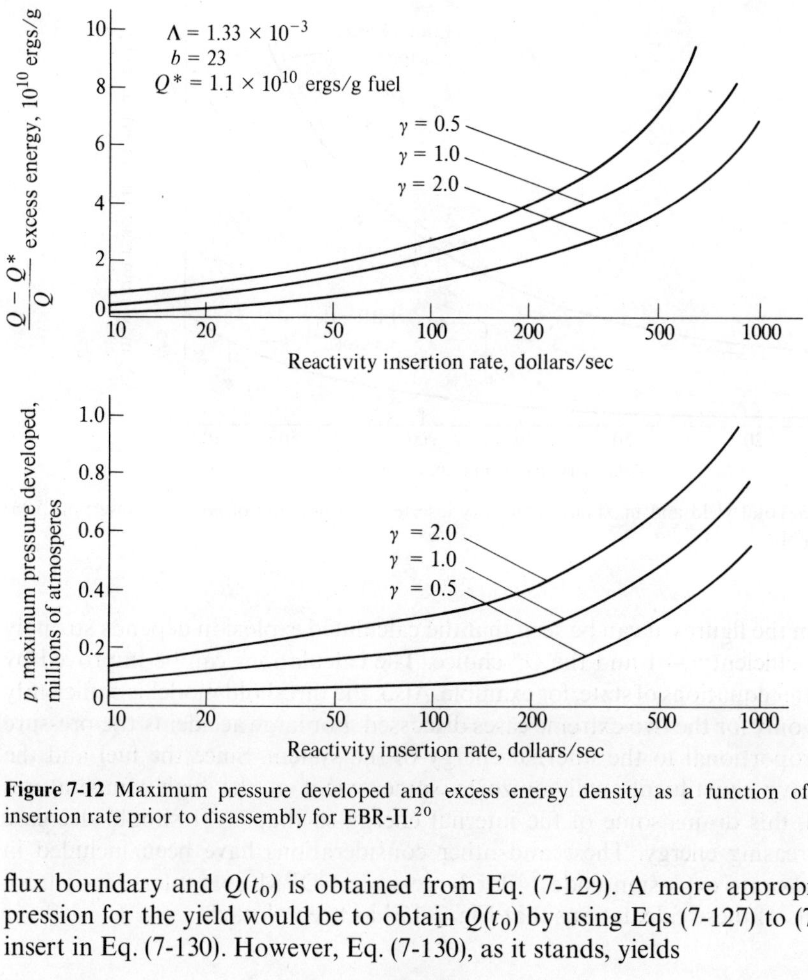

Figure 7-12 Maximum pressure developed and excess energy density as a function of reactivity insertion rate prior to disassembly for EBR-II.[20]

flux boundary and $Q(t_0)$ is obtained from Eq. (7-129). A more appropriate expression for the yield would be to obtain $Q(t_0)$ by using Eqs (7-127) to (7-129) to insert in Eq. (7-130). However, Eq. (7-130), as it stands, yields

$$Y \cong \tfrac{8}{15}\pi b^3 Q^* \, e^{(K^*/\Lambda)t_0} \tag{7-131}$$

or for the energy yield per unit volume

$$y = \frac{Y}{\frac{4}{3}\pi b^3} = \tfrac{2}{5} Q^* \, e^{(K^*/\Lambda)t_0} \tag{7-132}$$

Figures 7-12 and 7-13 depict the excess energy density $(Q - Q^*)/Q$, pressure developed, and total yield, all following disassembly as a function of the pre-expansion reactivity insertion rate. These are based on calculations using Eqs (7-127) to (7-129) following hypothetical accidents to both the Enrico Fermi and EBR-II reactors done on EDPM computers.†

† The Enrico Fermi reactor is a sodium-cooled fast reactor near Detroit, Michigan. The EBR-II, an experimental breeder reactor, is also a sodium-cooled fast reactor at the U.S. National Reactor Testing Station, near Arco, Idaho.

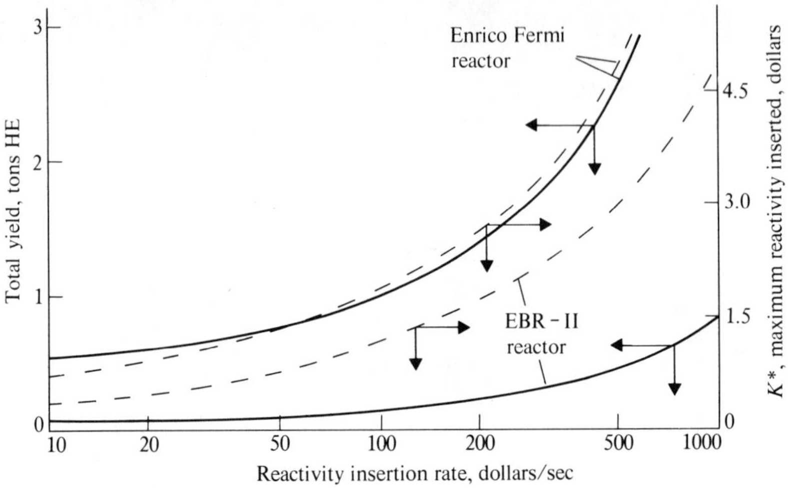

Figure 7-13 Total yield and maximum reactivity inserted as a function of reactivity insertion prior to disassembly ($\lambda = 2$).[21]

From the figures, it can be seen that the calculated explosion depends strongly on the coefficient $\gamma - 1$ and the Q^* choice. The calculations can be improved by using better equations of state, for example. Also, the threshold model is sufficiently accurate only for the two extreme cases discussed. For large accidents the pressure is not proportional to the internal energy of the system. Since the fuel and the other heavy constituents will ionize to some extent at the high temperatures involved, this drains some of the internal energy so that $\gamma - 1$ should decrease with increasing energy. These and other considerations have been included in some elaborate explosion codes[22] for large-scale EDPM computers. It is from the results of these calculations that Fig. 7-13 has been plotted.

7-6 SPACE–TIME KINETICS

As must be realized by now, the raw point kinetics model falls short of a proper description of reactor kinetics for many important situations. Large, relatively loosely coupled reactor kinetic behavior resulting from xenon induced oscillating flux tilts is a case in point. For a reasonably detailed and accurate space–time kinetic description of such a large reactor, rather sophisticated computer codes are a vital necessity. Attempts at "back of the envelope" computations to obtain detailed space–time kinetic descriptions is futile for essentially all real reactors. For these reasons, most of the following will consist of narrative descriptions, with adequate references, of the status of space–time kinetics computational methods, as opposed to taking a specific reactor undergoing control rod motion, for example, and investigating its space–time kinetics in depth. Hence, emphasis will lie in the areas of the computer oriented methodology. Part of the following is a paraphrasal of the work of W. M. Stacey JR.[23]

First, finite differencing methods for reducing the differential equations appropriate to the kinetics detail sought, are many and well known over the past three decades, and will be dealt with at this juncture in passing.

The most practical computationally reasonably accurate model of this type of reactor kinetics is that associated with the energy group diffusion approximation, which will immediately lead into what is called the modal expansion approximation. The general time-dependent group diffusion for the flux of the gth energy group can be written as

$$\nabla \cdot D^g(\bar{r},t)\nabla\phi^g(\bar{r},t) - \{\Sigma_a^g(\bar{r},t) + \Sigma_s^g(\bar{r},t)\}\phi^g(\bar{r},t) + \sum_{g' \neq g}^{G} \Sigma_s^{g'/g}(\bar{r},t)\phi^g(\bar{r},t)$$

$$+ (1-\beta)\chi_p^g \sum_{g'=1}^{G} \Sigma_f^{g'}(\bar{r},t)\phi^{g'}(\bar{r},t) + \sum_{i=1}^{6} \lambda_i \chi_i^g C_i(\bar{r},t) + Q^g(\bar{r},t) = \frac{1}{vg}\frac{\partial}{\partial t}\phi^g(\bar{r},t) \tag{7-133}$$

where the symbols have their usual meanings, and that $\Sigma_s^{g'/g}$ is the probability per second per unit flux that a neutron in group g' is scattered into group g, and $g = 1, 2, \ldots G$. G is the number of energy groups used in the particular approximation at hand.

The delayed neutron precursor concentrations satisfy

$$\frac{\partial}{\partial t}C_i(\bar{r},t) = -\lambda_i C_i(\bar{r},t) + \beta_i \sum_{g=1}^{G} v^g\Sigma_f^g(\bar{r},t)\phi^g(\bar{r},t) \tag{7-134}$$

Boundary conditions for the above system include the usual flux and current continuity conditions across all interfaces, and also that $\phi^g(\bar{R},t) + d\nabla\phi^g(\bar{R},t) = 0$, where d is the extrapolation distance, and $\bar{R}$ is the radial boundary vector of the reactor. All of the above can be written in matrix vector notation as

$$(\nabla \cdot \bar{D}\nabla - \bar{R}_a - \bar{R}_s + \bar{S} + (1-\beta)\bar{\chi}_p\bar{F}^T)\bar{\phi} + \sum_{i=1}^{6} \lambda_i\bar{\chi}_i\bar{C}_i + \bar{Q} = \bar{V}^{-1}\frac{\partial}{\partial t}\bar{\phi} \tag{7-135}$$

where:

G × 1 COLUMN VECTORS	*G × G DIAGONAL MATRICES*
$\bar{\phi}$; group fluxes	$\bar{D}$; D^g elements
$\bar{\chi}_p$; prompt fission group spectra	$\bar{R}_a$; Σ_a^g elements
$\bar{\chi}_i$; delayed neutron spectra	$\bar{R}_s$; Σ_s^g elements
$\bar{F}$; $v^g\Sigma_f^g$ components	$\bar{V}^{-1}$; $(v^g)^{-1}$ elements
$\bar{Q}$; group sources	$\bar{S}$; $\Sigma_g^{g'/g}$ (not diagonal)

The delayed neutron precursor equations are written analogously as

$$\dot{\bar{C}}_i = -\lambda_i\bar{C}_i + \beta_i F^T\phi; \qquad i = 1, \ldots 6 \tag{7-136}$$

where $\bar{F}^T$ is the transpose (row vector) of the column vector $\bar{F}$.

Modal Approximations

A representative type of modal expansion approximation is that of expanding the flux vector into known functions (modes), as

$$\bar{\phi}(\bar{r}, t) \cong \sum_{m=1}^{N} \bar{a}_m(t)\bar{\phi}_m(\bar{r}) \tag{7-137}$$

where $\bar{r}$ stands for x, y, and z, and since it is not possible to satisfy Eq. (7-135) exactly with an expansion of this sort, the approximation symbol is used therein. When this expansion is substituted into the matrix equations above, and the resulting flux matrix equation is premultiplied by $\overline{W}_m$, a group weighting function diagonal matrix, and the precursor vector equation premultiplied by $\overline{W}_m\bar{\chi}_i$, and both integrated over the reactor volume, then for each of N different weighting function matrices, the result, for $m = 1, 2, \ldots N$; yields the set of coupled ordinary differential equations.

$$\sum_{n=1}^{N} [-\overline{A}_{mn}(t)\bar{a}_n(t) + (1-\beta)\overline{F}^T_{mn}(t)\bar{a}_n(t)] + \sum_{i=1}^{6} \lambda_i \overline{C}_{im}(t) + \overline{Q}_m(t) = \sum_{n=1}^{N} \tau_{mn}\dot{\bar{a}}_n(t) \tag{7-138}$$

and for the delayed neutron precursors

$$\frac{\partial}{\partial t}\overline{C}_{i,m}(t) = -\lambda_i \overline{C}_{i,m}(t) + \beta_i \sum_{n=1}^{N} \overline{F}^T_{i,mn}(t)\bar{a}_n(t), \; i = 1, \ldots 6; \; m = 1, \ldots N \tag{7-139}$$

where

$$\overline{A}_{mn}(t) = \int \overline{W}_m(\nabla \cdot \overline{D}\nabla - \overline{R}_a - \overline{R}_s + \overline{S})\bar{\phi}_n \, d^3\bar{r} \tag{7-140}$$

$$\overline{F}^T_{mn}(t) = \int \overline{W}_m \bar{\chi}_p \overline{F}^T \bar{\phi}_n \, d^3\bar{r} \tag{7-141}$$

$$\overline{F}^T_{i,mn}(t) = \int \overline{W}_m \bar{\chi}_i \overline{F}^T \bar{\phi}_n \, d^3\bar{r} \tag{7-142}$$

$$\tau_{mn} = \int \overline{W}_m \overline{V}^{-1} \bar{\phi}_n \, d^3\bar{r} \tag{7-143}$$

$$\overline{C}_{i,m}(t) = \int \overline{W}_m \bar{\chi}_i \overline{C}_i \, d^3\bar{r} \tag{7-144}$$

$$\overline{Q}_m(t) = \int \overline{W}_m \overline{Q} \, d^3\bar{r} \tag{7-145}$$

The set of differential equations (7-138) and (7-139) is solved for the coefficient matrix $\bar{a}_n$, and the $\overline{C}_{im}$ matrix. Then the neutron flux is available from its expansion given in Eq. (7-137). The above differential equation solutions are those that satisfy the expansion equation (7-135) in a weighted integral calculus of variational sense, over the reactor; not at every point in or on the reactor. The choice of expansion functions (modes) and weights is important in the implementation of the modal expansion methods. This will be discussed in detail later.

Similarly to the preceding method is the space–time synthesis method. This is exemplified by a cylindrical reactor with annular rods coming down from the top, using, $\phi^g(r, z, t) = a^g_{unr}(z, t)\phi^g_{unr}(r) + a^g_{rod}(z, t)\phi_{rod}(r)$ where ϕ^g_{unr} is the flux in the bottom unrodded section of the reactor cylinder, and ϕ^g_{rod} is that in the top (rodded) section of the reactor.

Now Eqs (7-135) and (7-136) may be written as

$$\overline{H}\overline{\Psi} + \overline{Q} = \overline{T}\dot{\overline{\Psi}} \tag{7-146}$$

where $\overline{\Psi}$ is a column vector of group eigenfunctions (natural mode fluxes). If $\overline{\Psi}$ is assumed separable into space and time, where the natural mode fluxes ϕ_n are the spacial parts, then without the source $\overline{Q}$, one obtains a corresponding eigenvalue relation

$$\overline{H}\overline{\phi}_n = \lambda_n \overline{T}\overline{\phi}_n \tag{7-147}$$

where λ_n are the eigenvalues. The natural modes satisfy, in Dirac notation

$$\langle \phi_l^\dagger \,|\, T \,|\, \phi_m \rangle \equiv \langle l \,|\, \overline{T} \,|\, m \rangle = 0 \tag{7-148}$$

where the leftmost index signifies the adjoint mode vector, the rightmost index signifies the forward mode vector, and the adjoint fluxes $\phi_l^\dagger$ satisfy the eigenvalue relation

$$\overline{H}^\dagger \overline{\phi}_l^\dagger = \lambda_l \overline{T}\overline{\phi}_l^\dagger \tag{7-149}$$

Because of the orthogonality property, these equations are uncoupled, in that for each index

$$\langle n \,|\, \overline{T} \,|\, n \rangle \lambda_n \bar{a}_n(t) + \langle n \,|\, \overline{Q}' \rangle = \langle n \,|\, \overline{T} \,|\, n \rangle \dot{\bar{a}}_n(t) \tag{7-150}$$

where $\phi_n^\dagger$ is construed as the weighting function. Equation (7-150) implies that all the modes are uncoupled, since only a single mode index n is involved, which is referred to as the property of finality, Kaplan.[24]

The above is also related to what are called the omega modes, which are described in the following. In general if

$$\mathscr{L}\overline{\psi} = \frac{1}{v}\frac{\partial \overline{\psi}}{\partial t} \tag{7-151}$$

where $\mathscr{L}$ is, say, a transport operator, then if $\overline{\psi}$ is exponentially changing with time, that is, $\psi \sim \psi_r e^{r\omega_r t}$ then the omega modes ψ_r satisfy

$$\mathscr{L}\psi_r = \omega_r \psi_r \tag{7-152}$$

where ω_r are the eigenfrequencies of the omega modes. There also exists a bi-orthogonality relationship

$$\langle r \,|\, v^{-1} \,|\, m \rangle = 0 \tag{7-153}$$

which, however, does not evidently decouple the omega modes, that is, no finality exists in this case.

Nodal Approximations

The nodal approximation approach divides a reactor into cells of contiguous regions or nodes which are physical and neutronically coupled. Then equations like Eqs (7-133) and (7-134) are solved for each mode N_j, by using a group flux expansion similar to the preceding, of the form $\bar{\phi}(\bar{r},t) = \bar{\phi}_j(\bar{r})\bar{n}_j(t)$, where $\bar{\phi}_j$ is a diagonal matrix in the group modes $\phi_j^g(\bar{r})$ and $\bar{n}_j(t)$ is a column vector in the time-dependent group amplitudes $n_j^g(t)$. As in previous modal methods, equations for each mode are obtained by multiplying Eqs (7-135) by a suitable weighting function matrix $\bar{W}_j$, and Eq. (7-136) by $\bar{W}_j\bar{\chi}_i$, where $\bar{W}_j$ is a diagonal matrix of group weights $W_j^g(\bar{r})$. Following this by integration over a reactor node volume gives finally

$$\tau_j\dot{\bar{n}}_j(t) = \bar{\mathscr{L}}_j(t) - \bar{A}_j(t)\bar{n}_j(t) + (1-\beta)F_j^T(t)\bar{n}_j(t) + \sum_{i=1}^{6} \lambda_i \bar{C}_{i,j}(t) + \bar{Q}_j(t) \tag{7-154}$$

for each of the j nodes. Similarly for the delayed neutron precursors

$$\dot{\bar{C}}_{i,j}(t) = -\lambda_i \bar{C}_{i,j}(t) + \beta_i \bar{F}_{i,j}^T(t)\bar{n}_j(t) \tag{7-155}$$

where the definitions of the above matrices and vectors are very similar to those following Eqs (7-139), except that

$$\bar{\mathscr{L}}_j = \int d^2\bar{r}\, \bar{W}_j \bar{D}\nabla\bar{\phi}_j - \int d^3\bar{r}\nabla \bar{W}_j \cdot \bar{D}\nabla\bar{\phi}_j \tag{7-156}$$

Physically, $\bar{\mathscr{L}}_j$ is the net streaming of neutrons from node j within the reactor. For computational purposes

$$\mathscr{L}_j \approx \sum_i{}' \bar{\lambda}_{ij}(t)\{\bar{n}_i(t) - \bar{n}_j(t)\} \tag{7-157}$$

where $\bar{\lambda}_{ij}$ is a diagonal matrix whose elements λ_{ij}^g are coupling coefficients for neutrons of group g between the ith and jth node, and the summation prime indicates that the $i = j$ term is not included in the summation.

Collapsing of the individual groups to form new composite groups for reduction of the dimensionality of the system model can be accomplished for both modal and nodal methods. If all the groups are collapsed into one group (total collapsing), then $n_j^1 = n_j^2 = \cdots = n_j^G$ for each node. Then if the G Eqs (7-154) and (7-155) are added, the result is nearly the familiar point kinetics equations, namely

$$\frac{1}{v_j}\dot{n}_j(t) = \frac{K_j(t)-1}{v_j\Lambda_j(t)} n_j(t) + \sum_i{}' \lambda_{ij}(t)(n_i(t) - n_j(t)) + \sum_i \lambda_i C_{ij}(t) + Q_j(t) \tag{7-158}$$

$$\dot{C}_{i,j}(t) = -\lambda_i C_{i,j}(t) + \frac{n_j(t)}{v_j\Lambda_{i,j}(t)} \tag{7-159}$$

where

$$v_j^{-1} = \int d^3\bar{r}\, \overline{W}_j^T \overline{V}^{-1} \overline{\phi}_j$$

$$\lambda_{ij} = \sum_g \lambda_{ij}^g$$

$$K_j = \frac{1}{\beta} \int d^3\bar{r}\, \overline{W}_j^T \{\bar{\chi}_p \overline{F}^T - \overline{R}_a - \overline{R}_s + \overline{S}\} \overline{\phi}_j \Big/ \int d^3\bar{r}\, \overline{W}_j^T \bar{\chi}_p \overline{F}^T \overline{\phi}_j$$

$$\Lambda_j^{-1} = \beta v_j \int d^3\bar{r}\, \overline{W}_j^T \bar{\chi}_p \overline{F}^T \overline{\phi}_j$$

$$\Lambda_{i,j}^{-1} = \beta v_j \int d^3\bar{r}\, \overline{W}_j^T \bar{\chi}_i \overline{F}^T \overline{\phi}_j$$

$$Q_j = \int d^3\bar{r}\, \overline{W}_j^T \overline{Q}$$

$$C_{i,j} = \int d^3\bar{r}\, \overline{W}_j^T \chi_i C_i$$

and that $\overline{W}_j$ and $\overline{\phi}_j$ are column vectors with elements W_j^g and ϕ_j^g, respectively, K_j is the nodal reactivity in dollars, Λ_j is the nodal prompt generator time, and $\Lambda_{i,j}$ is an effective prompt neutron generation time.

An example of the nodal approach is given in Chap. 8 for two coupled core reactors, where each reactor is considered as a node with respect to their combined reactor kinetic behavior. One principal difficulty with the nodal method is obtaining the coupling coefficients λ_{ij}. This is discussed in Stacey.[23]

Point Kinetics

If only the total power or neutron level in the reactor as a whole is desired, then a more severe extrapolation than the preceding can be made. Let the group flux be represented by

$$\phi(\bar{r}, t) = \sum_l \phi_l(r, t) n_l(t) \tag{7-160}$$

and insert this into Eqs (7-135) and (7-136). Then as before, pre-multiply Eq. (7-135) by a diagonal weight matrix with elements $W^g(\bar{r}, t)$, and Eq. (7-136) by $\overline{W}\bar{\chi}_i$. Integrate over the whole reactor volume to obtain

$$\tau(t)\dot{\bar{n}}(t) = -\overline{B}(t)\bar{n}(t) + (1-\beta)\overline{G}^T\bar{n}(t) + \sum_i \lambda_i \overline{C}_i(t) + \overline{Q}(t) - \overline{E}(t)\bar{n}(t) \tag{7-161}$$

$$\dot{\overline{C}}_i(t) = -\lambda_i \overline{C}_i(t) + \beta_i \overline{G}_i^T \bar{n}(t) \tag{7-162}$$

where

$$\tau = \int d^3\bar{r}\, \overline{W}\overline{V}^{-1}\overline{\phi}$$

$$\overline{B}(t) = \int d^3\bar{r}\, \overline{W}(-\nabla \cdot \underset{\cdot}{D}\nabla + \overline{R}_a + \overline{R}_s - \overline{S})\overline{\phi}$$

$$\overline{G}^T = \int d^3\bar{r}\, \overline{W}\bar{\chi}_p \overline{F}^T\overline{\phi}$$

$$\overline{G}_i^T = \int d^3\bar{r}\, \overline{W}\bar{\chi}_m \overline{F}^T\overline{\phi}$$

$$\overline{Q} = \int d^3\bar{r}\, \overline{W}\overline{Q}$$

$$\overline{C}_i = \int d^2\bar{r}\, \overline{W}\bar{\chi}_i C_i$$

$$\overline{E}(t) = \int d^3\bar{r}\, \overline{W}\overline{V}^{-1}\dot{\phi}$$

$\overline{E}(t)$ is similar to the reactor invariant obtained in Sec. 6-3 where the fundamental bases are brought to bear in the derivation of the point model reactor equations. If the space shape function is constant in time then the above expansion is only approximate, but $\overline{E} = 0$. Or $\overline{\phi}$ can be renormalized so that $\overline{E}$ vanishes also. Ordinarily, when $\overline{W}$ is time dependent, $\overline{E} = 0$ implies the normalization on $\overline{\phi}$ such that

$$E = \frac{\partial}{\partial t} \int d^3\bar{r}\, \overline{W}\overline{V}^{-1}\overline{\phi} = 0 \tag{7-163}$$

Total group collapsing here implies that Eqs (7-158) and (7-159) are added on j, the group index, to give

$$\frac{1}{v(t)}\dot{\bar{n}}(t) = \frac{K(t) - 1}{v(t)\Lambda(t)}\bar{n}(t) + \sum_i \lambda_i \overline{C}_i(t) + \overline{Q}(t) \tag{7-164}$$

and

$$\dot{\overline{C}}_i(t) = -\lambda_i \overline{C}_i(t) + \frac{\bar{n}(t)}{v(t)\Lambda_i(t)} \tag{7-165}$$

where

$$v^{-1}(t) = \int d^3\bar{r}\, \overline{W}^T\overline{V}^{-1}\overline{\phi}$$

$$\Lambda^{-1}(t) = \beta v(t) \int d^3\bar{r}\, \overline{W}^T\bar{\chi}_p \overline{F}^T\overline{\phi}$$

$$\Lambda_i^{-1}(t) = \beta v \int d^3\bar{r}\, \overline{W}^T \bar{\chi}_i \overline{F}^T \bar{\phi}$$

$$K(t) = \frac{1}{\beta} \int d^3\bar{r}\, \overline{W}^T (\bar{\chi}_p \overline{F}^T + \nabla \cdot D \nabla - \overline{R}_a - \overline{R}_s + \overline{S}) \bar{\phi} \Big/ \int d^3\bar{r}\, \overline{W}^T \bar{\chi}_p \overline{F}^T \bar{\phi}$$

$$Q(t) = \int d^3\bar{r}\, \overline{W}^T \overline{Q}$$

$$\overline{C}_i(t) = \int d^3\bar{r}\, \overline{W}^T \chi_i C_i$$

and the normalization $0 = E = \int d^3\bar{r}\, \overline{W}^T \overline{V}^{-1} \dot{\bar{\phi}}$. Again the usual kinetics parameters have the same physical definitions as before.

The weighting functions W to be used above are usually obtained from the adjoint flux corresponding to static initial equilibrium. Then $E = 0$ is automatically satisfied and the point kinetics represents that for a fixed spacial flux but an amplitude varying $n(t)$. An improvement in the above is the so called adiabatic point kinetics model, which is more flexible in that the shape function can change discontinuously during the transient amplitude change where $E = 0$ still holds. This is accomplished by using eigenfunctions obtained from "equilibrium" configurations chosen during the transient to redetermine the point model kinetics parameters, such as the prompt generation time and reactivity. The adiabatic

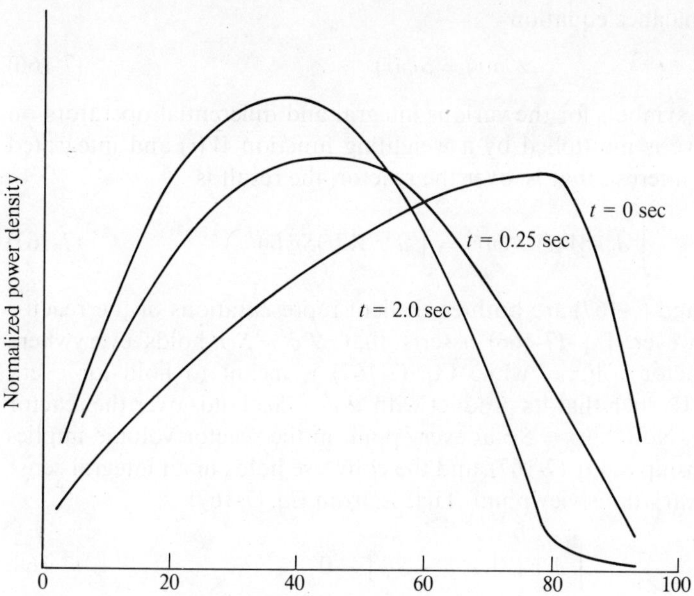

Figure 7-14 Power density distributions for slab reactor for two second reactivity ramp using an adiabatic computation. (*After Stacey.*[23])

model, of course, implies that the neutron flux distribution is in equilibrium with any sudden system changes, such as control rod motion. In so far as prompt neutrons are concerned this is a good assumption because control rod motion, for example, is very slow compared to the prompt generation time. However, for xenon induced spacial flux tilts, the relevant time constants are very long compared to the mean delayed neutron period (~ 10 sec), so that the delayed neutrons are not in equilibrium with the flux tilts, which former then fortuitously tend to retard the flux tilt changes. This effect increases with reactor size, that is as the flux tilt susceptibility becomes more pronounced. This is illustrated in Fig. 7-14, which depicts the neutron density in an eight foot high slab reactor using a two energy group diffusion model, Stacey.[23] From Fig. 7-14 it is seen that since the initial equilibrium power density differs from that at later times, the delayed neutrons must have retarded the power tilt appreciably. No retardation would imply that the equilibrium and dynamic power density flux shapes would remain coincident. A comparison of various computational methods for the same reactor is seen in Fig. 7-15, Stacey.[23]

Weighting Functions

Since weighting functions are used with almost all of the preceding methods an exposition of the theoretical aspects behind their use, as well as how to optimally choose them, is apropos here. In a general way, suppose that the flux, $\phi(\bar{r})$, satisfies a neutron balance equation

$$\mathscr{L}\phi(\bar{r}) = S\phi(\bar{r}) \tag{7-166}$$

where $\mathscr{L}$ and S are symbols for the various integral and differential operators on the flux. If the above is multiplied by a weighting function $\bar{W}(r)$ and integrated over the volume of interest, that is, over the reactor, the result is

$$\int d^3\bar{r}\; W(\bar{r})\mathscr{L}\phi(\bar{r}) = \int d^3\bar{r}\; W(\bar{r})S\phi(\bar{r}) \tag{7-167}$$

Equations (7-166) and (7-167) are both equivalent representations of the reactor flux behavior. However, Eq. (7-166) asserts that $\mathscr{L}\phi = S\phi$ holds everywhere throughout the reactor volume, while Eq. (7-167) is meant to hold for every weighting function W, such that its product with $\mathscr{L}\phi = S\phi$ holds over the reactor in an integral sense. Now, $\mathscr{L}\phi = S\phi$ at every point in the reactor volume implies the integral relationship of Eq. (7-167), and the converse holds in an integral sense from a calculus of variations viewpoint. That is, from Eq. (7-167)

$$\int d^3\bar{r}\; W[\mathscr{L}\phi - S\phi] = 0 \tag{7-168}$$

implies that if the approximation

$$\phi \simeq \sum_{m=1}^{N} a_m \psi_m(r) \tag{7-169}$$

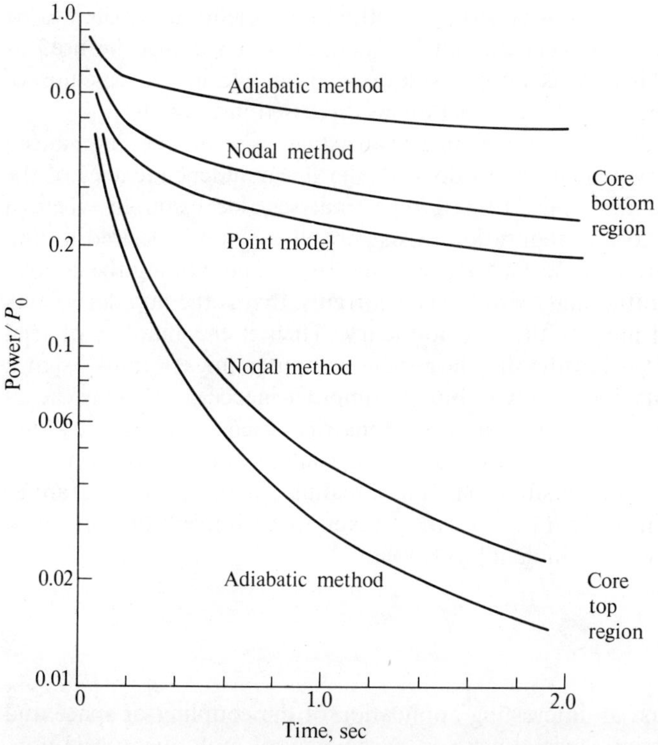

Figure 7-15 Comparison of various computational methods for the slab reactor. (*After Stacey.*[23])

were made in the integral, then

$$\sum_m a_m \int W[\mathscr{L}\psi_m - S\psi_m]\, d^3\bar{r} = \varepsilon \tag{7-170}$$

where ε is the error. So that if $\phi = \sum_{m=1}^{N} a_m \psi_m$ were the exact solution, then a_m could be chosen so that ε would vanish for all weighting functions W. However, the above expansion is only an approximation, so perhaps the integral can be made to vanish, or approach a minimum, for some class of functions W. If N such W's are chosen, this leads to N constraints which usually can be used to find the required a_m's.

To use the above method, which is to find the proper $\psi_m(\bar{r})$ to *synthesize* the above solution flux, judicious choices of these as well as the weighting functions must be made. Using the energy group approach, few trial functions, that is $\psi_m(\bar{r})$, are required to represent the fast flux since the latter is relatively smooth. For the thermal flux, each material (axial) zone should correspond to a trial function. For increased accuracy additional trial functions ψ_m can be used to span between different axial zones.

When choosing the weighting function, it is seen that many well-known approximation methods can be used. These include the method of moments, Gauss least squares, and collocation. In Galerkin's method the weights are chosen to be identical to the trial function. The adjoint flux in each zone can also be used as weighting functions. This stems from a stationary principle in the calculus of variations related to the importance function, as discussed in Chap. 6.

Because of the use of computer machinery, an artifact is sometimes introduced that can cause difficulty, which has to do with the linear independencies of the trial and/or weighting functions. In cell type analyses, for example where a particular control rod core configuration exists per cell, when it is desired to find the flux distribution through a section of a small cell region containing the control rod, the fast fluxes are rather flat. Also the cell currents, that is the flux derivatives may be small or vanishing at the cell boundary. Then if the number of trial functions that are used is inordinately large, the computer might "think" that a linear dependence exists due mainly to minor computer inaccuracies. This causes artificial singularities to be introduced into the matrices used in the energy group diffusion computation scheme, with concomitant roundoff and other computer errors. An indicator for such a situation can be manufactured in the program as part of the computations so that means can be taken to circumvent this difficulty. Such indicators are discussed in detail in Kaplan.[24]

7-7 NEUTRON BURSTS

It is well now to discuss an interesting application of the coupling of space and time kinetics to determine thermal neutron parameters of multiplying and non-multiplying media by using the neutron burst method.[25] The experiment consists in introducing pulses or bursts of fast or, if possible, slow neutrons into the medium under question and then counting the neutrons diffusing from the medium. One method of obtaining neutron bursts is to accelerate protons or deuterons onto a tritium target buried in the material to be tested. For example, the $T(d,n)He^4$ reaction produces 14-MeV neutrons which are available as bursts simply by modulating the deuteron beam. A second method is to obtain photo-neutron bursts by means of, for example, the $Be(\gamma,n)$ reaction. A beryllium target is inserted into the material to be tested, and the target is then irradiated by a modulated beam of γ rays obtained from an accelerator. The photoneutron energy is in the neighborhood of 2 MeV, which is near the mode of the fission spectrum as discussed in Chap. 1.

Since the medium parameters are obtained from the slowing-down and thermal neutron properties of the medium, it is necessary to solve the initial-value problem in the thermal flux given that it vanishes on the extrapolated boundaries. Initially, a burst of fast neutrons occurs in or near the test material. It is assumed that the resulting initial thermal flux distribution $\phi_0(\mathbf{r})$ occurs immediately. That is, the slowing-down time is neglected in comparison to the time scale implicit in the remainder of the discussion. This is a good approximation, since the slowing-down times are one to two orders of magnitude smaller than the thermal neutron diffusional times. Then, in a multiplying medium, the "initial" thermal flux following the burst begins to decay according to the diffusion equation

$$D\nabla^2\phi_0(\mathbf{r},t) + (\nu\Sigma_f p\overline{P}_\infty - \Sigma_a)\phi_0(\mathbf{r},t) = \frac{1}{v}\frac{\partial\phi_0}{\partial t}(\mathbf{r},t)\phi_0(\mathbf{r},0) = (\mathbf{r}) \tag{7-171}$$

where $\overline{P}_\infty$ is the Fourier transform of the slowing-down density, equivalent to the fast neutron nonleakage probability, discussed in Sec. 7-3. For mathematical convenience, the test material will be assumed to be in the shape of a parallelepiped of sides a, b, and c. The solution in Eq. (7-171) can be gotten easily by introducing three-dimensional finite Fourier transforms.[26] For example, the initial flux $\phi_0(\mathbf{r}) \equiv \phi_0(x, y, z)$ has such a transform defined as follows:

$$\overline{\phi}_0(u, v, w) = \int_0^a dx \int_0^b dy \int_0^c dz\, \phi_0(x, y, z) \sin\frac{u\pi x}{a} \sin\frac{v\pi y}{b} \sin\frac{w\pi z}{c} \tag{7-172}$$

with its inverse, the triple sum

$$\phi_0(x, y, z) = \frac{8}{abc}\sum_{u,v,w=1}^{\infty} \overline{\phi}_0(u, v, w) \sin\frac{u\pi x}{a} \sin\frac{v\pi y}{b} \sin\frac{w\pi z}{c} \tag{7-173}$$

Taking the finite Fourier transform of Eq. (7-171) with respect to the space variables only then gives the ordinary differential equation in $\overline{\phi}_0(u, v, w; t)$

$$\frac{d\overline{\phi}_0}{dt}(u, v, w; t) = -v[D\lambda_{uvw}^2 - (\nu\Sigma_f p\overline{P}_\infty - \Sigma_a)]\overline{\phi}_0(u, v, w; t) \tag{7-174}$$

where

$$\lambda_{uvw}^2 = \pi^2\left[\left(\frac{u}{a}\right)^2 + \left(\frac{v}{b}\right)^2 + \left(\frac{w}{c}\right)^2\right]$$

Its solution is simply

$$\overline{\phi}_0(u, v, w; t) = \overline{\phi}_0(u, v, w) \exp\left[-v(D\lambda_{uvw}^2 + \Sigma_a - \nu\Sigma_f p\overline{P}_\infty)t\right] \tag{7-175}$$

The inverse then gives the solution to Eq. (7-171) as

$$\phi_0(x, y, z; t) = \frac{8}{abc}\sum_{u,v,w=1}^{\infty} \overline{\phi}_0(u, v, w) \sin\frac{v\pi y}{b} \sin\frac{w\pi z}{c} \cdot \exp\left[-v(D\lambda_{uvw}^2 + \Sigma_a - \nu\Sigma_f p\overline{P}_\infty)t\right] \tag{7-176}$$

where $\overline{\phi}_0(u, v, w)$ is given by Eq. (7-172). Whatever the original flux distribution $\overline{\phi}_0(\mathbf{r})$ immediately following the burst happens to be, for long times following the burst the higher harmonics of this distribution have virtually disappeared. This leaves the fundamental mode which is the asymptotic form of Eq. (7-176), or

$$\phi_0(\mathbf{r}, t) \sim \frac{8}{abc}\overline{\phi}_0(1, 1, 1) \sin\frac{\pi x}{a} \sin\frac{\pi y}{b} \sin\frac{\pi z}{c} \exp\left[-v(D\lambda_{111}^2 + \Sigma_a - \nu\Sigma_f p\overline{P}_\infty)t\right] \tag{7-177}$$

with

$$\lambda_{111}^2 = \left(\frac{\pi}{a}\right)^2 + \left(\frac{\pi}{b}\right)^2 + \left(\frac{\pi}{c}\right)^2$$

and

$$\bar{\phi}_0(1, 1, 1) = \int_0^a \int_0^b \int_0^c \phi_0(x, y, z) \sin \frac{\pi x}{a} \sin \frac{\pi y}{b} \sin \frac{\pi z}{c}\, dx\, dy\, dz$$

The decay constant is given from Eq. (7-176) as

$$\alpha_{111} = v[D\lambda_{111}^2 + \Sigma_a(\text{mult}) - \nu\Sigma_f p\bar{P}_\infty] \tag{7-178}$$

where Σ_a(mult) is the total macroscopic cross section including the fuel. The modifier (mult), corresponding to a multiplying medium, has been added to delineate this fact. If the decay constant α_{111} can be measured as a function of the geometric buckling λ_{111}^2, then values of the parameters making up α_{111} can be determined. Therefore, following a burst, one counts, as a function of time, the neutrons emanating from the material being tested. When the curve of neutron count rate versus time has asymptotically become a straight line with negative slope showing that the higher-flux harmonics have died off, a measurement of

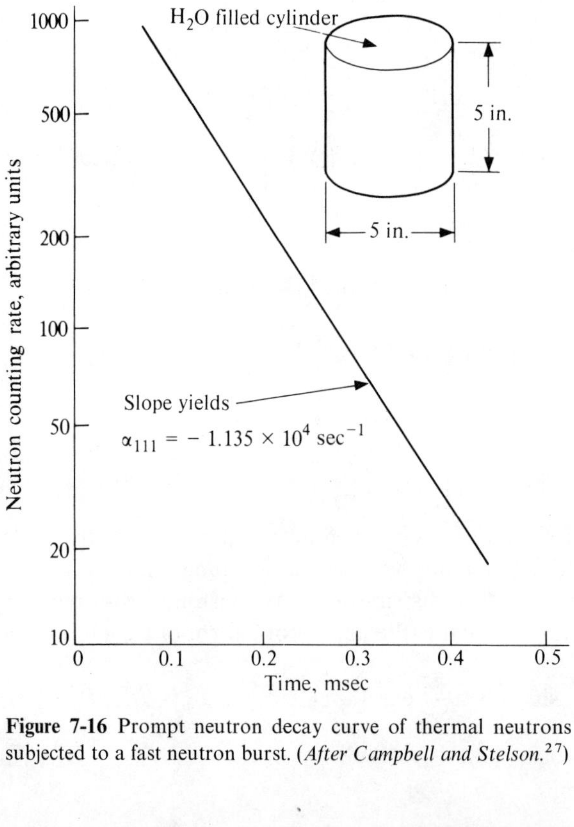

Figure 7-16 Prompt neutron decay curve of thermal neutrons diffusing from a cylinder of water subjected to a fast neutron burst. (*After Campbell and Stelson.*[27])

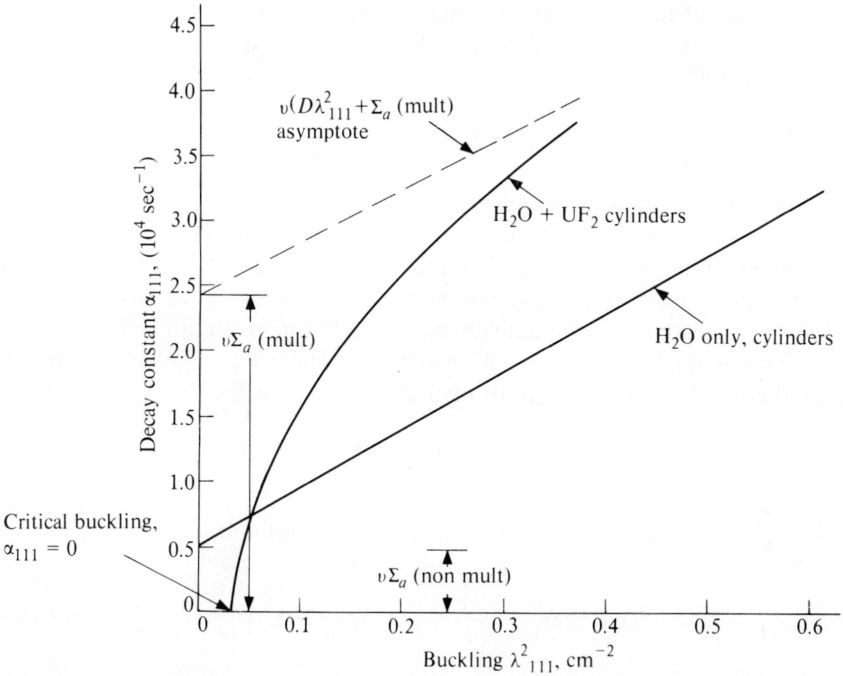

Figure 7-17 Prompt neutron decay constant as a function of buckling for thermal neutrons from H_2O-UF_2 cylinders following a burst of fast neutrons. (*After Meghreblian and Holmes.*[25])

the slope yields α_{111}. An example of such a curve is given in Fig. 7-16. This experiment is then repeated for various similar sizes of the same material so that a plot of α_{111} versus buckling λ_{111}^2 can be made. Such a plot is given in Fig. 7-17 for various cylinders of water with and without enriched uranyl fluoride in solution.

As can be seen from Eq. (7-178) in the limit of large buckling, which is tantamount to the limit of small size so that the multiplication diminishes in significance, the asymptotic decay constant is

$$\alpha_{111} \sim v[D\lambda_{111}^2 + \Sigma_a(\text{mult})] \tag{7-179}$$

which is plotted as a broken line in Fig. 7-17, where $v\Sigma_a$(mult) [and thereby Σ_a(mult)] is determined from its intercept. Dv is obtained from its slope. At the critical buckling $\alpha_{111} = 0$, which implies that the fundamental flux mode persists. Then from Eq. (7-178)

$$D\lambda_{111}^2 + \Sigma_a(\text{mult}) - v\Sigma_f p\overline{P}_\infty = 0 \tag{7-180}$$

which asserts that for criticality

$$k_{\text{eff}} \equiv \frac{v\Sigma_f p\overline{P}_\infty}{\Sigma_a(1 + L^2\lambda_{111}^2)} = 1 \tag{7-181}$$

For large blocks of material, the buckling is small so that the Fermi continuous slowing-down model is assumed valid. Then, with p assumed known, the fast nonleakage probability

$$\overline{P}_{\infty} \cong \exp\left(-\tau\lambda_{111}^2\right) \cong \left(\frac{\nu\Sigma_f p}{\Sigma_a}\right)^{-1} \qquad (7\text{-}182)$$

is determined by the above experiment.

It is assumed that the diffusion equation approximation is valid for the nonmultiplying situation as well. After the injection of a pulse of fast neutrons into the medium, the neutrons quickly become thermal, since the slowing-down times are short compared to the time during which the neutron diffuses at thermal energies. This is a good approximation for most materials of reactor interest. Then the thermal flux in the nonmultiplying medium satisfies

$$D\nabla^2\phi_0(\mathbf{r}, t) - \Sigma_a\phi_0(\mathbf{r}, t) = \frac{1}{v}\frac{\partial}{\partial t}\phi_0(\mathbf{r}, t) \qquad (7\text{-}183)$$

With $(\Sigma_a v)^{-1} = T_0$, the prompt neutron lifetime in an infinite medium, the above is rewritten as

$$Dv\nabla^2 n_0(\mathbf{r}, t) - \frac{1}{T_0} n_0(\mathbf{r}, t) = \frac{\partial}{\partial t} n_0(\mathbf{r}, t) \qquad (7\text{-}184)$$

where $\phi_0(\mathbf{r}, t) = vn_0(\mathbf{r}, t)$ and the neutron density is assumed separable as

$$n_0(\mathbf{r}, t) = \sum_m a_m\psi_m(\mathbf{r})n_m(t)e^{-t/T_0}$$

This, inserted into Eq. (7-184), yields

$$\nabla^2\psi_m(\mathbf{r}) + \lambda_m\psi_m(\mathbf{r}) = 0 \qquad m = 1, 2, \ldots \qquad (7\text{-}185)$$

$$n_m(t) = \exp\left[-\lambda_m \int_0^t vD(t')\,dt'\right] \qquad (7\text{-}186)$$

and the eigenfunctions vanish on the extrapolated boundary of the medium. This is the more usual method for solving the diffusion equation in this application, as opposed to the finite transforms used in the previous example. Then

$$n_0(\mathbf{r}, t) = \sum_m a_m\psi_m(\mathbf{r})e^{-\alpha_m t} \qquad (7\text{-}187)$$

with

$$\alpha_m t = \lambda_m \int_0^t vD(t')\,dt' + \frac{t}{T_0} \qquad (7\text{-}188)$$

The sequence of eigenvalues $\{\lambda_m\}$ are all negative in a nonmultiplying medium. As mentioned before, in a critical multiplying medium all the $\{\lambda_m\}$ are negative except one, which is zero. The latter corresponds to the persistent fundamental harmonic flux shape after all higher harmonics have been damped out. On the other hand, in a multiplying medium, the persistent harmonic is that with the algebraically largest λ.

Following a neutron burst, since $D = [3\Sigma_{tr}(v)]^{-1}$ is a function of the energy of neutrons, D assumes its asymptotic value D_0 as the system slows into the thermal energy neighborhood. Equation (7-188) then becomes

$$\alpha_m t = \lambda_m \tau + \frac{t}{T_0} \tag{7-189}$$

where τ is the neutron age[28] in the medium, assuming that continuous slowing down is applicable, i.e.

$$\tau = \int_0^t vD(t')\, dt' \tag{7-190}$$

After sufficient time has elapsed, which means that all of the neutrons have been thermalized and are now diffusing at thermal energies,

$$\alpha_1 = \lambda_1 v D_0 + \frac{1}{T_0} \tag{7-191}$$

since only the harmonic flux space mode corresponding to $m = 1$, the algebraically largest, still persists. Or

$$n_0(\mathbf{r}, t) \sim a_1 \psi_1(\mathbf{r}) e^{\alpha_1 t} \tag{7-192}$$

The eigenvalues $\{\psi_m\}$ are obtained from the solution to the boundary value problem in the Helmholtz equation (7-185). For a cube of side l, for example,

$$\lambda_1 = 3\left(\frac{\pi}{l}\right)^2 \tag{7-193}$$

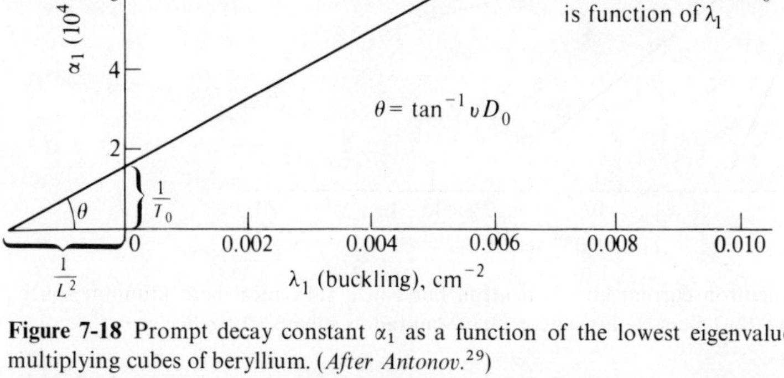

Figure 7-18 Prompt decay constant α_1 as a function of the lowest eigenvalue λ_1 for various non-multiplying cubes of beryllium. (*After Antonov.*[29])

Then by measuring α_1 experimentally for cubes of various sizes, as discussed earlier, a plot of α_1 versus λ_1 can be made. A plot for beryllium cubes is given in Fig. 7-18. From this plot, the parameters T_0, D_0, and L^2 can be obtained as shown. The y intercept is $1/T_0$; the absolute magnitude of the x intercept is $1/T_0D_0 = 1/L^2$; and the slope $\tan\theta = D_0v$. The experimentally determined curve is actually a parabola. This is because D_0 can be shown to be a decreasing function[30] of λ_1. As discussed in the earlier sections of this chapter, these same measurements can be made by using a source of periodically varying intensity.

This kind of an experiment easily lends itself to the measurement of the α, or e-folding time, of subcritical multiplying assemblies as well. Essentially, then, α is determined by introducing a short neutron pulse in a subcritical multiplying system and measuring the neutron (current) efflux therefrom as a function of time. The pulse neutron sources can be the same as discussed previously.

The experimental α is merely determined by measuring the slope of the curves after the initial pulse. For short pulse behavior, on the decay side, the neutron density $n \approx e^{-\alpha t}$, so that $\log n \approx -\alpha t$, which is plotted in Fig. 7-19 for spheres of various diameters.

Such experiments can also be used to measure control rod worths. This is similar to the danger coefficient method except that the transient kinetics are used (α's) as opposed to asymptotic kinetics (inhour equation) used in the latter method. Prior to the introduction of the neutron pulse, there are no fissions

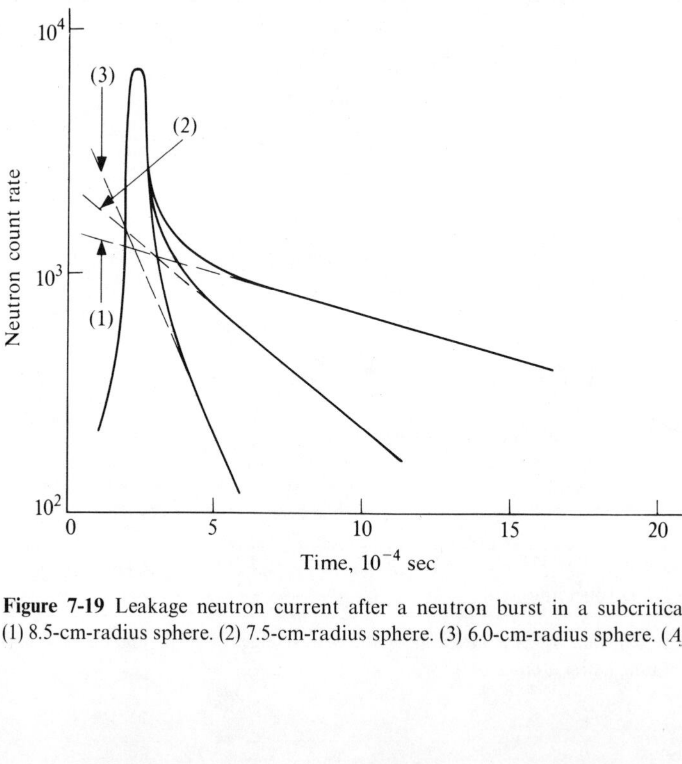

Figure 7-19 Leakage neutron current after a neutron burst in a subcritical bare uranium sphere. (1) 8.5-cm-radius sphere. (2) 7.5-cm-radius sphere. (3) 6.0-cm-radius sphere. (*After Bengston.*[28])

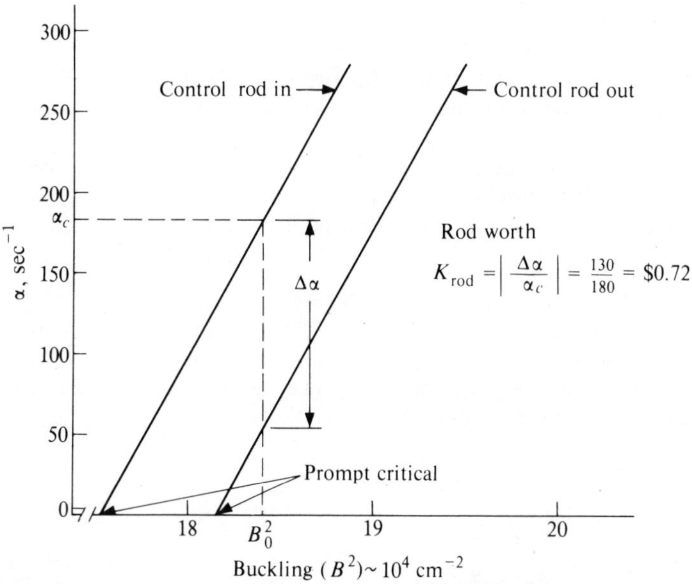

Figure 7-20 Variation of α near critical for bare parallelepipeds of graphite-moderated enriched uranium (C/U = 2,355). (*After Bengston.*[28])

occurring in the subcritical assembly, so that there are no delayed neutrons therein. Therefore, only $(1 - \bar{\beta})k_{\text{eff}}$ neutrons are produced per neutron absorbed; then

$$\dot{n}(t) = \frac{K - 1}{\Lambda} n(t) \tag{7-194}$$

so that

$$\alpha = \frac{K - 1}{\Lambda} \tag{7-195}$$

At "prompt" critical, $K = 1$ (\$1.00) and then $\alpha = 0$. At critical, $K = 0$ and then $\alpha_c = -1/\Lambda$. Combining these relationships yields

$$K = \frac{\Delta\alpha}{\alpha_c} \equiv \frac{\alpha_c - \alpha}{\alpha_c} \tag{7-196}$$

This relationship is used to determine control rod worths by measuring α for a series of subcritical assemblies of various sizes with and without a control rod absorber. The measurement is made only for weak absorbers (control rods of small reactivity) for an assembly near critical. The further limitations of this method are discussed in a problem at the end of the chapter.

PROBLEMS

7-1 Show from Eq. (7-1) that when the second time derivative term is neglected, the wavefront phenomenon disappears. Further, show that by neglecting the first time derivative term the residual wave property vanishes.

7-2 Make up a series of tables similar to Table 7-1 for the corresponding wave properties in moderators such as heavy water, light water, liquid sodium.

7-3 Obtain the three-dimensional inverse Fourier transform of the thermal flux as given in Eq. (7-22) as

$$\phi_0(\mathbf{r}) = \frac{1}{2\pi^2}\int_0^\infty \bar{Z}_0(\omega)\frac{\sin \omega r}{\omega r}\,\omega^2\,d\omega$$

where $\bar{Z}_0(\omega)$ is the Fourier transform of $\phi_0(\mathbf{r})$.

7-4 Derive the criticality relation from Eq. (7-26) and show that the three-dimensional Fourier transform of the slowing-down probability density is

$$\bar{P}(B_0^2) = 4\pi\int_0^\infty \bar{P}_\infty(r; E_{\text{th}})\frac{\sin B_0 r}{B_0 r}\,r^2\,dr$$

7-5 From Prob. 7-4 derive the general result: Eq. (7-28).

7-6 Reconstruct in detail the derivation leading up to the reactivity K introduced by an absorber as given by Eq. (7-38).

7-7 Trace the development yielding Eq. (7-50).

7-8 Calculate the characteristics of the $s = \mu = 0$ wave. Obtain the wavelength, velocity, and attenuation factor as functions of the excitation frequency by using an approximate form of the transcendental equation for the roots B_s. Apply this form to calculate B_0. *Hint:* The approximation to the transcendental equation involves an approximate form for $\bar{P}_\infty(B_s)$ for $B_s = B_0$ and B_0 is assumed very small. Use other approximations as required.

7-9 In a nuclear reactor accident, assume that the melted core is collapsing with an acceleration equal to that of gravity (32 ft/sec^2) and the reactivity introduced is given by $K = -\Delta r/r$, where r is the core radius (a fair approximation). Calculate the reactivity insertion rate during the first second $dK/dt = -(1/r)\,dr/dt$ in kilobucks per second. The original radius $r_0 = 50$ cm.

7-10 Using first-order perturbation theory, derive the reactivity change in the assembly to obtain $K = -4\int_{\text{core}} r_0 D\phi'^2 r\,dr / \int_{\text{core}} \nu\Sigma_f \phi^2 r^2\,dr$. r_0 is the outward material displacement.

7-11 In a nuclear reactor core meltdown accident analysis using the Bethe–Tait formulation, what is the justification for using merely perturbation theory to calculate the corresponding disassembly kinetic behavior?

7-12 Derive the expression for reactivity in terms of the volume integral of the pressure [Eq. (7-122)], following the derivation sketched in the text. Use the fact that $\rho_c/\rho \cong 1$, so that the fission importance in the core is

$$\zeta = \left[\frac{2}{5}\frac{\rho}{\rho_c}\frac{q^{-3}}{1 - 0.6q}\right]^{-1}$$

Eliminate q^{-3} from the reactivity expression by using ζ.

7-13 Obtain the more accurate expression for the excess energy density Q/Q^* generated in a fast-reactor accident, using the kinetics equations (7-27) for large values of maximum reactivity insertion K^*.

7-14 From the accident analysis given in the text, make a list of parameters to which the yield is sensitive (and insensitive). Discuss the reasons for the inclusion of each entry in the list.

7-15 Discuss the limitations of the α method for measuring control rod reactivity worth with regard to time dependencies, control rod–reactor similitude, etc.

REFERENCES

1. A. Weinberg and E. Wigner: "Physical Theory of Neutron Chain Reactors," University of Chicago Press, Chicago, 1958, chap. 12.
2. A. Weinberg and H. C. Schweinler: Oscillating Absorber in a Nuclear Reactor, *Phys. Rev.*, **74**: 851–863 (1948).
3. Weinberg and Wigner: *op. cit.*, p. 213.
4. V. Raieveskii and J. Horowitz: *Compt. Rend.*, **228**: 1992 (1954).
5. S. Glasstone and M. C. Edlund: "Elements of Nuclear Reactor Theory," Van Nostrand, Princeton, N.J., 1952, chap. 12.
6. Weinberg and Wigner: *op. cit.*, pp. 398–400.
7. L. D. Landau and E. M. Lifshitz: "Quantum Mechanics: Non-Relativistic Theory," Addison-Wesley, Reading, Mass., 1958, p. 17.
8. R. Bellman: "A Brief Introduction to Theta Functions," Holt, New York, 1961, pp. 7, 8.
9. D. Randall and D. S. St. John: Xenon Spatial Oscillations, *Nucleonics*, **16**(3): 82–87 (1958).
10. G. L. Gyorey: The Effect of Modal Interaction in the Xenon Instability Problem, *Nucl. Sci. Eng.*, **13**: 338–344 (1962).
11. G. S. Lellouche: Space Dependent Xenon Oscillations, *Nucl. Sci. Eng.*, **12**: 482–489 (1962).
12. J. Chernick: Dynamics of a Xenon Controlled Reactor, *Nucl. Sci. Eng.*, **8**: 233–248 (1960).
13. H. L. Garabedian and C. B. Leffert: A Time Dependent Analysis of Spatial Flux Distribution, *Nucl. Sci. Eng.*, **6**: 26–32 (1959).
14. H. Soodak (ed.): "Reactor Handbook," 2d ed., vol. 3, pt. A, Interscience, New York, 1962, chap. 5.
15. V. Z. Jankus: A Theoretical Study of Destructive Nuclear Bursts in Fast Power Reactors, *Argonne Natl. Labs. ANL* 6512, February, 1962.
16. H. A. Bethe and J. H. Tait: An Estimate of the Order of Magnitude of the Explosion when the Fast Reactor Core Collapses, *U.K. At. Energy Authority, RHM* (56)/113, April, 1956.
17. Weinberg and Wigner: *op. cit.*, chap. 16.
18. G. E. Hansen: Burst Characteristics Associated with Slow Assembly of Fissionable Materials, *Los Alamos Sci. Labs. LA* 1441, eq. 2-10.
19. W. R. Stratton et al.: Prompt Excursions in Fast Reactors, *Proc. U.N. Intern. Conf. Peaceful Uses At. Energy, 2nd, Geneva,* 1958, **12**: 196–205 (1958).
20. W. J. McCarthy et al.: *Proc. U.N. Intern. Conf. Peaceful Uses At. Energy, 2nd, Geneva,* 1958, **12**: 207–229 (1958).
21. McCarthy et al.: *op. cit.*, p. 227.
22. Stratton et al.: *op. cit.*, p. 197.
23. W. M. Stacey, JR: "Space–Time Nuclear Reactor Kinetics," Academic Press, N.Y., 1969.
24. J. B. Yasinsky and S. Kaplan: Anomalies Arising From the Use of Adjoint Weighting in a Collapsed Group-Space Synthesis Model, *Nucl. Sci. Eng.*, **31**: 354 (1968).
25. R. V. Meghreblian and D. K. Holmes: "Reactor Analysis," McGraw-Hill, New York, 1960, chap. 9.
26. I. N. Sneddon: "Fourier Transforms," McGraw-Hill, New York, 1951, chap. 3.
27. J. Campbell and R. Stelson: *Oak Ridge Natl. Labs. ORNL Rept.* 2076, March, 1956.
28. J. Bengston: Determination of Prompt Neutron Decay Constant, *Proc. U.N. Intern. Conf. Peaceful Uses At. Energy, 2nd, Geneva,* 1958, **12**: 63–71 (1958).
29. A. V. Antonov: Neutron Diffusion Studies Using Neutron Pulse Methods, *Proc. U.N. Intern. Conf. Peaceful Uses At. Energy, Geneva,* 1955, **5**: 3–12 (1955).
30. Antonov: *op. cit.*, p. 4.

CHAPTER

EIGHT

KINETICS OF REACTOR PROTOTYPES—I

8-1 INTRODUCTION

The preceding chapters have dealt with the kinetics of reactors without reference to a specific class or type of reactor. The earlier chapters plus the last chapter, taken together consist of a compendium on kinetics across the spectrum of reactors. In this, and the following chapter, a number of reactor archetypes throughout the relatively short history of the existence of nuclear reactors, will be considered from the point of view of their unique kinetics aspects. By the investigation of the kinetics problems inherent in these reactors, it is hoped that the base of understanding of this subject will be made that much more comprehensive. As is apparent, there is some unavoidable overlapping of reactor types. An immediate example is the all-metal fast-pulsed reactors, such as the Godiva family. The following discussion will attempt to consider the different kinds of reactors that can lead to unique and therefore interesting kinetics considerations, as opposed to the well-established kinetics and control knowledge of some of the earlier and better developed reactors. An example of the latter might be a thermal, graphite, or light water moderated, moderately enriched uranium, stationary reactor of medium power.

8-2 FAST REACTORS: IMPORTANT PARAMETERS

A fast reactor is one in which ordinarily no moderator is used, since moderating the neutrons, especially to thermal energies, is foreign to its operation. Therefore the coolant and, if possible, the associated structure should have relatively poor moderating properties, whereas the former should consist of an excellent heat transfer medium. Liquid metal coolants such as liquid sodium or sodium potassium eutectics immediately come to mind and are used in existing fast reactors. The fission neutrons are born fast, scatter (and are therefore moderated to some extent, largely through inelastic collisions), leak, and are captured all as essentially fast, or at least highly epi-thermal neutrons. The neutron spectrum of a typical large, fast power reactor has a maximum in the 200–400 keV region, with 50 percent points at approximately 100 keV and 3 MeV.[1]

The great interest in fast reactors, aside from an early one due to their similarity to certain types of nuclear weapons, arises because of their enhanced fuel breeding capabilities. Both $\nu(E)$ and $\eta(E)$ are increasing functions of the energy of the neutron that initiates the fission. Since η is defined as the number of fission neutrons produced per neutron captured in the fuel, it is desired that $\eta - 1$ be as large as possible for maximum breeding gain. It is made so by keeping the neutron spectrum as fast (hard) as possible. The "-1" in $\eta - 1$ is the neutron that must be subtracted to provide the link in the chain reaction.

A fast reactor requires a greater critical mass than its thermal counterpart, since for one reason the microscopic fission cross sections are orders of magnitude smaller, for example, at 100 keV, than at thermal energies. Also, there is no moderator, so that the fast reactor core is relatively small with a correspondingly poorer volume to surface ratio, which favors leakage. The three main constituents of a fast reactor are therefore the core fuel, the surrounding coolant, and the structure to hold it together, with a reasonable minimum of the latter. Even though the reactor minimum size is fixed by heat removal considerations, power densities are much higher than in thermal reactors.

There are two principal breeder reactor fuels with regard to their attractiveness for breeding. The first is the ^{239}Pu fueled fast reactor, breeding its ^{238}U reflector/blanket into more ^{239}Pu. The second is a ^{233}U fueled thermal reactor, breeding its ^{232}Th reflector/blanket into more ^{233}U. However, the Experimental Breeder Reactor (EBR), an early fast liquid metal cooled reactor (discussed later) is fueled with ^{235}U.

The grand scheme of the breeding reactor concept is to breed fissionable fuel from the plentiful amounts of fertile fuel, such as ^{232}Th and ^{238}U, while simultaneously producing useful power. In time then, a proliferation of breeder reactors could transmute enough fertile fuel to fissionable fuel, so that relatively cheap nuclear fuel would be no longer a world resource problem.

The differences in the kinetics, control, and stability between thermal and fast reactors are not always easily discernable. In general, however, the "coupling" of the kinetics of a fast reactor with the remainder of its nuclear engineering design and construction considerations is much closer than that of a thermal reactor. This will become evident in the following discussion. Below prompt critical, the kinetics characteristics of fast and thermal reactors are grossly similar, because they are both governed by delayed neutrons whose properties are largely the same in either type of reactor. Such differences are due to the type of fuel used, and not necessarily whether the reactor is fast or thermal.

A first major parametric difference is that the prompt neutron lifetime in fast reactors is from two to three orders of magnitude smaller than that for thermal reactors. Typical fast reactor prompt neutron lifetimes can range from microseconds to nanoseconds and less.

The expression for prompt neutron lifetime in an infinite reactor is proportional to $(v\Sigma_a)^{-1}$, where v is an average speed of the neutrons, and Σ_a is the total absorption cross section. It turns out that the large difference in prompt neutron lifetimes between fast and thermal reactors is due mainly to the large difference in their mean neutron speeds. This is because of the fact that even though the microscopic cross section decreases with energy from thermal to fast energies, the number density of fuel nuclei per cubic centimeter is increased in a fast reactor, because of the absence of moderator. Hence, Σ_a remains roughly constant in a thermal or fast reactor. For a thermal reactor, the bulk of neutrons have a mean energy of 0.025 eV. For a typical fast system, the corresponding mean neutron energy is in the 200 keV region. The square root of this ratio ($\sim 10^3$) gives the factor between fast and thermal prompt neutron lifetimes.

A second difference between thermal and fast reactors, assuming a plutonium-fueled fast reactor, is that β, the mean yield of delayed neutrons per fission neutron, in plutonium is about one-third that of ^{235}U, a common thermal reactor fuel. $\beta_{\mathrm{Pu}} \cong 0.0020$, $\beta_u \cong 0.0064$. However, β is not directly a function of whether the reactor is fast or thermal but is essentially a constant of the fuel material itself.† Table 8-1 compares delayed neutron parameters of ^{239}Pu and ^{235}U obtained from fast fission data.[2]

Another important difference between fast and thermal systems is that the absorption cross sections of most materials of interest in thermal reactors drops to negligible amounts in fast reactors. This certainly poses problems of control in the conventional manner of using control rods, as in thermal reactors. Also in the 100 keV region the total absorption cross section of the fuel is dominated by the inelastic cross section. The energy threshold for inelastic scattering is well within the fast reactor neutron spectrum for reactor materials of interest. An inelastic "scatter" (n, n') is the process whereby a neutron enters a given nucleus to form a compound nucleus. This event excites the now compound nucleus to a more energetic state, with the emission of a neutron at a lower energy than the one incident. The excited nucleus decays subsequently to a more stable state by emitting a gamma photon. Inelastic scattering[3] is a very complex process, and is one of the last of the nuclear scattering processes in this context to yield to satisfactory theoretical explanation. The experimental determination of inelastic cross sections is also difficult. Besides the fast neutron fission cross section, and the inelastic scattering process, neutrons are also absorbed by radiative capture (n, γ) in the heavy isotopes comprising a fast reactor core.

† The parameter β is modified by its effectiveness factor γ to give $\gamma\beta$, the effective β, as discussed in Sec. 6-6. γ is a function of the reactor importance, so that the effective β depends not only on the type of fuel but on the type of reactor as well. The latter is a comparatively insensitive dependence, however.

Table 8-1 Fast fission delayed neutron parameters of ^{235}U and ^{239}Pu

Mean life t_i (sec)	Decay constant λ_i (sec^{-1})	Delayed neutron fraction β_i
	^{235}U, $\beta = 0.0064$	
0.258	3.88	0.000168
0.715	1.40	0.000824
3.22	0.311	0.000263
8.65	0.116	0.00121
31.5	0.0317	0.00137
78.7	0.0127	0.000246
	^{239}Pu, $\beta = 0.0020$	
0.312	3.21	0.000073
0.793	1.26	0.000216
3.02	0.331	0.000687
7.50	0.133	0.000452
32.2	0.0311	0.000584
77.5	0.0129	0.000080

As already mentioned, the β of ^{239}Pu is about 30 percent of that of ^{235}U. Since prompt critical implies a $k_{\text{eff}} = 1 + \beta_{49}$ for ^{239}Pu, and $1 + \beta_{25}$ for ^{235}U, each ^{235}U reactivity dollar is worth about three ^{239}Pu dollars. This is merely another statement of the fact that the control regime between delayed critical and prompt critical for ^{239}Pu corresponds to about one third less margin than for ^{235}U. This is a second factor militating against easy control for fast (plutonium fueled) reactors. The first is that, through lack of absorbers for control rods compared to thermal reactors, a fast reactor is partially controlled by movement of fuel, such as reciprocating fuel rods, and especially through the doppler and other reactivity coefficients discussed later. Such control is limited to 2–3 percent reactivity, as opposed to 5–15 percent or more reactivity that normally resides in thermal reactors. Again, this is because above 100 keV, the absorption cross sections for reactor materials are in the vicinity of 1–10 barns, whereas their thermal counterparts are often thousands of barns. While the small reactivity control margin is dictated by β for the usual fast reactor fuels, this small range is not too alarming, because the xenon poison negative reactivity is not a factor in fast systems. Hence the additional reactivity to override xenon partially or totally is not needed as in a thermal reactor. This is, of course, due to the fact that the xenon cross section for neutron absorption is negligible for fast neutrons, and therefore no large amount of built-in reactivity need be provided for xenon absorption.

However, because of the potential danger of the very rapidly rising flux instabilities that could occur in a fast reactor, tight coupling is desired between the feedback coefficients of reactivity and reactor power. Some of this is provided by the fuel expansion itself, which figures strongly due to the high power densities, so that the fast reactor will be at least uniformly stable (bounded power excursions) in the sense of Liapunov, as discussed in Sec. 4-9.

In general, there is a much closer tie between core geometrical and material changes, and reactivity in fast reactors than in a thermal reactor. This is partially what is meant by the earlier statement that fast reactor kinetics and the corresponding core design is coupled much closer than in a thermal reactor. Any unique aspect of fast reactor kinetics does not necessarily lie in the delayed neutron behavior (except for the fact that the β of plutonium is 0.0020), which latter bears a strong similarity to that for thermal reactors, but in the more core criticality oriented considerations. As will be seen, these include the various reactivity coefficients, such as the doppler, sodium void, and fuel expansion coefficients of reactivity.

With respect to the latter in certain reactors, the fuel must be allowed to expand in a judicious manner. This is a consequence of the fact that in some of the EBRI cores, a large positive temperature reactivity coefficient was noted. This was found to be due to the longitudinal bowing expansion of the fuel elements toward the core central axis. The later cores had reinforced fuel elements to prevent bowing.

A further reactivity factor for a fast reactor, as mentioned above, is that neutron-medium interactions are influenced by the doppler effect. This is because neutron reaction rates depend on the relative velocity between the neutrons and the thermodynamically agitated fuel and other core material. This relative velocity is now represented by a probability distribution function centered about the energy of the incident neutron. This distribution function can be derived from the approximation that the core material nuclei act like a perfect gas at the equilibrium temperature corresponding to their interaction with the fast neutron. It is in the epi-thermal resonance region that the doppler effect is most noticeable. It effectively broadens the resonances of the neutron cross sections of, for example, ^{235}U and ^{238}U. This is discussed in detail in Sec. 3-6; it is similar to neutron thermalization where slow neutrons enter into thermal equilibrium with their moderator medium. The doppler reactivity coefficient is discussed in detail later.

8-3 FAST REACTORS: KINETICS EQUATIONS

Although the kinetics equations for fast reactors are almost formally identical to those of thermal systems, the parameters involved are quite different, as mentioned earlier. It is of interest to sketch briefly the derivation of the kinetics equation for fast systems to provide insight into the role played by these parameters. One begins by assuming that all of the neutrons exist in a single fast energy group. Then the corresponding time-dependent diffusion equation is assumed to hold. Since a fast reactor normally has no moderator, its core is frequently small with high fuel density. Its small size is not always a sufficient number of mean free paths in the dimensions of interest to maintain the validity of the diffusion approximation to the transport equation, as discussed in Sec. 1-6. Nevertheless, a diffusion equation for the fast flux can be written as

$$\frac{\lambda_{tr}}{3}\nabla^2\phi(\bar{r},t) + [(1-\beta)k_\infty - 1]\Sigma_a\phi(\bar{r},t) + \sum_i \lambda_i C_i(\bar{r},t) = \frac{1}{v}\frac{\partial\phi}{\partial t}(\bar{r},t) \qquad (8\text{-}1)$$

where λ_{tr} is the transport mean free path, that is, the macroscopic transport cross section $\Sigma_{tr} = \Sigma_s(1-\bar{\mu}_0)$ where $\bar{\mu}_0$ is the mean cosine of the scattering angle, averaged over the scattering cross section. In thermal reactors, the slow neutrons are scattered isotropically because of their low energy, so that $\bar{\mu}_0 \cong 0$. For fast reactors with their fast neutrons, higher modes of angular momentum scattering begin to be manifest, which implies an appreciable forward scattering component, which is much greater than in the thermal case, so that $\bar{\mu}_0 > 0$.

Also, the total absorption cross section is made up differently than in a thermal system, as discussed earlier, that is

$$\Sigma_a = \Sigma_c + \Sigma_f + \Sigma_{\text{inel}} \qquad (8\text{-}2)$$

where Σ_{inel}, the inelastic scattering cross section is the dominating term. Σ_c is the macroscopic capture cross section, which is primarily radiative capture (n,γ) at these energies, and Σ_f is the macroscopic fast fission cross section.

The delayed neutron precursor concentrations satisfy

$$\frac{\partial C_i}{\partial t}(\bar{r},t) = -\lambda_i C_i(\bar{r},t) + \frac{\beta_i k_p}{P}\Sigma_a\phi(\bar{r},t) \qquad (8\text{-}3)$$

The fast diffusion length $L_f = 1/3\,\Sigma_{tr}\,\Sigma_a$. To facilitate the development, the buckling B_0^2 is assumed to be that of the fundamental space flux mode which satisfies the appropriate Helmholtz equation $\nabla^2\phi + B_0^2\phi = 0$ for a near critical system.† All of this reduces Eq. (8-1) to

$$[(1-\beta)k_\infty - (1 + L_f^2 B_0^2)]\,\phi(\bar{r},t) + \frac{1}{\Sigma_a}\sum_i \lambda_i C_i(\bar{r},t) = \frac{1}{v\Sigma_a}\frac{\partial\phi}{\partial t}(\bar{r},t) \qquad (8\text{-}4)$$

with $\phi \equiv n(\bar{r},t)v$. Both $n(\bar{r},t)$ and $C_i(\bar{r},t)$ are assumed separable with the same space modes. This then gives the usual kinetics equation for the neutron density, $n(t)$, as

$$\dot{n} = \left(\frac{K-1}{\Lambda}\right)n + \sum_i \lambda_i c_i; \qquad n(0) = 1 \qquad (8\text{-}5)$$

where $\Lambda = \{\Sigma_a v(1 + L_f^2 B_0^2)\beta k_{\text{eff}}\}^{-1} \simeq (\beta v \Sigma_f v)^{-1}$, the usual invariant form of the prompt neutron lifetime. For the reasons already mentioned, the assumption of separability is generally none too good for small fast systems. However, it will be assumed as a basis for further analytical discussion. Eliminating the delayed

† This assumption, often made in thermal kinetics equations derivations is not actually required, and is not always justified as in startup, or when a reactor is excited by an external pulsed neutron source, cf. Usachev (33), Deverall (34).

neutron precursor concentration between Eqs (8-3) and (8-5), as was done in Sec. 1-6, results in the integral equation for the fast neutron density, namely,

$$n(t) = n_0 + \int_0^t G_f(t-\tau)K(\tau)n(\tau)\,d\tau \tag{8-6}$$

$G_f(t)$ is the zero power kernel for a fast reactor. It is formally the same as that for a thermal reactor. That is, the Laplace transform gives the zero power transfer function, from Eq. (8-6), as

$$\bar{G}_f(s) = \left[s\left\{ \Lambda + \sum_{i=1}^{6} \bar{a}_i/(s+\lambda_i) \right\} \right]^{-1} \tag{8-7}$$

where $\bar{a}_i = \bar{\beta}_i/\bar{\beta}$. The prompt neutron lifetime is relatively small in fast reactors, a typical $\Lambda \simeq 10^{-5}$, corresponding to a prompt neutron lifetime of about $5 \cdot 10^{-8}$ sec. So that Λ is usually neglected in the fast reactor transfer function. The validity of neglecting Λ is seen easily by considering the kinetics equation for the delayed neutron precursors

$$\dot{c}_i = -\lambda_i c_i + \bar{a}_i n/\Lambda; \qquad \dot{c}_i(0) = 0; \qquad i = 1, \ldots, 6 \tag{8-8}$$

If this is summed on i and added to Eq. (8-5) to eliminate $\sum_{i=1}^{6} \lambda_i c_i$, the result is

$$\Lambda \frac{d}{dt}\left(n + \sum_{i=1}^{6} c_i \right) = Kn \tag{8-9}$$

Near equilibrium which is the situation for transfer function analysis, $\Lambda \sum_{i=1}^{6} c_i \approx \bar{\tau} n$, where $\bar{\tau}$ is the mean life of the delayed neutrons in which c_i is approximated by its equilibrium concentration. Then, $d/dt(\Lambda n + \bar{\tau} n) \simeq \bar{\tau}\dot{n}$. The latter is a good approximation since $\Lambda \ll \bar{\tau}$. To illustrate, $\bar{\tau} \cong 12.8$ sec for ^{235}U while $\Lambda \simeq 10^{-5}$, so that Λ is assumed zero, or in Eq. (8-5) the left-hand side is assumed to vanish. The fast reactor zero power transfer function becomes

$$\bar{G}_f(s) = \{s\bar{f}(s)\}^{-1} = \left(s \sum_{i=1}^{6} \frac{a_i}{s+\lambda_i} \right)^{-1} \tag{8-10}$$

Of course, the above does not hold for highly transient situations, especially for super-prompt critical fast systems. In terms of $\bar{G}_f(s)$, neglecting Λ results in a poor representation of its high frequency characteristics, in that the high roll-off (corner) frequency is infinite. Now, for large but finite ω from Eq. (8-7) $|\bar{G}_{f0}(\omega)| \sim (1 + \omega^2\Lambda^2)^{-1/2}$ with the roll-off frequency satisfying $\omega_r \Lambda = 1$. As a matter of fact, with the above value of Λ of 10^{-5}, the roll-off frequency is approximately 16 kHz. It is difficult to produce such relatively high reactor kinetic frequencies by using mechanical oscillator transfer function experimental apparatus. Other means such as noise correlation measurements must be employed to probe such high frequency regions. Also, if Λ is neglected in transient calculations, discontinuities in n, such as the prompt jump or prompt drop situation discussed in Sec. 2-8 must be endured.

8-4 FAST REACTORS: TRANSFER FUNCTION RESONANCES

An excellent example of the use of transfer function analysis in the study of resonance kinetic behavior is that applied to the EBR Mark I, II, and III cores.[4] Even though such an application is not unique to fast reactors (the only differences between fast and thermal systems in this context are the magnitudes of the usual kinetics parameters below prompt critical), it played an important role in understanding the resonance kinetic behavior in the above reactor; behavior that is inimical to operating stability.

The EBR is a ^{235}U fueled (highly enriched uranium) fast reactor, unmoderated but with a natural uranium reflector, cooled with liquid NaK eutectic, and controlled by reciprocating fuel rods in the core. For the Mark I and II cores, the isothermal (ambient) temperature reactivity coefficient was about $\frac{1}{2}$ cent per degree C. It was soon discovered that in the Mark I core, a sudden decrease (increase) in coolant flow rate caused a relatively sudden power surge (drop), which corresponds to a prompt reactivity coefficient. It turned out to be due to the fact that the fuel elements bowed in toward the common longitudinal axis when overheated—for example, as caused by a coolant flow rate decrease—so that a supercritical configuration ensued. Further, there was evidence of an additional long time constant negative reactivity coefficient. The Mark III core fuel elements were strengthened[4] to prevent bowing, which eliminated the positive reactivity coefficient.

As discussed in Chaps 2–4, small excursions about equilibrium render the kinetics equation (8-6) linear, so that it can be rewritten as

$$\varepsilon(t) = \int_0^t G_{f0}(t - \tau)K(\tau)\, d\tau \tag{8-11}$$

where the small fluctuation about the equilibrium neutron density n_0 is $\varepsilon = (n - n_0)/n_0 \ll 1$. The reactivity is given by a control reactivity K_c, plus that due to the feedback reactivity power coefficient $H(t; n_0)$. Thus

$$K(t) = K_c(t) + \int_0^t H(t - \tau; n_0)\varepsilon(\tau)\, d\tau \tag{8-12}$$

Combining Eqs (8-11) and (8-12), taking the Laplace transform of the result, and letting $s \to i\omega$ gives the usual closed loop feedback relationship as

$$\bar{\varepsilon}(\omega)/\bar{K}_c(\omega) \equiv \bar{G}_f(\omega) = \frac{\bar{G}_{f0}(\omega)}{1 - \bar{G}_{f0}(\omega)\bar{H}(\omega; n_0)} \tag{8-13}$$

where $\bar{G}_f(\omega)$ is the transfer function at operating power n_0. The feedback power coefficient of reactivity $\bar{H}(\omega; n_0)$ can be investigated by measuring $\bar{G}_f(\omega)$ at various equilibrium power levels n_0, with known $\bar{G}_{f0}(\omega)$, the zero power no feedback transfer function, and plotting

$$-\bar{H}(\omega; n_0) = \frac{1}{\bar{G}_f(\omega)} - \frac{1}{\bar{G}_{f0}(\omega)} \tag{8-14}$$

in the complex plane. Figure 8-1 is the measured amplitude response $\bar{G}_f(\omega)$ for the Mark II core.

From Eq. (8-13) it is seen that power surge resonances occur when $1 - \bar{G}_{f0}(\omega)\bar{H}(\omega; n_0) = 0$, assuming (correctly) that the numerator $\bar{G}_{f0}(\omega)$ itself exhibits no poles in the positive half plane, which would contribute to increasing power resonances. Then for some frequency ω_0 and power level n_0, $\bar{H}(\omega_0; n_0) = \bar{G}_{f0}^{-1}(\omega_0)$. Figure 8-2 depicts the complex plane containing the polar vectors $-\bar{H}(\omega; n_0)$ and $\bar{G}_{f0}^{-1}(\omega)$, and the resultant $\bar{G}_f^{-1}(\omega)$ whose loci are plotted as a function of the frequency ω.

Since $\bar{G}_f^{-1}(\omega) \simeq 5 \cdot 10^{-3}$ in the neighborhood of the resonance, $|\bar{G}_f(\omega_0)|$, the amplitude of the transfer function at resonance, will correspond to a very high power as compared to a nominal non-resonant power. Because the resonant frequencies occur at small fractions of a hertz, the reactor will climb slowly to the resonance power, assuming that it has been excited at the resonant frequency. For example, if $f_{\text{res}} \simeq 0.03$ Hz, as in Fig. 8-1, the corresponding period is about 30 sec, so that there is sufficient time to take action to circumvent such a climb. This is opposed to the usual electronic or radio frequency applications, where resonant frequencies are measured in kilohertz or megahertz.

With the influx of data from the EBR experiments,[5] phenomenological models for $\bar{H}(\omega; n_0)$ could be postulated to fit the data. For a short time constant positive reactivity coefficient, and a long time constant negative reactivity coefficient, one model for $\bar{H}(\omega; n_0)$ is

$$-\bar{H}(\omega; n_0) \simeq \frac{n_0}{(1 + i\omega\tau_1)(1 + i\omega\tau_2)}; \qquad \tau_2 \gg \tau_1 \tag{8-15}$$

Another illustration of a model[4] for $\bar{H}$ is to consider that the fuel heats itself and the surrounding coolant in the core. Then the coolant is circulated out of the core to heat the peripheral structure. Let T_c and T_p be the mean temperature differences between the coolant inlet temperature and the core (plus coolant) and peripheral structure, respectively. A simple heat balance gives the pair of equations

$$\dot{T}_c(t) = an_0 - h_c T_c(t) \tag{8-16a}$$

$$\dot{T}_p(t) = bT_c(t - t_0) - h_p T_p(t) \tag{8-16b}$$

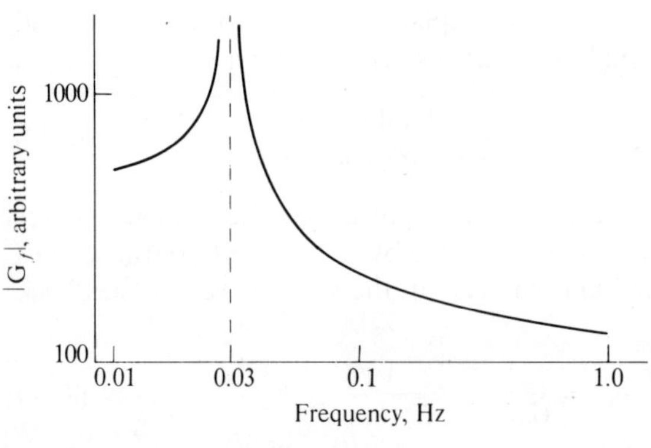

Figure 8-1 Amplitude response of EBR Mark II core at 550 kW, 95 g/min flow.[4]

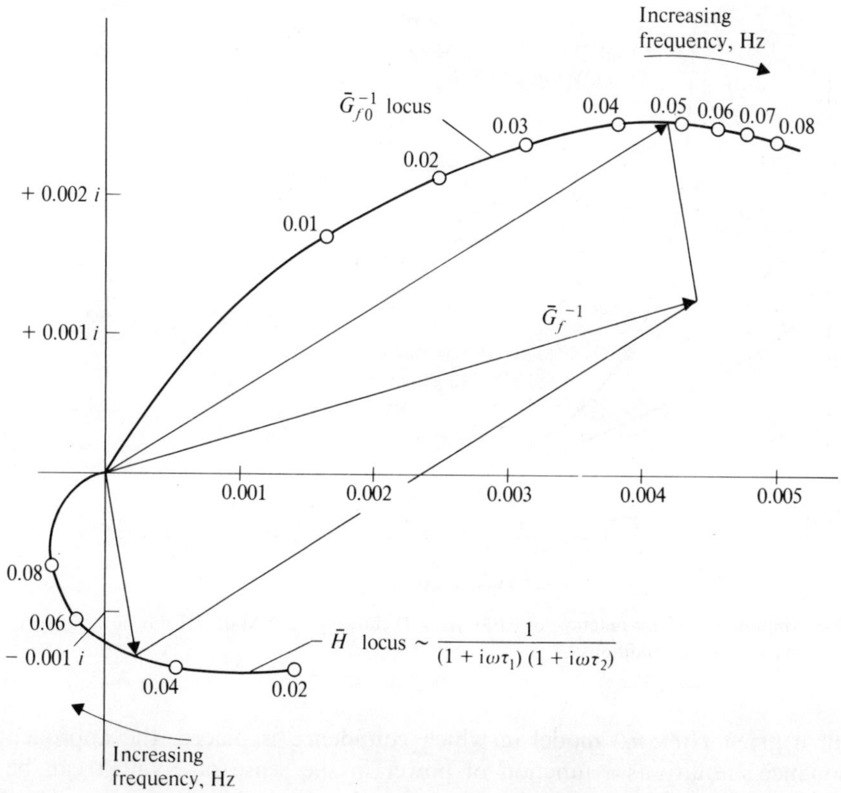

Figure 8-2 Plots of $-\bar{H}(\omega; n_0)$, $\bar{G}_{f0}^{-1}(\omega)$ for EBR Mark I core in the complex plane.[5]

Equation (8-16a) simply asserts that the core rate of temperature rise (fuel plus coolant) is given by the fission power generated less the heat loss rate from the coolant, whereas Eq. (8-16b) avers that the peripheral temperature rise is due to the rate of transport of heat from the core, via the coolant, to the periphery, taking time t_0, minus the heat loss rate from the latter. Eliminating T_c between these two equations by using Laplace transforms, and letting $s \to i\omega$ gives asymptotically

$$-\bar{H}(\omega; n_0) \simeq \bar{T}_p(\omega; n_0) \sim \frac{(ab/h_c h_p) n_0 \, e^{-i\omega t_0}}{(1 + i\omega\tau_p)(1 + i\omega\tau_c)} \tag{8-17}$$

where $\tau_c = h_c^{-1}$, and $\tau_p = h_p^{-1}$ are time constants for the core and periphery, respectively. Such a model seems to check the situation for the EBR Mark III core (rigidized fuel elements) as seen in Figs 3-3 and 3-4 (cf. Sec. 3-5). The Mark III core had ribbed fuel elements[5] which prevented bowing. The amplitude transfer functions of Mark II and III cores, compared under nearly identical operating conditions in Fig. 8-3, clearly reveal the elimination of the resonance which must have been due to fuel bowing.

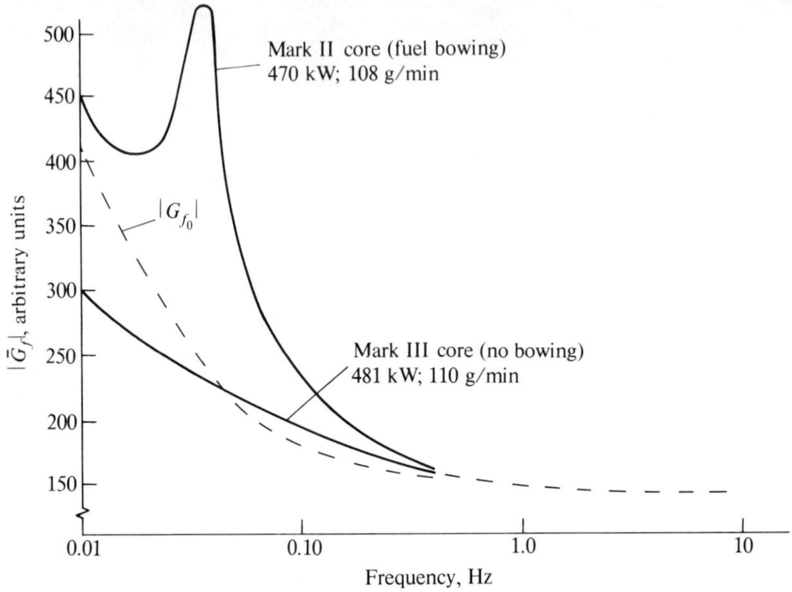

Figure 8-3 Amplitude transfer function of EBR Mark II (bowing) and Mark III (no bowing) cores under the same operating conditions.[4]

For a given $\overline{H}(\omega; n_0)$ model in which confidence is placed, the approach to resonance/stability as a function of power in the sense of Nyquist can be determined. From Eq. 8-13 it is seen that instability occurs when $\text{Re}[\overline{G}_{f0}(\omega_0)\overline{H}(\omega_0; n_0)] = 1$ and $\text{Im}[\overline{G}_{f0}(\omega_0)\overline{H}(\omega_0; n_0)] = 0$. Then a plot of $\text{Re}[\overline{G}_{f0}(\omega)\overline{H}(\omega; n_0)] < 1$, $\text{Im}[\overline{G}_{f0}(\omega)\overline{H}(\omega; n_0) = 0$ as a function of equilibrium power level n_0 can be made.† This can then be extrapolated to find the resonant power, where $\text{Re}[\overline{G}_{f0}(\omega)\overline{H}(\omega; n_0)] = 1$, $\text{Im}[\overline{G}_{f0}(\omega)\overline{H}(\omega; n_0)] = 0$, without risking the reactor. This is discussed further in Sec. 8-14 where an application is made to Boiling Moderator Reactors.

8-5 FAST REACTORS: REACTIVITY COEFFICIENTS

As mentioned earlier, the kinetics of a fast reactor in terms of the point model neutron density and associated delayed neutrons is very much like a thermal reactor. Of course, the idea of straightforward control using absorbing control rods does not play nearly as strong a role as it does for thermal reactors. This is the case, even though B_4C control rods are used as part of the control system

† $\text{Re}(\overline{G}_{f0}\overline{H}) = \text{Re}(1 - \overline{G}_{f0}/\overline{G}_f)$ can be plotted, which should give better accuracy, because $\overline{G}_{f0}$ is known more accurately than $\overline{H}$, and $\overline{G}_f$ is measured. This provides a function constraint with which to determine a fit to $\overline{H}$.

in the Fermi fast reactor. Hence, other reactivity influencing mechanisms assume a much more dominating role than for thermal reactors. These include reactivity coefficients and their behavior.

Again, the unique aspects of fast reactor kinetics are generally rather difficult to separate out from other reactor considerations, such as core criticality. That is, it can almost be said that fast reactor studies are fast reactor kinetics studies, and vice versa. This is as opposed to other type of reactors where, for example, point model kinetics are relevant but can be considered delineated from the other aspects of reactor behavior.

The pertinent fast reactor reactivity coefficients discussed below are due to the doppler effect, the sodium coolant voids, and the temperature–power coefficient. The physics of the doppler effect is discussed in detail in Sec. 3-6. Suffice it to say that when the reactor material temperature is changed, the relative velocity between the incident neutron and the material target nucleus, in the vicinity of a material cross section resonance becomes part of a velocity distribution. This has the effect of broadening the resonance, but lowering the resonance peak, keeping the mean cross section constant. Since both fission and capture resonances are broadened, the net doppler reactivity coefficient of a large fuel oxide or fuel carbide constituted fast reactor is a complicated affair. It has now been established by experiment and calculations that the net doppler reactivity coefficient in such reactors is negative.[7] This, of course, augurs well for the deterrence of fast reactor core melt-down accidents, and associated hazard considerations.

With the broadening of the resonances, the effective cross section over the resonances increases, while the average cross section over the resonance region stays constant, as mentioned. For the asymptotic case of when the temperature of the reactor material becomes very large, the resonances become smeared out, as the individual velocity probability distributions for each resonance merge together. The resonance fine structure "dissolves" into an average cross section. This is equivalent to the disappearance of the self-shielding factor, as discussed in Sec. 3-8. The latter also occurs when the fuel is infinitely dilute—that is, the antithesis of fuel lumping, hence the self-shielding effect vanishes.

Large, fast reactors using fuel in the form of oxide or carbide or other cermets, have a somewhat softer spectrum than pure metal fueled fast reactors. The ^{239}Pu fuel of the former reactors is well diluted with ^{238}U for breeding purposes, and will also contain ^{240}Pu. The fission cross section of such "even-even" isotopes is essentially zero in the resonance region, so that only their capture resonances are doppler broadened. This leads then to a net negative doppler reactivity coefficient, as mentioned. Hence, there is a built-in automatic quick acting negative reactivity feedback component. The relatively high concentration of ^{238}U required for breeding gain purposes fortuitously supplies the large negative reactivity coefficient, predominating over any positive doppler reactivity of further ^{239}Pu. Doppler coefficient magnitudes are of the order of a few tenths of a cent per degree centigrade. Since total temperature changes can vary over one or two thousand degrees C, the corresponding total doppler effect reactivity can amount to a few dollars, a not insignificant amount of

Table 8-2 Negative doppler reactivity effect coefficients computed for large PuO_2-$^{238}UO_2$ and Pu C-^{238}UC fueled fast reactors[6]

		300–2500 K		300–750 K		1500–2500 K	
Oxide fueled		Per cent $-\Delta k$	$-\Delta k/\Delta T$ (10^{-5} °C)	Per cent $-\Delta k$	$-\Delta k/\Delta T$ (10^{-5} °C)	Per cent $-\Delta k$	$-\Delta k/\Delta T$ (10^{-5} °C)
Pu/^{238}U	Na present						
1.9	Yes	2.63	1.15	1.09	2.40	0.59	0.59
1.9	No	1.44	0.65	0.63	1.40	0.29	0.29
1.7	Yes	1.81	0.79	0.77	1.73	0.40	0.40
1.7	No	0.92	0.40	0.39	0.87	0.18	0.18
Carbide fueled							
Pu/^{238}U	Na present						
1.9	Yes	1.95	0.89	0.82	1.82	0.40	0.40
1.9	No	1.11	0.48	0.47	1.05	0.20	0.20
1.7	Yes	1.30	0.59	0.54	1.20	0.29	0.29
1.7	No	0.63	0.29	0.27	0.60	0.12	0.12

plutonium dollars, for example. Table 8-2 depicts some typical calculated values for the above large, fast reactors.

The coefficients of reactivity due to the influences of the liquid sodium coolant in a large, fast reactor can be construed as made up of two components. One is the energy-dependent component, and the other is the purely physical component due simply to the presence, or absence, of the sodium, such as might occur in a hypothetical accident where an appreciable part of the sodium coolant is flushed away from the core. The physical aspect of the sodium void reactivity is discussed first. The very presence of the sodium enveloping core coolant will certainly inhibit leakage. Then it is appreciated that the relative absence of sodium due to density changes, voids, or removal, is a negative reactivity contribution to the sodium void reactivity coefficient.

With respect to the neutron spectrum aspects of the sodium void coefficient, a decrease in sodium due to voids, temperature-density changes, etc. will decrease the moderating capability of the reactor, and so shift the mean neutron energy to a higher value, that is, harden the spectrum. Then the net reactivity from this effect is an increase, due to the fact that $\nu(E)$ and $\eta(E)$ are increasing functions of the neutron energy. Thus, the spectral component of the sodium reactivity coefficient will be positive. If it is felt that the positive component is unduly large for the particular family of fast reactors of interest, then consideration can be given to reactor designs of odd geometries. For example, designing the core as a short squat cylinder or annulus to lower the volume to surface ratio, will then enhance the leakage, as a safety measure to counter the positive sodium reactivity. However, such geometries can lead to severe economic drawbacks, because too much leakage can be shown to be not very cost effective. Varying amounts of both the spectral component and leakage component of the sodium void reactivity can occur, depending on how the diminution of sodium voidage ensues.

The fast reactor temperature and power coefficients are analogous to those occuring in thermal reactors due to the relative expansion rates of fuel, moderator, and structure. An increase in reactivity resulting in an increase in reactor power, increases the temperature in the fuel, coolant, and reflector/blanket. The reactor core and blanket then expand ejecting some of the sodium coolant. This heat generation causes an outward propagation of material expansion as well as heat toward the reactor periphery. Whether or not the cumulative result is one of positive or negative reactivity depends on the particular reactor parameters. If the resultant reactivity is negative, the initial reactivity increase is then compensated, and the reactor will begin to assume an equilibrium state corresponding to a new higher power level. For the Fermi reactor, a sodium coolant density change, reactivity coefficient, is of the order of $K = (0.03/\beta)\,\Delta\rho/\rho$ dollars, where $\Delta\rho/\rho$ is the fractional change in the sodium density. On the other hand, for EBR II, the same reactivity coefficient is $K = -(0.35/\beta)\,\Delta L/L$ dollars, where $\Delta L/L$ is the fractional axial fuel expansion.[8]

8-6 SPECTRAL SHIFT CONTROL REACTORS

This reactor concept, due to Edlund et al.,[9,10] is a variation of the light water moderated reactor. Instead of control rod banks used to keep the reactor secularly critical, the moderator consists of an adjustable mixture of light and heavy water (H_2O—D_2O). Since D_2O is a poorer moderator than H_2O (its slowing down power is about 15 percent that of H_2O), the former causes the spectral shift control reactor (SSCR) neutron spectrum to maintain an appreciable neutron population segment in the epi-thermal resonance region, at the loss of some of the thermal neutron component. This then constitutes an equivalent negative reactivity. This moderator mixture spectrum is depicted in Fig. 8-4. It can be appreciated that the corresponding moderator mixture negative reactivity coefficient can be adjusted throughout the core life.

For economy of core fuel costs, it is desired that the core irradiation cycle time be maximized. This means that such cores are heavily loaded at the outset with fuel measured in terms of several critical masses, to compensate for the fuel burnup and fission product poison accumulation during the fuel cycle. Especially at the beginning of a fuel cycle, the corresponding excess reactivities due to fuel are measured in many dollars. Instead of using banks of control rods to supply the negative reactivity to compensate that represented by these large amounts of fuel, the water moderator in the SSCR is diluted to consist of mainly heavy water initially. Then, neutrons that normally would comprise part of the thermal population, are an additional part of the resonance population, thus constituting

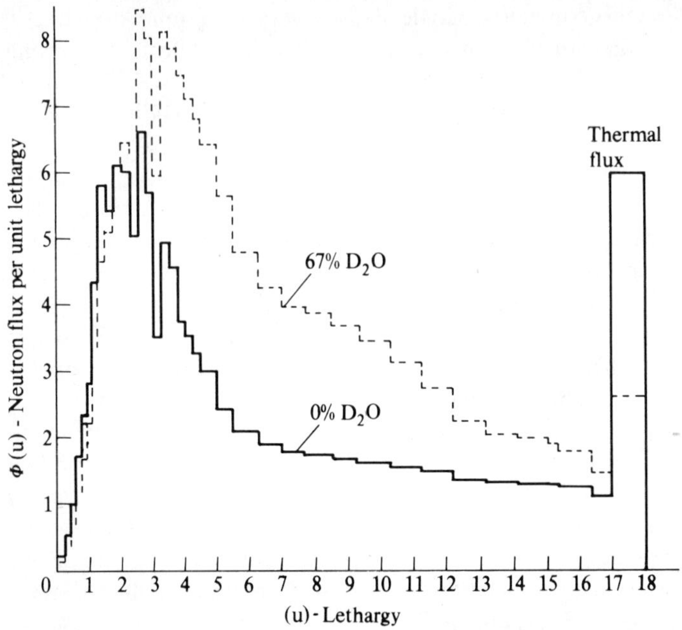

Figure 8-4 SSCR neutron spectrum for various fractions of D_2O.[9]

the required equivalent negative reactivity. Now, some of the neutrons are absorbed in the resonance region. With no control rods as such, they are absorbed mainly by the fuel, most of which is fertile, for typical light water reactor fuel enrichments of a few percent. In a sense, the fertile material acts like a homogeneous burnable poison. However, fertile fuel absorptions result in bred fissionable fuel, so that the neutrons are not non-productively absorbed, for example by poisons and/or actual control rods, but are productively absorbed to transmute fertile fuel to fissionable fuel. As the core burnup ensues, the moderator mixture is changed to shift the neutron spectrum towards thermal energies, by reducing the heavy water–light water ratio, thus increasing the moderator slowing down power. This results in an increasing positive reactivity to counter that lost due to fuel burnup.

A second advantage of the spectral shift reactor, compared to the garden variety pressurized water reactor (PWR), is that since there are no control rod banks inserted when the reactor is operating at full power, a much smoother flux distribution is manifest throughout the core. Then by proper core fuel distribution, a higher degree of flux flattening can be obtained in the SSCR compared to the PWR. In the SSCR it is possible to achieve a ratio of 2:1 or less maximum to average power, whereas in the latter case, ratios of 4:1 are commonplace. Such lower ratios allow more efficient core operation through minimizing heat flux gradients, thus allowing longer and more homogeneous core fuel irradiation. Among other consequences, this can be shown to be translated into a 70 per cent increase of power output for a given PWR system operated as an SSCR (Edlund[9]).

8-7 SPECTRAL SHIFT CONTROL REACTORS: TRANSFER FUNCTIONS

Since the moderator of the spectral shift reactor is a mixture of D_2O and H_2O, photoneutrons produced in the D_2O by fission product gamma rays play a role in the reactor kinetics behavior of this reactor. For reference, Table 1-3 in Chap. 1 gives the principal delayed photoneutron parameters for D_2O. The D_2O incident gamma ray energy threshold (2.23 meV) for the production of photoneutrons is well within the spectrum of fission product gamma rays in the reactor (0.1–6 meV). Besides the "prompt" gamma rays, i.e. the progeny of gamma emitters that are immediately born in fission, there are delayed gamma rays, of the required energy, that are progeny of the decay of fission product gamma ray delayed precursors. The latter are analogous to the usual delayed neutron precursors, that correspond to beta particle emitter precursors. All of the gamma rays can be construed to produce ultimately nine groups of delayed photoneutrons.

The concentration of the jth group of fission product parent delayed gamma precursors, P_j, satisfies

$$\dot{P}_j = -\lambda_{pj}P_j + a_j n/\Lambda \qquad j = 1,\ldots,9 \tag{8-18}$$

The production rate of the precursor parent is given in terms of the yield a_j.

To reiterate the nuclear decay chain for photoneutron production; it is that fission produces an excited parent nuclide concentration, P_j, with mean life λ_{pj}^{-1} for the emission of the precursor (daughter) nuclide itself. Prompt gamma rays are subsumed herein by using a sufficiently large value for λ_{pj}. The daughter nuclide, whose concentration is denoted by C_j, ultimately emits the gamma ray with mean life λ_j^{-1}. The daughter concentration then satisfies

$$\dot{C}_j = -\lambda_j C_j + \lambda_{pj}P_j \qquad j = 1,\ldots,9 \tag{8-19}$$

The decay of the daughter produces the jth type of gamma rays at the rate $\lambda_j C_j$. These interact instantaneously with the D_2O in the moderator to yield a $p\lambda_j C_j$ photoneutron production rate, where p is the fraction of D_2O in the moderator. Hence the corresponding neutron density satisfies

$$\dot{n} = \left(\frac{K-1}{\Lambda}\right)n + \sum_{i=1}^{6} \lambda_i c_i + p\sum_{j=1}^{9} \lambda_i C_j \tag{8-20}$$

The usual delayed neutron precursors satisfy

$$\dot{c}_i = -\lambda_i c_i + a_i n/\Lambda; \qquad i = 1,\ldots,6 \tag{8-21}$$

To obtain the SSCR transfer function, all concentrations are assumed to be in initial equilibrium, so that their initial rates $\dot{n} = \dot{c}_i = \dot{C}_j = \dot{P}_j = 0$. Using Laplace transforms to combine Eqs (8-18)–(8-21), with the above initial conditions, to eliminate all the concentrations except the neutron density, gives the kinetics integral equation for the SSCR neutron density as

$$n(t) = n_0 + \int_0^t G_{\text{SSR}}(t-\tau)K(\tau)n(\tau)\,d\tau \tag{8-22}$$

The Laplace transform of the kernel $G_{\text{SSR}}(t)$, is easily shown to be the SSCR transfer function using the methods of Sec. 3-6. It is

$$\bar{G}_{\text{SSR}}(s) = \left[s\left\{ \Lambda + \bar{f}(s) + p\bar{f}_j(s) + p \sum_{j=1}^{9} \frac{\lambda_j a_j}{(s+\lambda_j)(s+\lambda p_j)} \right\} \right]^{-1} \tag{8-23}$$

where $\bar{f}_j(s) = \sum_{j=1}^{9} a_j/(s+\lambda_j)$, is the delayed photoneutron decay function, analogous to that for the usual delayed neutrons, $\bar{f}(s)$, and that $1 - f_0 - f_{j0} = 0$. The latter simply asserts that the total delayed neutron fraction $\beta = \beta^i + p\beta^j$. β^i and β^j are the delayed neutron fraction corresponding to the delayed neutrons and delayed photoneutrons, respectively.

If instead of the "fourth order" system† above, that is, Eqs (8-18)–(8-21), the first two of these are added together to give

$$\dot{C}_j = -\lambda_j C_j + a_j n/\Lambda \tag{8-24}$$

where $\dot{P}_j \approx 0$ for the following reasons. It can be assumed to good approximation that λ_{pj} in Eq. (8-16) is sufficiently large that the rate of decay of the precursor parent nuclide is essentially equal to their production rate, that is, all delayed gamma rays are assumed to be born in fission, then $\dot{P}_j \simeq 0$. This results in a "third order" system given by Eqs (8-20), (8-21), and (8-24). Repeating the above procedure with this system of equations now results in the transfer function‡

$$\bar{G}_{\text{SR}}(s) = [s\{\Lambda + \bar{f}(s) + p\bar{f}_j(s)\}]^{-1} \tag{8-25}$$

so that the photoneutrons are now construed as an additional group of delayed neutrons.

With respect to the augmented prompt neutron lifetime, it is still $\Lambda = (\beta v \nu \Sigma_f)^{-1}$, changing only secularly with fuel burnup through the decrease of the macroscopic cross section Σ_f. However, the prompt lifetime l is given by

$$\frac{1}{l} = v\{p\Sigma_{a\text{D}} + (1-p)\Sigma_{a\text{H}} + \Sigma_c\} \tag{8-26}$$

where the three terms in the brackets are the effective macroscopic capture cross sections in the D_2O and H_2O moderator mixture, and fuel, respectively. Defining l_{D} as the prompt lifetime in an all D_2O moderated reactor, and l_{H} as that in an all H_2O moderated reactor, it easily follows that

$$\frac{1}{l} = \frac{p}{l_{\text{D}}} + \frac{1-p}{l_{\text{H}}} \tag{8-27}$$

The corresponding amplitude transfer function is obtained from Eq. (8-25) by letting $s \to i\omega$, to obtain

$$|\bar{G}_{\text{SR}}(i\omega)| = |[i\omega\{\Lambda + \bar{f}(i\omega) + p\bar{f}_j(i\omega)\}]|^{-1} \tag{8-28}$$

† The order of the system of equations is actually $N = I + J + 1$ where I is the number of delayed neutron groups, and J is the number of delayed photoneutron groups, and 1 is the neutron density equation, that is, $N = 16$, for $I = 6$ and $J = 9$.

‡ This is equivalent to letting $\lambda_{pj} \to \infty$ in the transfer function $\bar{G}_{\text{SSR}}(s)$ of Eq. (8-23).

As discussed in Sec. 3-6, the high corner angular frequency ω_{Hc}, is obtained from Eq. (8-28), for very large ω as

$$\omega_{Hc} = \Lambda^{-1} \simeq \frac{\beta}{l} = \beta\left(\frac{p}{l_D} + \frac{1-p}{l_H}\right) \tag{8-29}$$

showing the dependence on the heavy water fraction, p, in the moderator mixture. The low corner angular frequency ω_{Lc} is obtained from Eq. (8-28) for very small ω as

$$\omega_{Lc} = [\Lambda + \bar{\tau} + p\bar{\tau}_D]^{-1} \tag{8-30}$$

where $\bar{\tau}$ and $\bar{\tau}_D$ are the mean delayed neutron lifetimes in H_2O and D_2O, respectively, also dependent on the heavy water fraction p in the moderator.

To obtain an appreciation of the magnitudes of the corner frequencies $f_c = \omega_c/2\pi$ as a function of the D_2O fraction p, Table 8-3 is given below. Typical values for the parameters are: $l_H = 5 \cdot 10^{-4}$ sec, $l_D = 5 \cdot 10^{-2}$ sec, $\beta^i = 0.007$, $\beta^j = 0.001$ so that the total $\beta = 0.008$. $\bar{\tau} = 12.8$ sec, and $\tau_D = 1446$ sec.[11] Then from Eqs (8-29) and (8-30), respectively

$$f_{Hc} = 2.54(1 - 0.99p) \qquad \text{(Hz)} \tag{8-31}$$

$$f_{Lc} = \frac{1}{2\pi}\left(\frac{1}{0.16p + 16(1-p)} + 12.8 + 1446p\right)^{-1} \qquad \text{(Hz)} \tag{8-32}$$

Using the break-frequency method mentioned in Sec. 3-4, an all light water ^{235}U fueled reactor amplitude transfer function can be modified with additional asymptotes obtained from the data in Table 8-3 to reflect the fraction of D_2O in the moderator. Figure 8-5 depicts such a family of transfer functions. It shows the shift of the whole transfer function to lower frequencies with increasing D_2O in the moderator. This is a first approximation, in that the effect of fuel burnup with core life on the kinetics parameters has not been included in detail.

Table 8-3 SSCR transfer function low and high corner frequencies versus the fraction p of heavy water in the $D_2O : H_2O$ moderator

p	f_{Hc} (Hz)	ω_{Hc} (sec^{-1})	f_{Lc} (Hz)	ω_{Lc} (sec^{-1})
0	2.5400	15.96	0.01236	0.0777
0.0005	2.5387	15.95	0.01170	0.0735
0.0010	2.5375	15.94	0.01111	0.0698
0.0050	2.5274	15.88	0.00791	0.0497
0.010	2.5148	15.80	0.00582	0.0366
0.050	2.4143	15.17	0.00187	0.0117
0.10	2.2885	14.38	0.00101	0.0063
0.20	2.0371	12.80	0.00053	0.0033
0.50	1.2827	8.06	0.00022	0.0014
0.90	0.2769	1.74	0.00012	0.0008
0.95	0.1511	0.95	0.00012	0.0008
1.00	0.0254	0.16	0.00011	0.0007

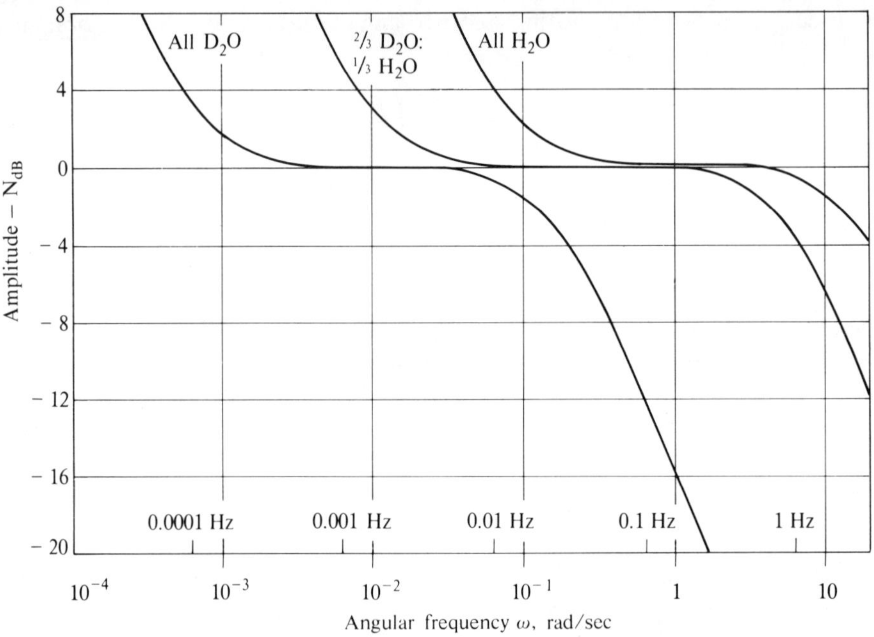

Figure 8-5 SSCR amplitude transfer function for various $D_2O:H_2O$ moderator ratios.

8-8 FAST-THERMAL COUPLED REACTORS

Coupled-core reactors were conceived very early in the post-World War II reactor technology.[12] The scheme was revived by R. Avery at Argonne National Laboratories. The discussion in this section is a paraphrasal of his work.[13] A coupled system implies that two assemblies (e.g., a thermal and a fast core) are physically coupled in the sense that some neutrons causing fissions in core A themselves originated from fissions in core B. Each core is subcritical if uncoupled from the other. The principal raison d'être of coupled-core reactors is that they combine the high breeding ratio of the fast reactor (since ν and η are largest for fast systems) with the relatively long prompt neutron lifetime of thermal reactors. The latter property obviously simplifies the control problem involved, because the thermal prompt neutron lifetime is often three orders of magnitude larger than the corresponding fast neutron lifetime.

The system operates with an almost critical fast core (for example, ^{239}Pu fuel) surrounded by a thick annular section which acts as a combination natural uranium (0.72 percent ^{235}U in ^{238}U) thermal core, fast core reflector, and breeding blanket. The beryllium shell acts as a moderator for the thermal core. The outer blanket of depleted uranium (from spent fuel) merely captures neutrons that have wandered away from the center of activity, and these neutrons convert ^{238}U to ^{239}Pu therein so that they do not go to waste, cf Fig. 8-6.

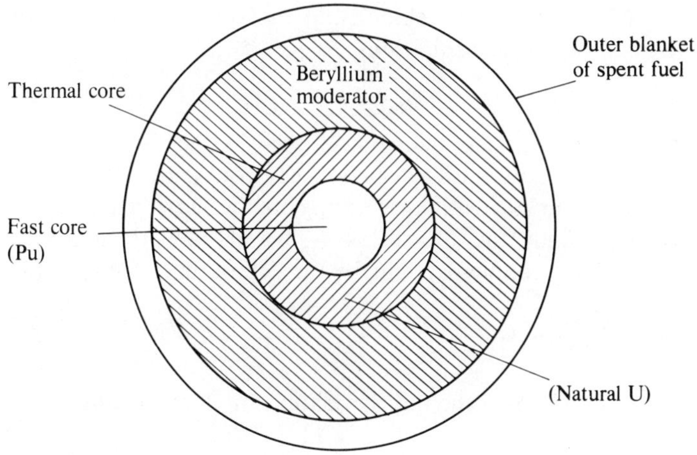

Figure 8-6 Fast-thermal coupled-core reactor configuration.

As can be seen, there are interesting kinetics analogies between coupled systems and ordinary reactors. The first is that coupled reactors are so constructed that the over-all system becomes critical only when all of the neutrons, especially the thermal ones, participate to maintain the chain reaction. The thermal neutrons act as a time bottleneck by virtue of their thermal prompt lifetime in a way that is quite similar to the manner by which the delayed neutrons act as a time bottleneck to lengthen the reactor time constant in ordinary thermal reactors. The second analogy has to do with the fact that the prompt neutron lifetime varies and therefore does not constitute a reactor invariant as described in Chap. 1. This is seen for the case where positive reactivity is added to speed up the response of the coupled-core reactor by making it critical on fast neutrons alone. This is analogous to a thermal reactor, where the period is shortened as the reactor approaches prompt critical. The coupled-core lifetimes are no more "constants of the motion" than the period in an ordinary reactor is.

In a sense, especially with respect to the time-dependent transport equation formulation, coupled-core reactors are a novel way of looking at a particular reactor configuration. The formulation developed for coupled-core systems, of course, apes the kinetics of ordinary reactors. For example, a fast-thermal coupled-core reactor is described by including the following parameters:

k_{fs} = expectation (mean) value that a fission neutron in the slow core will cause a next-generation fission in the fast core. (8-33)

l_{fs} = expected prompt neutron lifetime for a k_{fs} process. k_{fs} characterizes the coupling from the slow reactor to the fast, *not vice versa.* (8-34)

In general, $k_{fs} \neq k_{sf}$; also, k_{ss} and k_{ff} are the k_{eff} corresponding to the slow and fast cores, respectively.

The corresponding critical equation relationship that the k_{ij}'s must satisfy is obtained by first defining two source equations

$$S_f = S_{fs} + S_{ff} \qquad S_s = S_{ss} + S_{sf} \tag{8-35}$$

where S_f and S_s are the total fission neutron sources in the fast and slow cores, respectively. S_{fs} is defined as the total fission neutron source in the fast reactor caused by fissions from neutrons originating in the slow reactor. The other S_{ij} are defined in the obvious way. From Eqs (8-33) and (8-35), the following must hold for an overall critical coupled system, by the definition of the k_{ij}

$$\frac{S_{ss}}{S_s} = k_{ss} \qquad \frac{S_{sf}}{S_f} = k_{sf} \qquad \frac{S_{fs}}{S_s} = k_{fs} \qquad \frac{S_{ff}}{S_f} = k_{ff} \tag{8-36}$$

Combining Eqs (8-35) and (8-36) gives the matrix relationship for criticality

$$\begin{bmatrix} k_{ss} - 1 & k_{sf} \\ k_{fs} & k_{ff} - 1 \end{bmatrix} \begin{bmatrix} S_s \\ S_f \end{bmatrix} = 0 \tag{8-37}$$

Both S_s and S_f are assumed always positive, ensuring against the situation where one reactor is supercritical by itself. For nonzero S_s and S_f, the determinant

$$\begin{vmatrix} k_{ss} - 1 & k_{sf} \\ k_{fs} & k_{ff} - 1 \end{vmatrix} = 0 \tag{8-38}$$

or the criticality relationship is

$$(k_{ss} - 1)(k_{ff} - 1) = k_{sf}k_{fs} \tag{8-39}$$

That is, the system of equations represented in Eq. (8-37) is indeterminate with respect to the determination of S_s and S_f separately. Then only their ratio is available, which, from Eqs (8-37), is

$$\frac{S_s}{S_f} = \frac{k_{sf}}{1 - k_{ss}} = \frac{1 - k_{ff}}{k_{fs}} \tag{8-40}$$

All of this is valid even in the case where the coupling from one reactor to the other, in either direction, vanishes for whatever reason.

For the situation where the coupled system is not exactly critical, the fission source reproduces itself in the next generation by a factor $k = \nu/\nu_c$, where ν_c is the expected number of neutrons produced per fission at critical. Then Eq. (8-36) is rewritten, again from the definition of k_{ij}'s, as

$$kS_{ss} = k_{ss}S_s \qquad kS_{sf} = k_{sf}S_f \qquad kS_{fs} = k_{fs}S_s \qquad kS_{ff} = k_{ff}S_f \tag{8-41}$$

This together with Eq. (8-35) yields

$$\begin{bmatrix} k_{ss} & k_{sf} \\ k_{fs} & k_{ff} \end{bmatrix} \begin{bmatrix} S_s \\ S_f \end{bmatrix} = k \begin{bmatrix} S_s \\ S_f \end{bmatrix} \tag{8-42}$$

Now, let the overall reactivity K be given by $K = (k - 1)/k$. For the whole system close to criticality, that is, $(1 + K)^2 \cong 1 + 2K$, Eq. (8-42) yields

$$K \cong \frac{k_{sf}k_{fs} + (1 - k_{ff})(k_{ss} - 1)}{(1 - k_{ff}) - (k_{ss} - 1)} \tag{8-43}$$

The kinetics equations for the coupled-core reactors are very similar to the ordinary garden variety. They are a coupled system of ordinary differential equations in the neutron density and delayed neutron precursor concentrations. For the fast-thermal system, the equations for the corresponding neutron densities are

$$\dot{n}_{ss} = (1 - \beta)k_{ss}\left(\frac{n_{ss}}{l_{ss}} + \frac{n_{sf}}{l_{sf}}\right) - \frac{n_{ss}}{l_{ss}} + k_{ss}\sum_i \lambda_i C_{si} \tag{8-44a}$$

$$\dot{n}_{fs} = (1 - \beta)k_{fs}\left(\frac{n_{ss}}{l_{ss}} + \frac{n_{sf}}{l_{sf}}\right) - \frac{n_{fs}}{l_{fs}} + k_{fs}\sum_i \lambda_i C_{si} \tag{8-44b}$$

$$\dot{n}_{sf} = (1 - \beta)k_{sf}\left(\frac{n_{fs}}{l_{fs}} + \frac{n_{ff}}{l_{ff}}\right) - \frac{n_{sf}}{l_{sf}} + k_{sf}\sum_i \lambda_i C_{fi} \tag{8-44c}$$

$$\dot{n}_{ff} = (1 - \beta)k_{ff}\left(\frac{n_{fs}}{l_{fs}} + \frac{n_{ff}}{l_{ff}}\right) - \frac{n_{ff}}{l_{ff}} + k_{ff}\sum_i \lambda_i C_{fi} \tag{8-44d}$$

while those for the delayed neutron precursor concentrations are

$$\dot{C}_{si} = \beta_i\left(\frac{n_{ss}}{l_{ss}} + \frac{n_{sf}}{l_{sf}}\right) - \lambda_i C_{si} \tag{8-45a}$$

$$\dot{C}_{fi} = \beta_i\left(\frac{n_{fs}}{l_{fs}} + \frac{n_{ff}}{l_{ff}}\right) - \lambda_i C_{fi} \tag{8-45b}$$

with the usual equilibrium initial conditions implied for both systems of Eqs (8-44) and (8-45).

The next item is to obtain the overall inhour equation for the coupled reactors. As usual, let $n_{ij} \approx n_{ij}(0)\, e^{\lambda t}$ and $c_{ij} \approx c_{ij}(0)\, e^{\lambda t}$ for large time to determine the relationship between the asymptotic decay constant λ and the coupled-core parameters. These n_{ij} and c_{ij} are substituted into Eqs (8-44) and (8-45), which results in a simultaneous system of homogeneous algebraic equations. The condition for nontrivial (nonzero) solutions results in the vanishing of the determinant of the coefficient matrix. After some manipulation, the corresponding inhour equation is (see problem at end of the chapter)

$$(1 + \lambda l_{sf})(1 + \lambda l_{fs})\{1 - k_{ss} + \lambda[l_{ss} + \beta k_{ss}\bar{f}(\lambda)]\}\{1 - k_{ff} + \lambda[l + \beta k_{ff}\bar{f}(\lambda)]\}$$
$$= k_{sf}k_{fs}(1 + \lambda l_{ss})(1 + \lambda l_{ff})(1 - \lambda\bar{f}(\lambda))^2 \tag{8-46}$$

where $\bar{f}(\lambda) = \sum_i a_i/(\lambda + \lambda_i)$, the Laplace transform of the ordinary delayed neutron decay function with respect to λ as discussed in Chap. 2.

This development can be continued to obtain relationships analogous to those from ordinary kinetics equations for step function reactivity changes, and approximate solutions for certain time-varying reactivities. Some typical parameters for a coupled-core, fast-thermal system are now given to provide further insight into this kind of a reactor.[13]

$$\begin{gathered} k_{ff} = 0.96 \qquad k_{ss} = 0.17 \qquad k_{sf} = 0.11 \qquad k_{fs} = 0.30 \\ l_{ff} = 0.52\ \mu\text{sec} \qquad l_{ss} = 440\ \mu\text{sec} \qquad l_{sf} = 400\ \mu\text{sec} \qquad l_{fs} = 2.8\ \mu\text{sec} \end{gathered} \tag{8-47}$$

An overall prompt neutron lifetime l can be defined as

$$l^{-1} = -\frac{d}{dt} \log \text{(total neutron importance in entire coupled system)} \tag{8-48}$$

Using the set of kinetics equations adjoint to those in Eq. (8-44) and neglecting delayed neutrons, which does not detract from the generality of the development because only the prompt lifetime is involved, one can obtain

$$\begin{aligned} l = {} & \frac{(1-k_{ss})k_{ff}}{(1-k_{ff})+(1-k_{ss})} l_{ff} + \frac{(1-k_{ff})k_{ss}}{(1-k_{ff})+(1-k_{ss})} l_{ss} \\ & + \frac{(1-k_{ff})(1-k_{ss})}{(1-k_{ff})+(1-k_{ss})} (l_{sf}+l_{fs}) \end{aligned} \tag{8-49}$$

From the typical values above, to good approximation, l_{ff}, $l_{fs} \ll l_{ss}$, l_{sf}; $l_{ss} \cong l_{sf} = l_s$, which reduces the above expression for l to

$$l \cong \left(1 + \frac{1-k_{ss}}{1-k_{ff}}\right)^{-1} l_s \tag{8-50}$$

Again in a coupled-core fast-thermal breeder it is desired that the fast to slow power quotient S_f/S_s should be large for high-efficiency breeding and $k_{ff} \ll 1$, the latter being the k_{eff} for the fast core to provide easy and safe control. Since

$$\frac{S_f}{S_s} = \frac{k_{fs}}{1-k_{ff}} \tag{8-51}$$

it is seen that k_{fs} should be maximized to fulfill both of the above conditions. k_{fs} is, however, constrained by the fact that the thermal and fast portions of the physical coupled system must be sufficiently isolated to maintain the fast spectrum. The physical juxtaposition of the components in a fast-thermal system is in conflict with maintaining a high k_{fs}.

8-9 CIRCULATING FUEL REACTORS: ZERO-POWER KINETICS

A circulating fuel reactor is simply one in which the fuel slurry circulates through the reactor proper as well as the other components of the reactor–heat exchanger loop. Other components of the loop besides the reactor are the heat exchanger, pumps, fission product removal apparatus, etc. A principal reason for building circulating fuel reactors is that the fuel reprocessing is simplified. Also, for high-power thermal circulators, the xenon fission product poisons can be easily removed. The disadvantages are the obvious ones of fuel slurry corrosive effects, need for leak-tight loops, and tying up fuel in the part of the loop external to the reactor.[15]

The transfer function of the circulating reactor differs from that in stationary fuel systems owing to the fact that, for one thing, neutrons can be born in the loop outside the reactor and are therefore not immediately effective in the reactor itself. Initially, consider the space-independent zero-power kinetics equations obtained in Chaps. 1 and 2, but modified to account for the fact that the fuel circulates. These equations are, assuming initial equilibrium[16]

$$\dot{n}(t) = \frac{(K-1)n(t)}{\Lambda} + \sum_i \lambda_i c_i(t) \qquad n(0) = 1 \tag{8-52}$$

$$\dot{c}_i(t) = -(\lambda_i + \lambda_c)c_i(t) + \lambda_c e^{-\lambda_i/\lambda_l} c_i\left(t - \frac{1}{\lambda_c}\right) + \frac{a_i n(t)}{\Lambda} \qquad \dot{c}_i(0-) = 0,\ i = 1, 2, \ldots \tag{8-53}$$

where the second c_i on the right-hand side depends on the displaced time. As can be seen in this model, Eq. (8-52), involving neutron density remains the same as for the stationary fuel reactor. The two additional terms in Eqs (8-53) account for the fact that delayed neutron emitters are leaving and entering the reactor. $\dot{c}_i(t)$ is the rate of change of the concentration of the ith group of delayed neutron emitters *in the reactor*. The quantity $-\lambda_c c_i(t)$ is the flow rate of $c_i(t)$ out of the core. λ_c is the specific flow rate of slurry [and therefore of $c_i(t)$ for all i] out of the core in inverse seconds, as opposed to λ_i, the natural decay rate of the ith delayed neutron emitter. The term $\lambda_c e^{-\lambda_i/\lambda_l} c_i(t - 1/\lambda_c)$ is the immediate flow rate into the core of the ith delayed neutron emitters that were $1/\lambda_c$ sec away, in the external loop. However, these emitters have been reduced by a factor $e^{-\lambda_i/\lambda_l}$, since they decay exponentially with decay constant λ_i, having spent $1/\lambda_l$ sec in the loop exterior to the core. Certain assumptions are implied in the above model. These include (1) the fuel is thoroughly mixed in the reactor (homogeneous fuel concentration), (2) there is constant-rate one-dimensional laminar flow (slug flow), (3) fissions occur in the reactor core only and, (4) only the core itself is capable of criticality.

To obtain the aggregate forms of these kinetics equations, the Laplace transform is the method to be used as suggested by the form of Eqs (8-52) and (8-53). As in Chap. 1, the integral equation aggregate form is obtained analogously as

$$n(t) = \int_0^t G_c(t-\tau)[K(\tau)n(\tau) + H(\tau)]\, d\tau \tag{8-54}$$

where

$$\bar{H}(s) = \Lambda + \bar{M}(s),\ \bar{M}(s) = \sum_i \frac{a_i \alpha_i}{s + \lambda_i/\alpha_i + \lambda_c \bar{g}_i(s)} \tag{8-55a}$$

$$\bar{G}_c(s) = [s\Lambda + 1 - \bar{R}(s)]^{-1}, \; \bar{R}(s) = \sum_i \frac{a_i \lambda_i}{s + \lambda_i/\alpha_i + \lambda_c \bar{g}_i(s)} \tag{8-55b}$$

$$\alpha_i = \left[1 + \frac{\lambda_c}{\lambda_i}(1 - e^{-\lambda_i/\lambda_l})\right]^{-1} \tag{8-55c}$$

$$\bar{g}_i(s) = (1 - e^{-s/\lambda_c})\, e^{-\lambda_i/\lambda_l} \tag{8-55d}$$

That is, in the Laplace transform domain,

$$\bar{n}(s) = \bar{G}_c(s)[\mathscr{L}(Kn) + \bar{H}(s)] \tag{8-56}$$

$\mathscr{L}(Kn)$ is the Laplace transform of the product $K(t)n(t)$, assuming that it exists. For the ordinary stationary fuel reactor, $\lambda_c = \lambda_l = 0$. Then $\bar{g}_i(s)$ vanishes and $\alpha_i = 1$. $\bar{G}_c(s)$ becomes $\bar{G}(s)$, the usual zero-power transfer function, and the zero-power integral equation (1-47) ensues from Eq. (8-54).

To obtain the transfer function for the zero-power circulating reactor $\bar{G}_c(s)$, it is seen from Eqs (8-55) that near initial equilibrium (large s) $\bar{G}_c(s)\bar{H}(s)$ approaches s^{-1}. With $\bar{n}(s) \approx 1/s + \bar{\varepsilon}(s)$, where $\bar{\varepsilon}(s) \ll \bar{n}(s)$, near equilibrium as well, Eq. (8-56) becomes

$$\frac{1}{s} + \bar{\varepsilon}(s) \approx \frac{1}{s} + \bar{G}_c(s)\bar{K}(s) \tag{8-57}$$

and $\bar{\varepsilon}(s)$ has been neglected in comparison to $\bar{n}(s)$ on the right-hand side of Eq. (8-57). Then

$$\bar{\varepsilon}(s) \approx \bar{G}_c(s)\bar{K}(s) \tag{8-58}$$

Figure 8-7 depicts the corresponding zero-power transfer function approximation for one representative group of delayed neutrons with weighted mean decay constant $\bar{\lambda} = (\bar{\tau})^{-1}$. That is, for one group of delayed neutrons, Eq. (8-55b) gives

$$\bar{G}_{c1}(s) = \left[s\Lambda + 1 - \frac{\bar{\lambda}}{s + \bar{\lambda}/\alpha + \bar{g}(s)}\right]^{-1} \tag{8-59}$$

where α and $\bar{g}(s)$ are the one delayed neutron group counterparts of α_i and $\bar{g}_i(s)$, that is

$$\alpha = \left[1 + \frac{\lambda_c}{\bar{\lambda}(1 - e^{-\bar{\lambda}l\lambda_l})}\right]^{-1} \tag{8-60}$$

$$\bar{g}(s) = (1 - e^{-s/\lambda_c})\, e^{-\bar{\lambda}/\lambda_l} \tag{8-61}$$

As can be seen from Fig. 8-7, certain peaks or resonances appear. This is, of course, due to the coupling of the delayed neutron period with the loop circulation periods. Actually, using all of the delayed neutron groups would vitiate this effect somewhat, simply because of the multiplicity of the delayed neutron periods and their respective yields. Mixing effects of the fuel slurry, not taken into account above, would further diminish the resonance behavior of the transfer function.

The corresponding inhour equation can be obtained from Eq. (8-54) in the usual way by letting $n(t) \approx e^{\lambda t}$ therein. Equation (8-54) then gives for constant negative reactivity of amplitude $|K|$

$$1 = e^{-\lambda t} \int_0^t G_c(t - \tau) H(\tau)\, d\tau - |K| \int_0^t G_c(\tau)\, e^{-\lambda \tau}\, d\tau \tag{8-62}$$

The first integral above will vanish for very large time, as can be shown by using the final-value theorem of Laplace transforms. As time grows without bound, the

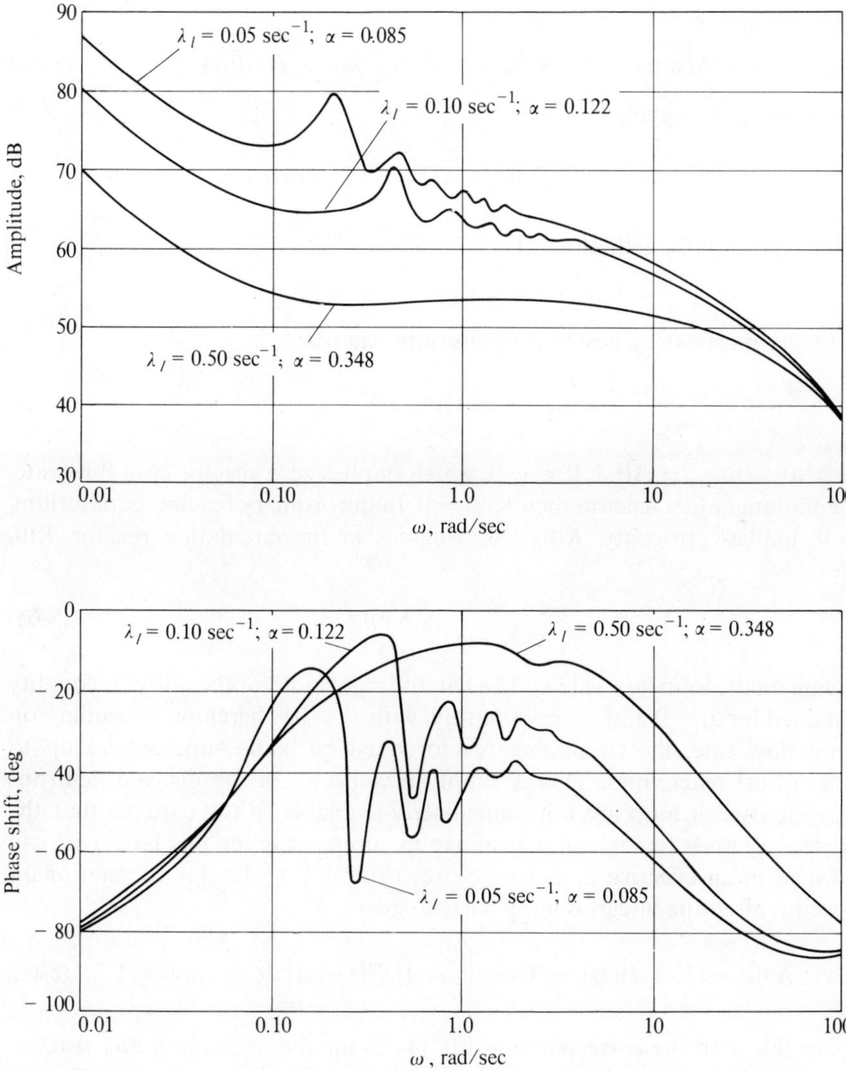

Figure 8-7 Amplitude and phase components of zero-power transfer for a ^{235}U circulating fuel reactor for various external loop residence times $1/\lambda_1 (B = 0.0064,\ \lambda = 0.07,\ \Lambda = 10^{-1}/B,\ \lambda_c = 1)$.

above yields the corresponding inhour equation

$$1 + |K| \bar{G}_c(\lambda) = 0 \tag{8-63}$$

where $\bar{G}_c(\lambda)$ is the Laplace transform of $G_c(t)$ with respect to the parameter λ. The algebraically largest (negative) root of Eq. (8-62) yields the "stable" inverse period for the circulating reactor.

For reasons to follow, it is of interest to obtain the integrodifferential equation form of the circulating reactor kinetics equations. By analogy to the derivation of Eq. (1-42) or, equivalently, by manipulating the Laplace transforms in Eqs (8-55) and (8-56), one obtains

$$\Lambda(s\bar{n}(s) - 1) = \mathrm{L}(Kn) + \bar{M}(s) - \bar{n}(s) + \bar{n}(s)\bar{R}(s) \tag{8-64}$$

Inverting yields the aggregate form

$$\Lambda\dot{n}(t) = [K(t) - 1]n(t) + M(t) + \int_0^t n(\tau)R(t - \tau)\, d\tau \tag{8-65}$$

At initial equilibrium, with $n(0) = 1$

$$0 = \Lambda\dot{n}(0) = K(0) - 1 + M(0) \tag{8-66}$$

so that there must exist a nonzero equilibrium reactivity

$$K(0) = 1 - M(0) = \sum_i a_i \alpha_i \tag{8-67}$$

where $M(0) = \lim_{s\to\infty} s\bar{M}(s)$. If $\lambda_c = 0$, which implies zero reactor core flow rate, i.e., the stationary fuel reactor, then $K(0) = 0$. In the ordinary reactor, equilibrium, $\dot{n}(0) = 0$, implies criticality, $K(0) = 0$; whereas in the circulating reactor $K(0)$ satisfies

$$1 - \sum_i \frac{a_i \lambda_i}{\lambda_i + \lambda_c} \le K(0) \le 1 \tag{8-68}$$

depending on the loop flow rate λ_l. The left and right sides of the above inequality are obtained for $\lambda_l = 0$ and ∞, respectively, with $\lambda_c \neq 0$. Therefore, depending on the loop flow rate, the circulating reactor must be made supercritical up to prompt critical reactivities. This is simply because all of the delayed neutrons born in the overall loop are not immediately available to the core, so that the core reactivity must be increased as above to compensate for this fact. This can be reflected in an effective β_i as well by rewriting Eq. (1-42), for the stationary fuel reactor, after one integration by parts to give

$$\Lambda\dot{n}(t) = (K - 1)n(t) + f(t) + \int_0^t n(\tau)|f'(t - \tau)|\, d\tau \qquad n(0) = 1 \tag{8-69}$$

Compare this with the corresponding Eq. (8-65) for the circulating fuel reactor. Initially, from Eq. (8-68)

$$\Lambda\dot{n}(0) = [K(0) - 1] + f(0) = K(0) \qquad f(0) = 1 \tag{8-70}$$

Equation (8-65) at initial equilibrium is

$$\Lambda\dot{n}(0) = [K(0) - 1] + M(0) \tag{8-71}$$

Making the identification between $f(0)$ and $M(0)$ yields an effective a_i as†

$$a_{i,\text{eff}} = a_i\alpha_i < a_i \tag{8-72}$$

8-10 CIRCULATING FUEL REACTORS: NEUTRONICS AND TEMPERATURE INTERACTION

It is now of interest to discuss the interaction of reactor heat production and temperature with the circulating reactor kinetics. In Sec. 4-5 (Example 3) a simple coupling of temperature and kinetics with no delayed neutrons was discussed. The power was assumed to be extracted by means of circulating fuel (or circulating coolant as well in this simple model) through the reactor at constant temperature. Here the next level in the hierarchy of such reactors as espoused by Ergen[14] and Welton[18] considers the fuel residence time in the circulating reactor core proper, using the same kinetics equation as in the earlier model. It is a special case of Eq. (4-57) with no delayed neutrons assumed ($\beta = 0$), that is, with $Q = \log n(t)$

$$\ddot{Q} + \omega^2(e^Q - 1) = \frac{\dot{K}_c}{\Lambda} + \omega^2 \int_0^\infty [e^{Q(t-\tau)} - 1]R(\tau)\,d\tau \tag{8-73}$$

and

$$R(t) = -\frac{\dot{H}(t)}{H(0)}, \qquad \int_0^\infty R(\tau)\,d\tau = 1, \qquad \text{and} \qquad \omega^2 = \frac{H(0)}{\Lambda} \tag{8-74}$$

$H(t)$ is the unit power reactivity coefficient; i.e., the power feedback reactivity K_p is given by

$$K_p(t) = \int_0^\infty H(\tau)n(t-\tau)\,d\tau \tag{8-75}$$

as discussed in Chap. 4. K_c is the control reactivity with the connotation of applied force in the Eq. (8-73) of Q motion.

As mentioned, the difference between this example and that in Sec. 4-5 is that the circulating fuel resides in the reactor core for a time $\lambda_c^{-1} = t_c$. The remainder of the assumptions is that the fuel is pumped into the core at constant inlet temperature (zero), laminar or slug flow is assumed, and temperature and neutron density are space-independent. Space-dependent considerations in circulating reactors are discussed by Fleck.[19] Following Welton,[18] let $T(t, y)$ be the temperature of fuel slurry at time t which has been in the reactor core for a

† This is not the effective $\bar{a}_i$ due to importance function considerations.

time $y \leq t_c$. Then with C being the core fuel heat capacity

$$CT(t, y) = \int_0^y n(t - \tau)\, d\tau \tag{8-76}$$

This equation simply asserts that the heat generated in the core is given by the integral of the neutron density (power) over the time y that a particular slug of fuel resides in the core. Letting α be a suitable reactivity temperature coefficient

$$K_p(t) = \frac{\alpha}{t_c} \int_0^\infty T(t, y)\, dy \tag{8-77}$$

Combining Eqs (8-76) and (8-77) results in

$$K_p(t) = \frac{\alpha}{Ct_c} \int_0^\infty dy \int_0^y n(t - \tau)\, d\tau \tag{8-78}$$

Since $0 \leq y \leq t_c$, the above is rewritten to give

$$K_p(t) = \frac{\alpha}{Ct_c} \int_0^{t_c} n(t - \tau)\, d\tau \int_\tau^{t_c} dy = \frac{\alpha}{C} \int_0^{t_c} \left(\frac{1 - \tau}{t_c}\right) n(t - \tau)\, d\tau \tag{8-79}$$

or

$$K_p(t) = \frac{\alpha}{C} \int_0^\infty \left(\frac{1 - \tau}{t_c}\right) n(t - \tau) h(t_c - \tau)\, d\tau \tag{8-80}$$

where $h(x)$ is the Heaviside unit function defined by

$$h(x) = \begin{cases} 1 & x \geq 0 \\ 0 & x < 0 \end{cases} \tag{8-81}$$

Making the identification from Eqs (8-74) and (8-75), the unit power reactivity coefficient is

$$H(t) = \frac{\alpha}{C} \left(\frac{1 - t}{t_c}\right) h(t_c - t) \tag{8-82}$$

and, within certain delta functions manifest through the process of differentiation

$$R(t) = \frac{h(t_c - t)}{t_c} \tag{8-83}$$

Inserting this expression for $R(t)$ into Eq. (8-73) gives the corresponding equation of Q motion when disturbed from initial equilibrium

$$\ddot{Q} + \omega^2(e^Q - 1) = \frac{\dot{K}_c}{\Lambda} + \frac{\omega^2}{t_c} \int_0^\infty (e^{Q(t-\tau)} - 1) h(t_c - \tau)\, d\tau \qquad Q(0) = \dot{Q}(0) = 0 \tag{8-84}$$

To investigate the stability of a circulating system so described, it is initially assumed that only small fluctuations from equilibrium ensue. Near equilibrium then, $e^Q \cong 1 + Q$, and assuming constant K_c results in the linearized version of the equation of motion

$$\ddot{Q} + \omega^2 Q = \frac{\omega^2}{t_c} \int_0^{t_c} Q(t - \tau)\, d\tau \tag{8-85}$$

Differentiating yields the differential equation with a lag term

$$\dddot{Q} + \omega^2 \dot{Q} + \frac{\omega^2}{t_c}[Q(t - t_c) - Q] = 0 \tag{8-86}$$

The stability behavior can be easily examined by letting $Q = e^{\lambda t}$ to obtain

$$\lambda^3 + \lambda\omega^2 - \frac{\omega^2}{t_c}(1 - e^{-\lambda t_c}) = 0 \tag{8-87}$$

a transcendental equation whose roots in general are complex. They are, however, shown to have negative definite real parts,[20] which guarantees local stability; i.e., the system is stable only for small fluctuations about equilibrium under constant reactivity excitation. A stronger stability criterion, due to Welton and discussed in Chap. 4, asserts that such a reactor as described by Eq. (8-84) is stable if the premultiplied Fourier sine transform of $R(t)$ is positive definite. That is, the system is stable, even for nonlinear fluctuations of Q, if

$$\int_0^\infty R(\tau)\omega \sin \omega\tau\, d\tau \geq 0 \tag{8-88}$$

For $R(t) = h(t_c - t)/t_c$ above, the system is stable, since

$$\frac{1}{t_c}\int_0^\infty h(t_c - \tau)\omega \sin \omega\tau\, d\tau = \frac{2}{t_c} \sin^2 \frac{\omega t_c}{2} \geq 0 \tag{8-89}$$

The delta function behavior of $R(t)$ at t_c, alluded to earlier, does not detract from the stability of this system, as is easily shown.

8-11 BOILING-LIQUID REACTORS: INTRODUCTION

Boiling-liquid-moderated reactors have a United States history beginning in 1953–1954, when the boiling reactor experiment (BORAX-I)† proved the feasibility of an operating boiling reactor at substantial power ‡ (BORAX-IV operated at 70 kW per liter of H_2O moderator). The main purpose of allowing the moderator to boil is that the fuel coolant-moderator boiling heat transfer rates are much

† The BORAX (boiling reactor experiment) series are large boiling-water-moderated and -cooled thermal reactors.

‡ The Dresden reactor, located near Chicago,[21] is the first large-scale boiling-water reactor producing commercial power (625 MW). It is fueled with 1.5 per cent enriched uranium oxide, moderated and cooled with light water, and controlled with 80 cruciform stainless-steel–boron control rods.

greater than nonboiling rates. This makes for greater reactor operating power densities. Nonboiling-water-moderated reactors must be operated at much higher moderator pressures and temperatures to get adequate boiling heat transfer rates. This requires thick reactor pressure vessels and the concomitant high-pressure valves and piping which are more expensive and unwieldy than their low-pressure counterparts. On the other hand, the boiling process must not become turbulent or effulgent, because this will deleteriously affect the reactor kinetic stability through making the boiling moderator neutronics properties vary over unduly large ranges. That is, even though excessive bubble void formation corresponds to an inherent negative reactivity temperature coefficient, such operation would strongly militate against normal stable control. Until control systems can be built to tolerate fulminating boiling, the power density is limited. There are other secondary advantages and disadvantages of boiling reactors.[22]

8-12 BOILING-LIQUID REACTORS: VOID REACTIVITY COEFFICIENTS

Among the usual reactivity coefficients due to moderator and fuel temperature, etc., that due to the moderator bubble voids is by far the largest. The amount of negative reactivity in the moderator voids can be as much as 10 dollars for a 70 kW per liter reactor, as shown in Fig. 8-8.

The void coefficient of reactivity can be defined as $dK/d\alpha$, where K is the reactivity and α is the steam void fraction.

$$\alpha = 1 - \frac{\text{water moderator density in boiling reactor}}{\text{water density at saturation conditions}} \tag{8-90}$$

or

$$\alpha = 1 - \text{water fraction contained in voids plus water in reactor}$$

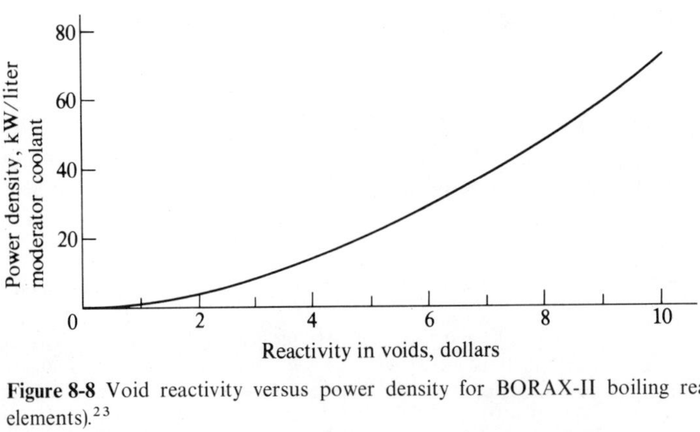

Figure 8-8 Void reactivity versus power density for BORAX-II boiling reactor (300 psig, 69 fuel elements).[23]

Table 8-4 Reactivity void coefficient terms for EBWR-I†

Term	(A)	(B)	(C)	(D)	(E)	(F)	(G)
Calculated value	+0.004	+0.110	−0.011	+0.016	~0	+0.208	−0.181

† After Thie.[24]

Consider a continuous slowing-down approximation for a reactor with a uniform void concentration. From the definition of reactivity

$$\frac{dK}{d\alpha} = \frac{d}{d\alpha}(1 - k_{\text{eff}}^{-1}) = \frac{1}{k_{\text{eff}}^2}\frac{d}{d\alpha}\left[\frac{k_\infty}{(1 + L^2B^2)\,e^{\tau B^2}}\right] \tag{8-91}$$

where $k_\infty = \eta\varepsilon fp$, the usual four-factor expression. Now

$$K = K(L^2, \tau, B^2, \varepsilon, \eta, f, p)$$

so that its total derivative with respect to α is[24]

$$\frac{dK}{d\alpha} = (1 - K)\left[-\frac{B^2}{1 + L^2B^2}\frac{dL^2}{d\alpha} - B^2\frac{d\tau}{d\alpha} - \left(\frac{L^2}{1 + L^2B^2} + \tau\right)\frac{dB^2}{d\alpha} + \frac{1}{\varepsilon}\frac{d\varepsilon}{d\alpha} + \frac{1}{\eta}\frac{d\eta}{d\alpha} + \frac{1}{f}\frac{df}{d\alpha} + \frac{1}{p}\frac{dp}{d\alpha}\right] \tag{8-92}$$

or labeling each term alphabetically

$$\frac{dK}{d\alpha} = (1 - K)[-(A) - (B) - (C) + (D) + (E) + (F) + (G)] \tag{8-93}$$

The terms (*A*) through (*G*) were calculated for the EBWR-I, and they are given in Table 8-4. The calculation yields $dK/d\alpha = -0.06(1 - K)$, $K < 1$, which asserts that the total void reactivity coefficient is negative. Of course, this follows from physical reasoning as well, since the presence of voids degrades the moderating properties of the liquid moderator, and to a sizable amount, as mentioned earlier.

Another type of void coefficient exists in homogeneous (nonboiling) liquid reactors. This is due to gas bubbles formed from radiolytic dissociation. An example of such a reactor is the kinetic experiment on water boilers (KEWB),[25] which consists essentially of a sphere of 11.5 liters of enriched uranyl sulphate solution (1890 g of ^{235}U). Owing to the radiolytic dissociation of the water solvent, the gas bubbles formed thereby result in a negative reactivity coefficient[25] of $\sim 3 \times 10^{-5}\,\Delta K$ per cubic centimeter, or $dK/d\alpha = -0.3\,\Delta K$ for a 10^4-cm^3 sphere. The KEWB reactor is discussed in detail in Sec. 9-15.

8-13 BOILING-LIQUID REACTORS: OSCILLATORY BEHAVIOR

Boiling-liquid reactors, because of hydrodynamic and neutronic coupling of various kinds, exhibit different types of oscillatory behavior. Each of these will be examined with regard to the kinetic stability of the reactor.

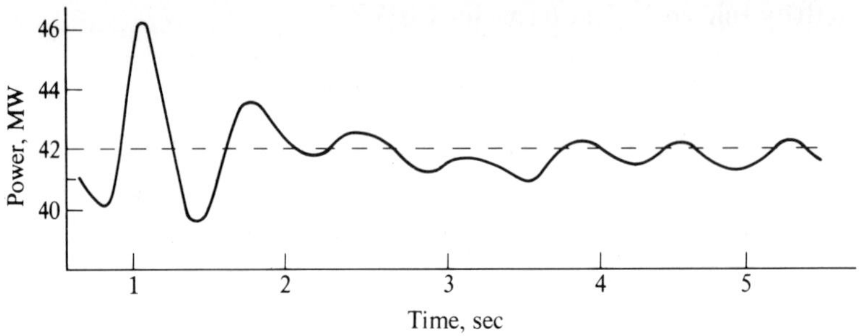

Figure 8-9 Boiling reactor (small amplitude) shugging oscillations after sudden partial withdrawal (2 in.) of central rod in EBWR (600 psi).[26]

Following Thie,[26] the first type is called "chugging," cf. Fig. 8-9. It is defined as oscillations whose amplitude is equal to or exceeds the average reactor power. These are quite drastic fluctuations in power and usually result in scramming the reactor. They can be caused by the quick ejection of a control-rod corresponding to a positive step change in reactivity that results in a prompt period of approximately 0.02 sec (BORAX-I).[21] For such fast periods, enough water moderator is ejected from the reactor to expose the core and cause a scram shutdown. When the chugging is such that the amplitude is constant or decreasing, the power or neutron level can be expressed by the following model[26]

$$n(t) = n_\infty \left[1 - \left(1 - \frac{n_0}{n_\infty} \right) e^{-\omega_m \zeta t} \cos \omega_m t \right] \tag{8-94}$$

where ω_m is the chugging angular frequency and ζ is the damping constant. If the transfer function of the boiling reactor, without feedback, is known at least well enough to discern the frequency at which it peaks, the decrement ζ in Eq. (8-94) can be obtained.

An expression for the transfer function can be obtained in terms of a small-amplitude initial positive step change in reactivity[27] (a ringing experiment) and cast in a form suitable to determine ζ. Consider Eq. (1-47), the zero-power integral equation of reactor kinetics for the neutron equilibrium level n_0 with an initial reactivity step of magnitude K_0. It is

$$n = n_0 + K_0 \int_0^t G(t - \tau) n(\tau)\, d\tau \tag{8-95}$$

For K_0 small enough, the neutron fluctuations about equilibrium are sufficiently small that the above equation can be linearized by letting $n = n_0$ in the integral term to obtain

$$n(t) = n_0 \left[1 + K_0 \int_0^t G(\tau)\, d\tau \right] \tag{8-96}$$

since $n \cong n_0$ for the small changes about equilibrium. After a long time, Eq. (8-96) gives the asymptotic neutron density as

$$n_\infty = n_0\left[1 + K_0 \int_0^\infty G(\tau)\,d\tau\right] \tag{8-97}$$

Subtracting the two equations gives

$$n(t) - n_\infty = -n_0 K_0 \int_t^\infty G(\tau)\,d\tau \tag{8-98}$$

The Fourier transform of the above, defined over the positive t axis, is

$$\int_0^\infty [n(t) - n_\infty]\,e^{-i\omega t}\,dt = -n_0 K_0 \int_0^\infty e^{-i\omega t}\,dt \int_t^\infty G(\tau)\,d\tau \tag{8-99}$$

Using Eq. (8-97) in Eq. (8-99), the result follows easily that

$$\overline{G}(\omega) = \frac{1}{n_0 K_0}\left\{(n_\infty - n_0) + i\omega \int_0^\infty [n(t) - n_\infty]\,e^{-i\omega t}\,dt\right\} \tag{8-100}$$

where $\overline{G}(\omega)$ is the Fourier transform of $G(t)$.† If Eq. (8-94) is used for $n(t)$ in the above expression for $\overline{G}(\omega)$, one obtains, for $\zeta \ll 1$ and at $\omega = \omega_m$, the chugging frequency

$$|\overline{G}(\omega_m)| \cong \frac{n_\infty - n_0}{2\zeta n_0 K_0} \tag{8-101}$$

It is, of course, implied that the maximum of $|\overline{G}(\omega)|$ occurs at the chugging frequency ω_m, that is, that the dominant resonant mechanism operating gives rise to the chugging phenomena. Therefore, determining $|\overline{G}(\omega_m)|$, a salient feature of $|\overline{G}(\omega)|$, from a ringing experiment enables one to calculate ζ by using Eq. (8-101). A typical, experimentally determined $|\overline{G}(\omega)|$ for the EBWR is depicted in Fig. 8-10.

On the other hand, the problem of determining ζ can be approached "backwards" as in the following development. Let the reactivity be represented by the empiricism $K = K_0\,e^{-\zeta\omega_m t}\sin\omega_m t$, where K_0 is the amplitude and ω_m is the chugging frequency. From Eq. (1-42), which is an integrodifferential equation for the neutron density given the reactivity function, but is merely an integral relationship for reactivity given the neutron density (for zero source therein)

$$\Lambda\dot{n} = Kn - \int_0^t n'(\tau)f(t-\tau)\,d\tau \tag{8-102}$$

Solving for K gives

$$K = \frac{\Lambda\dot{n} + \int_0^t n'(\tau)f(t-\tau)\,d\tau}{n(t)} \tag{8-103}$$

† $|\overline{G}(\omega)|$ will not correspond to $|\overline{G}_0(\omega)|$, the zero-power transfer function amplitude obtained by "linearizing" the reactor kinetics equations. This is because of the feedback mechanisms inherent in the reactor, which are part of $|\overline{G}(\omega)|$ when it is measured.

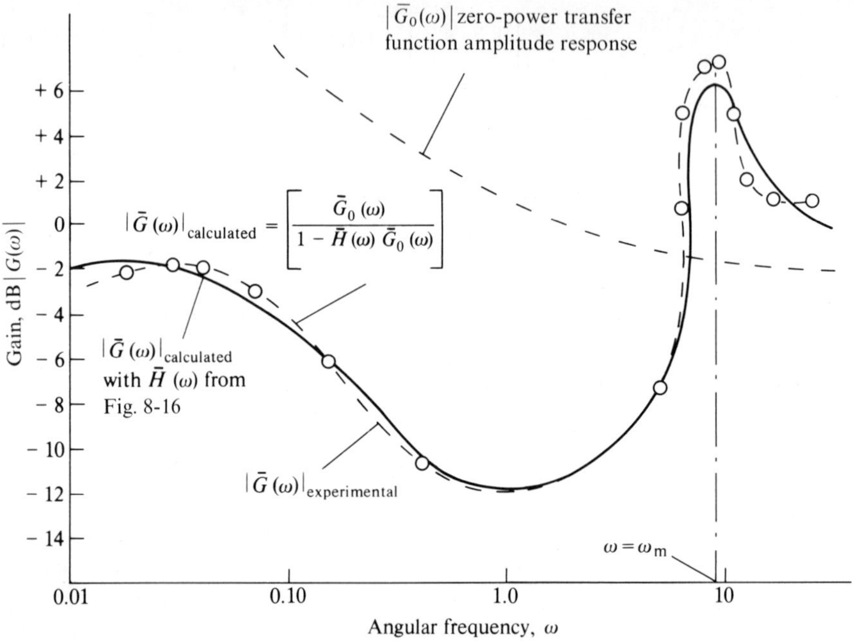

Figure 8-10 Experimental determined amplitude response $\{G(\omega)\}$ for EBWR at 40 MW.[1]

With experimentally obtained neutron level traces exhibiting chugging fed into the right-hand side of the above equation, programmed on EDPM machinery, a time plot of K can be obtained. Since K oscillates, its successive amplitudes (K_{max}) can be plotted as a function of time, from which the logarithmic decrement ζ can be obtained. Such a plot is shown in Fig. 8-11.

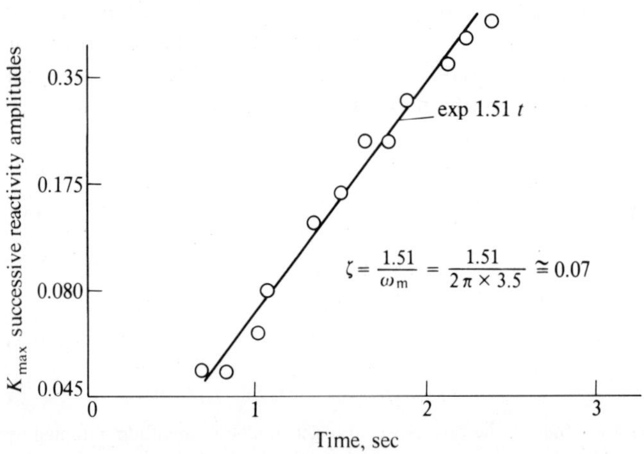

Figure 8-11 Successive reactivity amplitudes of an oscillating (chugging) reactivity versus time in BORAX-II (4 per cent reactivity in voids).[26]

Table 8-5 Logarithmic decrement of the reactivity $K = K_0 e^{-\zeta\omega_m t} \sin \omega_m t$ as a function of various reactor configurations†

Reactor	Moderator water pressure, psig	Power, MW	Reactivity, in voids, $	Chugging frequency Hz	ζ
SPERT-I‡	0	6	<3.3	2.13	−0.0180
BORAX-II (43-12)§	150	3.8	3.7	2.56	−0.0174
BORAX-II (46-12)	150	4	3.7	2.84	−0.0274
BORAX-II (46-12)	150	4	3.7	2.79	−0.0160
BORAX-II (46-12)	193	4.9	4.5	3.57	−0.0326
BORAX-II (69-3)	309	9.5	5.3	3.54	−0.0149
BORAX-II (69-3)	300	10.1	5.6	3.54	−0.0680

† After Thie.[26]

‡ SPERT-I is a boiling-water pulsed reactor, it is discussed in Sec. 9-13.

§ (43-12) indicates that the core is composed of 43 heavy fuel elements (157.3 g fuel) and 19 light elements (93.4 g fuel).

This technique[26] was used to determine the logarithmic decrement for a number of BORAX reactor configurations, as well as the SPERT-I reactor. The results are given in Table 8-5.

The cause of the chugging oscillation, as correlated by transfer function measurements of the oscillatory and ringing type, is primarily surface fluctuation of the boiling moderator water.[26] One explanation of the chugging oscillation[26] begins by considering a heated pulse of water generated in the nonboiling region of the moderator. This is known as the subcooled region. Assume that a pulse is generated just exterior to a fuel element. Then this heat pulse takes, on the average, about 0.5 sec to travel out into the subcooled water before rising. The pulse of heated water now takes about 0.25 sec to rise to the surface, where boiling is taking place. The additional heat represented in the pulse now causes a pulse of bubbles to form at the surface, which drops the power due to the bubble void. The bubbles, i.e., the steam and saturated water, pass out of the reactor in about 0.25 sec. Because the void "pulse" is making its influence less felt as it leaves the reactor, the power rises. As the power rises, the process can repeat itself with a period of about one second. Integrating the kinetics equations and the associated hydrodynamics equations, by using EDPM equipment, confirms the above model in a semiquantitative manner.[24]

8-14 BOILING-LIQUID REACTORS: TRANSFER FUNCTION ANALYSIS

The second major kind of boiling reactor oscillatory behavior is that due to the fluctuations of the steam voids owing to power being generated in the boiling zone. The first kind of oscillatory behavior, discussed in the preceding section,

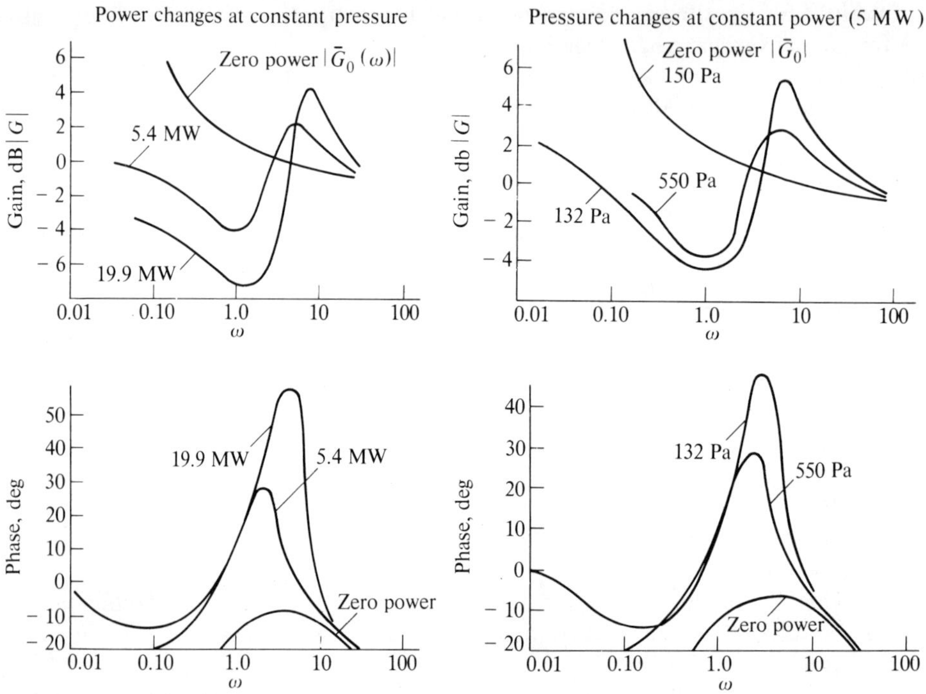

Figure 8-12 Experimental determined (open-loop) transfer functions for EBWR-I versus mean operating power level.[23]

is due to boiling boundary fluctuations because of the power generated in the nonboiling zone.

Transfer functions at various power levels were obtained from the EBWR-I to investigate the approach to instability as the power is increased. Some of these for EBWR-I are shown in Fig. 8-12.

The gain consists principally of two components. The first is that due to the usual delayed and prompt neutron decay parameters; the second is due to the presence of the internal boiling void feedback loops discussed above. Another, lesser, component is that due to pressure reactivity feedback, which causes the transfer function to rise at low frequencies. The pressure is coupled to the concentration of bubble voids, which, in turn, is inversely proportional to pressure. This was shown experimentally on BORAX-IV by an increase in the stability power upper limit with increasing moderator water pressure.

To determine the limit of stability as the power is increased, a phenomenological feedback model is first constructed. One such for EBWR is given in Fig. 8-16. It is essentially self-explanatory in the context of feedback control considerations. The procedure begins, as discussed in Chap. 4, by composing an $\bar{H}(s)$ from all of the neutronic, hydrodynamic, and thermodynamic parameters. This is an onerous, but not difficult, task because of the complexity of the interactions. Detailed excellent syntheses of $\bar{H}(s)$ for boiling reactors are given by Thie,[26] Akcasu,[28] and Beckjord.[29]

Part of the synthesis of $\bar{H}(s)$, composed of a number of subsidiary feedback loops, will be developed from the following list of transfer functions that include:

1. Fuel-to-moderator heat transfer/fuel fission power input.
2. Moderator void reactivity/moderator heat power input.
3. Moderator void reactivity/pressure rate input.
4. Moderator void reactivity/moderator boiling boundary height input.
5. Moderator boiling boundary height/moderator heat power input.
6. Moderator boiling boundary height/pressure rate input.
7. Moderator boiling boundary height/feedwater flow input.
8. Moderator water acceleration/net steam volume input.
9. Core net steam volume/core total steam volume input.
10. Moderator water temperature/pressure input.

To convey the essence of the analysis involved, transfer functions for the first two entries of the above list will be obtained.

For illustration consider a fuel-coolant interface, as shown in Fig. 8-13, where the fuel cladding is neglected. It is easily added when desired. The fission heat power, which is proportional to the neutron density $n(t)$ in the fuel element, is conducted to the coolant at temperature $T_2(t)$. For the fuel elements that interface the steam portion of the moderator, i.e., above the boiling boundary height, the film resistance at the interface,[28] $R = T(a, t)/\phi_{\text{heat}}(a, t)$. $T(a, t)$ and ϕ_{heat} are the temperature and the heat flux at the interface, respectively. For the corresponding situation below the boiling boundary (in the nonboiling region) the Dittus–Boelter equation[29] can be used to calculate the film resistance. These considerations provide interface conditions in order to integrate the heat conduction equations. For the illustrative problem, simpler boundary conditions will be employed.

The fuel temperature $T(x, t)$ is assumed to satisfy the one-dimensional heat

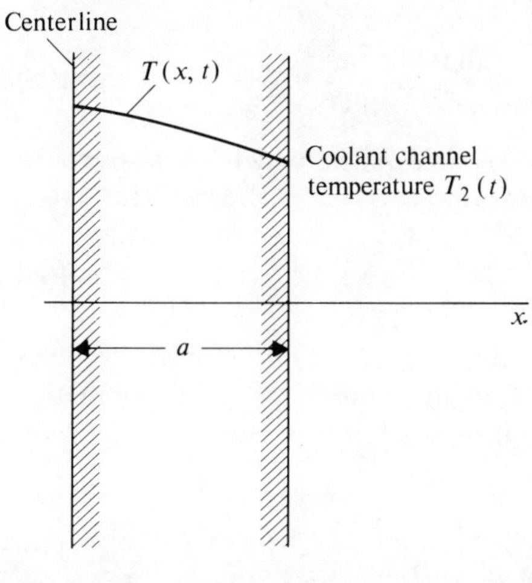

Figure 8-13 Section of fuel element slab immersed in moderator coolant.

conduction equation, cf. Iriarte,[30]

$$k \frac{\partial^2 T}{\partial x^2}(x, t) + bkn(t) = \rho c \frac{\partial T}{\partial t}(x, t) \tag{8-104}$$

ρ, c, and k are the fuel density, specific heat, and thermal conductivity, respectively. b is a proportionality constant such that $bkn(t)$ is the per unit volume heat power generated in the fuel element due to fission. The initial condition is taken to be a zero initial temperature $T(x, 0) = 0$. One boundary condition asserts that a temperature extremum (maximum) occurs at the center of the fuel element, by symmetry. This then implies $\partial T/\partial x\,(0, t) = 0$. The second boundary condition is that the heat flux conducted across the fuel coolant interface is proportional to the temperature difference between the fuel and coolant. This is expressed as $-k\,\partial T/\partial x\,(a, t) = H\{T(a, t) - T_2(t)\}$, where H is the film heat transfer coefficient.

Using Laplace transforms with respect to the time variable in Eq. (8-104) yields a simple second-order differential equation for the transform of the temperature:

$$\frac{d^2\bar{T}}{dx^2}(x, s) - \frac{s}{\kappa}\bar{T}(x, s) = -b\bar{n}(s), \qquad \kappa = \frac{k}{\rho c} \tag{8-105}$$

With the boundary conditions mentioned, the solution is

$$\bar{T}(x, s) = \frac{[s\bar{T}_2(s) - b\kappa\bar{n}(s)]\cosh(s/\kappa)^{1/2}x}{s[\cosh(s/\kappa)^{1/2}a + (1/h)(s/\kappa)^{1/2}\sinh(s/\kappa)^{1/2}a]} + \frac{\kappa b}{s}\bar{n}(s), \qquad h = \frac{H}{k} \tag{8-106}$$

Since the fuel heat flux output in the coolant $Q(t)$ is given by $H[T(a, t) - T_2(t)]$, then from Eq. (8-106) and $\bar{Q}(s) = H\{\bar{T}(a, s) - \bar{T}_2(s)\}$

$$\bar{Q}(s) = \bar{G}_{\text{fuel}}(s)\left[\bar{n}(s) - \frac{s}{\kappa b}\bar{T}_2(s)\right] \tag{8-107}$$

where the fuel heat power transfer function is

$$\bar{G}_{\text{fuel}}(s) = \frac{kb(\kappa/s)^{1/2}\sinh(s/\kappa)^{1/2}a}{\cosh(s/\kappa)^{1/2}a + 1/h(s/\kappa)^{1/2}\sinh(s/\kappa)^{1/2}a} \tag{8-108}$$

$\bar{G}_{\text{fuel}}(s)$ is analytic everywhere except as its simple poles, which correspond to the zeros of its denominator. Then the expansion

$$\bar{G}_{\text{fuel}}(s) = \sum_{r=0}^{\infty} \frac{G_r}{s - s_r} \tag{8-109}$$

holds, where the coefficients G_r are the residues at the infinite number of poles at $s = s_r$. The series can be truncated or, equivalently, numerator and denominator can be expanded in powers of s. Using the first two terms of such an expansion for the hyperbolic functions gives

$$\bar{G}_{\text{fuel}}(s) \cong \frac{ab\kappa[1 + (a^2/b\kappa)s]}{1 + (a/\kappa h)(1 + ah)s + (a^3/6h\kappa^2)s^2} \tag{8-110}$$

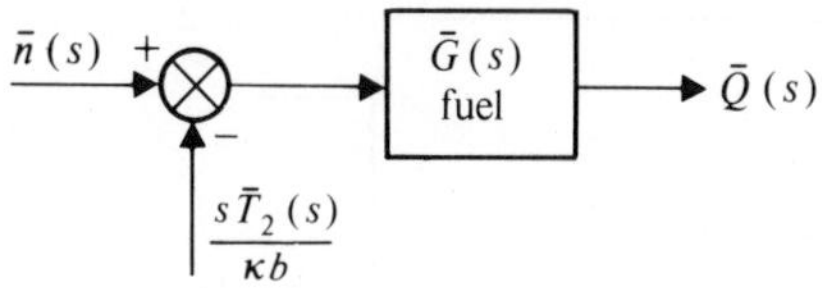

Figure 8-14 Boiling reactor transfer function fragment due to fuel heat power.

This constitutes a fragment, depicted in Fig. 8-14, of the overall transfer function of the boiling reactor.

To obtain the transfer functions for the moderator coolant characteristics, such as voids, steam volume, and pressure, equations to describe the hydrodynamics of the moderator are needed. They are obtained from the conservation equations of mass, momentum, and energy. For the one-dimensional slab situation, paraphrasing Akcasu,[28] the conservation of (water and steam) mass equation gives, for a volume element of thickness dx consisting of both steam volume $A_s\,dx$ and water volume $A_w\,dx$

$$\frac{\partial}{\partial x}(A_s\rho_s V_s + A_w\rho_w V_w) + \frac{\partial}{\partial t}(A_s\rho_s + A_w\rho_w) = 0 \tag{8-111}$$

where V_s and V_w are the steam and water velocities, respectively, and the other symbols have their usual meanings.

Assuming that the variation of kinetic and potential energies and pressure are negligible along the fuel coolant channel, the conservation of energy equation can be expressed in terms of saturated enthalpies h_s and h_w. Then

$$\frac{\partial}{\partial x}(A_s\rho_s V_s h_s + A_w\rho_w V_w h_w) + \frac{\partial}{\partial t}(A_s\rho_s h_s + A_w\rho_w h_w) = Q(x, t) \tag{8-112}$$

and Q is the heat power in the channel. With the further assumption of constant steam density and latent heat h_v, the above two conservation equations combine to yield

$$\frac{\partial}{\partial x}(A_s V_s) + \frac{\partial A_s}{\partial t} = \frac{1}{h_v}\left[Q(x, t) - \left(A_s\rho_s\frac{dh_s}{dp} + A_w\rho_w\frac{dh_w}{dp}\right)\dot{p}\right] \tag{8-113}$$

where the enthalpies are function of the pressure p only. For transfer function relationships, only small fluctuations about the equilibrium values of pertinent quantities are tolerated. Each variable is then expressed by its equilibrium value plus its fluctuation about equilibrium, for example, $Q = Q^0(x) + q(x, t)$, and $q(x, t) \ll Q^0(x)$. To terms through the first order, Eq. (8-113) then yields

$$\frac{\partial}{\partial x}(a_s V_s) + \frac{\partial a_s}{\partial t} = \frac{1}{h_v}\left[q(x, t) - \left(A_s^0\rho_s\frac{dh_s}{dp} + A_w^0\rho_w\frac{dh_w}{dp}\right)\dot{p}\right] \tag{8-114}$$

In the first term on the left-hand side, the correct quantity is

$$\frac{\partial}{\partial x}(a_s V_s^0 + v_s A_s^0) \approx \frac{\partial}{\partial x}(a_s V)$$

where V is the propagation velocity of the fluctuation.

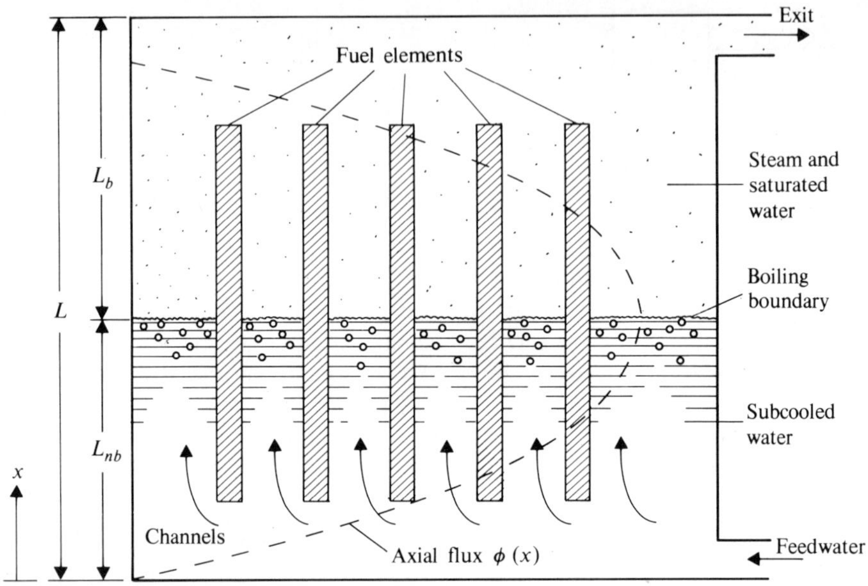

Figure 8-15 Boiling reactor core and moderator showing steam and water regions.

To obtain the moderator steam void heat power input transfer function, first assume that the pressure is constant so that $\dot{p} = 0$. Pressure variations are relegated to the pressure input transfer function (item 10 in the preceding tabulation). In the boiling region, power changes affect steam production, hence steam volume. In the nonboiling region, power changes produce more or less steam, thus shifting the height of the boiling boundary up and down (Fig. 8-15). With the above stipulations, Eq. (8-114) becomes simply

$$\frac{\partial}{\partial x}(a_s V^0) + \frac{\partial a_s}{\partial t} = \frac{q(x, t)}{h_v} \tag{8-115}$$

where $V \cong V^0(x)$, implying that the perturbational velocity is essentially the same as the mainstream velocity V^0. $a_s(x, 0) = 0$ and $a_s(L_{nb}, t) = 0$ are the initial and boundary conditions, respectively.† The latter implies that the boiling boundary height is constant with time. Boiling boundary height fluctuations correspond to transfer functions 5 to 7 in the preceding tabulation, and are not discussed herein. Let the thermal heat production follow the steady-state axial flux $\phi(x)$, so that $q(x, t) = q(t)\phi(x)/L$, where L is the core height and

$$\frac{1}{L}\int_0^L \phi(x)\, dx = 1$$

The Laplace transform with respect to time, with the above conditions, results

† L and L_{nb} are defined in Fig. 8-15.

in simply

$$\frac{d}{dx}[\bar{a}_s(x,s)V^0(x)] + s\bar{a}_s(x,s) = \frac{\bar{q}(x,s)\phi(x)}{Lh_v} \tag{8-116}$$

To calculate the effect of steam voids on the reactivity, a weighted steam volume must be used, since the void reactivity is a function of position in the channel. Because the voids affect the thermal diffusion length primarily, first-order one-energy group perturbation theory asserts that the correct weight is $|\nabla\phi|^2$. Then apropos of the foregoing, the weight $w(x) = \phi'^2(x)$. The reactivity "worth" of the voids $v(t)$ is the following integral over the steam volume in the channels, i.e.,

$$v(t) = \int_{L_{nb}}^{L} w(x)a_s(x,t)\,dx \tag{8-117}$$

Inserting the integral of Eq. (8-116) into the Laplace transform of Eq. (8-117) gives the sought-after void-power transfer function

$$\frac{\bar{v}(s)}{\bar{q}(s)} = \frac{1}{Lh_v}\int_{L_{nb}}^{L} dx\,\frac{\phi'^2(x)}{V^0(x)}\int_{L_{nb}}^{x} du\,\phi(u)\,e^{-s[t(x)-t(u)]} \tag{8-118}$$

where

$$t(x) = \int_{L_{nb}}^{x} \frac{du}{V^0(u)}$$

To achieve a more familiar looking transfer function, let $V^0(x) = V_0$, a constant velocity. Then with the axial flux $\phi(x) = (\pi/2)\sin(\pi x/L)$

$$\frac{\bar{v}(s')}{\bar{q}(s')} = \frac{\pi^2k^3}{8h_v\bar{V}_0(s'^2+k^2)}\left\{(1+\cos^3 kL_{nb})\frac{s'}{3k} + \frac{1}{k}(1-\tfrac{1}{3}\sin^2 kL_{nb})\sin kL_{nb}\right.$$
$$\left. + \frac{s'\sin kL_{nb} - k\cos kL_{nb}}{s'^2+4k^2}[s'\,e^{-s'L_{ab}} + \cos kL_{nb}(2k\sin kL_{nb} - s'\cos kL_{nb})]\right\} \tag{8-119}$$

with $s' = s/V_0$, $k = \pi/L$. If the boiling boundary is assumed to be at the core half-height, then $L_b = L_{nb} = L/2$ (in EBWR the average height of the boiling boundary $L_b = \frac{2}{3}L$). This results in the simplification

$$\frac{\bar{v}(s)}{\bar{q}(s)} = \frac{k^2\pi^2}{8h_v(s^2+\omega_0^2/4)}\left[\frac{\omega_0 s^2\,e^{-s\theta_b}}{2(s^2+\omega_0^2)} + \frac{s+2V_0}{3}\right] \tag{8-120}$$

where $\theta_b = L_b/V_0$ is the flow transit time through the boiling region and $\omega_0 = 2\pi V_0/L$.

Leaving the parameters undefined, because to discuss them would be to digress too far, an $\bar{H}(s)$ for the EBWR at 20 Mw and 300 psig is given as[11]

$$\bar{H}(s) = K_k\frac{[V-(\sigma/\eta)(b+d)][1+s(\gamma V-\sigma d)T_r]/[\eta V-\sigma(b+d)]}{[1+s(\gamma/\eta)T_r](1+sT_h)(1+sT_{pv})(1+s\tau)} \tag{8-121}$$

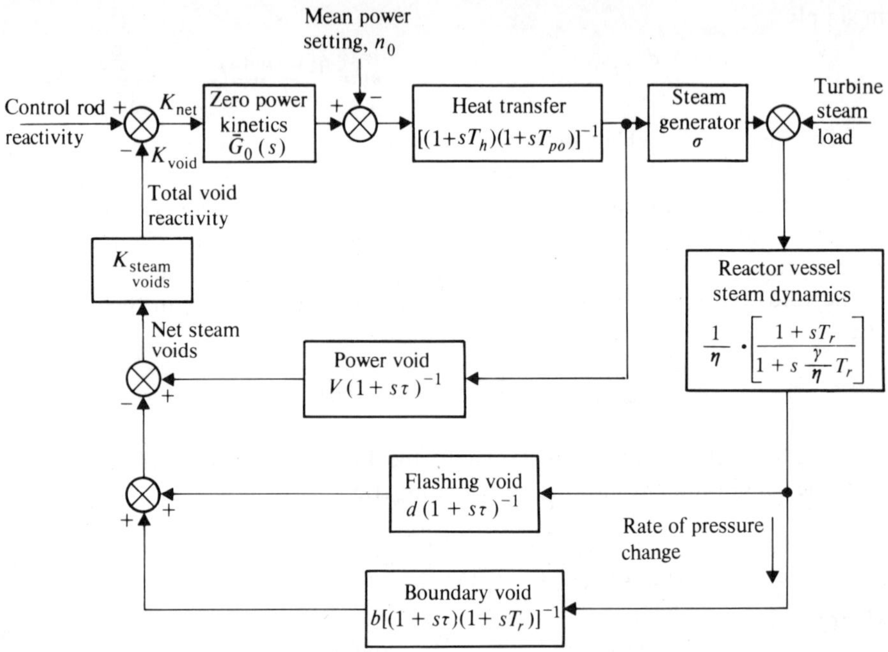

Figure 8-16 Feedback model of EBWR reactor.[23]

Analyzing and fitting the parameters appearing in $\bar{H}(s)$ in Fig. 8-16 for the EBWR finally yields[11]

$$\bar{H}(s) \cong \frac{0.94(1+9.6s)}{(1+0.075s)(1+0.17s)(1+0.23s)(1+29s)} \tag{8-122}$$

Again, this technique of synthesizing an $\bar{H}(s)$, with the exception of void and boiling water–steam boundary considerations, is not unique to boiling-water reactor systems.

$H(t)$ is given by the inverse Laplace transform of the above $\bar{H}(s)$. From the partial fraction expansion of $\bar{H}(s)$

$$H(t) = \sum_{r=1}^{4} h_r e^{-s_r t} \tag{8-123}$$

where the s_r are the roots of the denominator of Eq. (8-122). The coefficients h_r are given as

$$h_r = 0.94\,\frac{1+9.60s_r}{D'(s_r)} \tag{8-124}$$

where $D'(s_r)$ is the derivative of the denominator of Eq. (8-122) evaluated at s_r. From Fig. 8-17 the closed loop transfer function $\bar{G}(\omega)$ is

$$\bar{G}(\omega) = \frac{\bar{G}_0(\omega)}{1+\bar{G}_0(\omega)\bar{H}(\omega)} \tag{8-125}$$

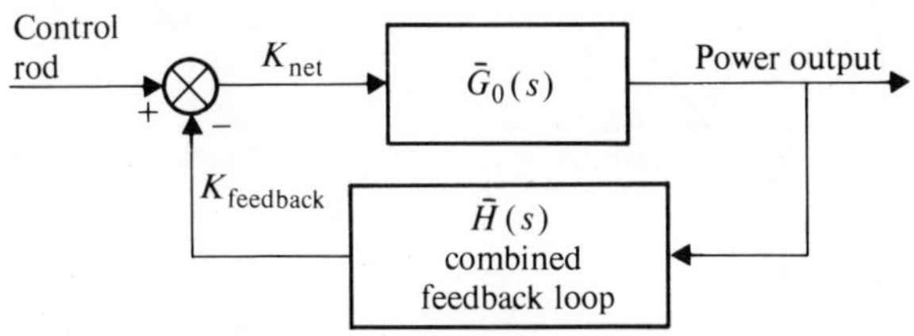

Figure 8-17 Combined schematic of EBWR model as given in Fig. 8-16.

where $i\omega$ has been substituted for s. Solving for $\bar{H}(\omega)$, one obtains

$$\bar{H}(\omega) = \bar{G}^{-1}(\omega) - \bar{G}_0^{-1}(\omega) \tag{8-126}$$

Therefore, by measuring $\bar{G}(\omega)$ experimentally at a given power level and by knowing $\bar{G}_0(\omega)$, the zero-power transfer function, one can determine $\bar{H}(\omega)$. There are practical difficulties associated with this technique. For example, unless the difference between $\bar{G}^{-1}$ and $\bar{G}_0^{-1}$ can be obtained with reasonable accuracy, the method will not produce a creditable $\bar{H}(\omega)$. The $\bar{H}(\omega)$ for EBWR [Eq. (8-122)] was obtained with good precision, but that for BORAX-IV was only approximate. A major difficulty is that the method is, of course, only as good as the postulated model. However, once the model is fixed, it is usually a straightforward procedure to calculate the needed model parameters.

Further, $\bar{H}(\omega)$ is also desired as a function of reactor operating power. $\bar{G}_0(s)\bar{H}(s)$ plots are made as a function of power and the gain margin noted, as in Fig. 8-18. That is, if $\bar{G}_0(s)\bar{H}(s)$ encircles the point $(-1, 0)$ in the complex $\bar{G}_0\bar{H}$ plane, the system is unstable. This is the essence of the Nyquist stability

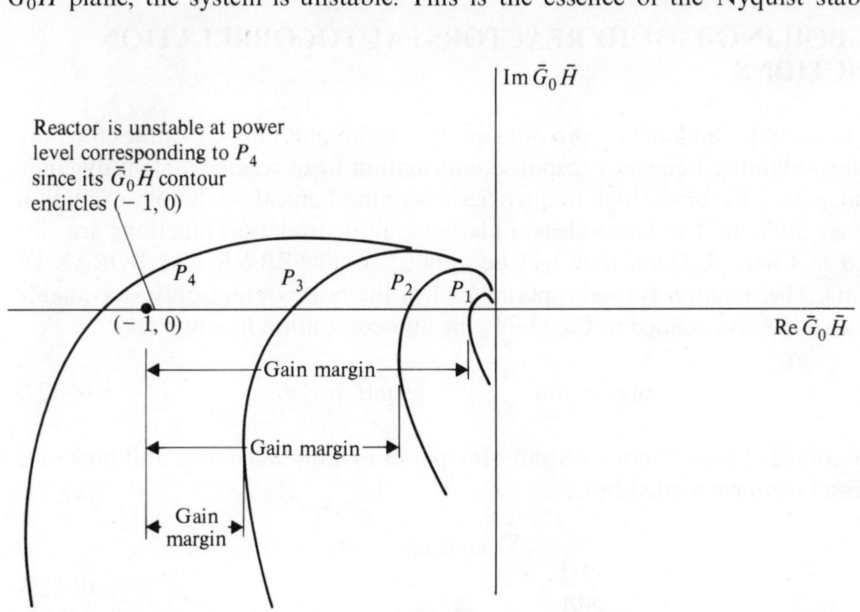

Figure 8-18 $\bar{G}_0(s)\bar{H}(s)$ for various power levels $(0 < P_1 < P_2 < P_3 < P_4)$.

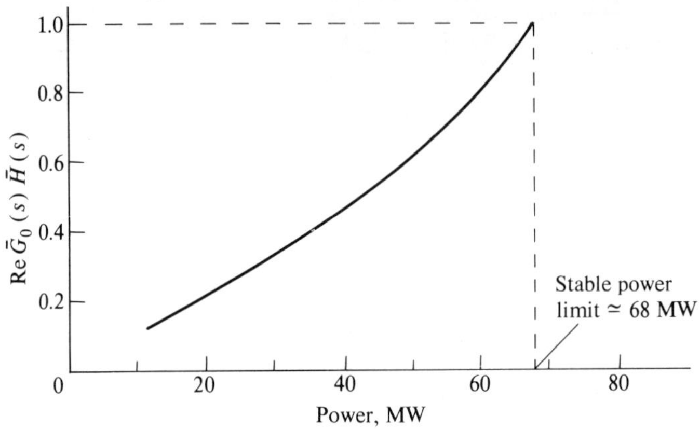

Figure 8-19 Re $\overline{G}_0(s)\overline{H}(s)$ for EBWR as function of power level.

criterion (Sec. 3-5). Then the approach of the real part of $\overline{G}_0\overline{H}$ to $(-1,0)$ as a function of power can be seen by inspection. The gain margin as a function of power can be obtained more concisely by plotting Re $\overline{G}_0(s)\overline{H}(s)$ with power level. Re $\overline{G}_0(s)\overline{H}(s)$ is the component of the $\overline{G}_0(s)\overline{H}(s)$ "vector" along the real axis in the $\overline{G}_0(s)\overline{H}(s)$ plane. If Re $\overline{G}_0(s)\overline{H}(s) \geq -1$ when Im $\overline{G}_0(s)\overline{H}(s) = 0$, the system is unstable. Such a plot is given in Fig. 8-19 for the EBWR reactor, and a hypothetical $\overline{G}_0\overline{H}$ is sketched in Fig. 8-18. In Fig. 8-19, it is seen that the stability limit for the EBWR occurs at 68 MW.

8-15 BOILING-LIQUID REACTORS: AUTOCORRELATION FUNCTIONS

The autocorrelation function provides another technique, and sometimes the only one, for obtaining frequency response information from reactor system dynamic output, especially for the high frequencies where mechanical oscillatory excitation becomes difficult. The basic ideas underlying autocorrelation functions are discussed in Chap. 3. Here, they will be applied to the EBWR and BORAX-IV reactors. The function is used especially when the noise output only is available from a system. As defined in Eq. (3-79), the autocorrelation function is

$$\phi(\tau) = \lim_{T\to\infty} \frac{1}{2T} \int_{-T}^{+T} n(t)n(t+\tau)\,dt \tag{8-127}$$

For a sample of reactor noise output of N pieces of data taken at Δ unit intervals, $\phi(\tau)$ can be approximated by

$$\frac{\phi(\tau)}{\phi(0)} \cong \frac{\sum_{r=1}^{N} n(r\Delta)n(\tau + r\Delta)}{\sum_{r=1}^{N} n^2(r\Delta)} \tag{8-128}$$

which is easily hand- or machine-computed. Wiener's theorem [Eqs (3-84), (3-86)] asserts that the spectral density $|\bar{n}(\omega)|^2$ and the autocorrelation are Fourier cosine transform mates, i.e.,

$$|\bar{n}(\omega)|^2 = \int_0^\infty \phi(\tau) \cos \omega\tau \, d\tau \tag{8-129}$$

$$\phi(\tau) = \int_0^\infty |\bar{n}(\omega)|^2 \cos \omega\tau \, d\omega$$

Further, if $\bar{K}_n^2(\omega)$ is the spectral density of the reactivity noise input, then

$$|\bar{n}(\omega)|^2 = |\bar{G}(\omega)|^2 \, |\bar{K}_n(\omega)|^2 \tag{8-130}$$

where $\bar{G}(\omega)$ is the transfer function of the linear system, which has this noise as input. Taking the square root of Eq. (8-130) yields

$$\bar{n}(\omega) = |\bar{G}(\omega)| \, |\bar{K}_n(\omega)| \tag{8-131}$$

Therefore, if $|\bar{n}(\omega)|$ is found by using autocorrelation techniques, $|\bar{G}(\omega)|$ is obtained if $\bar{K}_n(\omega)$ is known. Usually $\bar{K}_n(\omega)$ is assumed constant (white noise), or at least it is only slowly changing over the ω range of interest with respect to $|\bar{G}(\omega)|$. Within a constant, then, $\bar{n}(\omega) \cong |\bar{G}(\omega)|$, so that the square root of the spectral density gives the amplitude response portion of the transfer function. In essence, this is a method for determining the amplitude response $|\bar{G}(\omega)|$ from pile noise measurements. Variations of this method using cross-correlation techniques are discussed in Chap. 3. The curves in Fig. 8-20 are autocorrelation

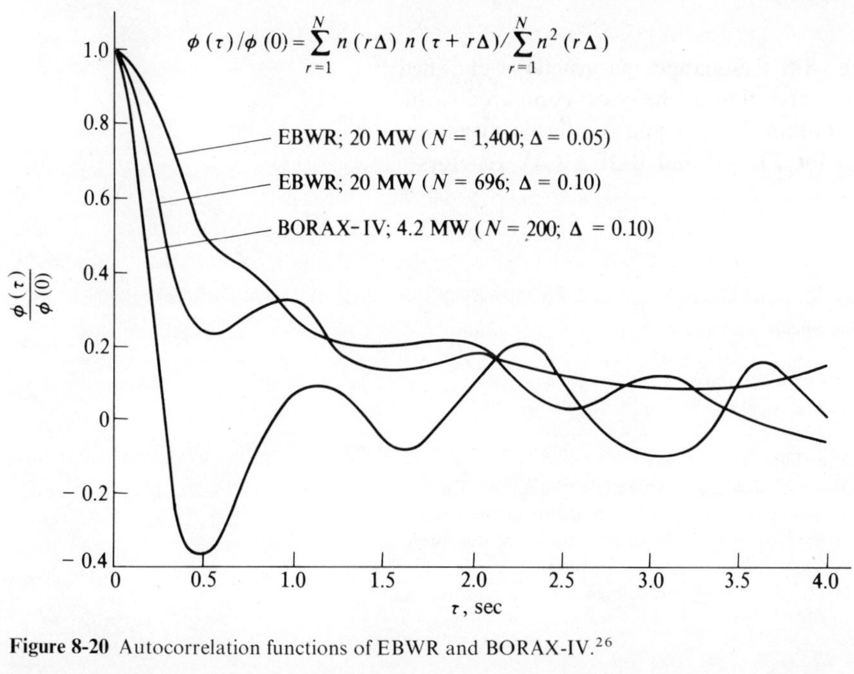

Figure 8-20 Autocorrelation functions of EBWR and BORAX-IV.[26]

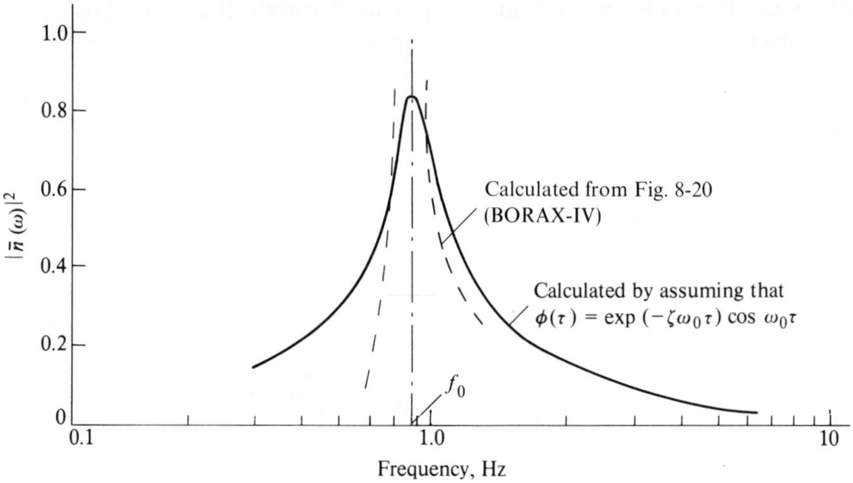

Figure 8-21 Spectral density (Fourier cosine transform) of $\phi(\tau)/\phi(0)$.

functions obtained from neutron level noise traces from EBWR and BORAX-IV. If, as is often done in radar and communication noise theory applications, the autocorrelation function is stipulated empirically by

$$\phi(\tau) = e^{-\zeta\omega_0\tau}\cos\omega_0\tau \tag{8-132}$$

where ω_0 is a postulated resonance frequency, and Eq. (8-132) is inserted in Eq. (8-127), the result is

Table 8-6 Resonance parameters obtained from correlation techniques compared with those obtained from pile oscillator measurements for EBWR and BORAX-IV reactors†

	ω_0, sec^{-1}	$\Delta\omega_0$ ‡
BORAX-IV:		
From $\phi(\tau)/\phi(0)$ plots	5	2.01
From pile oscillator	5	2.76
EBWR:		
From $\phi(\tau)/\phi(0)$ plots	7.54	—§
From pile oscillator	6.28	—

† After Thie.[26]

‡ $\Delta\omega_0 = Q\omega_0 = 2\omega_0\zeta$, where Q is the figure of merit used in radio engineering as an indicator of the sharpness of a resonance. It is similar to Γ in the Breit Wigner cross section resonance formula.

§ The $\Delta\omega_0$ was too broad to measure with reasonable precision.

$$|\bar{n}(\omega)|^2 = \frac{\omega_0}{2}\left[\frac{1}{(\omega_0\zeta)^2 + (\omega + \omega_0)^2} + \frac{1}{(\omega_0\zeta)^2 + (\omega - \omega_0)^2}\right] \tag{8-133}$$

This spectral density exhibits the resonance behavior of the usual dispersion type of formula as seen in Fig. 8-21. That is, it delineates the power spectrum near the resonance ω_0. It has a maximum at $\omega = \omega_0$, approaches a constant value as ω approaches zero, and decreases like ω^{-2} for large, increasing ω. The experimentally determined autocorrelation function, like the ringing experiment, reveals periodicities (resonances), as shown in Table 8-6, cf. Fig. 8-22.

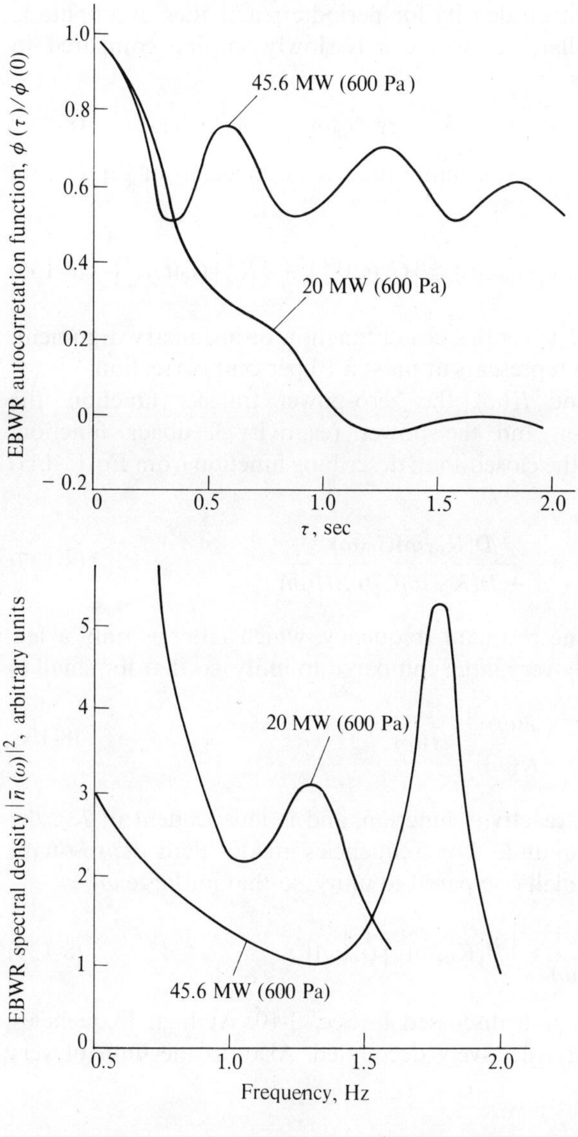

Figure 8-22 Autocorrelation and spectral density of EBWR at two different values of power.

8-16 BOILING-LIQUID REACTORS: DESCRIBING FUNCTIONS

The describing functions discussed in Sec. 3-10 are applied here, Smets,[31] to a simplified model of a boiling-water reactor in order to aid in a description of the resonance behavior of its transfer function. Let the zero-power transfer function be represented by one delayed neutron group whose decay constant $\lambda = 0.08\ \text{sec}^{-1}$. Then with $\Lambda = 10^{-2}$, corresponding to a prompt neutron lifetime of $\sim 10^{-4}$ sec,

$$\bar{G}_0(\omega) = \left[s\left(\Lambda + \frac{1}{s+\lambda}\right)\right]^{-1}_{s=i\omega} = \frac{100(i\omega + 0.08)}{i\omega(i\omega + 100)} \tag{8-134}$$

The approximation to the neutron density for periodic reactivities of amplitude appreciably less than one dollar, i.e., where n is slowly varying compared to $\dot{n}/n$ as discussed in Sec. 2-10, is

$$n(t) \cong \exp\left[K_0 \,|\bar{G}_0(\omega)|\right] \sin(\omega t + \delta) \qquad \delta = \arg \bar{G}_0(\omega) \qquad K_0 < \$1.00 \tag{8-135}$$

The corresponding describing function amplitude is obtained from Eq. (3-136) as

$$|\bar{D}(K_0, \omega)| = \frac{2}{K_0} I_1(K_0\,|\bar{G}_0(\omega)|) \cong |\bar{G}_0(\omega)|\,[1 + \tfrac{1}{8}K_0^2\,|\bar{G}_0(\omega)|^2] \tag{8-136}$$

where $K_0\,|\bar{G}_0(\omega)| \le 0.9$, and $I_1(x)$ is the Bessel function of imaginary argument, so that the second term above represents at most a 10 per cent correction.

Now $\bar{G}_0(\omega)$, $\bar{D}(K_0, \omega)$, and $\bar{H}(\omega)$, the zero-power transfer function, the individual describing function, and the power reactivity feedback function, respectively, combine to yield the closed-loop describing function from Eq. (3-147) as

$$\frac{\bar{n}(\omega)}{\bar{K}(\omega)} = \frac{D(K_0, \omega)\bar{G}_0(\omega)}{1 + \bar{D}(K_0, \omega)\bar{G}_0(\omega)\bar{H}(\omega)} \tag{8-137}$$

For frequencies lower than the resonant frequency, which latter is only a few hertz, $|\bar{G}_0(\omega)|\,|\bar{H}(\omega)|$ is usually very large compared to unity, so that for small ω

$$\frac{\bar{n}(\omega)}{\bar{K}(\omega)} \cong \bar{H}(\omega)^{-1} \tag{8-138}$$

the reciprocal of the feedback reactivity function, and is independent of K_0, the input oscillating reactivity amplitude. For frequencies greater than the resonant frequency, $|\bar{G}_0(\omega)|\,|\bar{H}(\omega)|$ is small compared to unity, so that for large ω

$$\frac{\bar{n}(\omega)}{\bar{K}(\omega)} \approx |\bar{D}(K_0, \omega)|\,|\bar{G}_0(\omega)| \tag{8-139}$$

the open-loop describing function discussed in Sec. 3-10. At high frequencies, therefore, the feedback loop is effectively decoupled. Also, in the limit of very

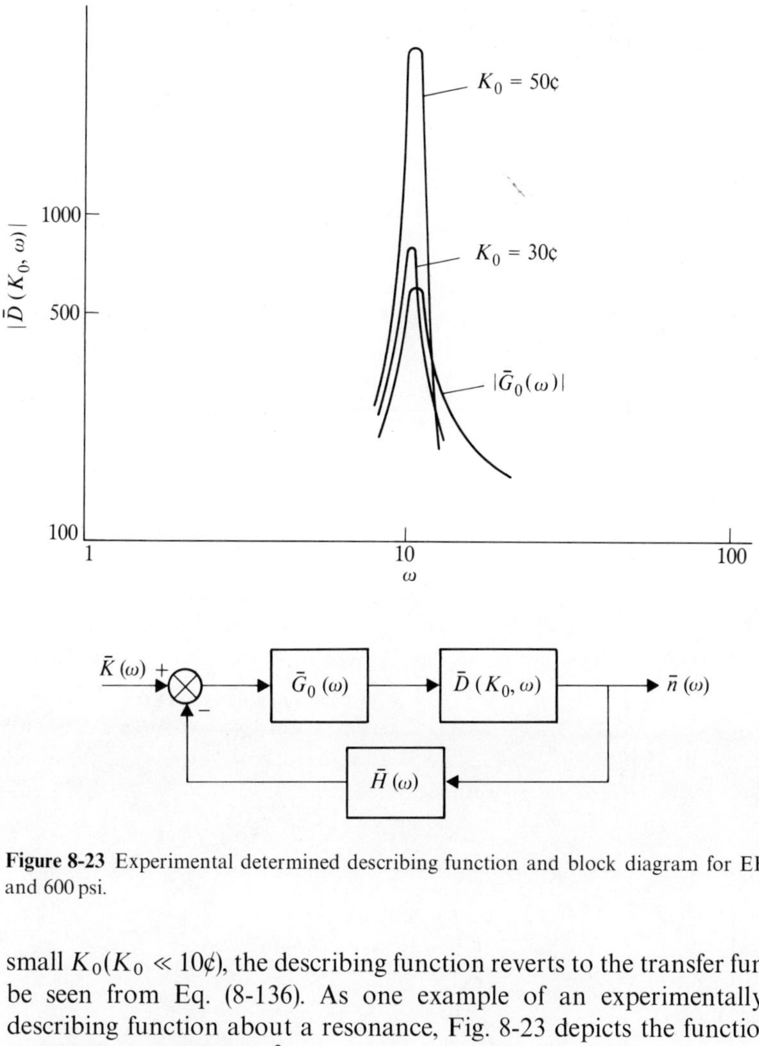

Figure 8-23 Experimental determined describing function and block diagram for EBWR at 50 MW and 600 psi.

small $K_0(K_0 \ll 10¢)$, the describing function reverts to the transfer function, as can be seen from Eq. (8-136). As one example of an experimentally determined describing function about a resonance, Fig. 8-23 depicts the function for EBWR at 50 MW and 600 lb/in^2.

The simplified model for a boiling water reactor uses $\overline{G}_0(\omega)$ from Eq. (8-134) and a power reactivity feedback function given by[31]

$$\overline{H}(\omega) = \frac{K_2}{(i\omega + 0.01)(i\omega + 0.5)} \tag{8-140}$$

to include thermal and bubble void lag time constants only. A plot of $\overline{G}_0(\omega)\overline{H}(\omega)$ in the Argand plane (Nyquist diagram) will yield its gain margin for stability. That is, how close does Re $\overline{G}_0(\omega)\overline{H}(\omega)$ approach $(-1, 0)$ for $0 \le \omega \le \infty$? As discussed in Sec. 4-7, as $\overline{G}_0(\omega)\overline{H}(\omega)$ approaches $(-1, 0)$, the amplitude of $\overline{G}_0(\omega)\overline{H}(\omega)$ will manifest a resonance peak. A very high resonance peak augurs

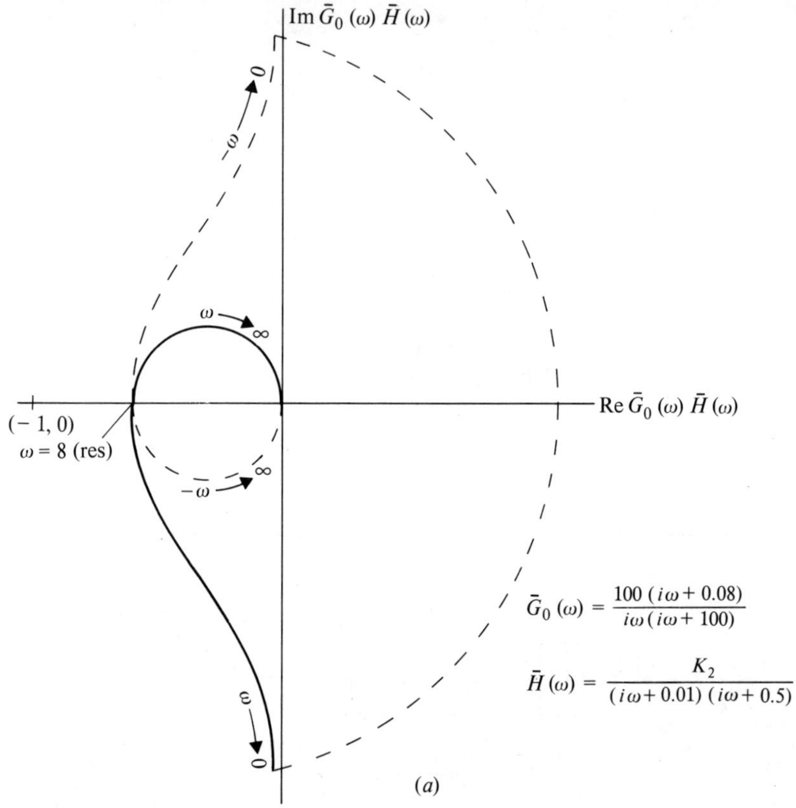

(*a*)

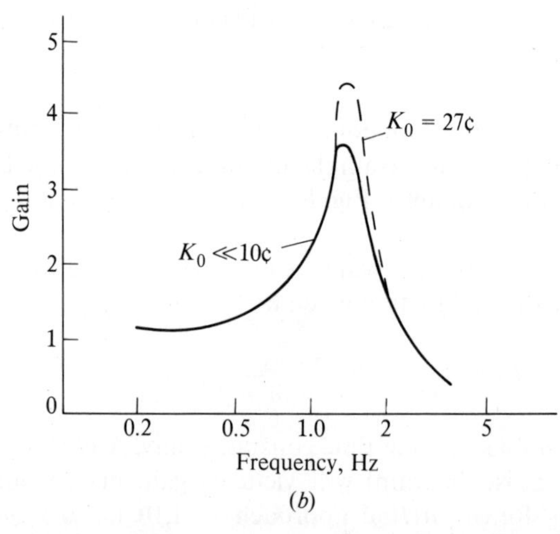

(*b*)

Figure 8-24 (*a*) Nyquist diagram for simplified boiling water reactor. (*b*) describing function amplitude for simplified water reactor.

Table 8-7 Describing function amplitudes for $|\bar{D}_1|$ and $|\bar{D}_2|$ at $\omega = 8\ \text{sec}^{-1}$ for simplified boiling water reactor

K_0, \$	$\lvert\bar{D}_1\rvert$	$\lvert\bar{D}_2\rvert$
0.10	1.001	1.008
0.27	1.01	1.07
0.30	1.01	1.07
0.50	1.03	1.24
0.80	1.08	2.08
1.00	1.13	∞

instability at the corresponding frequency. From the upper plot of Fig. 8-24, this angular frequency ≈ 8 radians per sec.[32] The units of the above curves are arbitrary because K_0 can be fixed by convenience in this simplified model (assuming that the reactor equilibrium power, $n_0 = 1$). The lower plot of Fig. 8-24 depicts the calculated closed-loop transfer function (describing function) for $K_0 \ll 10¢$, while the upper plot corresponds to $K_0 = 27¢$.

Table 8-7 is calculated for two describing function amplitudes derived in Sec. 3-10 at the resonance $\omega = 8$, namely, $|\bar{G}_0(8)| \cong 1$

$$|\bar{D}_2(K_0, 8)| = \frac{2}{K_0^2}\left[(1 - K_0^2)^{-1/2} - 1\right]$$

$$|\bar{D}_1(K_0, 8)| = \frac{2}{K_0}\, I_1(K_0)$$

By using the entries from Table 8-7, estimates can easily be made for the manner by which $|\bar{D}_1|$ or $|\bar{D}_2|$ enhances the $\bar{G}_0(\omega)$ resonance. This is done, as a problem at the end of the chapter, for the closed-loop EBWR reactor depicted in Fig. 8-23 by using the above $\bar{G}_0(\omega)$ and $\bar{H}(\omega)$ from Fig. 8-23. The closed-loop describing function analysis proceeds exactly as for transfer functions, with the overall describing function for the closed loop given by Eq. (8-137).

PROBLEMS

8-1 From its foregoing development, derive the kinetics equations (8-5) and (8-6).

8-2 Starting from the product form of the zero power fast reactor transfer function $\bar{G}_p(s) = \prod_{i=1}^{6}(s + \lambda_i)/\Lambda s \prod_{r=1}^{6}(s + s_r)$, where s_r are the nonzero roots of $\Lambda + \sum_{i=1}^{6}(a_i/s + \lambda_i) = 0$, show that its asymptotic, i.e., high frequency, form is $\bar{G}_p(\omega) = 1/(1 + i\omega\Lambda)$.

8-3 Enumerate and discuss the relative kinetic parameter advantages and disadvantages of a ^{232}Th-^{233}U thermal breeder with that of the ^{238}U-^{239}Pu fast breeder.

8-4 Why does the doppler reactivity coefficient vanish for the case of infinite fuel dilution?

8-5 For the Spectral Shift Control Reactor (SSCR), obtain the transfer function expression of Eq. (8-23) for the general case where both prompt and delayed gamma photoneutron precursors are included.

8-6 For a given fraction, p, of heavy water in the SSCR moderator mixture, show that the relationship between l, l_D, and l_H, where the foregoing are the prompt neutron lifetimes in the moderator mixture, pure heavy water moderator, and pure light water moderator, respectively, is given by Eq. (8-27).

8-7 Using the break-frequency technique as described in Sec. 3-4, construct the amplitude transfer functions for $\frac{1}{4}:\frac{3}{4}$, $\frac{1}{2}:\frac{1}{2}$, $\frac{3}{4}:\frac{1}{4}$, D_2O/H_2O moderator mixtures for the SSCR. Use the amplitude transfer function expression which construes the photoneutrons merely as an additional number of delayed neutron groups.

8-8 For the fast-thermal coupled reactor, derive the corresponding inhour equation (8-46).

8-9 One method for simplifying the fast-thermal coupled core kinetics equations, is to impose the approximations that l_{ff}, $l_{fs} \ll l_{sf}$, l_{ss}, or in a stronger manner let l_{ff} and l_{fs} approach zero. Then show that the coupled core kinetics Eqs (8-44) and (8-45) reduce to

(a) $$\frac{\dot{n}_{fs}}{l_{fs}} = k_{fs}(1-\beta)\left(\frac{n_{sf}}{l_{sf}} + \frac{n_{ss}}{l_{ss}}\right) + k_{fs}\sum_i \lambda_i c_{fi}$$

(b) $$\frac{\dot{n}_{ff}}{l_{ff}} = \left\{k_{ff}(1-\beta)\left[k_{fs}(1-\beta)\left(\frac{n_{sf}}{l_{sf}} + \frac{n_{ss}}{l_{ss}}\right) + k_{fs}\sum_i \lambda_i c_{si}\right] + k_{ff}\sum_i \lambda_i c_{fi}\right\}\Big/\{1 - k_{ff}(1-\beta)\}$$

(c) $$\dot{n}_{sf} = \frac{k_{sf}(1-\beta)k_{fs}(1-\beta)}{1-k_{ff}(1-\beta)}\left(\frac{n_{sf}}{l_{sf}} + \frac{n_{ss}}{l_{ss}}\right) - \frac{n_{sf}}{l_{sf}} + \frac{k_{sf}}{1-k_{ff}(1-\beta)}\sum_i \lambda_i c_{fi} + \frac{k_{fs}(1-\beta)k_{sf}}{1-k_{ff}(1-\beta)}\sum_i \lambda_i c_{si}$$

(d) $$\dot{n}_{ss} = k_{ss}(1-\beta)\left(\frac{n_{sf}}{l_{sf}} + \frac{n_{ss}}{l_{ss}}\right) - \frac{n_{ss}}{l_{ss}} + k_{ss}\sum_i \lambda_i c_{si}$$

(e) $$\dot{c}_{fi} = \frac{\beta_i k_{fs}(1-\beta)}{1-k_{ff}(1-\beta)}\left(\frac{n_{sf}}{l_{sf}} + \frac{n_{ss}}{l_{ss}}\right) + \frac{\beta_i k_{ff}}{1-k_{ff}(1-\beta)}\sum_i \lambda_i c_{fi} + \frac{\beta_i k_{fs}}{1-k_{ff}(1-\beta)}\sum_i (\lambda_i c_{si}) - \lambda_i c_{fi}$$

(f) $$\dot{c}_{si} = \beta_i\left(\frac{n_{sf}}{l_{sf}} + \frac{n_{ss}}{l_{ss}}\right) - \lambda_i c_{si}$$

8-10 For the circulating fuel reactor, obtain the corresponding zero-power integral equation (8-54).

8-11 Derive the circulating fuel reactor transfer function $\overline{G}_c(s)$, from the long time behavior following an initial disturbance of the above zero power integral equation.

8-12 From the inhour equation of the circulating fuel reactor, obtain the stable period following an initial negative step function reactivity insertion, and compare with that from a stationary fuel reactor.

8-13 For a boiling liquid moderator reactor where the reactivity K is a function of $L^2, B^2, \ldots f$, and p, show that its total derivative with respect to the steam void fraction α is given by Eq. (8-92).

8-14 Show that for the boiling liquid moderator reactor, its zero power transfer function as obtained from a ringing experiment is given as

$$\overline{G}(\omega) = \frac{1}{n_0 K_0}\left[(n - n_0) + i\omega \int_0^\infty (n - n_\infty)\, e^{-i\omega t}\, dt\right]$$

8-15 From the boiling liquid moderator reactor amplitude transfer function $|\overline{G}(\omega_m)|$ at its maximum frequency (the chugging frequency), show that the logarithmic decrements appearing in Eq. (8-94) is given to good approximation by $\zeta \simeq (n_\infty - n_0)/(2n_0 K_0 |\overline{G}(\omega_m)|)$.

REFERENCES

1. L. J. Koch and H. C. Paxton: Fast Reactors, *Ann. Revs. Nucl. Sci.*, **9**:453 (1959).
2. G. R. Keepin and T. F. Wimett: Reactor Kinetic Functions: A New Evaluation, *Nucleonics*, **16** (10):86 (1958).
3. A. M. Weinberg and E. Wigner: "The Physical Theory of Chain Reactors," University of Chicago Press, Chicago, 1959, pp. 100–105.
4. F. W. Thalgott et al.: Stability Studies on EBR-I, *Proc. U.N. Intern. Conf. Peaceful Uses At. Energy, 2nd, Geneva*, 1958, **12**:242–265 (1958).
5. R. R. Smith: *Proc. Conf. Transfer Function Measurements and Reactor Stability Analysis, Argonne Natl. Labs. ANL-6205*, May 1960.
6. M. G. Bhide and H. H. Hummel: Doppler Coefficient of Fast Reactors, *Argonne Natl. Labs. ANL-6601*, 1962.
7. R. Avery: Fast Reactor Physics, "Fast Reactors and the University," M. Becker, Ed.: Session B-I, p. 11, 1968.
8. W. J. McCarthy and D. Okrent: Fast Reactor Kinetics, "The Technology of Nuclear Reactor Safety," Vol. I, p. 538. T. J. Thompson, J. G. Beckerley, eds.
9. M. C. Edlund: The Spectral Shift Control Reactor, "Small and Medium Power Reactors," Intern. At. Energy Agency, Vienna, 1961.
10. M. C. Edlund and G. K. Rhode: Spectral Shift Control, *Nucleonics*, **16**: No. 5, pp. 80–81 (May, 1958).
11. J. A. Deshong: "Analysis of Experimental Power Reactivity Feedback Transfer Functions for a Natural Circulation Boiling Water Reactor," *Argonne Natl. Labs. ANL-5850*, 1958.
12. J. R. Menke: Monsanto Chem. Corp., M-3753, Monp-412, 1947.
13. R. Avery: Theory of Coupled Reactors, *Proc. U.N. Intern. Conf. Peaceful Uses At. Energy, 2nd, Geneva*, 1958, **12**:182–191 (1958).
14. W. K. Ergen: Kinetics of the Circulating Fuel Reactor, *J. Appl. Phys.*, **25** (6): 703–711 (1954).
15. C. F. Bonilla: "Nuclear Engineering," McGraw-Hill, New York, 1957, chap. 14.
16. J. MacPhee: The Kinetics of Circulating Fuel Reactors, *Nucl. Sci. Eng.*, **4**:588 (1958).
17. M. A. Schultz, "Control of Nuclear Reactors and Power Plants," 2nd edn., McGraw-Hill, New York, 1961, p. 129.
18. T. A. Welton: "Kinetics of Stationary Reactor Systems," *Proc. U.N. Intern. Conf. Peaceful Uses At. Energy, Geneva*, 1955, **5**, 377–388 (1958).
19. J. A. Fleck: Theory of Low Power Kinetics of Circulating Fuel Reactors, *Brookhaven Natl. Labs. BNL-334* (*T-57*), April 1955; *BNL-1933*, April 1954; *Nucleonics*, **12**(11):1954.
20. R. V. Meghreblian and D. K. Holmes: "Reactor Analysis," McGraw-Hill, New York, 1960, p. 609.
21. A. W. Kramer: "Boiling Water Reactors," Addison-Wesley, Reading, Mass., 1958, p. 58.
22. C. F. Bonilla: *op. cit.*, p. 721.
23. J. M. Harrer: et al., *Progr. Nucl. Energy, Ser. II, Reactors*, **2** (1961), chap. 14.
24. J. A. Thie: Theoretical Statics and Kinetics of Boiling Reactors," *Proc. U.N. Intern. Conf. Peaceful Uses At. Energy, 2nd, Geneva*, 1958, **11**:440–446 (1958).
25. D. L. Hetrick, et al.: Experimental Studies on the KEWB Reactor, *Proc. U.N. Intern. Conf. Peaceful Uses At. Energy, 2nd, Geneva*, 1958, **11**:447–456 (1958).
26. J. A. Thie: Dynamic Behavior of Boiling Reactors, *Argonne Natl. Labs., ANL-5849*, May 1959, p. 7.
27. H. A. Bethe: Atomic Power Development Associates, Rept. APDA-117.
28. A. Z. Akcasu: "Theoretical Feedback Analysis in Boiling Water Reactors," *Argonne Natl. Labs., ANL-6221*, Oct. 1960.
29. E. S. Beckjord and G. M. Ray: Performance Characteristics of Large Boiling Water Reactors, *Adv. Nucl. Sci. Techn.*, **1** (1962). E. J. Henley and H. Kouts, eds.
30. M. Iriarte, JR: *Nucl. Sci. Eng.*, **7**:26–32 (1960).

31. H. B. Smets: The Describing Function of Nuclear Reactors, *IRE Trans. ·Nucl. Sci.* NS-6:8–12 (1959).
32. H. A. Sandmaier: Non-Linear Treatment of Large Perturbations in Power Reactor Stability, *Nucl. Sci. Eng.*, **6**:85–92 (1959).
33. L. N. Usachev: Equation for Importance of Neutrons: Reactor Kinetics and Perturbation Theory, *Proc. U.N. Intern. Conf. Peaceful Uses At. Energy, Geneva*, 1955, **5**:503–570 (1955).
34. L. J. Deverall: Solution of the Time Dependent Thermal Neutron Diffusion Equation, *Nucl. Sci. Eng.*, **4**:495–498 (1958).

CHAPTER

NINE

KINETICS OF REACTOR PROTOTYPES—II

9-1 NUCLEAR ROCKETS: INTRODUCTION

With the simultaneous advent of nuclear reactor and missile development during the post-World War II decades, it was natural to begin to investigate a combination of the two: a nuclear reactor rocket.[1] The nuclear rocket provides the next level in the hierarchy of rocket performance, the first level being that of conventional liquid or solid chemical propelled rockets. The direction here is to describe briefly the fundamentals of rocketry and to proceed quickly to the kinetics aspects of interest with regard to the nuclear rocket. First, then, it is necessary to establish the specific reasons for the very consideration of a reactor rocket.

9-2 NUCLEAR ROCKETS: PERFORMANCE COMPARED WITH THAT OF CHEMICAL ROCKETS

One key measure of performance of a rocket is the velocity (speed) it possesses in space when the rocket engine is cut off. This is called the burnout velocity, and it usually corresponds to the rocket velocity when all of the propellant has been expended. Some required mission burnout velocities are given in Table 9-1.

A second key parameter, better related to the propulsion capabilities, is called the specific impulse. To aid in defining this term, first consider a rocket in free space, of mass m and velocity with respect to the ground v, from which an element of gas Δm has just been ejected. Assume that no external forces are acting on this

Table 9-1 Required mission burnout velocities†

Mission	Required burnout velocity for single-stage rocket, ft/sec	Available propulsion plant burnout velocity capability, ft/sec	Description
V-2, 200-mile range	5,500		
IRBM, 1500-mile range	15,000		
ICBM, 5000-mile range	22,000	20,000	Medium-performance chemical rocket (Minuteman II)
Satellite	27,000		
Escape, interplanetary	35,000	30,000	High-performance chemical rocket
		44,000	Medium-performance nuclear rocket (3000°F)
		55,000	High-performance nuclear rocket (4500°F)

† After Grey.[2]

system, so that linear momentum is conserved. Then[3]

$$p = mv + (v - v_e)\,\Delta m = \text{const} \tag{9-1}$$

where v_e is the velocity of ejection of Δm relative to the rocket. Now

$$\dot{p} = 0 = \frac{d}{dt}\left[mv + (v - v_e)\,\Delta m\right] \tag{9-2}$$

and $(d/dt)\,\Delta m = -\dot{m}$, since the material has been ejected through loss of rocket mass. $\dot{v}_e = 0$, since the exhaust velocity is assumed constant. Equation (9-2) in the limit of zero Δm becomes

$$m\dot{v} = -\dot{m}v_e = F \tag{9-3}$$

where F is the reaction force or rocket thrust.

The rocket engine operates differently from an ordinary engine in that the latter (e.g., airplane engine) usually propels its load at constant speed. A rocket engine is used to accelerate a payload to a burnout velocity and operates ideally at constant thrust. Therefore, the change in momentum is a more important measure for a rocket than, for example, its power output.

The quantity called specific impulse, I_{sp}, is used to rate rocket propulsion capability. It is defined as the thrust per unit weight rate of propellant flow, i.e.,

$$I_{sp} = \frac{F}{\dot{m}g}\left(\frac{\text{pounds}}{\text{pounds/sec}} = \text{seconds}\right) \tag{9-4}$$

Inserting I_{sp} into Eq. (9-3) gives

$$m\dot{v} = \dot{m}gI_{sp} - mg \tag{9-5}$$

where mg, the gravity force, has been added at this juncture. Dividing by m and integrating from zero time to t_b, the time of burnout, gives from the above equation

$$v_b = gI_{sp} \ln \frac{m_b}{m_0} - gt_b + v_0 \tag{9-6}$$

From Eqs (9-3) and (9-4), $-v_e = gI_{sp}$, so that

$$v_b = v_e \ln \frac{m_0}{m_b} - gt_b + v_0 \tag{9-7}$$

As can be appreciated from Table 9-1, it is desired to maximize the burnout velocity to achieve maximum rocket performance. Looking at Eqs (9-6) and (9-7) and assuming that $v_0 = 0$, as it usually is for a single-stage rocket, the means for maximizing v_b are limited. One can (1) minimize t_b, (2) maximize m_0/m_b, or (3) maximize I_{sp}. Minimizing t_b is limited, since the rocket structure can stand accelerations of only a few g's. Besides, the gt_b term is dwarfed by the first term for the rather high burnout velocities available today, and surely for those available in the future. The ratio of takeoff mass to burnout mass, m_0/m_b, is limited as well, to correspond to about 90 percent payload. For all practical purposes then, $m_0/m_b \cong 10$. For Eq. (9-2)

$$v_b \cong 2.3v_e \tag{9-8}$$

neglecting v_0 and gt_b.

The richest area of promise for increasing the thrust and thereby the burnout velocity is to increase the specific impulse. An important way to estimate I_{sp} is to obtain it as a function of the thermodynamic properties of the propellant as it burns in a thrust chamber or duct, delivering thrust. The thrust for a constant exit velocity of propellant gases in a duct of cross-sectional exit area A_e is[4]

$$F = m\dot{v}_e + (p_e - p_a)A_e \approx \dot{m}v_e \tag{9-9}$$

where p_a is the atmospheric pressure. The pressure thrust is usually much smaller than the momentum thrust so that the former (second term above) is neglected. Assuming isentropic flow, the ideal exit velocity of the propellant gases is derived from the conservation of energy in terms of Bernoulli's equation of fluid flow in the thrust chamber. It is the well-known expression[5]

$$v_e = \left[2\left(\frac{\gamma}{\gamma - 1}\right) RT_c \left(1 - \frac{p_e}{p_c}\right)^{(\gamma-1)/\gamma} \right]^{1/2} \tag{9-10}$$

where γ is the ratio of specific heats and p_c and T_c are, respectively, the pressure and temperature in the thrust chamber. From the definition of specific impulse and Eqs (9-9) and (9-10) it is seen that

$$I_{sp} = \frac{F}{\dot{m}g} \approx (RT_c)^{1/2} \tag{9-11}$$

In appropriate units R is inversely proportional to the molecular weight of the

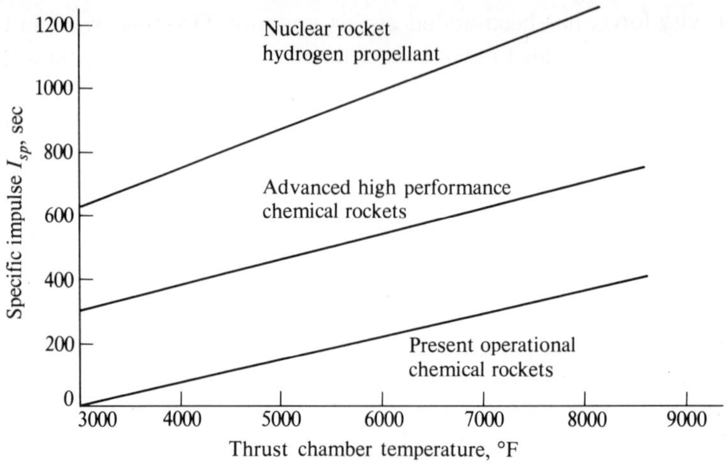

Figure 9-1 Specific impulse as a function of temperature for nuclear and chemical rockets.

propellant gases M so that

$$I_{\text{sp}} = \text{const}\left(\frac{T_c}{M}\right)^{1/2} \tag{9-12}$$

Therefore, to maximize I_{sp} it is desired to maximize T_c/M. This can be done by heating the propellant gas so that the thrust chamber temperature is as high as possible and by using light propellant gases. The nuclear rocket uses H_2 and/or H as propellant so that M is as small as possible. Figure 9-1 compares the specific impulse of nuclear rockets with that of their chemical counterparts.

The advantage of a nuclear rocket is that of an almost unlimited energy supply from a low-molecular-weight fuel. That is, the thrust chamber temperature T_c is limited only by the yield strength of the rocket structural material. The lack of high specific impulse capability in present-day chemical rockets is overcome by staging. In other words, a multistaged rocket is used, where the payload of the $(N-1)$st stage is the Nth stage. For a two-stage rocket

$$\begin{aligned} v_b = gI_{\text{sp}} \ln \frac{m_b}{m_0} + v_0 &= gI_{\text{sp}}\left[\ln\left(\frac{m_b}{m_0}\right)_1 + \ln\left(\frac{m_b}{m_0}\right)_2\right] \\ &\cong 2gI_{\text{sp}} \ln\left(\frac{m_b}{m_0}\right) \end{aligned} \tag{9-13}$$

if the mass ratio per stage is constant. The effective specific impulse is thereby doubled by two stages.

9-3 NUCLEAR ROCKETS: REACTOR PROTOTYPE DESIGN

There is basically little difference in the organization of and basic kinds of components needed for a chemical or a nuclear rocket. Of course, the outstanding

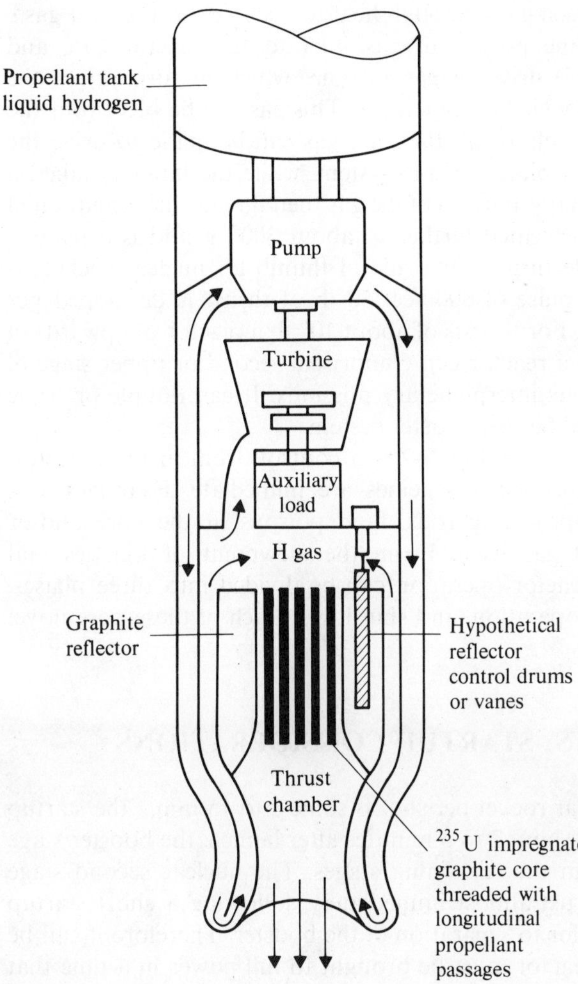

Figure 9-2 Nuclear rocket engine.[9]

difference is that the chemical rocket supplies thrust from the direct molecular decomposition of its propellant, while the nuclear rocket acts as a nuclear energy exchanger by heating a working fluid (gas), to expand it at tremendous flow rates to provide thrust. Figure 9-2 depicts a prototype rocket engine heating liquid hydrogen from cryogenic temperatures to 3000°F or higher and discharging it as a gas to provide thrust.

The rocket is straightforward in operation. Liquid hydrogen is pumped from the propellant tank through the walls of the thrust chamber, thereby cooling them as well as the reflector. The requisite power densities are about 4000 kW per liter of reactor, of which about 6 per cent, or about 250 kW per liter, emanates as γ photons. A large part of the γ and other particle energy is dissipated in the reflector and immediately adjacent structure, which thereby vaporizes the liquid hydrogen coolant. The hydrogen gains enough heat to become a rather hot gas.† A small portion bled from the propellant gas inlet to the reactor core and exhausted separately (Fig. 9-2) drives a gas turbine, which in turn drives the propellant pump and auxiliary electric generator. This gas can be bled from the nozzle end of the reactor as well, or all the inlet gas can be made to drive the turbine. The former is called a bleed turbine system, while the latter is called a topping turbine system. The major portion of the gas then threads the longitudinal reactor tubes, where it is superheated further to about 3000°F and is expanded through the nozzle to provide thrust. One rule of thumb for nuclear rockets is that for a median specific impulse of 800 sec, 50 lb of thrust is developed per megawatt of power generated. For thrusts of about 10^6 lb a reactor of upwards of 20,000 MW is required. Such a reactor can comprise a second or upper stage of a multistage system for various interplanetary missions. It can double or triple the payload of the all-chemical Saturn booster system.

The reactivity control shown in Fig. 9-2 is a control drum in the reflector. It is merely suggestive, because other schemes are immediate. Examples are: control by conventional reciprocating rods, fluid poisons in the core and/or reflector, and the propellant gas itself. From the viewpoint of kinetics and control, the nuclear rocket reactor operation can be divided into three phases. These are startup, full thrust operation, and shutdown. Each of these pose novel and formidable problems.

9-4 NUCLEAR ROCKETS: STARTUP CONSIDERATIONS

Since the useful life of a nuclear rocket per shot is some 2 to 15 min,‡ the startup problems can be horrendous. About 2 to 3 minutes after launch, the booster stage separates and falls away from the remaining stages. The nuclear second stage must begin to supply full thrust almost immediately, following a short startup period probably occurring prior to separation of the booster. Therefore it can be appreciated that the rocket reactor must be brought to full power in a time that is small compared to its total burning time. This must be accomplished in meticulous fashion without propellant density inhomogeneities, which could represent disruptive positive reactivity jumps to dangerously disturb the startup period rate of change. A simple calculation illustrates this well.[8] The fractional amount of propellant used to warm up the reactor to full power is given by

† Calculations have shown[6] that, for high thermodynamic efficiency operation, the gas will be insufficiently heated upon entering the reactor proper. Then fuel can be added to the reflector, making it part of the reactor, to supply the needed heat to the gas on its way to the reactor thrust chamber. That is, the propellant makes two passes through the reactor. The first pass is up through the peripheral portion of the reactor, and the second is down through the central portion and discharge through the nozzle.

‡ Since a nuclear rocket will be much more expensive than a chemical rocket, it is desired to arrange to reuse the rocket. The usual duration of a chemical rocket, e.g., an Intercontinental Ballistic Missile, to burnout of the last stage, is about 5 min.

$$F_b = \frac{P_0}{Pt_b}\int_0^T n(t)\,dt \tag{9-14}$$

The numerator is proportional to the amount of propellant consumed to warm up the reactor in time T. Once at full power P, the reactor is assumed to provide thrust at constant power P until burnout time t_b. The amount of propellant consumed is proportional to Pt_b, the denominator. Actually, the total amount of propellant consumed is the sum of the numerator and denominator. The numerator is considered small with respect to the denominator, and so it is neglected, by comparison, where it would appear in the denominator. For a 1 per cent propellant warm-up allowance ($F_b = 0.01$), a power rise of eight orders of magnitude ($P/P_0 = 10^8$) from critical, and a burnout time of 300 sec, Eq. (9-15) becomes, with $\bar{\lambda} = (\bar{\tau})^{-1}$,

$$3 \times 10^8 = \int_0^T \exp\left(\frac{\bar{\lambda}mt^2}{2}\right) dt \tag{9-15}$$

where a linear increase in reactivity is assumed and delayed neutrons are considered in an asymptotic fashion (cf. Sec. 2-11), since the reactor is prompt critical or better over most of its startup interval. This implies the kinetics equation $\dot{n} = \bar{\lambda}mtn$, where m is the reactivity insertion rate. The solution is $n = \exp(\bar{\lambda}mt^2/2)$, which is the integrand of the above equation.

For a $\bar{\lambda}$ of 0.10 sec^{-1}, which is the right order of magnitude for the reactor in Fig. 9-2, and a warm-up time T of 60 sec, which is very generous from the propellant consumption standpoint, the reactivity insertion rate m is of the order of 10 cents per second, a not inconsiderable rate.

As Mohler and others[4] have shown using a mean square error criterion optimal control, the rocket reactor can proceed to operating power with essentially constant reactivity. The maximum value of the reactivity is limited mainly by the maximum thermal stress of the reactor materials. Since the reactor power level changes radically over a relatively short time period, as already discussed, an appropriate more slowly changing control variable is the logarithmic derivative, that is, the "alpha" of the power level. So letting $\alpha = \dot{n}/n$ in the point model integro-differential equation (1-42) without a source, results in an expression for alpha, namely

$$\Lambda\alpha(t) = K(t) - \int_0^t \alpha(\tau)\, e^{-\int_\tau^t \alpha(u)\,du} f(t-\tau)\,d\tau \tag{9-16}$$

Since the above reactivity is essentially constant, $\alpha(t)$ is relatively slowly varying. So that in the above, $\exp - \int_\tau^t \alpha(u)\,du \simeq \exp - \alpha_0(t-\tau)$ where $\alpha_0 = K_0/\bar{\tau}$ for reasonably large K_0 near one dollar, which is very nearly the positive root of the inhour equation $K_0\bar{G}(\alpha_0) = 1$. Then Eq. (9-16) becomes

$$\Lambda\alpha(t) = K(t) - \int_0^t \alpha(\tau)\, e^{-\alpha_0(t-\tau)} f(t-\tau)\,d\tau \tag{9-17}$$

with Laplace transform

$$\bar{\alpha}(s) = \bar{K}(s)/\{\Lambda + \bar{f}(s + \alpha_0)\} \tag{9-18}$$

Inverting Eq. (9-18) gives the control variable $\alpha(t)$ as

$$\alpha(t) = K_0 G_{\alpha_0}(t) + \int_0^t G_{\alpha_0}(t - \tau) K'(\tau)\, d\tau \tag{9-19}$$

where $G_{\alpha_0}(t)$ is defined through its Laplace transform

$$\bar{G}_{\alpha_0}(s) = [s\{\Lambda + \bar{f}(s + \alpha_0)\}]^{-1} \tag{9-20}$$

For a one delayed neutron group approximation to the above, Mohler (*op. cit.*) uses a control variable $q(t)$, with the corresponding control trajectory equation, to obtain the above near constant reactivity control, which is

$$\lambda \dot{q}(t) = -\dot{K}/(1 - K) \tag{9-21}$$

where, with λ as the one delayed neutron group decay constant,

$$q(t) = \frac{1}{\lambda} \ln \frac{(1 - K)n}{(1 - K_0)n_0}$$

and with α assumed negligible by comparison in its derivative.

If a ramp reactivity startup from a subcritical condition with a source ΛR neutrons per second is considered, Eq. (1-42) can be used, where the reactivity is modified by a temperature reactivity coefficient according to the Fuchs model discussed in Sec. 9-10 on pulsed reactors. The typical aggregate form of the kinetics equations then becomes

$$\Lambda \dot{n} = \left[mt - \alpha_T \int_0^t n(\tau)\, d\tau \right] n - \int_0^t n'(\tau) f(t - \tau)\, d\tau + \Lambda R \qquad n(0) = 0 \tag{9-22}$$

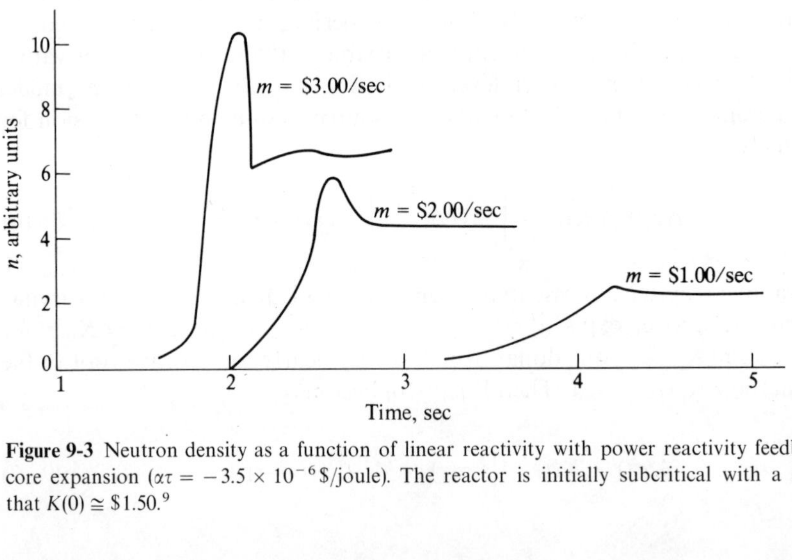

Figure 9-3 Neutron density as a function of linear reactivity with power reactivity feedback due to core expansion ($\alpha\tau = -3.5 \times 10^{-6}$ \$/joule). The reactor is initially subcritical with a source such that $K(0) \cong \$1.50$.[9]

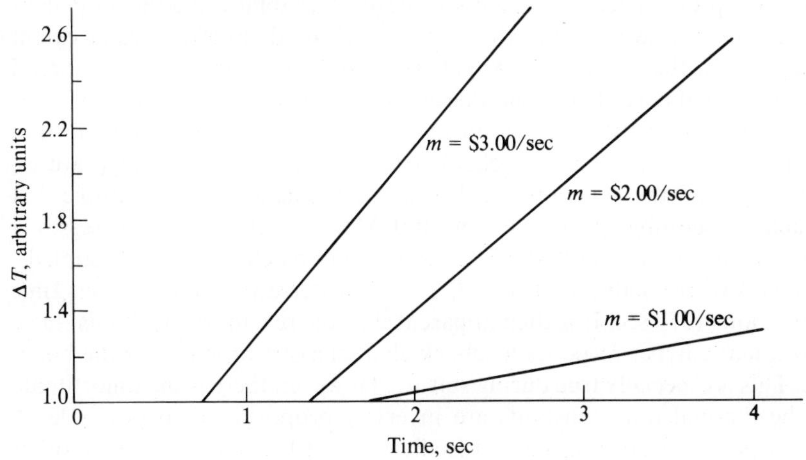

Figure 9-4 Reactor temperature rise for linear reactivity with power reactivity feedback due to core expansion. The reactor is initially subcritical with a source such that $K(0) \cong -\$1.50$.

where m is the input reactivity rate, in dollars per second, and α_T is an appropriate reactivity coefficient. For more accurate and ready computation on digital computing machinery, Laplace transforms can be used to convert this equation to its integral form

$$n(t) = \int_0^t G(t-\tau)\left[\Lambda R + \left(m\tau - \alpha_T \int_0^\tau n(x)\,dx\right) n(\tau)\right] d\tau \tag{9-23}$$

so that no derivatives need be approximated by their difference quotients. Solutions for the type of reactivity depicted above are given in Figs 9-3 and 9-4. The accompanying temperature rise in the core, depicted in Fig. 9-4, is obtained simply from

$$C(T - T_0) = \int_0^t n(\tau)\,d\tau \tag{9-24}$$

where C is a heat capacity constant of the reactor core.

9-5 NUCLEAR ROCKETS: FULL-THRUST OPERATION

At full-thrust (power) operation the rocket reactor differs from the conventional thermal reactor in a number of ways with respect to its kinetic operation. First, the reactor is not truly a thermal reactor because the operating core temperature of about 3000°F corresponds approximately to a lowest neutron energy, which is an order of magnitude higher than that in the ordinary room-temperature thermal reactor. This decreases the prompt neutron lifetime, since the "thermal" velocity of the neutrons is now approximately tripled. Second, the hot core, presumably graphite because of its high-temperature durability, becomes relatively porous to fission product decay fragments, some of which diffuse into the propellant channels and are thus swept out of the reactor. This holds true to a lesser extent for the mobility of the fuel as the propellant erodes the channels which thread the fuel-impregnated core. Third, the thermal-lag time constants of the fuel, that is, the time required for the temperature to reach $(1 - e^{-1})$ of its equilibrium value, must be short in the nuclear rocket reactor owing to the high temperatures, power densities, and the generally quick response demanded. For example,[10] a rocket reactor operating at 4000°F and 100 MW per cubic foot possesses a thermal time constant of about 2 sec. The Calder–Hall reactors, also "gas cooled," operate at 11 kW per cubic foot at 350°F with a corresponding thermal time constant of about 1350 sec. It is then apparent that the reactor control must have a highly automatic hyper-response feedback characteristic probably of the bang-bang type. This is especially true during startup. However, there is the unfortunate fact that the thermal time constants are inversely proportional to power level. Then for low power levels, this form of feedback must be supplanted by another form, e.g., gas nozzle temperature. A nuclear rocket enjoys no natural feedback loop between propellant power output and flow rate or pressure, closed by propellant combustion, such as exists in a chemical rocket. Then the controller is somewhat more difficult to synthesize in a nuclear rocket in that certain semi-artificial loops must be provided for adequate closed-loop system response. Some of these characteristics are reflected in the transfer function, as will now be shown.

The delayed neutron precursor diffusion loss, mentioned above, appears in an augmented one-dimensional equation for the precursor concentration $c_i(x, t)$, with time in units of Λ, as[10]

$$\frac{\partial c_i}{\partial t}(x,t) = -\lambda_i c_i(x,t) + a_i n(t) + D\frac{\partial^2 c_i}{\partial x^2}(x,t) \quad \frac{\partial c_i}{\partial t}(x,0) = 0 \quad i = 1, 2, \ldots \tag{9-25}$$

The boundary conditions are $c_i(x, 0) = 0$, since no precursors exist prior to startup because of the nascent condition of the reactor.

$$c_i(0, t) = c_i(a, t) = 0$$

(a is the principal axial dimension), merely means that the fuel is so dispersed that the precursor concentration will sensibly vanish at the ends of the core. The last term on the right-hand side of the above equation accounts for the precursor diffusion loss. In a rocket reactor, the heat energy is transferred to the propellant over the length of the reactor. For detailed analysis, the reactor is usually compartmentalized into a series of contiguous regions along its length. Each region is assumed homogeneous throughout its volume with respect to its neutronic, thermodynamic, etc., characteristics. The region coupling is characterized by input to the next region and output from the preceding region. This is exactly analogous to the treatment of transmission lines using lumped parameters. In rough calculations, the entire reactor is treated as a single region, as in the following solution of Eq. (9-25).

The Laplace transform of Eq. (9-25) with respect to time and the initial condition $c_i(x, 0) = 0$ yields the ordinary differential equation in $\bar{c}_i(x, s)$, the Laplace transform of $c_i(x, t)$:

$$\frac{d^2\bar{c}_i(x,s)}{dx^2} - \frac{s+\lambda_i}{D}\bar{c}_i(x,s) = -a_i\bar{n}(s) \tag{9-26}$$

The boundary conditions as transformed are $\bar{c}_i(0,s) = \bar{c}_i(a,s) = 0$. The solution of the above equation is

$$\bar{c}_i(x,s) = \frac{a_i\bar{n}(s)}{s+\lambda_i}\left\{1 - \frac{2\sinh\left[\left(\frac{s+\lambda_i}{D}\right)^{1/2}\frac{a}{2}\right]\cosh\left[\left(\frac{s+\lambda_i}{D}\right)^{1/2}\left(x-\frac{a}{2}\right)\right]}{\sinh\left(\frac{s+\lambda_i}{D}\right)^{1/2}a}\right\} \tag{9-27}$$

Since the reactor is considered as a single region, the $\bar{c}_i(x, s)$ is averaged over the length a of the reactor to give

$$\bar{c}_i(x,s)_{\text{avg}} = \frac{a_i\bar{n}(s)}{s+\lambda_i}\left\{1 - \frac{2}{a}\left(\frac{D}{s+\lambda_i}\right)^{1/2}\tanh\left[\left(\frac{s+\lambda_i}{D}\right)^{1/2}\frac{a}{2}\right]\right\} \tag{9-28}$$

If the ordinary kinetics equations (1-39) and (1-40) are "linearized" by assuming small fluctuations about the equilibrium level, the result is, when Eq. (9-25) is used for the delayed neutron precursor concentration and time in seconds is reintroduced

$$\bar{\varepsilon}(s) = \frac{\bar{K}(s)}{s(\Lambda+\bar{f}(s)) + \frac{2}{a}\sum_i \frac{a_i\lambda_i}{s+\lambda_i}\left(\frac{D}{s+\lambda_i}\right)^{1/2}\tanh\left(\frac{s+\lambda_i}{D}\right)^{1/2}\frac{a}{2}} \tag{9-29}$$

where $\varepsilon(t) = n - 1$ is the deviation from the equilibrium level and

$$\bar{f}(s) = \sum_i \frac{a_i}{(s+\lambda_i)}$$

the Laplace transform of the delayed neutron decay function. The corresponding zero-power transfer function $\bar{G}_D(s)$ is then, cf. Fig. 9-5,

$$\bar{G}_D(s) = \left[s(\Lambda+\bar{f}(s)) + \frac{2}{a}\sum_i \frac{a_i\lambda_i}{s+\lambda_i}\left(\frac{D}{s+\lambda_i}\right)^{1/2}\tanh\left(\frac{s+\lambda_i}{D}\right)^{1/2}\frac{a}{2}\right]^{-1} \tag{9-30}$$

As can be seen, if the precursor diffusion constant D vanishes, the result is the ordinary zero-power transfer function $\bar{G}_0(s)$. The comparison between $\bar{G}_D(s)$ and $\bar{G}_0(s)$ can be easily sketched by examining their behavior for small, medium, and large s.

For very small s

$$\bar{G}_0(s) \cong [s(\Lambda + \bar{f}(0)]^{-1} \cong \frac{1}{s\bar{\tau}} \tag{9-31}$$

Medium s implies $1/\Lambda \gg s \gg \lambda_i$ for all i, so that

$$\bar{G}_0(s) \cong 1 \tag{9-32}$$

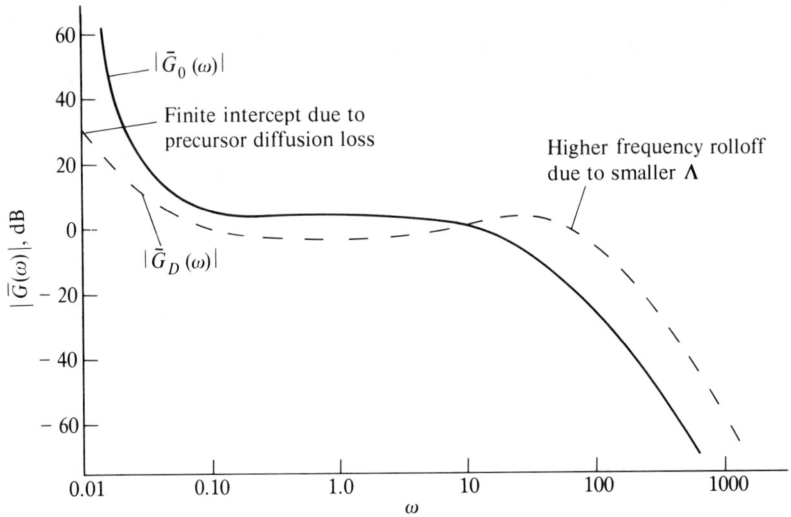

Figure 9-5 Amplitude response of zero-power transfer with and without the delayed neutron precursor loss.

For large s

$$\bar{G}_0(s) \cong (1 + s\Lambda)^{-1} \tag{9-33}$$

while for very large s

$$\bar{G}_0(s) \approx (s\Lambda)^{-1} \tag{9-34}$$

Similarly, $\bar{G}_D(s)$ takes on the corresponding forms

$$\bar{G}_D(s) \cong \left[s(\Lambda + \bar{\tau}) + \frac{2}{a} \sum_i a_i \left(\frac{D}{\lambda_i} \right)^{1/2} \tanh \left(\frac{\lambda_i}{D} \right)^{1/2} \frac{a}{2} \right]^{-1} \tag{9-35a}$$

$$\bar{G}_D(s) \cong \left(1 + \frac{2D^{1/2}}{a\bar{\tau}} s^{-3/2} \right)^{-1} \tag{9-35b}$$

$$\bar{G}_D(s) \cong \left(s\Lambda + 1 + \frac{2D^{1/2}}{a\bar{\tau}} s^{-3/2} \right)^{-1} \tag{9-35c}$$

$$\bar{G}_D(s) \cong (s\Lambda + 1)^{-1}, (s\Lambda)^{-1} \tag{9-35d}$$

For an appreciation of the interaction of the novel components in a nuclear rocket reactor system, examine the schematic diagram in Fig. 9-6. Consider a multistage rocket with the reactor rocket as the second stage. Just prior to burnout of the booster, or first stage, the programmer initiates a thrust demand on the basis of its startup program. The thrust error signal is obtained from core exit pressure, or temperature, and is injected into the thrust controller. On the basis of the required thrust, the scheduler initiates a power demand from the propellant flow demand according to the required operating temperature.

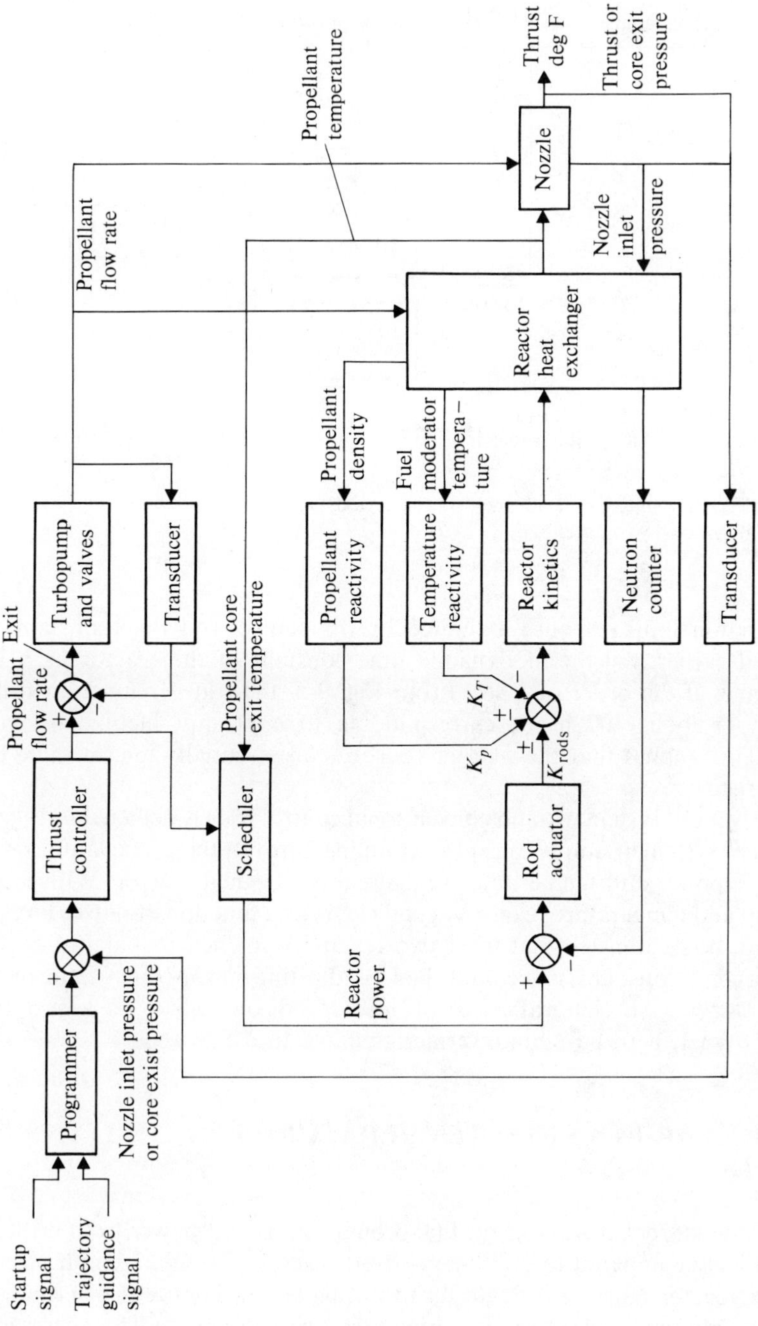

Figure 9-6 Schematic diagram of contemplated nuclear rocket system.[4]

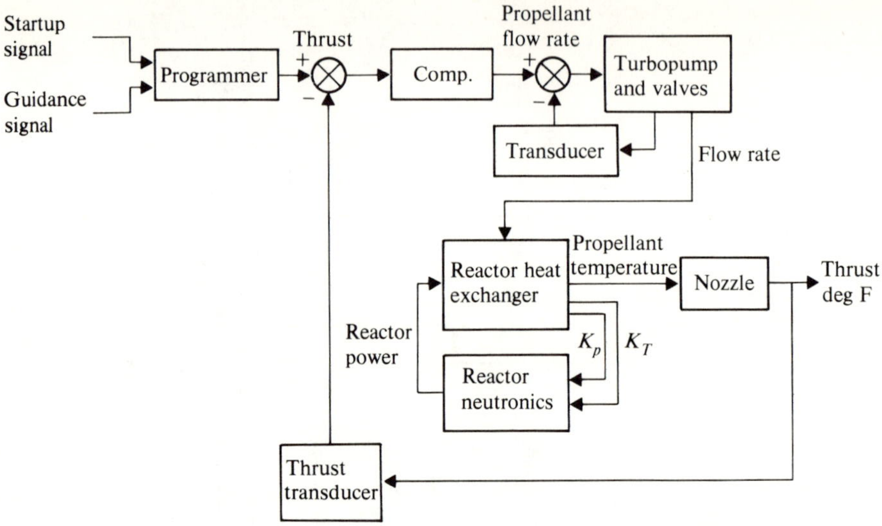

Figure 9-7 Schematic diagram of advanced nuclear rocket system controlled only by temperature and propellant reactivity balance.

The power loop response is limited by the control rod time constants. As mentioned earlier, the thermodynamic time constant of the reactor as a heat exchanger is of the order of 1 sec. From Fig. 9-5, the transfer function roll-off frequency is about 100 hertz, corresponding to a prompt lifetime of $\beta/\omega \cong$ 10 μsec. This implies that the nuclear rocket is an essentially intermediate (epithermal) reactor.

An advanced system with no conventional control rods is depicted in Fig. 9-7. The control system attempts to exploit coupling between the propellant flow and the reactor power and temperature in a suitably designed reactor. With known propellant and temperature reactivity coefficients as a function of power level, the system can be so adjusted that these two reactivity coefficients balance at rated operating conditions. This is accomplished by adjusting poison (rods) in the reactor prior to startup. An elaboration of the use of poison rods is to adjust them remotely (by telemetry) to obtain vernier guidance thrust as well.

9-6 NUCLEAR ROCKETS: TEMPERATURE REACTIVITY EFFECTS

The operational rocket reactor must be brought up to full power from criticality rather quickly compared to stationary-based reactors. To again emphasize this point, the reactor must be brought up to full power, of hundreds of megawatts, from essentially negligible power levels in about one-half minute. This corresponds to about 10 to 100 MW per second. For an operating temperature of 3000°F, this is a 100°F per second temperature increase. Neutronically this corresponds to about a 2-sec period (*e*-folding time). Therefore, it is apparent that temperature reactivity effects will play a major role in the kinetics and control aspects of a nuclear rocket. This will induce certain novel temperature, density, etc., reactivity effects which will merely be mentioned. This is principally because adequate realistic solutions of the associated problems stemming from these effects have not been forthcoming as yet.

As the temperature of the critical reactor is increased from, say, 300°K to 10 times this value, the neutron thermalization energy increases from 0.025 to about 0.20 eV. In ^{235}U, a feasible rocket reactor fuel, the epithermal fission and total absorption cross sections possess resonances that begin about 1 eV and range roughly to 100 eV. If it is assumed that thermal neutrons come into equilibrium with the thermal motion of the fuel and moderator nuclei at temperatures of 3000°F and that the thermal energy spectrum of the neutrons is given by the Maxwell–Boltzmann distribution, then an appreciable portion of these neutrons will find themselves in the epithermal resonance region. Therefore, the calculations made for temperature reactivity coefficients should take into detailed account the effect of these resonances.

A further complication of neutron thermalization is due to the reflector.[11] For high core fuel density, the thermal flux will peak strongly in the reflector. Since the reflector is anterior to the core, its temperature will be somewhat less than the core temperature. Then, thermal neutrons in the reflector that scatter into the core will be moderated upward because of the elevated temperature of the core.

The next item is the effect on the reactivity of the hot propellant hydrogen gas. Its behavior is not quite that of a perfect gas at the operating temperatures and pressures of a nuclear rocket. Therefore, an empirically derived equation of state should be used similarly to the way equations are used in the analysis of reactor accidents (cf. Chap. 7). One simple way to couch the propellant reactivity[4] is in terms of an appropriate reactivity coefficient. Let the propellant reactivity be expressed as $K_p = \alpha_p m_p$, where α_p is the reactivity coefficient in dollars per mole of propellant. m_p is the number of moles of propellant in the system. Then $K_p = \alpha_p PV/ZRT$, where P, V, and T are the pressure, volume, and temperature of the propellant. Z is the compressibility factor to reflect the actual propellant density, and R is the universal gas constant.

Since the propellant gas is a good moderator and absorber as well, its passage through the reactor will result in both positive and negative reactivity effects. This is because most of the neutrons are thermalized in the reflector, and therefore the predominant effect of the propellant flowing through the reflector on thermal neutrons that scatter *into* the reflector will be that of an absorber. With respect to the core neutrons, however, the propellant represents a positive reactivity because it will moderate some of the core epithermal neutrons to thermal energies and thereby increase the resonance escape probability. As mentioned, this can be dangerous when, for some reason, an amount of hydrogen propellant insufficiently vaporized enters the core. This corresponds to a positive step reactivity insertion which may cause a prompt rise in the neutron density of unmanageable proportions.

9-7 NUCLEAR ROCKETS: SHUTDOWN

As can be imagined, the shutdown (engine cutoff) of a reactor rocket also poses some almost intractable problems, especially when many and diverse post-shutdown requirements must be satisfied. For example, for a nuclear rocket that forms either the booster or upper stage of a multistage rocket, it is imperative that the reactor thrust drop to a negligible amount very quickly in order that it separate in space from the stage to follow. Unfortunately, it is difficult to satisfy this requirement, since even scramming the system causes it to drop promptly to only $(1 + K)^{-1}$ of its operating level. Then it proceeds to drop further, approximately on its asymptotic period of about 80 sec. Not only this, but the other post-shutdown neutron-producing reactions, such as would occur with a beryllium reflector, for example, would further complicate the shutdown.

Another troublesome factor, at these very high powers, is that the fission products keep the reactor thermally and radioactively hot as Fig. 9-8 shows. Still another problem of operating power-producing equipment in space is that the only mechanism available for cooling is by radiation. For example, satellite reactors must be built around an efficient radiator for the system cannot dissipate its inherent heat losses otherwise. For a rocket reactor that is to be reused in space, means must be obtained to cool the reactor engine to a temperature at which it can safely dissipate its heat losses through radiation.[13] Perhaps this can

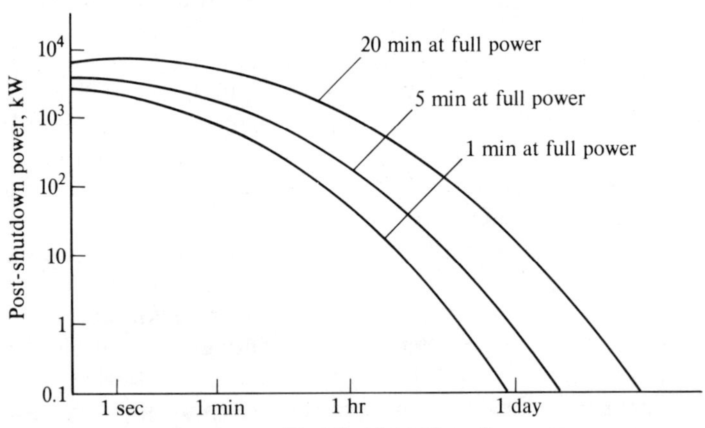

Figure 9-8 Post-shutdown *B*- and *γ*-generated heat power for a 100-MW reactor (based on a $T^{-1.2}$ decay law).[12]

be done with a liquid propellant reserve. A second, smaller reactor could supply the prime mover for the power needed to pump the post-propellant coolant reserve. If the rocket reactor is expendable, then simple means can be found to fragment it so that the fragments behave as bodies with poor reentry characteristics and thereby burn up upon reentering the atmosphere.

9-8 GASEOUS CORE/CAVITY REACTORS

The analytical and experimental relative successes of the various nuclear rocket program efforts through the years, concurrently prompted other workers to propose the gaseous core/cavity reactor (rocket). This, usually cylindrical, reactor contains a core of swirling gaseous fuel, for example, $^{235}UF_6$, in which jets of propellant gas are introduced. The propellant gas in diffusing through the fissioning fuel gas, extracts the fission energy through scattering from the fission fragments. The now very hot propellant gas leaves the "combustion" chamber through the nozzle in the rear, thus providing the required thrust.

The gaseous fuel reactor does not require the solid fuel core bearing structure as in other reactors, and so is not limited by reactor material thermal stresses. This makes for ultimate specific impulse ratios over that of the earlier discussed nuclear reactor rockets, of about 2–5.[14]

The scheme of the gaseous fuel reactor is to retain the heavy fissionable fuel gas within the combustion chamber-reactor vessel, while simultaneously transferring fission energy to a very light propellant gas (e.g. hydrogen) that is diffusing roughly perpendicularly to, and through, the swirling fuel gas. The principal method by which it is anticipated to accomplish the above is to exploit the very large mass ratio of the fuel gas to the propellant gas. One way by which this can be done is to establish a radial vortex flow of the fuel gas so the centripetal force generated thereby will keep the fuel gas in place in the combustion chamber. The very light propellant gas will only minimally participate in the radial vortex flow because of its relatively small mass, however, acquiring the fission fragment kinetic energy, and then rapidly diffusing out of the nozzle to provide the thrust force. Efforts in this regard[15–17] have shown that vortices must be generated in a sophisticated manner to obtain an essentially laminar vortex flow, or else the vortex can degenerate into turbulent flow thus destroying the two-flow gas regime. Herein, it will be assumed that the requisite flow of the gases is obtained, and attention will be directed toward the reactor kinetics of such a system. In this sense then, consider travelling in a coordinate system that is rotating with the mean vortex angular velocity of the gaseous fuel (considered as one large vortex). Then this system devolves to a "cavity reactor." That is, it consists of a central gaseous fuel core cavity surrounded by a low Z moderator/reflector, whether solid or gaseous itself. The effects of the propellant gas are neglected in this model, but can be construed as a reactivity coefficient dependence given in a later development. Again, the cavity is not a void, but contains the fuel gas at relatively low density compared to the moderator/reflector density. Typical gaseous fuel densities are in the vicinity of a fraction of molecular gas densities at standard conditions.[18,19]

For the cavity reactor kinetics model, it is assumed that:

(*a*) Core:

The low density gaseous fuel core is essentially transparent to fast neutrons. That is, the core gas can weakly scatter fast neutrons, but does not moderate or capture them. Thus all fission neutrons born, almost immediately penetrate into the moderator/reflector. Those that happen to be scattered into the core from the moderator/reflector, merely traverse the core region back into the former.

Slow neutrons can interact with the core gas with the usual cross sections. Some produce fissions, whereas others pass through the core, and perhaps ultimately leak from the reactor.

(*b*) Moderator/Reflector:

Fast neutrons are slowed to thermal in the moderator/reflector only. Following thermalization, slow neutrons that migrate to the core take an average of l_f seconds to do so. That is, there is an additional diffusion time l_f for thermal neutrons to diffuse back to the core. l_f is construed as a delay time, in addition to the normal prompt neutron lifetime—to absorption. Delayed neutrons are manifest in the moderator/reflector only.

With these assumptions, neutron kinetics balance rate equations can be written for the neutron density in the core, n_c, that in the moderator/reflector, n_r, and the usual six groups of delayed neutron precursors concentration c_i. Then, in the core

$$\dot{n}_c = -\frac{n_c}{T_c} + \lambda_r n_r(t - l_f) \tag{9-36}$$

where T_c is the prompt generation time in the core. $\lambda_r = T_r^{-1}$ is the neutron decay rate in the moderator/reflector. Equation (9-36) states that the core balance rate is made up of core neutrons leaving at a rate T_c^{-1} to enter the moderator/reflector, plus moderator/reflector neutrons arriving at a rate λ_r, but with a mean delay time l_f needed to diffuse into the core.

In the moderator/reflector

$$\dot{n}_r = -\lambda_r n_r(t - l_f) + (1 - \beta)\frac{kn_c}{T_c} + \sum_{i=1}^{6} \lambda_i c_i \tag{9-37}$$

where k is the usual effective reproduction factor. The moderator/reflector neutron balance rate is comprised of those neutrons leaving at a rate of λ_r per second, but taking l_f seconds to get to the core, plus prompt core neutrons born in fission in the core, together with the delayed neutron precursor concentration rate $\lambda_i c_i$ to provide delayed neutrons. For completeness, the delayed neutron precursor concentration satisfies

$$\dot{c}_i = -\lambda_i c_i + \beta_i k n_c / T_c; \qquad i = 1, \ldots 6 \tag{9-38}$$

in the usual fashion.

Initial equilibrium is assumed, indicated by the subscript zero, which implies from Eq. (9-36), where $n_r(0 - l_f) = n_r(0) = n_{r0}$, that

$$\lambda_r n_{r0} = n_{c0}/T_c = n_{c0}/\beta\Lambda_c \tag{9-39}$$

where Λ_c is the augmented neutron lifetime in the core, that is, $\Lambda_c = T_c/\beta k$. It is seen from Eqs (9-37) and (9-38) evaluated at zero time, that Eqs (9-39) are merely reiterated, where $k(0) = 1$.

With the reactivity in dollars, that is, $k = (k - 1)/\beta k$, used to eliminate k, the above balance equations are rewritten, respectively, as

$$\dot{n}_c = -\left(\frac{1}{\beta} - K\right)\frac{n_c}{\Lambda_c} + \lambda_r n_r(t - l_f) \tag{9-40}$$

$$\dot{n}_r = -\lambda_r n_r(t - l_f) + \left(\frac{1}{\beta} - 1\right)\frac{n_c}{\Lambda_c} + \sum_{i=1}^{6} \lambda_i c_i \tag{9-41}$$

$$\dot{c}_i = -\lambda_i c_i + a_i n_c/\Lambda_c; \qquad i = 1, \ldots 6 \tag{9-42}$$

Eliminating the delayed neutron precursor concentrations c_i, reduces these equations to the neutron densities in the core and moderator/reflector, respectively, namely

$$\dot{n}_c = -\left(\frac{1}{\beta} - K\right)\frac{n_c}{\Lambda_c} + \lambda_r n_r(t - l_f) \tag{9-43}$$

$$\dot{n}_r = -\lambda_r n_r(t - l_f) + \frac{n_c}{\beta\Lambda_c} - \frac{1}{\Lambda_c}\int_0^t n_c'(\tau) f(t - \tau)\, d\tau \tag{9-44}$$

Now, two transfer functions for this reactor will be obtained. They are (*a*) the core transfer function $\bar{G}_c(s)$, which relates the neutron response in the core to a small reactivity disturbance, and (*b*) the moderator/reflector transfer function $\bar{G}_r(s)$, which relates a like response in the moderator/reflector to a small reactivity disturbance. This is done simply, to first order in the perturbational sense, by letting $n_r(t) = n_{r0} + \varepsilon_r(t)$, $n_c(t) = n_{c0} + \varepsilon_c(t)$, where ε_r and ε_c are small compared to their corresponding steady state neutron densities. The reactivity K is already assumed to be a first order perturbational quantity. Inserting these quantities in Eqs (9-43) and (9-44), and retain first order terms only, yields the transfer function relationships for the core and moderator/reflector, respectively. That is

$$\bar{\varepsilon}_c(s) = \bar{G}_c(s)\bar{K}(s) \tag{9-45}$$

where the core transfer function is given by

$$\bar{G}_c(s) = \frac{\beta n_{c0}}{s} \cdot \frac{(s + \lambda_r e^{-sl_f})}{1 + \beta[\Lambda_c^s + \lambda_r e^{-sl_f}(\Lambda_c + \bar{f}(s))]} \tag{9-46}$$

and

$$\bar{\varepsilon}_r(s) = \bar{G}_r(s)\bar{K}(s) \tag{9-47}$$

where the moderator/reflector transfer function is given by

$$\bar{G}_r(s) = \frac{\beta n_{r0}}{s} \cdot \frac{\lambda_r(1 - \beta s \bar{f}(s))}{1 + \beta[\Lambda_c s + \lambda_r e^{-s l_f}(\Lambda_c + \bar{f}(s))]} \tag{9-48}$$

The zero order terms from Eqs (9-43) and (9-44) merely yield the initial condition relationships, of which $n_{r0}/n_{c0} = (\beta\Lambda_c\lambda_r)^{-1} = T_r/l_c$ is used above. l_c is the prompt generation time, that is, $l_c = (vv\Sigma_f)^{-1}$.

To roughly sketch these transfer functions, the upper and lower corner angular frequencies, ω_{Hc} and ω_{Lc} must be calculated. For a graphite moderator/reflector and a $^{235}UF_6$ gaseous core, with a ^{235}U atomic number density of $10^{19}/\text{cm}^3$, Table 9-2 depicts the pertinent parameters. The transfer functions are sketched from their asymptotes given in Table 9-2 using the break-frequency technique mentioned in Sec. 3-4 and are depicted in Fig. 9-9.

First, it is seen that the amplitude transfer functions are separated in level by about 17 dB. This implies that the amplitude gain of the moderator/reflector transfer function is 17 dB over that of the core. This is because the ratio (n_{r0}/n_{c0}) of equilibrium neutron levels in the moderator/reflector to that of the core is about a factor of 50, as seen from Table 9-2. That $n_{r0}/n_{c0} = 50$ holds is simply because the neutron lifetimes are in this ratio, due to the fact that the neutrons, on the average, spend 50 times more of their life in the moderator/reflector than in the core. This can be seen in Fig. 9-10 where it is seen that the flux peaks in the moderator/reflector, as compared to the core flux which is relatively flat due to the low density of the core fuel.

Second, the low frequency characteristics of both transfer functions are the same, except for the 17 dB gain separation. This is because the low corner

Table 9-2 Cavity reactor parameters

$^{235}UF_6$ core	Graphite moderator/reflector
$\rho_{25} = 0.004\ \text{g/cm}^3$	$\lambda_r = v\Sigma_a = 70.4\ \text{sec}^{-1}$
$\Sigma_f = 6.13 \cdot 10^{-3}\ \text{cm}^{-1}$	$\Sigma_a = 3.2 \cdot 10^{-4}\ \text{cm}^{-1}$
$\Lambda_c = (\beta vv\Sigma_f)^{-1} = 0.042\ \text{sec}$	$T_r = \lambda_r^{-1} = 0.014\ \text{sec}$
$l_c = \beta\Lambda_c = 2.9 \cdot 10^{-4}\ \text{sec}$	$n_{r0}/n_{c0} = T_r/l_c = 48.28\ (16.8\ \text{dB})$
$\lim_{\omega\to\text{large}} \lvert\bar{G}_c(\omega)\rvert \simeq \beta n_{c0}/\sqrt{1 + (\beta\Lambda_c\omega)^2} \simeq \frac{n_{c0}}{\omega\Lambda_c}$	$\lim_{\omega\to\text{large}} \lvert\bar{G}_r(\omega)\rvert \sim \frac{\beta n_{r0}\lambda_r}{\omega} \Big/ \sqrt{1 + (\beta\Lambda_c\omega)^2} \sim \frac{n_{r0}\lambda_r}{\omega^2\Lambda_c}$
$\omega_{\text{Hc}} = \Lambda_c^{-1} = 23.8\ \text{sec}^{-1}$	$\omega_{\text{Hc}} = \sqrt{\lambda_r/\Lambda_c} = 40.9\ \text{sec}^{-1}$
$f_{\text{Hc}} = \omega_{\text{Hc}}/2\pi = 3.79\ \text{Hz}$	$f_{\text{Hc}} = \omega_{\text{Hc}}/2\pi = 6.52\ \text{Hz}$
$\lim_{\omega\to\text{small}} \lvert\bar{G}_c(\omega)\rvert \simeq \frac{\beta n_{c0}}{\omega} \cdot \frac{\lambda_r}{1 + \beta\lambda_r\bar{\tau}}$	$\lim_{\omega\to\text{small}} \lvert\bar{G}_r(\omega)\rvert \simeq \frac{\beta n_{r0}}{\omega} \cdot \frac{\lambda_r}{1 + \beta\lambda_r\bar{\tau}}$
$\omega_{\text{Lc}} = \beta\lambda_r/(1 + \beta\lambda_r\bar{\tau}) \simeq 0.0674\ \text{sec}^{-1}$	$\omega_{\text{Lc}} = \beta\lambda_r/(1 + \beta\lambda_r\bar{\tau}) \simeq 0.0674\ \text{sec}^{-1}$
$f_{\text{Lc}} = \omega_{\text{Lc}}/2\pi = 0.0107\ \text{Hz}$	$f_{\text{Lc}} = \omega_{\text{Lc}}/2\pi = 0.0107\ \text{Hz}$

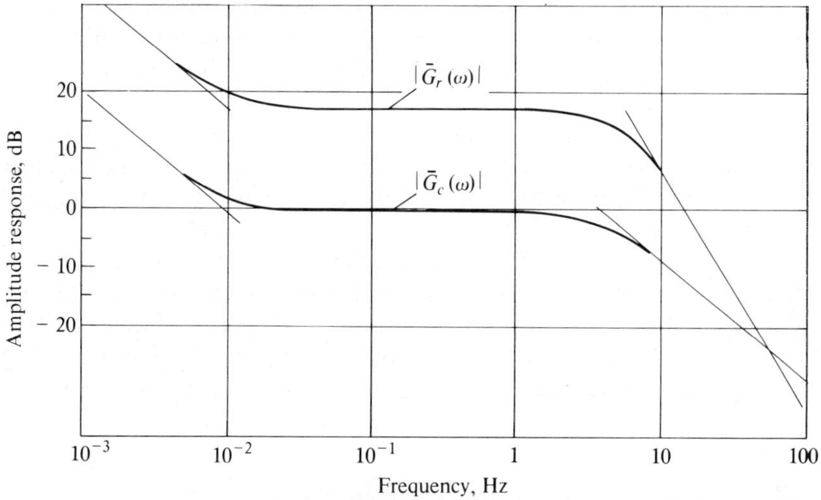

Figure 9-9 Cavity reactor core and moderator/reflector amplitude transfer functions for $^{235}U_6$ gas core and graphite moderator/reflector.

frequency depends on moderator/reflector parameters only, and only indirectly on the core parameters through the delayed neutron precursor constants.

Third, the high frequency characteristic of the moderator/reflector transfer function falls off twice as fast (12 dB per octave) as that of the core. The asymptotes provide only a rough sketch of the amplitude transfer functions, so that any rapidly varying fine structure that might lie in the intermediate frequency range, due to the phase shifts manifest by the delay time l_f is not obtained. The delay

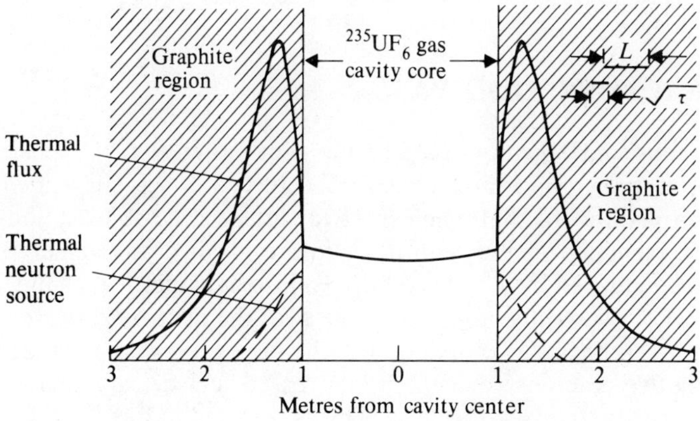

Figure 9-10 Thermal flux in cavity reactor fueled by $^{235}U_6$ gas, externally moderated and reflected by graphite.[19]

time l_f is of the order of a quarter that of the neutron lifetime in the moderator/reflector.[20] So that l_f is about 3 milliseconds as seen from Table 9-2. It is appreciated that this corresponds to a very small phase shift in the frequency regime involved herein.

If one assumes that the core and moderator/reflector are coupled, in the sense that the total transfer function of the cavity reactor is given by the product $\bar{G}_c(s)\bar{G}_r(s)$, then it is seen from Fig. 9-9 that the corresponding sum of the two logarithmic curves constitute a reactor that acts as a "sharp" low pass filter, which falls off at -18 dB per octave at high frequencies. This implies that any small disturbances in the system (e.g. weak shock waves in the core gas), having frequency components higher than the composite high corner frequency $f_c = (1/2\pi)(\lambda_r/\Lambda_c^2)^{1/3} = 5.44$ Hz, are sharply cut off by this reactor. The preceding formulations, and transfer function asymptote calculations are left as exercises for the reader.

9-9 PULSED REACTORS: INTRODUCTION

Pulsed reactors are, as the name implies, reactors whose neutron flux or density varies in a heavily transient fashion during their normal operating mode. That is, extremely large neutron pulses, but of short duration, are generated in such systems. There are many reasons for building such reactors; they range from neutron pulse research to novel power applications. With a small duty cycle, defined as the ratio of pulse duration to its period, such reactors deliver peak power in the thousands of megawatts and an average power that is but a minute fraction of the maximum power. This discussion will be limited to four representative reactors. These are the Godiva, TREAT, TRIGA, and SPERT reactors. As mentioned before, these classifications overlap; for example, the Godiva can be also classed as a fast reactor.

9-10 PULSED REACTORS: GODIVA

The Godiva reactor consisted† merely of a 93.5 per cent enriched ^{235}U sphere weighing about 53 kg sectioned into three pieces.[21] Each section was suspended, and the whole was so juxtaposed that when the two outer retractable sections were brought near to each side of the central section, criticality was achieved. Neutron bursts could be induced by a ^{235}U rod driven into a diametrical channel of the center section. Following introduction of the rod, the neutron level rose rapidly to a peak and then began to drop, because the heat generated expanded the fuel, and the negative reactivity due to the decreased fuel density thereby dropped the neutron density to a very low level. This is depicted in Fig. 9-11. For ultra fast bursts, an inertial time lag manifests itself between fission heating, and the expansion of the metal fuel.

† Godiva-I was dismantled ca. 1953 following an accidentally large burst.

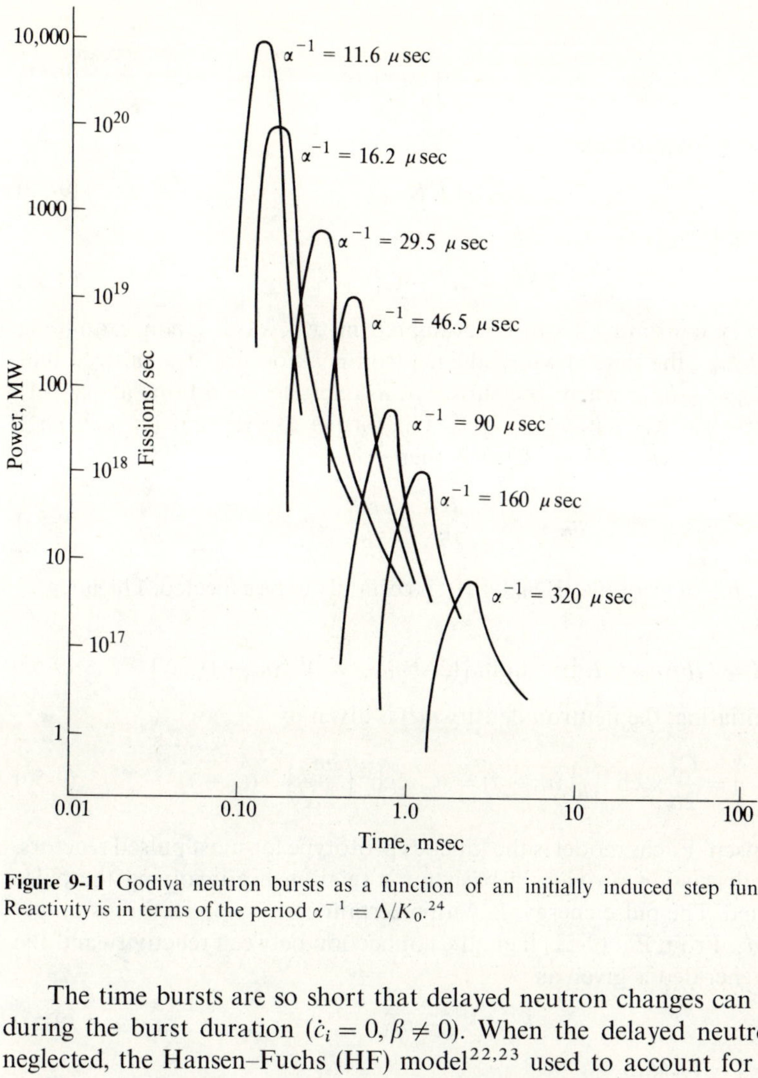

Figure 9-11 Godiva neutron bursts as a function of an initially induced step function reactivity. Reactivity is in terms of the period $\alpha^{-1} = \Lambda/K_0$.[24]

The time bursts are so short that delayed neutron changes can be neglected during the burst duration ($\dot{c}_i = 0, \beta \neq 0$). When the delayed neutrons are thus neglected, the Hansen–Fuchs (HF) model[22,23] used to account for the negative reactivity due to core expansion upon heating yields

$$\Lambda\dot{n} = Kn \equiv [K_0 - \alpha T(t)]n \tag{9-49}$$

where K_0 is the initially introduced reactivity step, in dollars, α is the negative reactivity temperature coefficient in dollars per degree, and T is the temperature in the ^{235}U assembly. Since $CT = \int_0^t n(\tau)\,d\tau$, where C is the thermal capacity of the fuel in appropriate units, the integral is proportional to the energy generated up to time t. Then Eq. (9-49) can be rewritten to give

$$\Lambda\dot{n} = \left[K_0 - b\int_0^t n(\tau)\,d\tau\right]n, \qquad n(0) = n_0 \tag{9-50}$$

where $b = \alpha/C$. Let

$$K = K_0 - b \int_0^t n(\tau)\,d\tau \tag{9-51}$$

to insert in Eq. (9-50) to obtain

$$\Lambda \ddot{K} = K\dot{K} \tag{9-52}$$

A first integral is

$$2\Lambda \dot{K} = K^2 - C_1^2 \tag{9-53}$$

The arbitrary constant C_1 can be evaluated in two ways. Upon examining the situation at $t_{\max}$, the time at which the neutron peak occurs, it is noticed that $K_m = 0$ and $\dot{K}_m = -bn_m$, where the subscript m refers to conditions at $t_{\max}$. At $t = 0$ when $K(0+) = K_0$, $\dot{K}(0+) = -bn_0$. The former asserts that $C_1^2 = 2\Lambda b n_m$; the latter gives $C_1^2 = K_0^2 + 2\Lambda b n_0$. Clearly then

$$n_m = n_0 + \frac{K_0^2}{2b\Lambda} \cong \frac{K_0^2}{2b\Lambda} \tag{9-54}$$

since $n_0 \ll K_0^2/2b$. For example, $n_m/n_0 \cong 10^9$ is certainly to be expected. The integral of Eq. (9-53) is

$$K = (2b\Lambda n_0 + K_0^2)^{1/2} \tanh\left[(2\Lambda b n_0 + K_0^2)^{1/2}(t_m - t)/2\Lambda\right] \tag{9-55}$$

By differentiating, the neutron density $n(t)$ is given by

$$n = \frac{C_1^2}{2b}\operatorname{sech}^2\frac{C_1}{2}(t_m - t) = n_m \operatorname{sech}^2\left(\frac{bn_m}{2\Lambda}\right)^{1/2}(t_m - t) \tag{9-56}$$

As the Hansen–Fuchs model is the kinetics prototype for most pulsed reactors, the preceding discussion provides a basis for a relationship involving the pulse energy generated. The pulse energy, E, within certain conversion factors is simply $E(t) = \int_0^t n(\tau)\,d\tau$. From Eq. (9-51) then, the connection between reactivity and the pulse energy generated is given as

$$K = K_0 - bE \tag{9-57}$$

By integrating Eq. (9-56)

$$E(t) = \sqrt{\frac{2n_m\Lambda}{b}}\left[\tanh\sqrt{\frac{bn_m}{2\Lambda}}\,t_m - \tanh\sqrt{\frac{bn_m}{2\Lambda}}\,(t_m - t)\right] \tag{9-58}$$

and the total energy in the pulse is therefore

$$E(\infty) = \sqrt{\frac{2n_m\Lambda}{b}}\left[1 + \tanh\sqrt{\frac{bn_m}{2\Lambda}}\,t_m\right] \tag{9-59}$$

From Eq. (9-56) it is seen that the pulse is symmetric about the maximum neutron density value. Hence, half the energy is generated prior to the maximum, and half beyond the occurrence of the maximum. Insofar as this model describes the physical situation, then $E(\infty) = 2E_m$, where E_m is the energy generated in the pulse up to the maximum neutron density. So that from Eqs (9-57), (9-58), and (9-59), as Nordheim first pointed out

$$E(\infty) \cong 2E_m = 2K_0/b \tag{9-60}$$

So that the energy developed in the pulse is proportional to the initially inserted reactivity K_0.

Following the neutron density pulse peak, there is a residual level, due in part to the manifestation of the delayed neutrons, which is now investigated. Consider the Hansen–Fuchs model modified to include one group of delayed neutrons, namely, where λ is the one group effective decay constant,

$$\Lambda \dot{n} = \left(K_0 - b \int_0^t n(\tau)\, d\tau \right) n(t) - \int_0^t n'(\tau)\, e^{-\lambda(t-\tau)}\, d\tau \quad ; \quad \Lambda \dot{n}_0 = K_0 n_0 \tag{9-61}$$

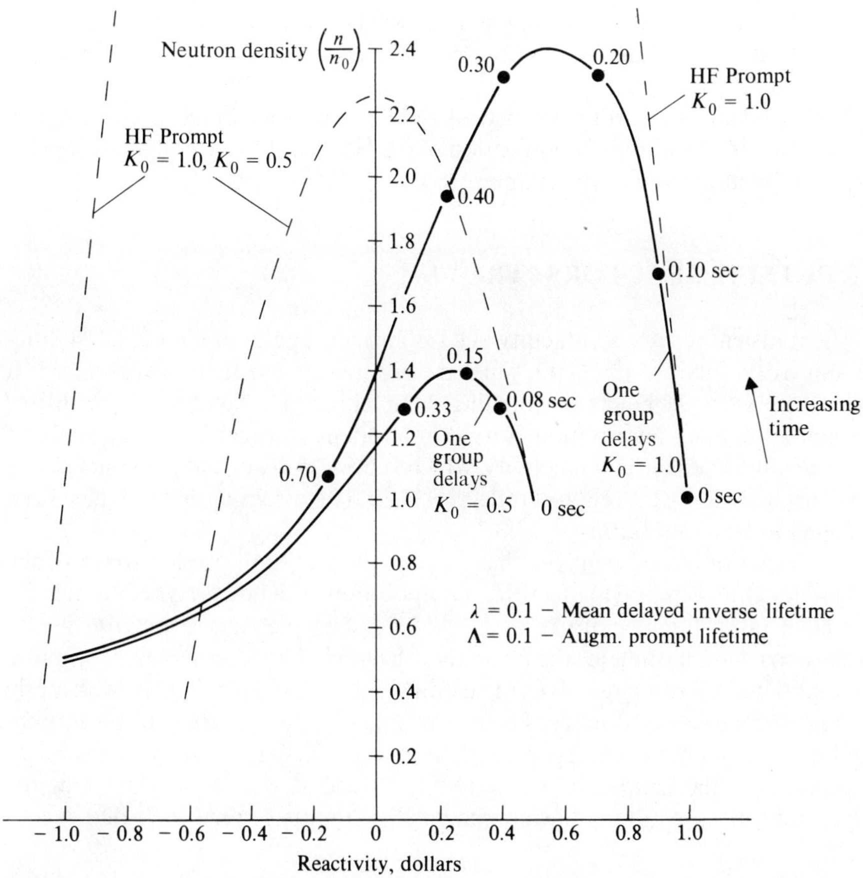

Figure 9-12 Hansen–Fuchs model modified for one group of delayed neutrons, compared with that with no delayed neutrons.

Since the reactivity $K(t) = K_0 - b\int_0^t n(\tau)\,d\tau$, and that $\dot{K} = -bn$, transforming the above equation to the neutron density-reactivity phase plane yields, after eliminating the rightmost integral by differentiation

$$\Lambda bnn' + (K - 1)n = \frac{\lambda}{2b} K^2 + C \tag{9-62}$$

where the prime denotes differentiation with respect to the reactivity K, and that $\lambda\Lambda \ll 1$, and so is neglected by comparison. In a fashion similar to the no delayed neutron Hansen–Fuchs model, the above constant is evaluated both at the initial time, and at the time at which the neutron density peak occurs. This gives

$$-C = \frac{\lambda}{2b} K_0^2 + n_0 = \frac{\lambda}{2b} K_m^2 + (1 - K_m)n_m \tag{9-63}$$

Notice that due to the effect of the delayed neutrons, the neutron density peak occurs for a finite reactivity K_m, as opposed to zero reactivity in the no delayed neutron case. All of this is plotted in Fig. 9-12, where the neutron density is depicted as a function of the reactivity K, for no delayed neutrons, and one group of delayed neutrons.†

The problem of introducing instead only the equilibrium concentration of delayed neutrons as another modification of the Hansen–Fuchs model is discussed as an exercise at the end of the chapter.

9-11 PULSED REACTORS: TREAT

The transient reactor test facility (TREAT) is a graphite-moderated, graphite-uranium oxide mixture ($^{235}U_3O_8$) fueled core, air-cooled thermal reactor.[25] It was built to test fuel elements by melting them with a high flux ($\simeq 10^{16}$ neutrons/cm^2/sec) pulse. Such information is used in accident and reactor safety analysis. Its temperature coefficient of reactivity, which is principally due to neutron heating of the reactor, is -2.4 cents per degree C.[25] The prompt lifetime has been calculated to be about $0.10\beta_{25}$.

Two types of pulsed behavior have been studied; both use a variant of the HF kinetics model employing EDPM‡ computations.[26] The first type initiates the pulse by a ramp function up to $K_{max} = \$2.67$. The negative reactivity due to thermal expansion terminates the pulse very sharply. The second, or flat-top-pulse type, is initiated by a step reactivity. The quickly rising neutron density is abruptly curtailed by reinserting the rods. The neutron density then flattens to an equilibrium level. To maintain the flat-topped level, the reactivity is then increased to compensate for the temperature reactivity decrease as well as the fact that the delayed neutrons are not manifest immediately. In units of time such that $\Lambda = 1$,

† Equation (9-62) is numerically integrated to obtain the curves, using the HP-65 hand calculator program Math 1-36A, a third order Runge Kutta algorithm.

‡ e.g., Argonne National Laboratories, RE-129 Code.

the sharply peaked pulse type of neutron density for a ramp reactivity of 5.34 dollars per second for 0.5 sec, until $K_{max} = \$2.67$ is attained, corresponds to

$$\dot{n} = \left[mt - b\int_0^t n(\tau)\,d\tau \right] n(t), \qquad 0 < t < 0.5 \tag{9-64}$$

$$\dot{n} = -bn(t)\int_0^t n(\tau)\,d\tau, \qquad t \geq 0.5 \tag{9-65}$$

Before examining the solutions to this set of equations, it is of interest to compute the constant b.[27] The heat absorbed per unit volume of the reactor up to time t is given by the usual expression for the number of fissions per second per cubic centimeter:

$$\int_0^t \Sigma_f \phi(\tau)\,d\tau = \Sigma_f v \int_0^t n(\tau)\,d\tau \tag{9-66}$$

Σ_f is the macroscopic fission cross section, and the monoenergetic flux $\phi = nv$. Since the prompt neutron lifetime is given by $l \cong (\nu v \Sigma_f)^{-1}$, $v\Sigma_f = (\nu l)^{-1} = (0.10\beta\nu)^{-1}$, using the value of prompt lifetime given in the above discussion: and ν, the mean number of neutrons per fission of ^{235}U, is 2.5. The specific heat $C = 0.53$ cal/cm^3/deg C. One fission corresponds to 7.5×10^{-12} cal of heat energy, so that the fuel temperature change ΔT is given by

$$\Delta T = \frac{\int_0^t n(\tau)\,d\tau}{0.10\beta\nu C} = \frac{10}{7.6 \times 10^{-3}} \frac{7.5 \times 10^{-12}}{0.53 \times 2.5} \int_0^t n(\tau)\,d\tau \cong 0.8 \times 10^{-8} \int_0^t n(\tau)\,d\tau \tag{9-67}$$

The reactivity change due to temperature $K_T = (K_T/\Delta T)\,\Delta T$. The reactivity temperature coefficient is $K_T/\beta\,\Delta T \cong 0.025$ (dollars per degree C). Finally

$$K_T = -b\int_0^t n(\tau)\,d\tau \qquad b \cong 10^{-10} \tag{9-68}$$

To find the solutions to Eqs (9-64) and (9-65), the former is differentiated to give

$$\ddot{n} = \frac{\dot{n}^2}{n} + n(m - bn) \tag{9-69}$$

Letting $\dot{n} = p$ reduces this to

$$\frac{d}{dn}\left(\frac{p^2}{n^2}\right) = 2b\left(\frac{r}{n} - 1\right) \qquad p(0) \equiv \dot{n}(0) = 0 \qquad n(0) = 1 \tag{9-70}$$

where

$$r = \frac{m}{b} = \frac{0.025/0.5}{10^{-10}} = 5 \times 10^8$$

for an initial reactivity insertion rate of about 8 dollars per second.[27] The above initial conditions result in the integral of this equation as

$$\int_1^n \frac{dx}{x[r\ln x-(x-1)]^{1/2}}=(2b)^{1/2}t \tag{9-71}$$

The above integral will be approximated for three different portions of the neutron density excursion. These are the small time initial change, the region around the peak level n_m, and the asymptotic neutron density level.

CASE 1 SMALL TIME NEAR THE ORIGIN Transform the integral to

$$\int_0^{\log n} \frac{dx}{[rx-(e^x-1)]^{1/2}}=(2b)^{1/2}t \tag{9-72}$$

Then approximate the exponential in the integrand by $e^x = 1 + x + x^2/2$ to give, with $r - 1 \cong r$,

$$\int_0^{\log n} \frac{dx}{[2rx-x^2]^{1/2}}=b^{1/2}t \tag{9-73}$$

This is easily integrated to obtain

$$n(t) \cong \exp\left(2r\sin^2\frac{b^{1/2}t}{2}\right) \tag{9-74}$$

which for very small time becomes

$$n(t) \cong e^{mt^2/2} \tag{9-75}$$

the well-known immediate neutron density rise for a ramp reactivity. This is equivalent to neglecting the integral term in Eq. (9-64), which corresponds to such short times that the neutron level is well on its way toward the peak value n_m, while the thermal lags have prevented the temperature reactivity effects from occurring as rapidly.

CASE 2 BEHAVIOR NEAR n_m If Eq. (9-70) is integrated again, but in terms of conditions near the neutron density pulse peak, that is, $n(t_m) = n_m$; $\dot{n}(t_m) = 0$, the result is

$$\int_n^{n_m} \frac{dx}{x[r\ln(x/n_m)-(x-n_m)]^{1/2}}=(2b)^{1/2}(t-t_m) \tag{9-76}$$

Near n_m the logarithm in the integral is approximately given by $\log(x/n_m) \cong (x - n_m)/n_m$. With this approximation, the above simplifies to

$$\int_n^{n_m} \frac{dx}{x(x-n_m)^{1/2}}=\left[2b\left(\frac{r}{n_m}-1\right)\right]^{1/2}(t-t_m) \tag{9-77}$$

$$n = n_m \operatorname{sech}^2\left[\frac{b}{2}(n_m-r)\right]^{1/2}(t-t_m) \tag{9-78}$$

which is very nearly the same as the expression obtained for the neutron density due to an initial step function of reactivity, as given in Eq. (9-56). This implies that near the peak, the effect of the form of the initiating reactivity has largely been vitiated.

CASE 3 NEUTRON LEVEL FOLLOWING THE PEAK This corresponds to Eq. (9-65) which, when differentiated, gives

$$\ddot{n} = \frac{\dot{n}^2}{n} - bn^2, \dot{n}(t_m) = 0, \qquad n(t_m) = n_m \tag{9-79}$$

This is the same for the case of the HF model for an initial step function of reactivity. The solution follows from the development given in Sec. 9-10 and is

$$n = n_m \operatorname{sech}^2 \left(\frac{bn_m}{2}\right)^{1/2} (t - t_m) \tag{9-80}$$

Figure 9-13 portrays the results of EDPM computation for the cases above.

9-12 PULSED REACTORS: TRIGA

The training, research, isotope production reactor (TRIGA) is a light-water-cooled zirconium-hydride-moderated reactor.[31] The fuel is 20 per cent enriched ^{235}U impregnated into the ZrH. The reactor is constructed of combined fuel-moderator elements which make for a large, prompt, negative temperature reactivity coefficient (~ -1.5 cents, per degree C). The Flash reactor is a TRIGA-type reactor designed for pulse operation. The principal difference is that the fuel-moderator combination is graphite-uranium instead of ZrH-uranium. The former has a greater high-temperature structural stability as well as a greater heat capacity than the ZrH-^{235}U mixture. The pulse peak flux is 6×10^{17} fast neutrons/cm^2/sec, an order of magnitude over that obtainable in the Godiva reactor.[29] The pulse peak occurs so rapidly that thermodynamically the reactor remains in an adiabatic state for the duration of the pulse. As discussed earlier in the analysis of the TREAT reactor, the form of the pulse is essentially independent of the exact manner of reactivity insertion, i.e., of the control rod movements.

The allowable period between pulses (or flashes) is a function of the cooling capability of the reactor. The reactor is designed to remain intact even if the cooling water does not circulate. For a peak pulse power of 8×10^4 MW, or a generated energy in the pulse of 800 MW sec (900 deg F temperature rise), the pulse half-width is about 8 msec. A cooling system of 50 kW capacity would allow a 4-hr period, corresponding to a mean reactor power of 50 kW.

A novel safety aspect of the ZrH moderator is that at room temperature (300 deg K) the bound proton of the H atom acts like an Einstein harmonic oscillator with energy levels spaced at $\frac{3}{2}h\nu$ ergs apart. For ZrH, $h\nu = 0.13$ eV,[30] so that for neutrons with energies greater than 0.13 eV, these protons moderate the neutrons essentially as if the former were free particles. Below 0.13 eV, however, the protons behave like a bound system, so that their moderating effectiveness is greatly impaired. Moderation below 0.13 eV takes place in the reactor by virtue of the water coolant only. However, neutrons below 0.13 eV in energy can be moderated upward,[31] by the bound protons, in multiples of 0.13 eV. This

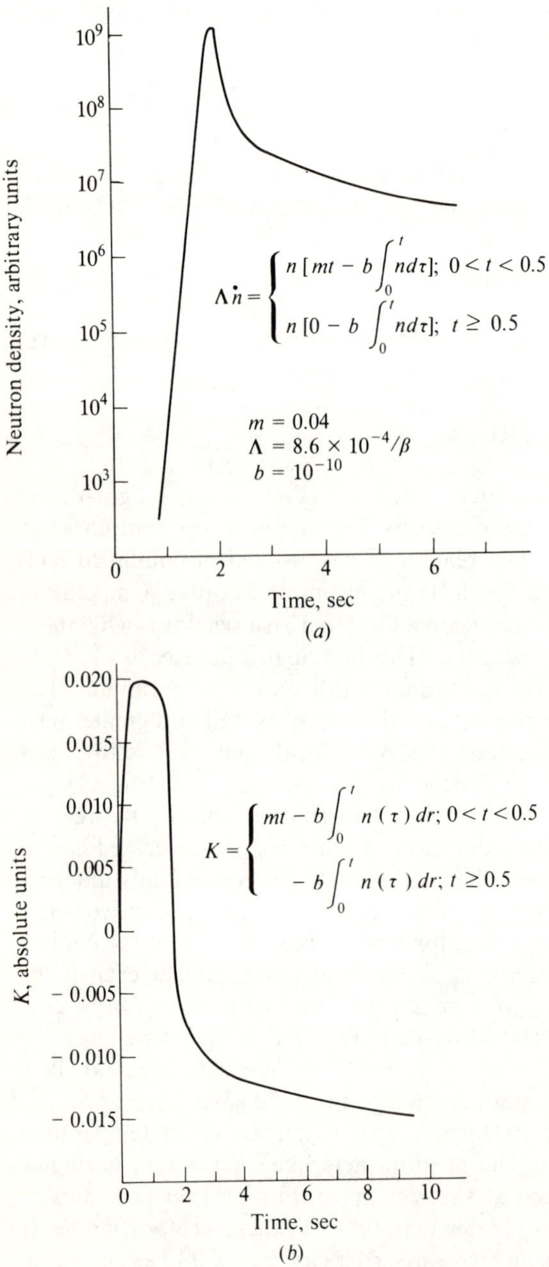

Figure 9-13 (*a*) TREAT reactor pulse as a function of time. (*b*) TREAT reactivity as a function of time.[25]

corresponds to the bound protons being in excited states and falling to stable states by giving their energy to the neutrons. At room temperature, $h\nu/kT \gg 1$ for this phenomenon, and in this temperature region the distribution function giving the occupation number of excited states varies as $e^{-h\nu/kT}$. As the temperature of the ZrH increases, the occupation number density of the excited states increases thereby increasing the up-scattering cross section. At the onset of a neutron density pulse, the fuel-moderator mixture becomes heated promptly, with an immediate increase in the number of neutrons being moderated up and out of the thermal region to energies where the fission cross section is relatively low. This causes a prompt drop in the thermal ηf corresponding to a safety negative reactivity temperature coefficient.

Again, the HF model is used; for an initial step function reactivity it becomes

$$\dot{n} = \left[\lambda - b \int_0^t n \, d\tau \right] n(t) \qquad n(0) = 1 \tag{9-81}$$

in time units of Λ. λ is the step function reactivity in units of inhours as measured by using the inhour equation (Chap. 2).

For the TRIGA reactor[31] $b = 2 \times 10^{-5}$ watts^{-1} sec^{-2}: $\Lambda = 0.01$; $\lambda = 102$ sec^{-1}. Plotted in Figs 9-14 and 9-15 are some neutron and temperature excursions for the TRIGA and Flash reactors computed on EDPM machinery. With the

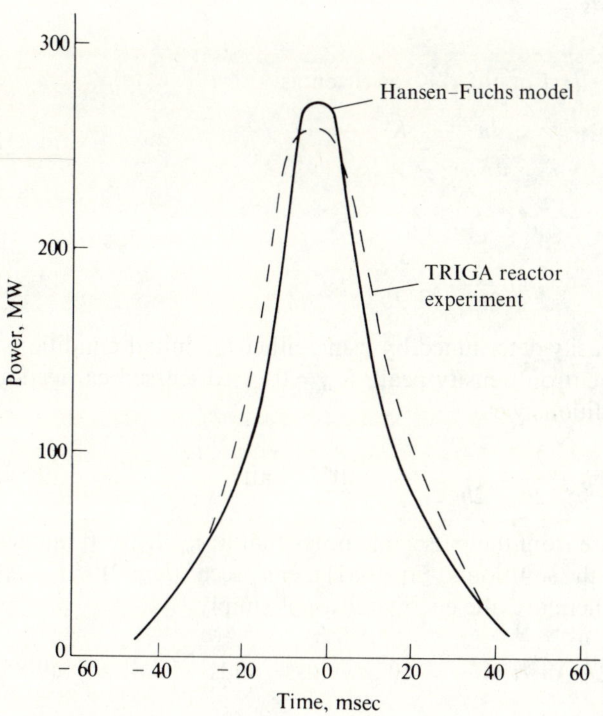

Figure 9-14 Power level as a function of time for 2-dollar step reactivity for TRIGA reactor.[31]

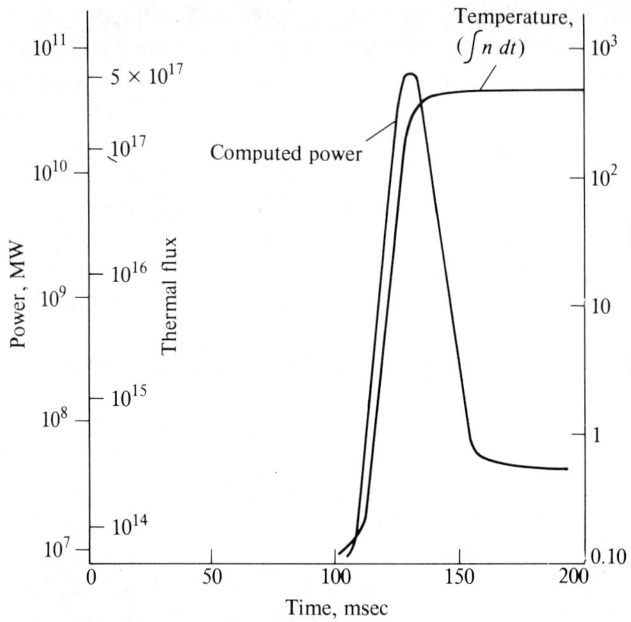

Figure 9-15 Power level as a function of time for ramp reactivity of 67.00 dollars per second for 0.10 sec for TRIGA-Flash reactor.[32]

reactivity $K = \lambda - b \int_0^t n(\tau)\,d\tau$, Eq. (9-81) is rewritten as

$$\frac{dn}{dK} = -\frac{K}{b} \tag{9-82}$$

with the integral

$$n = c_1 - \frac{K^2}{2b} \tag{9-83}$$

The arbitrary constant is easily determined by using either the initial conditions, $n(0) = 1$, or those at the neutron density peak, $K_m = 0$. As discussed earlier for Godiva reactor, these conditions yield

$$n_m = n_0 + \frac{\lambda^2}{2b} \cong \frac{\lambda^2}{2b} = 2.5 \times 10^4 \lambda^2 \text{ watts} \tag{9-84}$$

To obtain the energy release from the pulse, one notes that at t_m, $K_m = 0$ implies $\int_0^{t_m} n\,d\tau = \lambda/b$ and also that the solution to Eq. (9-81) $n = n_m \operatorname{sech}^2 (bn_m/2)^{1/2}(t - t_m)$ is symmetrical about t_m. Therefore, the energy release is simply

$$E = 2\int_0^{t_m} n\,d\tau = \frac{2\lambda}{b} = 10^5 \lambda \text{ watt-seconds} \tag{9-85}$$

The corresponding temperature rise is

$$2b\int_0^{t_m} n(\tau)\,d\tau = 2\lambda°\text{C} \tag{9-86}$$

For the ZrH moderator another variant of the HF model suggests itself. If the occupation number of bound proton states in ZrH increases with absolute temperature T as $e^{-h\nu/kT}$, then the temperature-dependent negative reactivity, due to protons pumping thermal neutrons out of the thermal energy region, should also change in the same manner. Let the ambient temperature of the moderator prior to the pulse be zero; then the reactivity K_T can be expressed as

$$K_T = \alpha_T\, e^{-h\nu/kT} \tag{9-87}$$

where α_T is a suitable reactivity coefficient. Since the heat energy produced is related to the temperature $\int_0^t n(\tau)\,d\tau = CT$, where C is the thermal heat capacity, then

$$K_T = \alpha_T \exp -\left[A\Big/\int_0^t n(\tau)\,d\tau\right] \tag{9-88}$$

with $A = h\nu C/k \equiv T_0 C$. For the proton oscillators whose level spacing is 0.13 eV, $T_0 = h\nu/k = 1500$ K. With $C = 2.5$ W-sec/cm^3/deg C, $A = 3.75$ kW-sec/cm^3. $\alpha_T \cong 10^{-4}$ per degree K[33] or 1 cent per degree K. Then the kinetics equation is simply

$$\dot{n} = \left\{K_0 - \alpha_T \exp -\left(A\Big/\int_0^t n\,d\tau\right)\right\} n \qquad n(0) = 1 \tag{9-89}$$

For $E = \int_0^t n\,d\tau$

$$\ddot{E} = (K_0 - \alpha_T\, e^{-A/E})\dot{E}, \qquad E(0) = 0 \tag{9-90}$$

In the same manner as before, integrating this equation yields

$$\int_0^E \frac{dx}{1 + x[K_0 - \alpha_T E_2(A/x)]} = t \tag{9-91}$$

where $E_2(y) = \int_1^\infty e^{-yz}\,dz/z^2$, the well-tabulated Milne function[34] used extensively in neutron and photon transport problems.

9-13 PULSED REACTORS: SPERT

The special power excursion reactor tests (SPERT) reactor is an enriched ^{235}U-fueled (U-Al-fueled elements), thermal, water-moderated and -cooled system. It is used to provide information on the kinetic behavior of enriched fuel, water-moderated and reflected reactors, especially with regard to suddenly induced reactivities.

There are[35] two kinds of control rod(s) in this reactor. These are a bank of shim rods and a central transient rod. With the transient rod partially extending through the core bottom, the reactor is made critical with the calibrated shim rods. Then the transient rod is drawn up into the core, rendering the reactor subcritical. The shim rods are then repositioned to compensate for the amount of negative reactivity introduced by the transient rod. All this takes place at a very low power of a few watts. The transient rod is then partially ejected from the core leaving it supercritical by the amount of reactivity compensated by the shim rod bank. The power rise, almost immediately, becomes exponential with a reciprocal period α. Peak powers of 1300 MW are attainable. The power keeps rising until the thermodynamic and hydrodynamic shutdown mechanisms accumulate enough negative reactivity to compensate for that in the initial rise. The power passes through a maximum, then drops to some low asymptotic level. The system then automatically shuts itself down.

To fit the experimental data, an empirical form of the HF model was postulated to include a lag time T between the neutron level and the onset of the shutdown mechanism. The kinetics equation is then

$$\dot{n} = \left[\alpha_0 - b_1 \left(\int_0^{t-T} n \, d\tau\right)^r\right] n \qquad n(0) = 1 \tag{9-92}$$

where $\alpha_0 = K_0/\Lambda$, the inverse period corresponding to the initial reactivity insertion K_0; $b_1 = b/\Lambda$; and r and T are obtained empirically from experiment. Letting the energy production up to time t be given by $\phi(t) = \int_0^t n \, d\tau$, we write[36]

$$\ddot{\phi}(t) = \begin{cases} \alpha_0 \dot{\phi}(t) & 0 < t < T \\ [\alpha_0 - b_1 \phi^r(t - T)]\dot{\phi}(t) & t \geq T \end{cases} \tag{9-93}$$

For fast transients (prompt periods less than 20 msec) the above equation holds strongly, with respect to neglecting delayed neutrons, since the corresponding step reactivities are greater than prompt critical, that is, $K_0 > \$1.00$. The neutron level increases to many orders of magnitude over the initial level, so that the last few neutron generations contribute the overwhelming bulk of the neutrons. The subsequent neutron level is then essentially independent of the initial neutron level and how the initial reactivity was introduced.

In the region $0 < t < T$, Eq. (9-93) gives the neutron density as $n = e^{\alpha_0 t}$. For $t \geq T$ a first integral of the same equation gives

$$\dot{\phi}(t) = 1 + \alpha_0 \phi - b_1 \int_0^{t-T} \phi^r(x) \, d\phi(x + T) \tag{9-94}$$

a nonlinear difference differential equation for $\phi(t)$ which must be integrated numerically as it stands. Its asymptotic form is

$$\dot{\phi}(t) \sim \alpha_0 \phi - b_1 \int_0^{t-T} \phi^r(x) \, d\phi(x + T) \tag{9-95}$$

There are two cases of interest that render these equations analytically tractable. The first is for a zero time lag ($T = 0$) of the shutdown mechanism. The second is for a time lag which is long compared to the width of the neutron pulse. For $T = 0$, Eq. (9-95) reduces to

$$\dot{\phi}(t) \cong \alpha_0 \phi - \left(\frac{b_1}{r+1}\right) \phi^{r+1}(t) \tag{9-96}$$

Its integral is, using the fact that $\phi(t_{\max}) \equiv \phi_m = (\alpha_0/b_1)^{1/r}$ from Eq. (9-93),

$$\phi(t) = \left[\frac{(1+r)\alpha_0}{b_1(1 + r\,e^{-\alpha_0 r(t-t_m)})}\right]^{1/r} \tag{9-97}$$

The derivative of this equation gives the neutron density in terms of

$$n(t_{\max}) \equiv n_m = \frac{r\alpha_0^{1+1/r}}{(1+r)b_1^{1/r}}$$

as

$$n(t) = n_m \left[\frac{1+r}{1 + r\,e^{-\alpha_0(t-t_m)}}\right]^{(r+1)/r} e^{-\alpha_0 r(t-t_m)} \tag{9-98}$$

For small times, but not too small in the sense of the above asymptotic forms of these equations, prior to the onset of the pulse, $0 \ll t \ll t_m$ so that $r\,e^{-\alpha_0 r(t-t_m)} \gg 1$ reduces the above to

$$n(t) \cong n_m \left(\frac{1+r}{r}\right)^{(1+r)/r} e^{\alpha_0(t-t_m)} \tag{9-99}$$

For times long after the pulse, the denominator of Eq. (9-98) approaches unity, so that

$$n(t) \sim n_m(1+r)^{(1+r)/r}\,e^{-\alpha_0 r(t-t_m)} \tag{9-100}$$

It is seen, therefore, that the pulse rises on the inverse period α_0 and descends with inverse period $-r\alpha_0$.

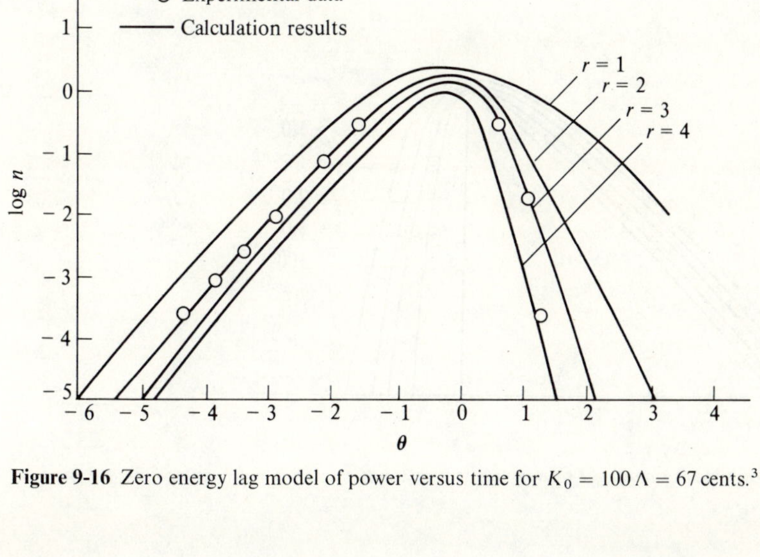

Figure 9-16 Zero energy lag model of power versus time for $K_0 = 100\,\Lambda = 67$ cents.[37]

Choosing t_m as the origin by letting $\theta = t - t_m$, Eq. (9-98) yields

$$n(\theta) = n_m \left(\frac{1 + r}{1 + r\, e^{-\alpha_0 r \theta}} \right)^{(1+r)/r} e^{-\alpha_0 r \theta} \tag{9-101}$$

This is plotted in Fig. 9-16 for several values of r.

The situation for the lag time, which is long compared to the neutron pulse width, is considered by examining Eq. (9-92) and looking strictly for the asymptotic solutions. That is

$$\frac{\dot{n}}{n} \sim \begin{cases} \alpha_0 & t < T \\ \alpha_0 - b_1 \left[\int_0^{t-T} n(\tau)\, d\tau \right]^r & t \geq T \end{cases} \tag{9-102}$$

For times prior to the manifestation of the shutdown mechanism ($t < T$), the above gives $n \sim e^{\alpha_0 t}$. In other words, the neutron level has been on an exponential rise for many neutron generations, so that a good approximation to the asymptotic solution for $t > T$ is obtained by inserting $e^{\alpha_0 t}$ into the integral of Eq. (9-102), in the sense of successive approximations to its solution. This gives for $e^{\alpha_0 (t-T)} \gg 1$

$$\frac{\dot{n}}{n} \sim \alpha_0 - \frac{b_1}{\alpha_0^r} e^{r\alpha_0 (t-T)} \qquad t > T \tag{9-103}$$

By integrating this, again with $e^{r\alpha_0 (t-T)} \gg 1$, one obtains

$$n = \exp\left(\alpha_0 t - \frac{b_1}{r\alpha_0^{r+1}} e^{r\alpha_0 (t-T)} \right) \tag{9-104}$$

As before, the time at which the pulse peak occurs and the neutron density (power) maximum are

$$t_m = T + \frac{1}{r\alpha_0} \ln \frac{\alpha_0^{r+1}}{b_1} \qquad n_m = \left[\frac{\alpha_0^{r+1}}{b_1} \right]^{1/r} e^{\alpha_0 T - 1/r} \tag{9-105}$$

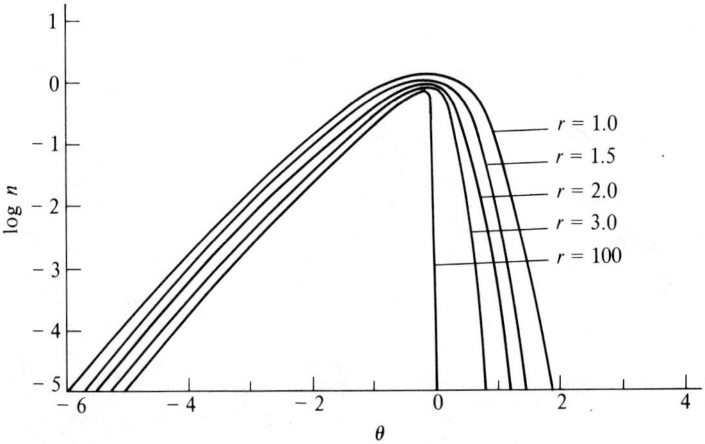

Figure 9-17 Long time lag model of power versus time for $K_0 = 100\,\Lambda = 67$ cents.[37]

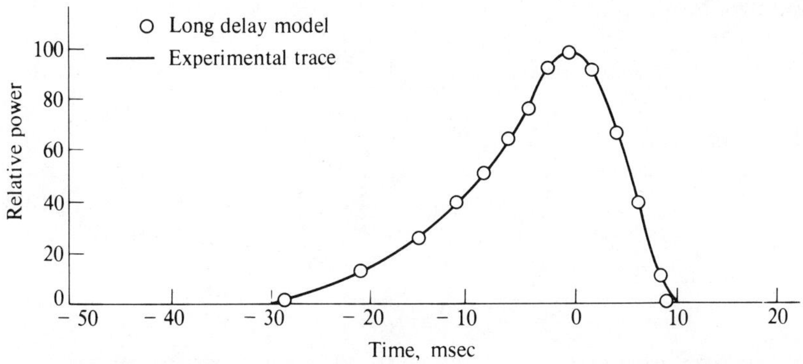

Figure 9-18 Comparison of experimentally obtained transients with long delay model ($r = 2$) $K_0 = 133$, $\Lambda = 88$ cents.[37]

Further, placing the origin at t_m by letting $\theta = t - t_m$, Eqs (9-104) and (9-105) combine to give

$$n = n_m \exp\left(\alpha_0\theta - \frac{1}{r}e^{r\alpha_0\theta - 1}\right) \tag{9-106}$$

This is plotted in Fig. 9-17 for various r. It is seen then from Figs 9-16 and 9-17 that the zero-lag model fits the data best for $r = 2$ while the long time delay model exponent is $r = 1.5$. The agreement for the combined case of letting $r = 2$ in the long-time-delay model is quite good, as seen from Fig. 9-18. To ascertain the value of T, the delay time for the long delay time model, it is found experimentally[36] that the peak power n_m and the initial reactivity step K_0 in the SPERT reactor are well approximated by

$$n_m = a\alpha_0^m \tag{9-107}$$

where a and m are constants obtained by fitting this expression to the data. Comparing this with the long delay model peak power expression from Eq. (9-105) by

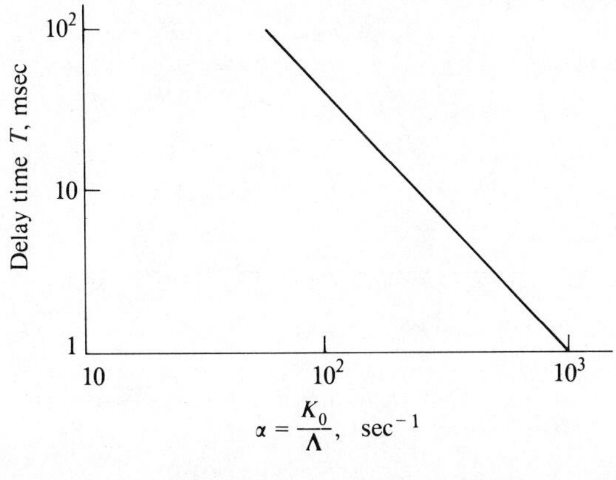

Figure 9-19 Long delay time as a function of the initially induced step reactivity $K_0 (r = 2)$.[36]

equating the two, results in

$$a\alpha_0^m = \left(\frac{\alpha_0^{r+1}}{b_1}\right)^{1/r} e^{\alpha_0 T - 1/r} \tag{9-108}$$

Certainly $T = T(\alpha_0)$ and

$$\alpha_0 T = \frac{1}{r} + \ln ab_1^{1/r} + \left(m - \frac{r+1}{r}\right) \ln \alpha_0 \tag{9-109}$$

This is plotted in Fig. 9-19.

9-14 PULSED REACTORS: SPERT SHUTDOWN REACTIVITY MECHANISMS

For initial step function (or ramp) reactivities in the SPERT reactors, the resulting power levels (neutron density) consist of heavily transient pulses as depicted in Fig. 9-20. The pulse form per se results from the various internal reactivity shutdown mechanisms, such as thermal expansion of the fuel and/or core and

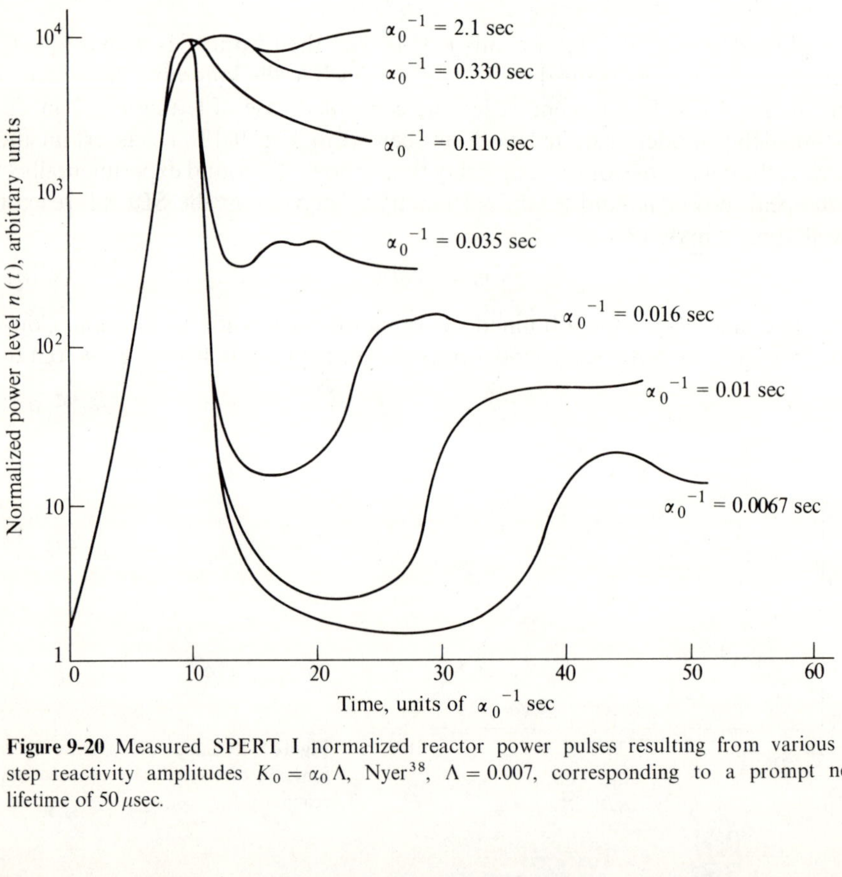

Figure 9-20 Measured SPERT I normalized reactor power pulses resulting from various initial step reactivity amplitudes $K_0 = \alpha_0 \Lambda$, Nyer[38], $\Lambda = 0.007$, corresponding to a prompt neutron lifetime of 50 μsec.

moderator steam void formation. Internally generated reactivities such as the foregoing are collectively termed the reactivity compensated by the reactor, cf. Fig. 9-21.

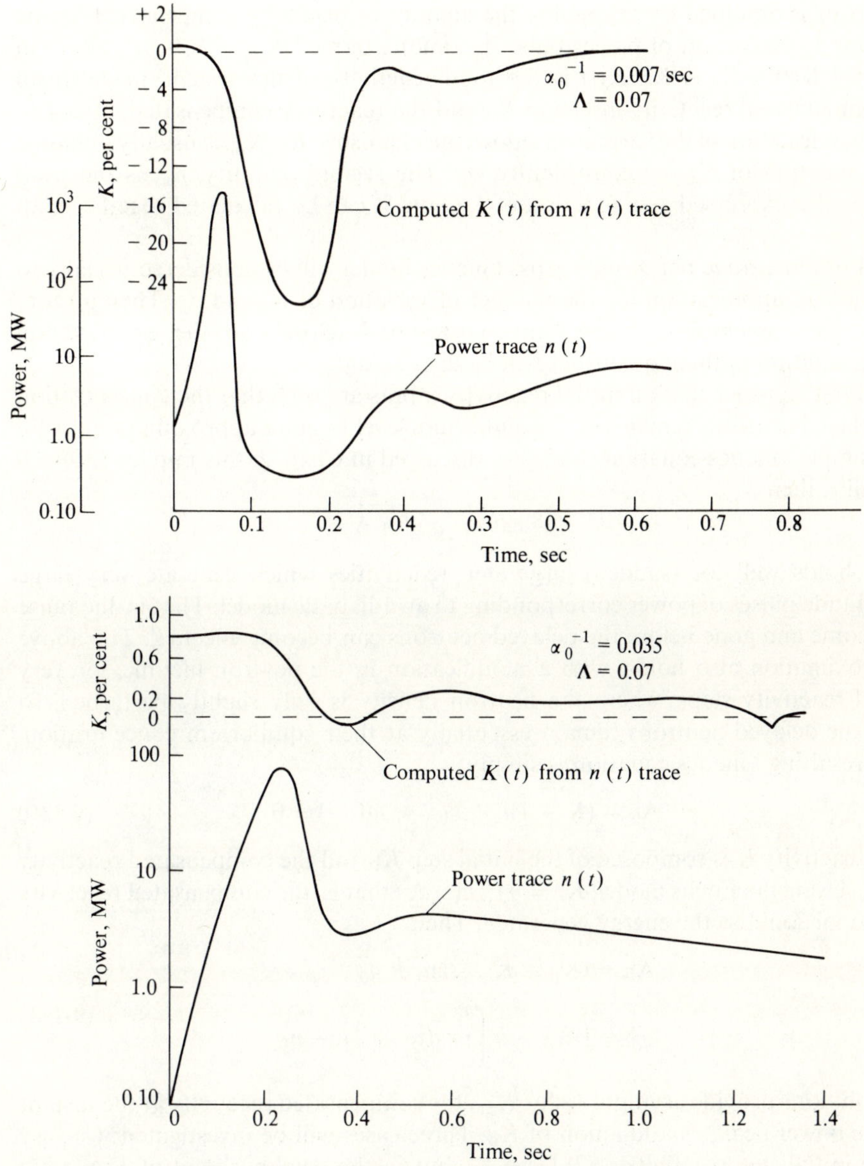

Figure 9-21 Computed reactivities resulting from experimentally obtained SPERT I power traces, $n(t)$, obtained from digital computer solutions of zero-power kinetics equation

$$K(t) = \left(\Lambda \dot{n} + \int_0^t n'(\tau) f(t-\tau)\, d\tau \right) \Big/ n(t)$$[36]

To aid in the determination of the nature of the self-shutdown mechanisms in SPERT systems following an initial insertion of, for example, a step reactivity of amplitude K_0, knowledge of the subsequent reactivity behavior and power pulse maxima are two important quantities that are usually desired. Reactivity behavior is obtained by examining the amount of reactivity compensated by the reactor at the instant of power pulse maximum, termed $K_{cm} = K_c(t_{\max})$, where in general $K(t) = K_0 - K_c$. $K(t)$ is the total reactivity of the system, made up of the initial (step) reactivity insertion K_0 and the reactivity compensated, owing to the manifestation of the various shutdown mechanisms, K_c. K_{cm} is usually obtained as a function of K_0, or equivalently, α_0. The second quantity, n_m, is the peak power also expressed as a function of K_0, and it can be calculated together with K_{cm}.

For illustrative purposes, a gross kinetics model will be considered initially to provide an appreciation for the manner of variation of K_c and n_m. Then a more elaborate semiempirical model that corresponds closely to the experimental determination of these quantities will be discussed.

First consider that all initial reactivity inputs are such that the concentration of delayed neutrons remains at its equilibrium level to good approximation. From the simple kinetics equation solutions, discussed in Chap. 2, this implies $\dot{c}_i(0) = 0$ for all i; then

$$c_i \equiv c_i(0) = a_i n(0)/\Lambda$$

This holds well for (sudden) large step reactivities which generate very large amplitude pulses of power corresponding to an adiabatic model. That is, the pulse has come and gone before the delayed neutrons can become effectual. The above approximation also holds, with a modification in the neutron lifetime, for very small reactivity steps, where the neutron density is only slightly perturbed, so that the delayed neutrons remain essentially at their equilibrium concentration. The resulting kinetics equation is simply

$$\Lambda \dot{n} = (K - 1)n + n_0 \qquad \dot{n}(0-) = 0 \tag{9-110}$$

The reactivity K is composed of the initial step K_0 and the compensated reactivity $-K_c$. Using the Fuchs model, $K_c = b\int_0^t n(\tau)\,d\tau$; that is, the compensated reactivity is proportional to the energy generated. Then†

$$\begin{aligned}\Lambda \dot{n} &= (K_0 - K_c - 1)n + n_0 \\ &\equiv \left(K_0 - b\int_0^t n\,d\tau - 1\right)n + n_0\end{aligned} \tag{9-111}$$

Its solutions provide asymptotes for K_{cm}, the compensated reactivity at the instant of the power peak, as a function of K_0. Three cases will be investigated: (1) very large initial step reactivities, (2) large K_0 but in the neighborhood of $K \cong 1$, (3) very small $K_0 \ll 1$.

† Note that solutions to Eq. (9-111) can be obtained in closed form, as given in problem 9-15 at the end of the chapter.

Let $K = K_0 - b\int_0^t n\,d\tau$, $10^{-10} \leq b \leq 10^{-8}$, so that the above equation becomes

$$\Lambda b \frac{dn}{dK} = -(K-1) - \frac{n_0}{n} \tag{9-112}$$

CASE 1 $K_0 \gg 1$ For very large initial step reactivities, $K_0 \gg 1$, the power pulse has an amplitude order of magnitude larger than the initial power n_0. Then

$$\Lambda b \frac{dn}{dK} \approx -K \tag{9-113}$$

with an integral

$$\Lambda bn = -\frac{K^2}{2} + c_1 \tag{9-114}$$

The arbitrary constant can be evaluated to sufficiently good approximation for the purposes at hand for initial conditions at $t = 0+$, which gives $\Lambda bn_0 = -K_0^2/2 + c_1$, and for those at the power maximum implying $\Lambda bn_m = c_1$, since Eq. (9-113) asserts that $K_m = 0$. Elimination of c_1 gives

$$n_m = n_0 + \frac{K_0^2}{2\Lambda b} \cong \frac{K_0^2}{2\Lambda b} \tag{9-115}$$

Since $K_m \equiv K_0 - K_{cm} = 0$, the reactivity compensated at the power peak is the initial step inserted ($K_{cm} = K_0$).

CASE 2 $K \cong 1$ For large reactivity, but near prompt critical, $K \cong 1$. Equation (9-112) becomes

$$\Lambda b \frac{dn}{dK} = -\frac{n_0}{n} \tag{9-116}$$

then

$$\frac{\Lambda bn^2}{2} = -n_0 K + c_2 \tag{9-117}$$

Again, the conditions at equilibrium and at power peak yield

$$(\Lambda b/2)n_0^2 = -n_0 K_0 + c_2$$

and $(n_0/2)n_m^2 \cong -n_0 + c_2$, respectively. Eliminating c_2, with $n_m^2 \gg n_0^2$ gives

$$n_m = \left[\frac{2n_0}{\Lambda b}(K_0 - 1)\right]^{1/2} \tag{9-118}$$

Equation (9-112) at the instant of the power peak asserts that

$$K_{cm} = K_0 - 1 + n_0/n_m \approx n_0/n_m$$

so that from the above

$$K_{cm} = \left[\frac{\Lambda b n_0}{2(K_0 - 1)}\right]^{1/2} \tag{9-119}$$

CASE 3 $K_0 \ll 1$ For small reactivities, $n_0/n \cong 1$, so that Eq. (9-112) now gives

$$\Lambda b \frac{dn}{dK} \cong -K \tag{9-120}$$

which is the same as the asymptotic form of the kinetics equation for very large reactivities, but for different reasons. However, for small K_0 the delayed neutrons are roughly in equilibrium with the prompt neutrons which change together asymptotically with a mean time constant given by $\bar{\tau}$, the mean life of the delayed neutrons. This is easily seen by examining the asymptotic form of the aggregate kinetics equation

$$\Lambda \dot{n}(t) = Kn - \int_0^t n'(\tau) f(t - \tau)\, d\tau \tag{9-121}$$

for slowly changing $n(t)$. Letting the upper limit in the above integral become infinite and assuming $n'(\tau) \cong \dot{n}(t)$, which can then be factored from the integral as discussed previously, Eq. (9-121) becomes

$$(\Lambda + \bar{\tau})\dot{n} = Kn$$

where $\bar{\tau} = \int_0^\infty f(x)\, dx$. Since $\Lambda \ll \bar{\tau}$

$$\bar{\tau}\dot{n} = Kn \tag{9-122}$$

describes the long-term behavior for small K_0. Equation (9-120) should then be amended to look like

$$\bar{\tau} b \frac{dn}{dK} = -K \tag{9-123}$$

Then the integral is of the same form as in Case 1, where $K_0 \gg 1$. The development is then formally the same as in Case 1 but with $\bar{\tau}$ instead of Λ, so that, with b very small

$$n_m \cong \frac{K_0^2}{2b\bar{\tau}} \tag{9-124}$$

and $K_{cm} = K_0$, also as in Case 1. In sum, Cases 1, 2, and 3 provide three asymptotes for both n_m and K_{cm} as a function of K_0. For very large and small K_0, $n_m \sim K_0^2$ and $K_{cm} \sim K_0$. In the vicinity of prompt critical ($K \cong 1$), $n_m \sim (K_0 - 1)^{1/2}$ and $K_{cm} \sim 1/(K_0 - 1)^{1/2}$. This is sketched in Fig. 9-22. The asymptotes in Fig. 9-22 correspond well with empirical fits to the data. For example, for large K_0 and $K_0 \cong 1$, the data for n_m give exponents of 1.7 and 0.8, respectively.[38] The corresponding exponents in Fig. 9-22 are 2 and 0.5, which are, of course,

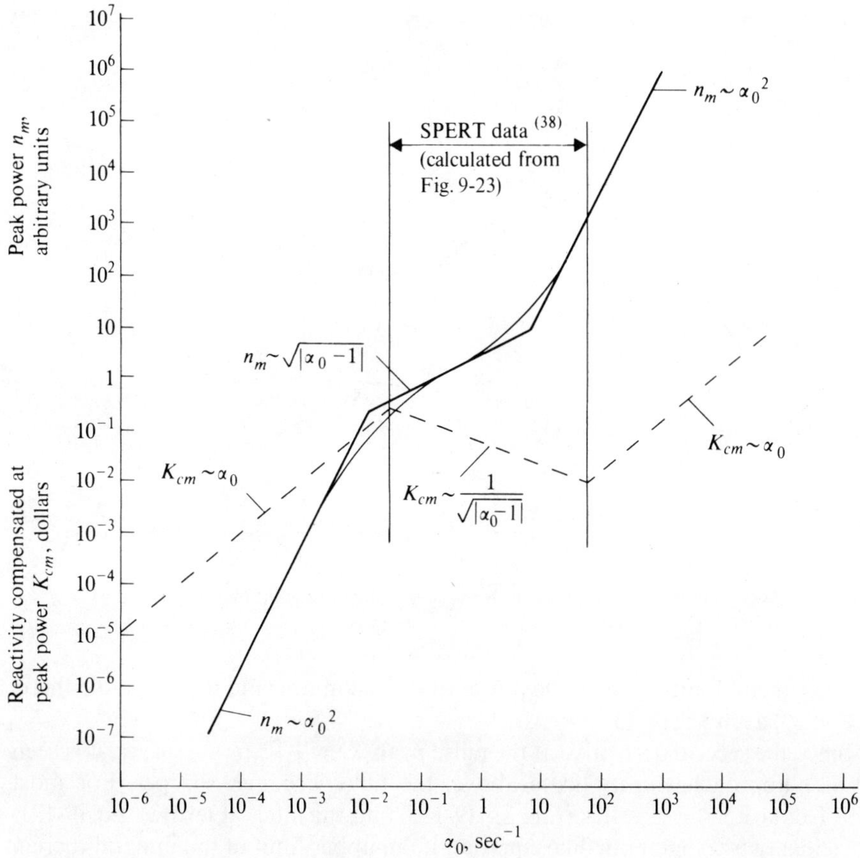

Figure 9-22 Asymptotes for the peak power n_m and the reactivity compensated at peak power K_{cm} as a function of the initial step reactivity $K_0 = \alpha_0 \Lambda$.

due to the gross model employed. However, it is important to note that this model delineates the fact that the inflections in the curves of n_m and K_{cm} are inherent in the kinetics, and are not merely due to the fact that empirical neutron pulses are substituted into the kinetics equations to produce such reactivity behavior, as below.

To gain further insight into the reactivity behavior, which we do by including delayed neutrons in detail because they cause the inflections in the curves of Fig. 9-22, a two-term exponential fit to experimentally determined power pulses is inserted into the kinetics equation (9-130) to obtain K_{cm} and n_m as functions of α_0. Let the power (neutron density) be represented by[38]

$$n(t) = n_m\left[c\, e^{\alpha_0(t-t_m)} - (c-1)\, e^{[\alpha_0 c/(c-1)](t-t_m)}\right] \tag{9-125}$$

where t_m is the time of the power maximum. The parameter c is adjusted to fit

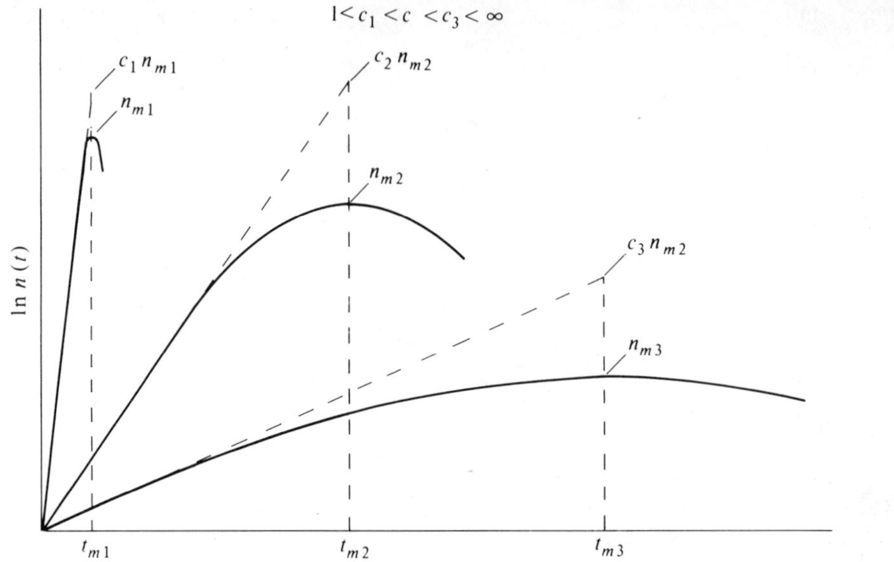

Figure 9-23 Power pulses and extrapolations for various values of the fitting parameter c.

the experimental initial slope, the power at its maximum, and its derivative there, $1 < c < \infty$, as in Fig. 9-23.

Since the second derivative at the pulse peak $\ddot{n}_m = -[c/(c-1)]n_m\alpha_0^2$, it is seen that as c approaches unity (from above) the pulse becomes sharper. For most SPERT data $1.5 \leq c \leq 2$. Inserting Eq. (9-125) into the integral term of Eq. (9-121) will yield a two-term inhour-like equation if the upper limit of the integral therein is extended to infinity. This is justified for most of the α_0 range of interest, since it is assumed that the pulse occurs "long" after the system has been on its initial rise. The result is

$$K \equiv K_0 - K_c = \frac{1}{n}\left\{\Lambda\dot{n} + \alpha_0 c n_m\left[e^{\alpha_0(t-t_m)}\bar{f}(\alpha_0) - e^{[\alpha_0 c/(c-1)](t-t_m)}\bar{f}\left(\frac{\alpha_0 c}{c-1}\right)\right]\right\} \tag{9-126}$$

where $\bar{f}(\alpha_0)$ is the Laplace transform of the delayed neutron decay function $f(t)$, with respect to the parameter α_0. During the initial power increase, the first term of Eq. (9-125) is very large compared to the second term. This implies that during this period, Eq. (9-126) should read

$$K_0 - K_{c0} \cong \frac{1}{n}[\Lambda\dot{n} + \alpha_0 c n_m e^{\alpha_0(t-t_m)}\bar{f}(\alpha)] = \alpha_0[\Lambda + \bar{f}(\alpha_0)] \tag{9-127}$$

which is essentially the ordinary inhour equation, since $K_{c0} \ll K_0$. $K_{c0}(t)$ is the reactivity being compensated on the initial rise. At the instant of power maximum $(t = t_m)\dot{n} = 0$, so that Eq. (9-126) becomes

$$K_0 - K_{cm} = \alpha_0 c\left[\bar{f}(\alpha_0) - \bar{f}\left(\frac{\alpha_0 c}{c-1}\right)\right] \tag{9-128}$$

Combining Eqs (9-127) and (9-128) and neglecting K_{c0} in comparison to K_0 gives

$$K_{cm} = \alpha_0\left[\Lambda + \sum_{i=1}^{6} \frac{a_i\lambda_i}{(\alpha_0 + \lambda_i)[\alpha_0 c/(c-1) + \lambda_i]}\right] \qquad \alpha_0 = \frac{K_0}{\Lambda} \tag{9-129}$$

The reactivity compensated at peak power, K_{cm}, is plotted as a function of α_0 in Fig. 9-24. For illustration, c is very large, corresponding to a rounded burst of power. Plots are obtained for $\Lambda \cong 10^{-1}$ (heavy-water-moderated thermal reactor, $\Lambda \cong 10^{-2}$ (light-water- or graphite-moderated thermal reactor), and $\Lambda \cong 10^{-4}$ (fast reactor). It is seen that K_{cm} has a pronounced dip, especially for the short-lifetime reactors. For the case depicted in Fig. 9-24, $c \gg 1$, corresponding to a relatively broad power pulse following an initial step reactivity insertion. For such pulses, the figure shows that the peak power reactivity is principally compensated by (whatever internal reactor mechanisms such as core fuel expansion and moderator expulsion exist) delayed neutrons for small α_0. This simply implies that for small α_0 the power pulse is small, broad, and of duration corresponding to delayed neutron mean time constants. As α_0 increases, the domination of the compensated reactivity due to delayed neutrons decreases (dip in curve of K_{cm})

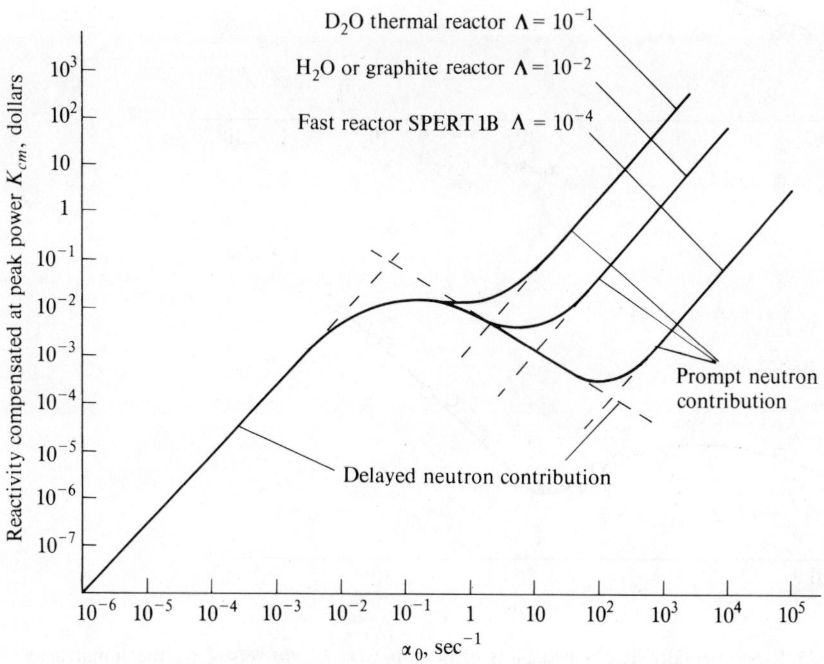

Figure 9-24 Reactivity compensated at peak power versus initial step reactivity $K_0 = \alpha_0\Lambda$ for various prompt neutron lifetimes corresponding to reactor prototypes.[38]

because the power pulse peak appears too soon and, for sufficiently large α_0, is relinquished to the prompt neutrons. For very large α_0, the pulse amplitude is large and appears very early, so that delayed neutron contributions are negligible. Without too much license α_0 (sec^{-1}) can be considered as a frequency-like variable, and Fig. 9-25 is the "frequency response" of K_{cm} with respect to α_0. For the more sluggish reactors, like the heavy-water-moderated one, the prompt lifetime is of the order of magnitude of some of the delayed neutron lifetimes, so that the dip is less pronounced.

The slopes of the K_{cm} curves depend on prompt lifetime, delayed neutron characteristics, and the neutron pulse shape. Similar curves can be obtained for ramp inputs as well. The above analysis does not take into account any reactor

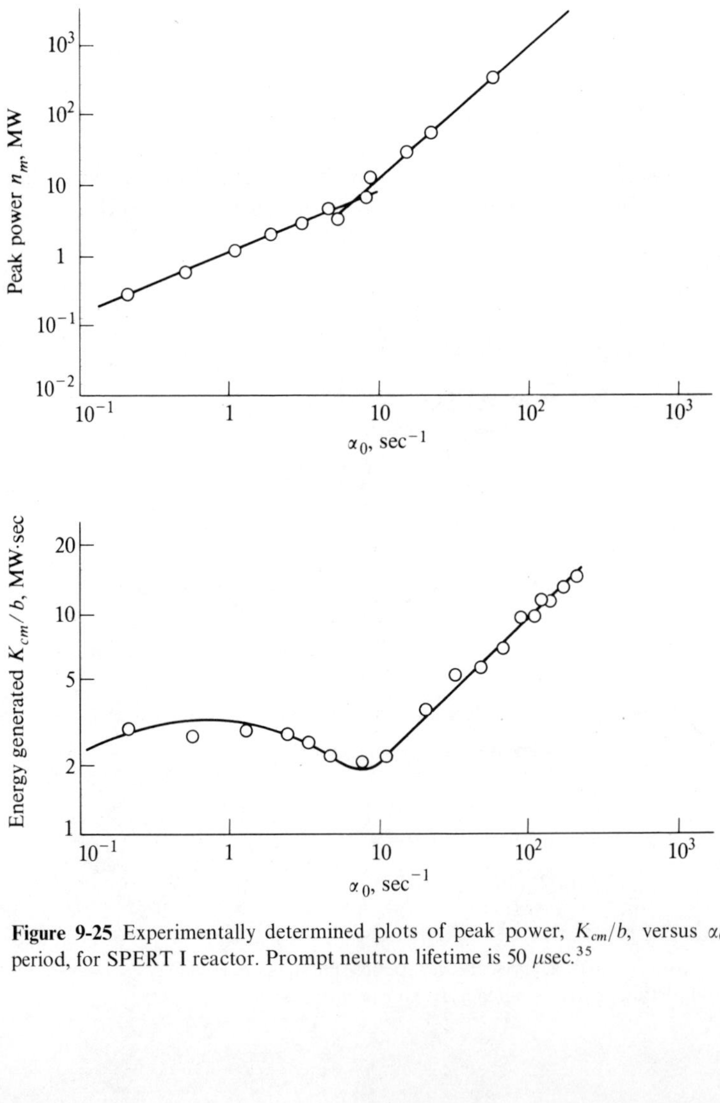

Figure 9-25 Experimentally determined plots of peak power, K_{cm}/b, versus α_0, the initial inverse period, for SPERT I reactor. Prompt neutron lifetime is 50 μsec.[35]

energy-dissipating mechanisms such as cooling. These can be accommodated by adding an energy equation of state to the existing kinetics equations.[40,41]

The peak power n_m can be obtained as a function of α_0 from the above very easily. Since

$$K_{cm} = b \int_0^{t_m} n(\tau)\, d\tau$$

by definition, the two-term $n(t)$ empiricism given by Eq. (9-125) and inserted into the integral gives for very large c and t_m

$$n_m \cong \frac{\alpha_0}{2b} K_{cm} \tag{9-130}$$

Then from Eq. (9-138)

$$n_m = \frac{\alpha_0^2}{2b}\left[\Lambda + \sum_{i=1}^{6} \frac{a_i \lambda_i}{(\alpha_0 + \lambda_i)^2}\right] \tag{9-131}$$

For $\alpha_0 \gg 1$, $n_m \sim \Lambda\alpha_0^2/2b$, which is equivalent to the large α_0 asymptote developed in the earlier model, and similarly for the small α_0 asymptote. In Fig. 9-22, the n_m asymptotes are paired together by using the above expression for n_m. Again, even this model becomes crude, especially for small α_0 where deviations from the adiabatic state of affairs occur through energy dissipation.

9-15 PULSED REACTORS: KEWB

The KEWB (Kinetic Experiment Water Boiler) built by the Atomics International Division of Rockwell International, was a small highly enriched homogeneous reactor, consisting of a solution of uranium sulphate in ordinary water. The solution partially filled a one foot diameter stainless steel spherical vessel, topped by a large necked structure, all surrounded by a graphite reflector. The necked structure included the control rods. A diametrical (glory) hole threaded the vessel through the neck, and formed part of a pneumatic tube. An absorber pneumatically ejected from the hole initiated the power pulse from a controlled critical solution of the reactor fuel.

Very early, following production of the pulse, the reactor is immediately shutdown due to the thermal expansion of the fuel solution, and the dissociation of the water due to the intense activity from the fission products. The dissociation of the water forms gas bubbles which constitute a void reactivity feedback. Subsequent to this epoch, the fuel solution itself is ejected forcefully up into the neck of the vessel, and slowly falls back through a baffle system.

Paraphrasing Dunenfeld and Gamble,[42] Hetrick and Gamble,[43] and Hetrick,[44] the initial prompt neutron shutdown model can be formulated as in the following.

The prompt neutron reactor kinetics satisfies a simple equation allowing for the fact that delayed neutrons are absent initially. It is

$$\Lambda \dot{n} = (K - 1)n; \qquad n(0) = n_0 \tag{9-132}$$

The reactivity, K, consists of an initial positive insertion K_0 causing the pulse, but decreased by a temperature–reactivity feedback coefficient (θ), and a radiolytic gas void feedback coefficient (b). So that the reactivity is given by

$$K = K_0 - \theta T - bV \tag{9-133}$$

The reactor temperature satisfies

$$\dot{T} = An \tag{9-134}$$

representing the prompt temperature feedback due to the fission fragment heat production. The reactor bubble void volume V satisfies

$$\dot{V} = cnE \tag{9-135}$$

which is a rudimentary form for the manner in which the water is radiolytically decomposed to form gas bubble voids. That is, the bubble formation rate is proportional to the reactor power (neutron density), while the gas concentration in solution is proportional to the energy generated $E = \int_0^t n\, d\tau$. c is simply a constant of proportionality.

Eliminating the neutron density, n, from among Eqs (9-132)–(9-135). using $n = \dot{E}$, yields

$$\Lambda \ddot{E} = (K_0 - 1 - A\theta E - \tfrac{1}{2}bcE^2)\dot{E} \tag{9-136}$$

Integrating once yields a cubic relationship between the neutron density and the energy generated, namely

$$\Lambda n = E\left(K_0 - 1 - \frac{A\theta E}{2} - \frac{bcE^2}{6}\right) \tag{9-137}$$

The neutron density, or power, pulse peaks when $\dot{n} = 0$, which provides an expression for the energy generated up to the peak time, E_p, as given by the positive root of the right-hand side of Eq. (9-136). For an assumed large initial reactivity K_0, that root is well approximated by

$$E_p \simeq \sqrt{\frac{2\alpha\Lambda}{bc}} \tag{9-138}$$

where the initial inverse period $\alpha = (K - 1)/\Lambda$.

From Eq. (9-137) the corresponding peak power level for large α, n_p, is given approximately by

$$n_p \simeq \tfrac{2}{3}\sqrt{2\Lambda/bc}\,\alpha^{3/2} \tag{9-139}$$

This exponential dependence between the initial inverse period and peak power is well borne out by experiment, Hetrick.[45] For a somewhat more complicated model of the KEWB void reactivity, cf. Hetrick, *op. cit.*

9-16 MICROREACTORS

With the imminent development of very high power megajoule output lasers, as well as extremely high current density relativistic energy electron beam generators, detailed sophisticated computer studies and experiments have been conducted to investigate the feasibility of using these beams to compress and confine millimeter size spherical pellets. These are made of fusionable or fissionable isotopes to achieve a net overall energy gain to constitute a new type of nuclear reactor. Such a concentrically layered pellet can be construed as a "one-shot" fusion or fission microreactor depending on the pellet material.

In the case of pellets made of light isotopes only, that is fusionable materials, a number of studies have been made at the Lawrence Livermore Laboratories, Nuckolls et al.,[50] using a precisely timed and shaped laser energy pre-pulse to first ablate the pellet peripheral material. The resulting ultra-fast vaporization and outward ablation of the pellet material in nanoseconds or less (*a*) forms an atmosphere of pellet material gas about the pellet, (*b*) from the conservation of momentum plus the arrival of the main laser pulse, a very strong spherical inward directed hypervelocity shock front is formed in this gas to compress the remainder of the pellet. This strong shock front implodes the pellet material to thermonuclear pressures of millions of atmospheres, and temperatures of millions of degrees absolute. At the point of maximum implosion the supercompressed fermi-degenerate pellet plasma has a density of several thousand grams per cubic centimeter, and (*c*) ignites in a thermonuclear reaction which then forms an outward travelling spherical burn (fuel burnup) shock front. DT pellets are favored since the DT reaction has a high cross section at these temperatures, as opposed to pure DD for example. The above is by way of introduction to fusion microreactors. The laser induced fusion exploits inertial confinement of the fusionable plasma material instead of the magnetic confinement characteristic of most fusion reactors.

The microreactor kinetics lies in the temporal neutronic supercriticality of the fissionable or fission/fusion pellet. That is, similar studies have also been made, Ligou et al.,[47] of light isotope tamped fissionable pellets, such as ^{6}LiD reflected ^{239}Pu, as well as vice versa, that is, uranium and thorium tamped DT pellets, Seifritz et al.[51] For the light isotope tamped fission pellet the tamper acts as (*a*) an ablator to contribute to the initial inward shock momentum, (*b*) a neutron reflector, and (*c*) a source of copious amounts of very fast (14 MeV) neutrons to enhance the fast fission of the compressed heavy isotope fuel thereby increasing the burn efficiency of the system. In the reverse case, that is, the heavy isotope high Z tamper, the latter not only acts as an inertial pusher material with subsequent fission energy production, together with the DT pellet core in a "bootstrapping" action, but more important its high Z, that is, its corresponding high electron density provides a very effective absorber of the fast electron energy for the case of electron beam generator driven implosion. The corresponding principal interaction (photoelectric and compton) cross sections vary as Z^r where r is appreciably larger than unity, typically $2 \leqq r \leqq 5$.

As can be appreciated, certain aspects of microreactor kinetics share a commonality with reactor core meltdown accidents. The latter are usually studied in detail using multigroup codes that couple material hydrodynamics, thermodynamics, and neutronics/kinetics. The same computational apparatus can also be used to study microreactor kinetics.

Since the pellet is of small size compared to most mean free paths of interest, geometrical zonal considerations can be neglected. So that following the hydrodynamic implosion phase, the adiabatic or point kinetics model should be adequate to describe the kinetics during the burn phase.

The prompt neutronics as expressed through the Rossi $\alpha(t)$ is simply

$$\dot{n} = \alpha(t)n(t) \tag{9-140}$$

with solution

$$n(t) = n_0 \exp \int_0^t \alpha(\tau)\, d\tau \tag{9-141}$$

where n_0 is an initial neutron density fiducial discussed later. The identification can be made that $\alpha(t) = K/\Lambda$, where the reactivity-lifetime quotient is now $\rho/l(t)$. ρ is the reactivity in dimensionless units (e.g. per cent), and l is the prompt neutron generation time in seconds. The excess k, that is, $k - 1$, can be used in the definition of α, since k is not far from unity at this point. As the times involved are on the order of fractions of microseconds, the delayed neutron kinetics are neglected throughout this discussion.

As is evident, $\Lambda = l/\beta k$ is no longer invariant since Σ_f changes radically through N_f, in a very short time, as the fuel is explosively consumed. This type of fuel burn is the antithesis of that in a power reactor where the fuel concentration changes only secularly through burnup. In the above detailed computer studies the dependent variable is usually the angular density $\phi(\bar{r}, \bar{\Omega}, t)$ as discussed in Chap. 1. It can be assumed separable so that $\phi(\bar{r}, \bar{\Omega}, t) = \Phi_0(r, \Omega) \exp\left(\int_0^t \alpha\, d\tau\right)$. The usual transport equation is then symbolically written as

$$L\phi(\bar{r}, \bar{\Omega}, t) = \frac{1}{v}\frac{\partial \phi}{\partial t}(\bar{r}, \bar{\Omega}, t) \tag{9-142}$$

where L is the corresponding integro-differential transport operator. With the separable neutron density, Eq. (9-142) becomes

$$L\Phi_0(\bar{r}, \bar{\Omega}) = \frac{\alpha}{v}\Phi_0(\bar{r}, \bar{\Omega}) \tag{9-143}$$

With suitable pellet boundary conditions this becomes an eigenvalue problem for the eigenvalues α. Equation (9-143) is then solved numerically for a series of steady state "time slices" during the pellet compression and subsequent burn phases. This procedure is characteristic of the general nature of the space and time iteration procedures used in such codes. This provides a set of $\{\alpha\}$ for each time slice thus producing a time-dependent $\alpha(t)$ in accordance with the pellet material dynamics.

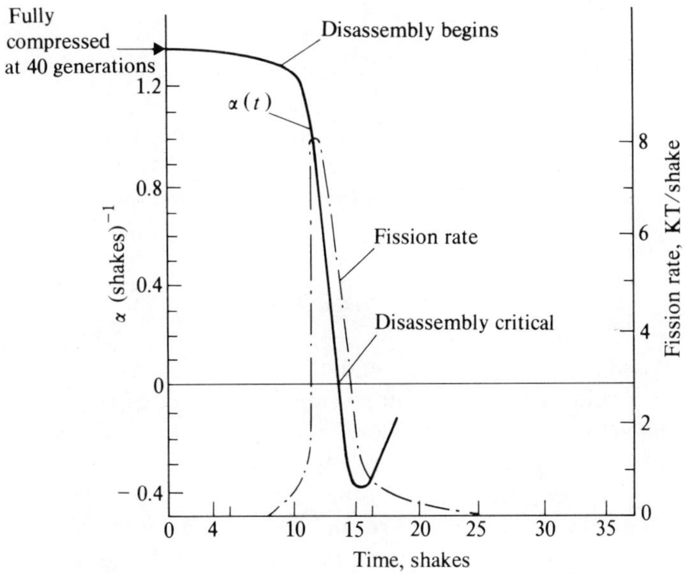

Figure 9-26 Rossi α for 24 KT S device as calculated using the HENRE code.[46] 1 shake = 10^{-8} sec.

Again, all of this assumes the validity of time and space separability and its implications as discussed in Chap. 6.

An example is given by the explosion of an S device, a ^{239}Pu core surrounded by high explosive material, Sandmeier et al.[46] A coupled hydrodynamics-neutronics code HENRE generated a time-dependent α, a segment of which is given in Fig. 9-26.

As already mentioned for a supercompressed pellet tamped or not, following the hydrodynamic implosion phase and at the onset of the burn yield phase but prior to disassembly, the point kinetics model represents well the time-dependent neutronics/kinetics. Besides a modified Hansen–Fuchs model to be discussed below, one type of point model, Ligou,[47] features the very rapid fast neutron density increase followed by a rapid drop due simply to burnup feedback as the fuel is exceedingly rapidly burned. That model is represented by the following two coupled equations in the neutron density, and the fuel number density, N_f. The neutron density satisfies

$$\dot{n} = Kn/\Lambda + S_0 \equiv \sigma_a v(k-1)nN_f + S_0 \tag{9-144}$$

where $k = \nu\sigma_f/\sigma_a$ is the reproduction factor for an all metal imploded fuel system. S_0 is an assumed constant neutron source originating from an implanted source or due to fusion neutrons from the compressed DT pellet. The other symbols have the usual meanings as given in Chaps 1 and 2. Similarly, the fuel density satisfies

$$\dot{N}_f = -\Sigma_a \phi = -\sigma_a v n N_f \tag{9-145}$$

with an integral

$$N_f = N_0 \exp - \sigma_a v \int_0^t n(\tau)\, d\tau \tag{9-146}$$

where N_0 is the imploded fuel number density prior to the burn phase and subsequent disassembly. Equation (9-146) is inserted into Eq. (9-144) to yield an equation in the neutron density explicitly. It is, where $\alpha(t)$ is given as seen in Fig. 9-26,

$$\dot{n} = \alpha n\, e^{-\sigma_a v \int_0^t n(\tau)\, d\tau} + S_0; \qquad \alpha = N_0 \sigma_a v(k-1) = \frac{k-1}{l_0} \tag{9-147}$$

Its integration is discussed as a problem at the end of the chapter.

An alternative description of the post implosion burn kinetics is a more easily integrated modified Hansen–Fuchs model. This modification features two modes of reactivity feedback in the pellet microreactor. One feedback mechanism is the negative reactivity energy feedback. The other is that due to the explosive burnup of the fuel. These features are evident in the following two coupled equations in the neutron density n and the fuel number density N_f. The neutron density is given by

$$\dot{n} = Kn/\Lambda = \left(K_0 - b\int_0^t \Sigma_f \phi\, d\tau\right)\frac{n}{l} = \left(K_0 - b_1 \sigma_f v \int_0^t nN_f\, d\tau\right)\sigma_a v n N_f \tag{9-148}$$

All initial conditions are reckoned from the onset of the burn phase. The initial reactivity K_0 is herein assumed dimensionless. The fuel loss rate due to burnup and non-fission absorption is given by

$$\dot{N}_f = -\Sigma_a \phi = -\sigma_a v n N_f; \qquad N_f(0) = N_0 \tag{9-149}$$

As before, integrating Eq. (9-149) gives $N_f = N_0 \exp - \sigma_a v \int_0^t n(\tau)\, d\tau$ which when substituted into Eq. (9-148) yields an equation in the neutron density only. It is, after integrating the bracket in Eq. (9-148),

$$\dot{n} = R_a N_0 \left[K_0 - r\left(1 - e^{-R_a \int_0^t n(\tau)\, d\tau}\right)\right] \cdot n\, e^{-R_a \int_0^t n(\tau)\, d\tau} \tag{9-150}$$

where the reaction rates for total absorption and fission are $R_a = \sigma_a v$ and $R_f = \sigma_f v$, respectively, and $r = b_1 R_f N_0 / R_a$. Now let $Z = \exp - R_a \int_0^t n\, d\tau$ in Eq. (9-150), and integrating once gives the result

$$\dot{Z}/Z = R_a N_0 \left[\frac{r}{2} Z^2 + (K_0 - r)Z\right] + c_1 \tag{9-151}$$

with c_1 as the constant of integration.

As in the Hansen–Fuchs model there are two fiducials that are apparent. The first is that at the initial time $t = 0$, the initial conditions are satisfied. The second is at $t = t_m$, the time at which the neutron density maximum or peak occurs. Here $n(t_m) = n_m$, and $\dot{n}(t_m) = 0$. This allows the determination of certain relations between the important parameters. For example, from Eq. (9-150), $\dot{n}(t_m) = 0$ implies that

$$K_0 = r(1 - Z_m) \qquad (9\text{-}152)$$

where $Z_m = \exp - R_a \int_0^{t_m} n \, d\tau$. Now, evaluating the constant c_1 in Eq. (9-151) at each of these times gives for (*a*) at $t = t_m$, using Eq. (9-152)

$$c_1 = \frac{R_a N_0}{2} \left[r Z_m^2 - \frac{2n_m}{N_0} \right] \qquad (9\text{-}153)$$

while (*b*) at $t = 0$

$$c_1 = -\frac{R_a N_0}{2} \left[r(1 - 2Z_m) + \frac{2n_0}{N_0} \right] \qquad (9\text{-}154)$$

Equating these two relationships and using Eq. (9-152) results in an expression for Z_m as

$$Z_m = 1 - \frac{2}{N_0 K_0}(n_m - n_0) \simeq 1 - \frac{2n_m}{N_0 K_0} \qquad (9\text{-}155)$$

since $n_m \gg n_0$.

Using c_1 as evaluated in Eq. (9-154), Eq. (9-151) now becomes with Eq. (9-152)

$$\dot{Z}/Z = \frac{\alpha_0}{2(1 - Z_m)} \left[(Z - Z_m)^2 - \left(1 - Z_m + \frac{n_0}{N_0 K_0} \right)^2 + \frac{n_0}{N_0 K_0} \right] \qquad (9\text{-}156)$$

where the initial Rossi alpha is given by

$$\alpha_0 = R_a N_0 K_0 = \frac{K_0}{l_0}$$

The quantity

$$\frac{n_0}{N_0 K_0} = \frac{n_0 R_a}{\alpha_0}$$

is seen to be very small so that it is neglected in Eq. (9-156). Then the latter becomes

$$\dot{Z}/Z \simeq \frac{\alpha_0}{2(1 - Z_m)} [(Z - Z_m)^2 - (1 - Z_m)^2] \qquad (9\text{-}157)$$

To evaluate Z_m it is seen that $N_f = N_0 Z$ from the integral of Eq. (9-149). Also within the bracket of Eq. (9-150), the energy generated is proportional to $r(1 - Z)$. So that the ratio, ε, of energy generated up to the neutron density peak to the total energy generated, is given by

$$\varepsilon = (1 - Z_m)/(1 - Z_\infty) \qquad (9\text{-}158)$$

A burnup efficiency η can now be defined as the ratio of fuel burned to that available initially at the onset of burn. This is seen to be

$$\eta = 1 - \frac{N_\infty}{N_0} = 1 - Z_\infty \qquad (9\text{-}159)$$

where N_∞ is the number density of fuel nuclei remaining after the burn phase. From Eqs (9-158) and (9-159), Z_m becomes

$$Z_m = 1 - \varepsilon\eta \tag{9-160}$$

For illustration, a plausible assumption is that half the pellet energy release takes place prior to the occurrence of the neutron density peak. Then $\varepsilon = \frac{1}{2}$. Also a companion assumption is made that half the pellet fuel is ultimately burned so

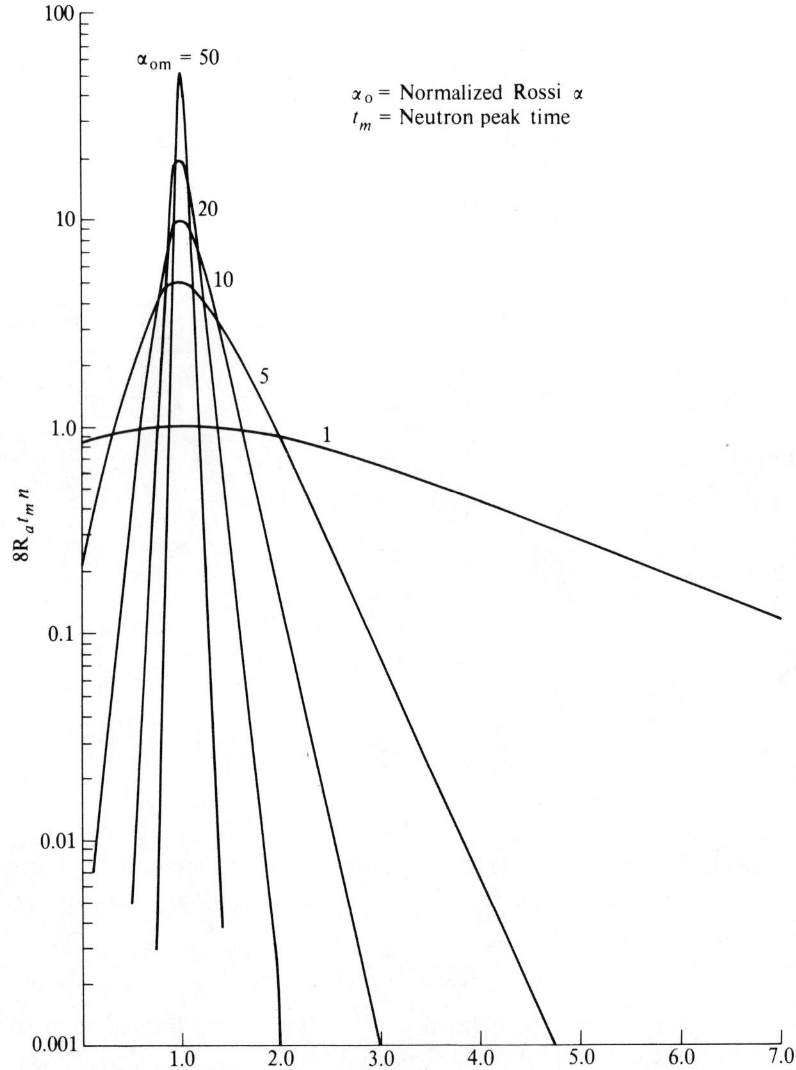

Figure 9-27 Modified Hansen–Fuchs model, normalized neutron density versus normalized time for various values of $\alpha_0 m = \alpha_0 tm$.

that $\eta = \frac{1}{2}$ as well. This gives $Z_m = 1 - \frac{1}{2}\cdot\frac{1}{2} = 3/4$. Then Eq. (9-157) yields

$$\dot{Z}/Z = \alpha_0(2Z - 1)(Z - 1) \tag{9-161}$$

Integrating and evaluating the constant of integration at t_m gives

$$Z = \tfrac{1}{2}[1 + \{1 + 3\exp - \alpha_0(t_m - t)\}^{-1/2}] \tag{9-162}$$

Since $n = -\dot{Z}/R_a Z$, the above finally results in the expression for the neutron density as

$$n = 12n_m\, e^{-\alpha_0(t_m - t)}/[(1 + 3\,e^{-\alpha_0(t_m - t)})^{3/2} + (1 + 3\,e^{-\alpha_0(t_m - t)})] \tag{9-163}$$

This is plotted in Fig. 9-27 as a one parameter of curves for the parameter $\alpha_0 t_m$ where $n_m = N_0 K_0/8$ from Eq. (9-155) and that $Z_m = 3/4$. It is seen from the figure that, although it is assumed that equal amounts of energy are released before and after t_m, the curves have the characteristic shape about the peak. That is, a sharp initial rise followed by a relatively gradual drop. The slopes and peak magnitudes depend on the parameter $\alpha_0 t_m$.

If the burn phase is so efficient that essentially all the pellet fuel is consumed, the corresponding Z_m for equal burn energy generated before and after the neutron density maximum yields the following integral for the neutron density (problem at the end of the chapter):

$$\ln\left(\exp R_a \int_0^t n\,d\tau - 1\right) + \exp R_a \int_0^t n\,d\tau = \alpha_0(t_m - t) + 2 \tag{9-164}$$

Also note Figs 9-28 and 9-29.

9-17 SPACE BORNE REACTORS

Space borne reactors are intended to be used as prime movers for electrical generating equipment for space needs, such as for satellites. The compactness, reliability, and longevity characteristics of such reactors make them a major candidate for medium to large space borne power sources, for example, from 3 kW and up. These systems in the USA belong to the SNAP (Systems for Nuclear Auxiliary Power) family. The even numbered series (SNAP 2, 8, 10, etc.) employ a turbo-electric generator as the electric power source, whereas the odd numbered series do not employ a reactor, but use thermo-electric elements only, to generate electricity directly. The latter, though even more compact and containing no moving parts, are limited by their relatively low power output due to the very low efficiency of the thermo-electric process. The discussion herein will dwell on the SNAP 2 and SNAP 8 reactor turbo-alternator systems as reactors typical of the family.

The SNAP 2 system specifications include a capability for yielding 3 kW of 2 kHz a.c. power for one year of unattended operation in space. The SNAP 2 reactor weighs about 750 pounds with a space radiation (radiator) heat dump of 110 ft^2 of surface area. It is a near-thermal (1000°F) reactor with zirconium hydride core elements impregnated with 10 per cent by weight of ^{235}U. It has a beryllium reflector and is cooled by liquid sodium potassium alloy (NaK) eutectic threading the core interstices. It delivers about 50 kWt (kilowatts thermal), most of which is used to drive the remainder of the reactor-alternator system. The primary NaK loop transfers heat to a secondary mercury liquid-vapor loop by means of a boiler heat exchanger. The mercury vapor drives a turbine, which in turn drives a 3 kW 2 kHz alternator, which supplies electrical power for use throughout the remainder of the space vehicle. The SNAP 2 system is shown in Figs 9-30 and 9-31.

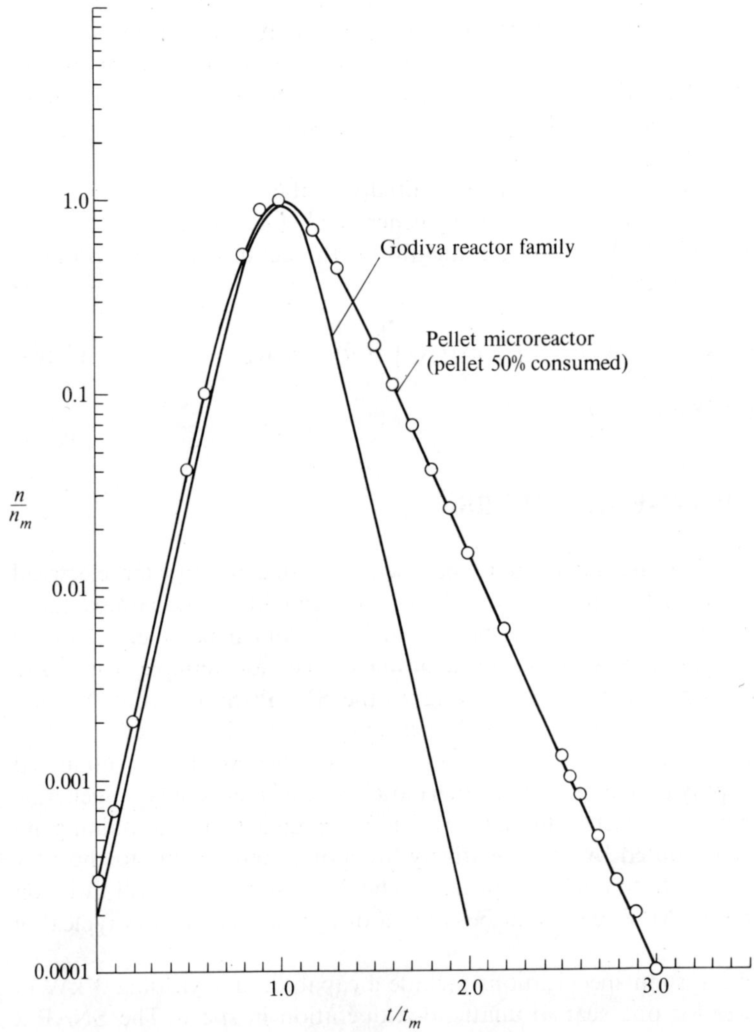

Figure 9-28 Neutron density pulse comparison of Godiva type all metal reactor and all metal pellet microreactor.

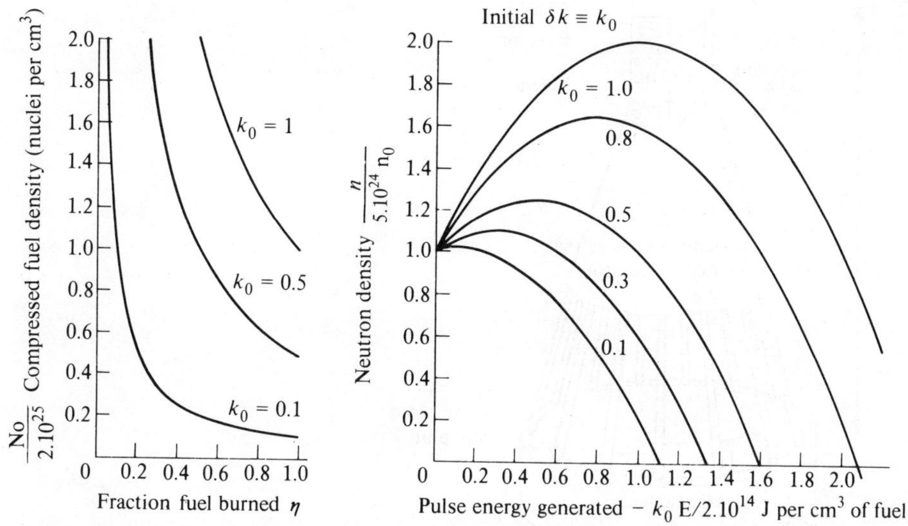

Figure 9-29 Hansen–Fuchs model modified for compressed pellet fuel burn.

The SNAP 8 reactor system is basically a larger version of the SNAP 2, as the output power is much larger, being about 600 kWt nominally with a coolant outlet temperature of 1300°F. The output power can be increased to one megawatt.

The two principal and interesting aspects of this type of reactor from the kinetics and control viewpoint are orbital startup and movable reflector control. The orbital reactor startup constitutes a precision control problem. One reason for this specification is that after determining that the system has established a successful orbit, most of the control signals must be telemetered from the ground. Aside from the task of monitoring the system to ascertain whether it is operating normally, the startup is complicated by the extreme temperatures of space. An

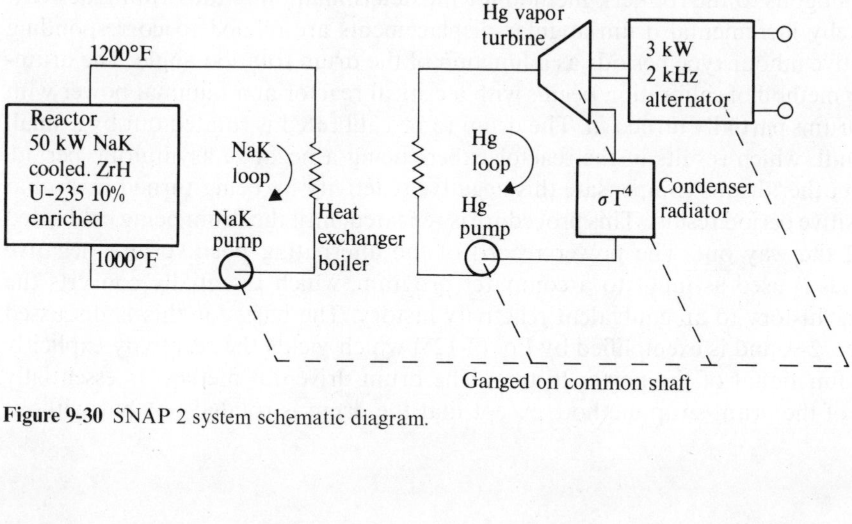

Figure 9-30 SNAP 2 system schematic diagram.

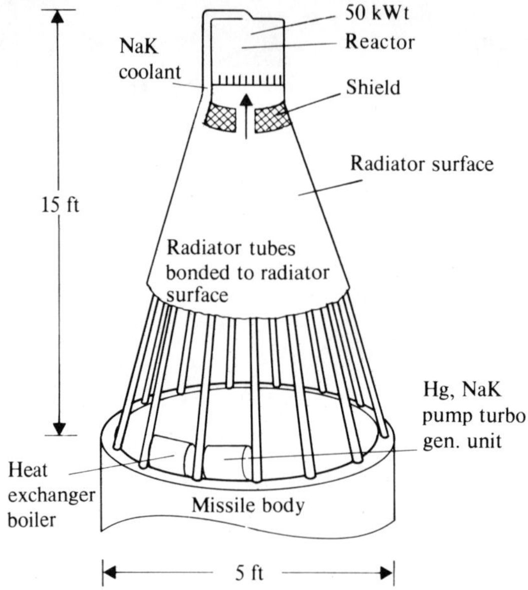

Figure 9-31 SNAP 2 system mounted on a missile nose cone.

automatic sequential controller will first sense that the system is preheated to the point where adequate mercury pump inlet pressure will ensue. Then the reactor will be slowly brought up to power.

The reactor is controlled by semicircular movable drums which constitute the main portion of the beryllium reflector. In SNAP 2 scram control is accomplished by dropping away sections of the reflector/drums as shown in Fig. 9-32, whereas for SNAP 8 a battery of torsional springs, one per drum, provide scram control where the drums are only free to rotate about their axes as in Fig. 9-33.

The experimental determination of the control drum reactivity worth was made by three different, but somewhat similar, methods.[48] They are the bump-period, the drum-swap, and the drum-drive out methods. The bump-period method is analogous to the rod-jerk method for the determination of the drum reactivity, whereby incremental drum angular displacements are related to corresponding negative inhour type periods as a function of the drum rotation angle. The drum-swap method of calibration begins with a critical reactor at a minimal power with the drums partially turned in. The drum to be calibrated is rotated out by a small amount, which results in the reactor experiencing a negative asymptotic period. Then other drums compensate this negative reactivity by being turned in so that a positive period results. This procedure is repeated until the drum being calibrated is all the way out. The power record of the alternating positive and negative periods is used as input to a computer program, which essentially converts the power history to an equivalent reactivity history. The basis for this is discussed in Sec. 2-9, and is exemplified by Eq. (2-125) which yields the reactivity explicitly as a functional of the power history. The drum drive-out method is essentially that of the drum-swap method, except that the drum is rotated out but without

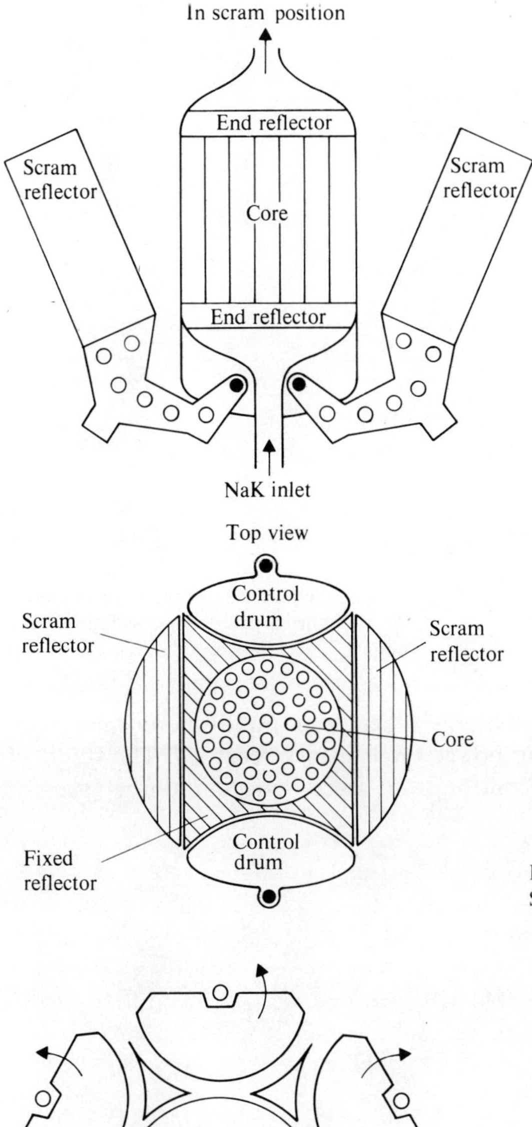

Figure 9-32 Elevation and plan view of SNAP 2 reactor.

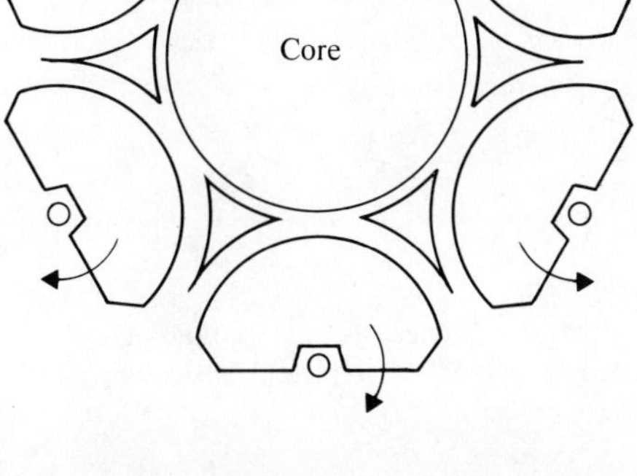

Figure 9-33 SNAP 8 reactor top view showing control drums.[48]

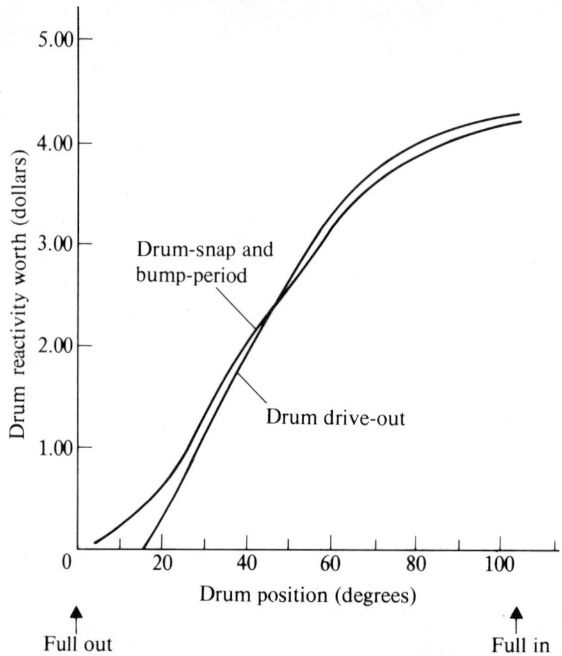

Figure 9-34 Comparison of control drum calibration methods for the SNAP reactor.[49]

compensation of the reactivity loss by the other drums, until the drum is fully withdrawn with the corresponding power level history recorded. The result of these three methods for a typical control drum reactivity calibration is given in Fig. 9-34.

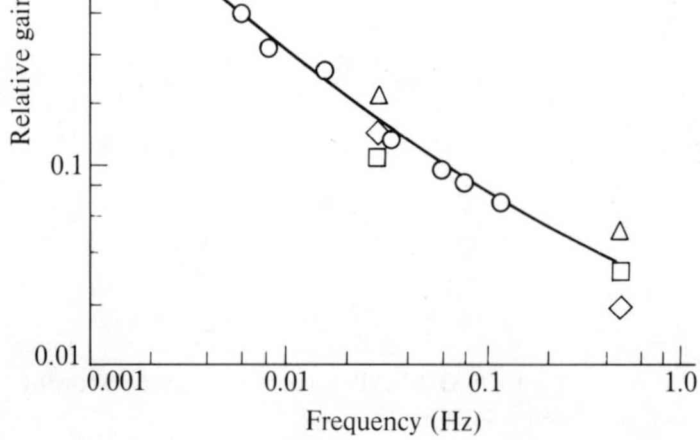

Figure 9-35 Amplitude plot of the SNAP 8 reactor transfer function.[49]

The transfer function of the SNAP 8 reactor was obtained by oscillating the drum about a given equilibrium power level. This was done for a number of corresponding equilibrium levels up to the rated power of 600 kWt. The transfer function amplitude plot is shown in Fig. 9-35, where it is seen that the mid-frequencies are of the order of 0.03 Hz. Part of the contribution to the very low frequency transfer function signature of this type of reactor, is the delayed photo-neutron contribution to the normal delayed neutrons. This is due to the large beryllium, material component of the reflectors.

PROBLEMS

9-1 For a boiling reactor where the reactivity $K = K(L^2, \tau, B^2, \ldots, f, p)$, show that its total derivative with respect to the steam void fraction α is given by Eq. (9-3).

9-2 Show that the zero-power transfer function obtained from a ringing experiment can be obtained as

$$\bar{G}(\omega) = \frac{1}{n_0 K_0}\left[(n - n_0) + i\omega \int_0^\infty (n - n_\infty)\, e^{-i\omega t}\, dt\right]$$

9-3 From the amplitude of the transfer function at its maximum frequency (the chugging frequency) $|\bar{G}(\omega_m)|$ show that the logarithmic decrement ζ in Eq. (9-5) is given approximately by

$$\zeta \cong \frac{n_\infty - n_0}{2n_0 K_0 |\bar{G}(\omega_m)|}$$

9-4 Convert the pulse periods in Fig. 9-18 to equivalent reactivities, given that $\beta = 75 \times 10^{-4}$ and the prompt lifetime is 1 μsec. Graph n_m as a function of $K_0^2/2b$ and compare with Fig. 9-18.

9-5 Using the $\bar{D}_2(K_0, \omega)$ function, plot the closed-loop describing function amplitude for the EBWR reactor. Let $\bar{H}(\omega)$ be given by Eq. (9-52) and use the one delayed neutron group approximation for $\bar{G}_0(\omega)$. Compare the calculations with the value determined experimentally as shown in Fig. 9-17.

9-6 Plot the curves of the asymptotic neutron density with and without considering delayed neutrons in an approximate way. Use the fractional exponent form given in Eq. (9-72).

9-7 Obtain the solution to the proton oscillator kinetics equation

$$\dot{n} = \left\{K_0 - \alpha_T \exp - \left(A \Big/ \int_0^t n\, d\tau\right)\right\} n; \qquad n_0 = 1$$

given by the integral in Eq. (9-100). Obtain approximations to the solutions for large and small $E = \int_0^t n\, d\tau$.

9-8 For the same model as in Prob. 9-7 but with a ramp reactivity mt, obtain solutions for small E by successive approximation techniques. Letting $\dot{n}/n$ be vanishingly small in the derivative of the equation in Prob. 9-7, obtain solutions for n near n_m, the peak of the pulse. One of these solutions is

$$n \cong \frac{A}{t[\ln (mt/\alpha)]^2}$$

9-9 For the curves of Figs 9-26 and 9-27 compute the equivalent step reactivity K_0 inserted initially. The prompt neutron lifetime is 50 μsec, $\beta = 0.007$.

9-10 Show that the relationship for the envelope of the peaks n_m and the time t_m at which they occur for the HF model is

$$\frac{n_m}{n_0} = \cosh^2\left(\sqrt{\frac{bn_m}{2}}\, t_m\right)$$

9-11 For the microreactor pellet that is totally burned, show that the energy generated as expressed by the time integral of the neutron density is given implicitly by the expression

$$\ln\left(\exp R_a \int_0^t n\,d\tau - 1\right) + \exp R_a \int_0^t n\,d\tau = \alpha_0(t_m - t) + 2$$

Assume that the same amount of energy is generated before and after the neutron density pulse peak.

9-12 In a comparison of the Godiva and microreactor kinetics, using Eq. (9-163) for the neutron density of the latter and the HF model for the former, calculate the neutron density "variances" of both. Use $\sigma^2 = \int_0^\infty (t - t_m)^2 n(t)\,dt / \int_0^\infty n(t)\,dt$, where t_m is the peak pulse time. Show that for the Godiva reactor the "standard deviation" is given by $\sigma_g \simeq (\sqrt{2}\,at_m)^{-1}$, where $a = \sqrt{bn_m/2\Lambda}$, and for the microreactor that $\sigma_m \simeq \sqrt{2}(\alpha_0 t_m)^{-1}$, where $\alpha_0 = K_0/l_0$. Assume that $n_m/n_0 \cong 10^8$ and that $\alpha_0 t_m \cong \ln 2\sqrt{n_m/n_0}$. Plot the curves given in Fig. 9-28

9-13 Using the HF model modified for compressed pellet fuel burnup, obtain the required expressions to plot the curves given in Fig. 9-29.

9-14 For the HF model with ramp input reactivity, derive the solution in the neutron density-reactivity plane, and plot a one parameter family of (closed) curves, where the parameter is $n_m/n_{\min}$. n_m is the neutron density at minimum period and $n_{\min}$ is the minimum neutron density value. The HF model is given by $\dot{n} = \left(mt - b\int_0^t n\,d\tau\right)n$, where the reactivity $K = mt - b\int_0^t n\,d\tau$.

9-15 Consider the HF model with a constant reactivity input, and with delayed neutrons approximated by their initial equilibrium concentration, that is the constant source model. The corresponding equations from Chap. 2 are $\Lambda\dot{n} = (K - 1)n + n_0$, where the reactivity is given by $K = K_0 - b\int_0^t n\,d\tau$. Show that the neutron density is given by the expression,

$$n = 2\Lambda \frac{d^2}{dt^2} \ln J_{2/3}\left[\frac{\Lambda^2}{\sqrt{2n_0}}\left(\frac{n_0 t}{\Lambda^2} + \frac{c}{\Lambda}\right)^{3/2}\right]$$

where $c = n_0 + (K_0 - 1)^2/2\Lambda$. *Hint:* Derive the differential equation which the reactivity satisfies, which ultimately results in the form $W'' - xW = 0$, which has the pertinent solution $W = AJ_{2/3}(x^{3/2})$, where $W = \exp - \int_0^t (K - 1/2\Lambda)\,d\tau$, and

$$x = \left(\frac{\Lambda^4}{2n_0^2}\right)^{1/3} \cdot \left(\frac{n_0 t}{\Lambda^2} + \frac{c}{\Lambda}\right)$$

9-16 Integrate the Ligou microreactor model given in Eq. (9-147) for constant α. Obtain for the neutron density

$$n = -\frac{1}{R_a}\frac{d}{dt}\ln\frac{d}{dt}\ln w$$

where

$$w = 1 - \frac{\alpha}{b} e^{a^2} \sqrt{\frac{\pi}{2}}\,[\operatorname{erf}(a + bt) - \operatorname{erf}(a)]$$

where, $a = (R_a n_0 + \alpha)/\sqrt{2R_a S_0}$, $b = \sqrt{R_a S_0/2}$, and $R_a = \sigma_a v$. In the n, N_f plane, plot the solutions for the $-\alpha$ case using the method of isoclines. Specifically, depict the situation when $S_0 = 0$ for times $t \geqq t_0 > 0$.

9-17 For the KEWB prompt kinetics model, integrate the appropriate relationship to obtain the energy generated as a function of time. Make reasonable approximations for the parameters involved to obtain an explicit expression for the energy, which when differentiated will yield the neutron density as a function of time. Obtain and plot this expression for the neutron density with time.

REFERENCES

1. R. Serber: The Use of Atomic Power for Rockets, *Proj. RAND RAD-2*, Douglas Aircraft Company, July, 1946.
2. J. Grey: Fundamentals of Rocket Propulsion, *Nucleonics*, **16**: 62–65 (1958).
3. H. Seifert (ed.): "Space Technology," Wiley, New York, 1959, chaps 3, 14.
4. R. R. Mohler: Stability and Control of Nuclear Rocket Propulsion, *IRE Trans. Auto. Control*, **AC-2**(2): 86–96 (1962).
5. M. J. Zucrow: "Aircraft and Missile Propulsion," vol. II, Wiley, New York, 1958, chap. 10.
6. D. F. Vanica: *Nucl. Propulsion Conf.*, Monterey, Calif., Oct., 1962.
7. G. T. Seaborg: Nuclear Power and Space, address at *Internatl. Symp. Aerospace Nucl. Propulsion*, Oct. 23–27, 1961.
8. Seifert, *op. cit.*, chap. 7.
9. D. J. Blake and W. H. Esselman: Nuclear Rocket Engine Dynamics Study, *IRE Trans. Nucl. Sci.*, **NS-9**(1): 78–81 (1962).
10. B. R. Felix and R. J. Bohl: Dynamic Analysis of a Nuclear Rocket Engine System, *Am. Rocket Soc. J.*, **29**: 853–862 (1959).
11. J. E. Perry, JR, and R. R. Mohler: Nuclear Rocket Engine Control, *Nucleonics*, **19**(4): 80–84 (1961).
12. R. W. Bussard and R. D. DeLauer: "Nuclear Rocket Propulsion," McGraw-Hill, New York, 1958, chap. 6.
13. H. P. Smith and A. H. Stenning: Open Loop Stability and Response of Nuclear Rocket Engines, *Nucl. Sci. Eng.*, **11**: 76–84 (1961).
14. R. V. Meghreblian: Gaseous Fission Reactors for Booster Propulsion, *ARS J.*, pp. 13–21 (Jan., 1962).
15. M. L. Rosenzweig, W. S. Lewellen, and J. L. Kerrebrock: Feasibility of Turbulent Vortex Containment in the Gaseous Fission Rocket, *ARS J.*, pp. 873–883 (July, 1961).
16. J. L. Kerrebrock and R. V. Meghreblian: Vortex Containment for the Gaseous Fission Rocket, *J. Aerospace Sci.*, **28**: 710–724 (Sept., 1961).
17. H. Weinstein and R. G. Ragsdale: "A Coaxial Flow Reactor—A Gaseous Reactor Concept," ARS 15th Ann. Mtg., Wash. DC, Dec. 5–8, 1960, ARS Preprint 1518-60.
18. G. Safanov: The Criticality and Some Potentialities of Cavity Reactors, *RAND Corp. Doc. No. RM-1835*, 17 July, 1955, p. 15, *ASTIA Doc. No. AD 112410.*
19. G. Safanov: Externally Moderated Reactors, *RAND Corp. Doc. No. R-316* (July, 1957).
20. W. A. Podney and H. P. Smith: Prompt Neutron Kinetics of a Spherical Cavity Reactor, *Nucl. Sci. Eng.*, **29**: 373–380 (1967).
21. J. A. Thie: Theoretical Statics and Kinetics of Boiling Reactors, *Proc. U.N. Intern. Conf. Peaceful Uses At. Energy, 2nd, Geneva*, 1958, **11**: 440–446 (1958).
22. G. E. Hansen: *Los Alamos Sci. Labs. Rept. LA* 1441, 1952: K. Fuchs, *LA* 496, 1946.
23. J. H. Bowen: "Nuclear Reactors for Power Generation," Newnes, London, 1958, chap. 6.
24. T. F. Wimett and S. D. Orndoff: Application of Godiva II Pulses, *Proc. U.N. Intern. Conf. Peaceful Uses At. Energy, 2nd, Geneva*, 1958, **10**: 449–460 (1958).
25. G. H. Freund et al.: Treat Reactor, *Proc. U.N. Intern. Conf. Peaceful Uses At. Energy, 2nd, Geneva*, 1958, **10**: 461–475 (1958).
26. D. Okrent et al.: Treat Reactor Kinetics, *Argonne Natl. Labs. ANL* 6174, 1960.
27. L. J. Koch et al.: Hazard Summary Report for EBR-II, *Argonne Natl. Labs. ANL* 5719, p. 204, 1957.
28. D. L. Hetrick et al.: Experimental Studies on the KEWB Reactor, *Proc. U.N. Intern. Conf. Peaceful Uses At. Energy, 2nd, Geneva*, 1958, **11**: 447–456 (1958).
29. C. F. Bonilla: "Nuclear Engineering," McGraw-Hill, New York, 1957, p. 721.
30. A. W. McReynolds: Neutron Thermalization, *Proc. U.N. Intern. Conf. Peaceful Uses At. Energy, 2nd, Geneva*, 1958, **16**: 297–313 (1958).
31. R. S. Stone: Transient Behavior of Triga, *Nucl. Sci. Eng.*, **6**: 249–255 (1959).

32. A. W. McReynolds et al.: Pulsed Solid Homogeneous Research Reactors, *Proc. U.N. Intern. Conf. Peaceful Uses At. Energy, 2nd, Geneva*, 1958, **10**: 476–486 (1958).
33. McReynolds: Ref. 32, p. 301.
34. K. M. Case, F. deHoffman, and G. Placzek: "Introduction to Theory of Neutron Diffusion," vol. 1, Los Alamos Scientific Laboratories, GPO, 1953.
35. F. Schroeder et al.: Transient Behavior in Spert, *Nucl. Sci. Eng.*, **2**: 96–115 (1957).
36. S. G. Forbes: Reactor Shutdown Model, *Quart. Progr. Rept.*, Phillips Petroleum Corp., Atomic Energy Division, January–March, 1958, IDO 16452 (TID 4500), p. 38.
37. W. E. Nyer and S. G. Forbes: Spert-I Safety Studies, *Proc. U.N. Intern. Conf. Peaceful Uses At. Energy, 2nd, Geneva*, 1958, **11**: 470–480 (1958).
38. W. E. Nyer et al.: Experimental Investigation of Reactor Transients, *AEC Res. Develop. Rept. IDO* 16285, April, 1956.
39. S. G. Forbes: Dependence of K_{cm} versus α_0 on Reactor Parameters, in *AEC Res. Develop. Rept. IDO* 16539, November, 1959.
40. F. Schroeder et al.: Power Excursion Studies, in 1960 *Idaho Conf. Reactor Kinetics, IDO* 16791, pp. 37–56, October, 1960.
41. G. W. Griffing and L. I. Deverall: Kinetics Studies on the Spert-I Reactor, *AEC Res. Develop. Rept. IDO* 16397, April, 1958.
42. M. S. Dunenfeld and D. P. Gamble: Analytic Solutions to KEWB Kinetics Equations for Short Period Transients, *Trans. Amer. Nucl. Soc.*, **2**(2): 62 (1959).
43. D. L. Hetrick and D. P. Gamble: Transient Reactivity During Power Excursions in a Water Boiler Reactor, *Trans. Amer. Nucl. Soc.*, **1**(2): 48 (1958).
44. D. L. Hetrick: "Dynamics of Nuclear Reactors," Univ. of Chicago Press, 1971, pp. 219–220.
45. D. L. Hetrick: *op. cit.*, pp. 170–171.
46. H. A. Sandmeier, G. E. Hansen, and S. A. Dupree: Time Dependent Escape of Neutrons and Photons from a Nuclear Explosion and Generated EMP Fields, *Los Alamos Sci. Labs. Doc. No. LA-DC-11512*, 1970: *Nucl. Sci. Eng.*, **48**: 343–352 (1972).
47. J. Ligou: Kinetics of Highly Compressed Fissionable Pellets, *Nucl. Sci. Eng.*, **62**: 151–160 (1977).
48. L. M. Fead, L. D. Felten, and V. L. Rooney: SNAP 8 Experimental Reactor Operations and Test Results, *Atomics International Doc. No. NAA-SR-10903*, May, 1973.
49. L. D. Felten and H. D. May: Summary of SNAP 8 Developmental Reactor (S8DR) Operations, *Atomics International Doc. No. AI-AEC 13071*, June, 1973.
50. J. Nuckolls, J. Emmett, and L. Wood: Laser-induced Thermonuclear Fusion, *Physics Today*, August, 1973, pp. 46–53, 1973.
51. W. Seifritz: Three-phase Pellets for Laser Fusion Reactors, Swiss Federal Institute for Reactor Research, CH-5303, Würenlingen, Switzerland, July, 1976; *Trans. Amer. Nucl. Soc.*, **21**: 18–19 (1975) (same title).

CHAPTER

TEN

STATISTICAL KINETICS

10-1 INTRODUCTION

This chapter considers, initially, the reactor kinetics of startup from weak sources, which are those that imply appreciable neutron density fluctuations in the early phases of reactor startup. This, and certain of the topics herein are somewhat specialized, even in the context of earlier chapters. Nevertheless, it behooves the serious reader to study the methods involved in obtaining solutions to these kinds of problems. At minimum, this is meant as exercising one's capacities in aspects of analysis that are not usually experienced within the mainstream of reactor kinetics, or nuclear reactor theory for that matter. For example, a much neglected skill is that of finding solutions to first-order partial differential equations. A good share of course work and applications in the milieu of reactor theory has to do with second-order partial differential equations, and the corresponding eigenvalue problems of critical mass, etc., while their first order counterpart is hardly dealt with in passing.

Other examples such as those from functions of a complex variable will be manifest throughout this chapter. An attempt is made to enlighten the uninitiated reader by expanding on both the mathematical as well as the physical ideas involved, in order to ease the road to understanding the subject matter of this chapter. Certain of the more classical statistical techniques for dealing with neutron density fluctuations in zero power reactors, such as the Rossi α method, are dealt with excellently in books by Pacilio,[1] Uhrig,[2] and Williams.[3]

During reactor startup it is desired that adequate monitoring of the power level be maintained, when the reactor is still substantially subcritical. With less than adequate monitoring, unwanted power excursions could take place unnoticed, particularly when the reactivity is increasing rapidly between delayed and prompt critical, that is, for $0 \leq K \leq 1$.

If the reactor startup source happens to be weak in the sense discussed below, the early phases of the startup will be characterized by appreciable random fluctuations in the neutron density, and thereby the power level. Such random fluctuations can be inimical to a smoothly controlled and operating system. There is ample evidence that such fluctuations occur as will be seen later. In the present context, such early startup fluctuations provide the initial conditions for the later average neutron density and power level kinetics.

The following is a development of statistical approaches for the calculation of quantities important for the measurement of early phase safe startup power levels. As already mentioned, this is especially important when the reactor state is in the vicinity of delayed critical, that is, near the vicinity of zero K, the normal operating reactivity regime. Near delayed critical, an important simplification in the statistical approach can be made, after Hurwitz et al.[4] It is that the prompt neutron lifetime can be assumed zero in the following sense. For a slightly subcritical reactor, $K \lesssim 0$, a single neutron injected into this reactor will cause a prompt neutron fission chain reaction that will persist on the order of more than 100 generations. To digress, this is easily seen from the fact that for step K near to zero, the corresponding neutron density response is given by $n(t) \sim \exp(-t/\Lambda)$ as noted from Eq. (2-11) for $K \simeq 0$, $G(t) \simeq e^{-t/\Lambda}/\Lambda$ for small values of time, and $R(t) = \delta(t)$, the single neutron injected. The corresponding mean life $\bar{t}$ then satisfies $\bar{t}/\Lambda = 1$, or $\bar{t}/l \simeq 1/\beta \simeq 140$ so that $\bar{t}$ is of the order of 100 lifetimes in duration. Now, during this time of about 100 generations (about 1 millisecond) there is virtually no change in any mechanical insertion of reactivity. Also, during this time the delayed neutron precursor densities will change mostly due only to their production from the above prompt neutron chains, as opposed to their loss by decay because of their relatively long decay times. In these terms then, the neutron lifetime can be assumed zero, so that such chains will occur instantaneously following a fission, thus producing precursors in "zero" lifetime.

A quantity of interest will be the asymptotic time-dependent probability distribution of the number of delayed neutron precursors, that are produced by one single instantaneously occurring prompt neutron fission chain. The zero neutron lifetime approximation will not be invoked at the outset, but at a later stage.

Following Hurwitz et al., loc. sit, let $P(n, m, t)$ be the probability that at time t, after the injection of a neutron into the reactor, there are n neutrons present (both prompt and delayed), as well as m delayed precursors. The delayed neutron precursors are assumed to consist of a single group that is an approximate representation of the usual six groups of delayed neutron precursors, with mean decay constant $\bar{\lambda} \simeq 0.1\ \text{sec}^{-1}$.

Now, the $\lim_{t\to\infty} P(n, m, t) = 0$ in a reactor that is delayed critical when all neutron sources are hypothetically turned off, since all of the neutrons remaining in such a distribution would ultimately disappear, as would all the decaying precursors. However, the asymptotic precursor distribution construed as $\lim_{t\to\infty} P(0, m, t) = P(0, m, \infty)$ is asserted to be finite and nonzero. $P(0, m, \infty)$ is defined to be the asymptotic time probability distribution of m delayed precursors. It is asymptotic in time with respect to the instantaneous single prompt neutron chain that produced them, but not asymptotic with respect to the delayed precursor decay times, as well as times over which the reactivity might change. Then, let $P(0, m, \infty) \equiv p(m, t)$ be a function of time t at which the particular prompt neutron chain occurred and which instantaneously created the m precursors. That is, the neutron chain occurs instantaneously, creating precursors instantaneously, hence there is zero neutron population in that sense prior to time t.

10-2 MATHEMATICAL PRELIMINARIES

The aforementioned approach proceeds by deriving an ordinary differential-difference equation in the quantity $P(n, m, t)$. Using the device of a generating function, to be defined, transforms this equation to a first-order partial differential equation in the generating function. The desired parameters are then extracted from the moments of the generating function.

The derivation of the differential-difference equation and how it is handled by a generating function, is illustrated by an example from queueing theory. Suppose that a single pump gasoline station can service μ cars per minute, and that the automobile arrival rate is λ cars per minute. Then the following equation can be written for $P(m, t)$, the probability that there are m cars in the system, that is, including those in the queue and being serviced. In the interval between t and $t + \Delta t$ the above probability can be expressed as the sum of four compound probabilities, namely

$$P(m, t + \Delta t) = (1 - \mu\Delta t)(1 - \lambda\Delta t)P(m, t) + (1 - \lambda\Delta t)\mu\Delta t P(m + 1, t) + (1 - \mu\Delta t)\lambda\Delta t P(m - 1, t) + \lambda\Delta t \cdot \mu\Delta t P(m, t) \tag{10-1}$$

The first term on the right of Eq. (10-1) is the probability that nothing happens in time Δt. That is, given the probability that there are m cars in the system, $1 - \lambda\Delta t$ is the probability that no car arrives during Δt, where $\lambda\Delta t$ is the arrival probability of one car in Δt. Similarly, $1 - \mu\Delta t$ is the probability that no car service is completed during Δt. Multiple arrivals and multiple service probabilities are considered to be zero. Hence, the first term is the probability that the system stays in "state m." The second term is; given the probability of $m + 1$ cars in the system, that no car arrives during Δt, but that one car's service is completed (and car leaves) during Δt thus bringing the system from the state $m + 1$ to state m. The third term is, given that the probability that there are but $m - 1$

cars in the system, there is one arrival during Δt, but no car is serviced in Δt. Hence, this term is the compound probability of transition from state $m - 1$ to m.

Only transitions from states but one car removed from m at time t are considered. Even if the gasoline station could accommodate them, $m - 2$, $m + 2$, $m - 3$, $m + 3$, etc. transitions to m can be shown to be $0(\Delta t^2)$ and higher order, and so the corresponding probabilities are considered negligible.

Taking the limit of vanishing Δt in Eq. (10-1), and neglecting terms of order higher than Δt, results in the differential-difference equation

$$\frac{dP}{dt}(m, t) = -(\lambda + \mu)P(m, t) + \lambda P(m - 1, t) + \mu P(m + 1, t); \qquad m > 0 \tag{10-2}$$

The initial condition is given by $P(m, 0) = \delta_{m1}$, where δ_{m1} is zero for $m \neq 1$, and 1 for $m = 1$. That is, initially, one car is assumed to be in the system with probability unity. Secondly, it is assumed that the system is never empty of cars, that is, $P(0, t) \equiv 0$.

Now, introducing a generating function G for the $P(m, t)$, which is a power series in a dummy variable v with a particular kind of coefficient. Using the above condition on $P(m, 0)$, let

$$G(v, t) = \sum_{m=0}^{\infty} P(m, t)v^m; \qquad G(v, 0) = v \tag{10-3}$$

Multiplying Eq. (10-2) through by v^m and summing on m yields the first-order partial differential equation in $G(v, t)$

$$\frac{\partial G}{\partial t}(v, t) = -(\lambda + \mu)G(v, t) + \lambda vG(v, t) + \mu/vG(v, t) \tag{10-4}$$

Since v occurs only as a parameter, a general solution is easily given by integration. It is $G(v, t) = F(v) \exp - \{\lambda(1 - v) - \mu(1 - v)/v\}t$. The initial condition $G(v, 0) = v$ asserts that $F(v) = v$, so that the solution of interest is

$$G(v, t) = v \exp - (1 - v)(\lambda - \mu/v)t \tag{10-5}$$

Now, an interesting parameter, such as the mean number of cars, $\bar{m}$, in the system is seen from Eqs (10-3) and (10-5) to be

$$\bar{m} = \frac{\partial G}{\partial v}(1, t) = 1 + \lambda - \mu \tag{10-6}$$

which is independent of time in this case. If $\lambda = \mu$, then $\bar{m} = 1$. So that if cars serviced (and departing) are balanced by cars arriving, there will always be on the average, only one car in the system if there was only one initially. This is exactly true since the variance of m, $\sigma_m^2 = \overline{m^2} - (\bar{m})^2 = 0$ for $\lambda = \mu$. As is now apparent other statistics of the system are available from moments of $G(v, t)$.

10-3 PROMPT NEUTRON/DELAYED NEUTRON PRECURSOR DISTRIBUTION

Turning to the problem at hand, $P(n, m, t)$ defined earlier satisfies an equation, analogous to the previous illustrative example, in that it describes a balance of probabilistic contributions to the (n, m) state of the neutron and precursor system.

$$P(n, m, t + \Delta t) = \left(1 - \frac{n\Delta t}{l}\right)P(n, m, t) + (n + 1)\frac{\Delta t}{l}\left(1 - \frac{k_p}{\nu_p}\right)P(n + 1, m, t)$$

$$+ \sum_{\nu=0}^{\infty} (n + 1 - \nu)\frac{\Delta t}{l} \cdot \frac{k_p}{\nu_p}(1 - \beta\nu_p)p(\nu)P(n + 1 - \nu, m, t)$$

$$+ \sum_{\nu=0}^{\infty} (n + 1 - \nu)\frac{\Delta t}{l} \cdot \frac{k_p}{\nu_p} \cdot \beta\nu_p p(\nu)P(n + 1 - \nu, m - 1, t) \tag{10-7}$$

where l, k_p, ν_p, β, and $p(\nu)$ are the prompt neutron lifetime or prompt generation time, prompt multiplication constant, mean number of prompt neutrons produced per fission, mean ratio of delayed to prompt neutrons, and the probability of neutrons produced per fission, respectively. The initial condition on the probability distribution is $P(n, m, 0) = 1$ for $n = 1$ and $m = 0$, and zero otherwise. This corresponds to one neutron being injected into a neutronless reactor at zero time.

Now $n\Delta t/l = nv\,\Sigma_T\,\Delta t$ is the interaction probability of n neutrons in time Δt, where Σ_T is the total macroscopic cross section for all interaction processes, and v is the neutron speed. So that $1 - (n\Delta t/l)$ is the non-interaction probability of the n neutrons. Hence the first term on the right-hand side of Eq. (10-7) is the probability that nothing happens in time Δt, given the system probability of being in the state (n, m). The second term is, given the system probability of being in the $(n + 1, m)$ state, that it will interact, $(n + 1)(\Delta t/l)$, but not by fission, $1 - k_p/\nu_p$, since $k_T/\nu_p = \Sigma_f/\Sigma_T$ is the fission probability per neutron. Therefore, non-fission interactions, that is, absorptions can transform the system from the $(n + 1, m)$ state to the (n, m) state.

The third term is, given the probability that if the system is in any of the myriad $(n + 1 - \nu, m; \nu = 0, 1, 2, 3, \ldots)$ states, that the interaction probability in Δt is $(n + 1 - \nu)\Delta t/l$, and that a fission ensues, k_p/ν_p, accompanied by no delayed precursors, $1 - \beta\nu_p$, where $\beta\nu_p$ is the probability of producing a single precursor. The fission produces ν prompt neutrons with probability $p(\nu)$. This term summed over all states $(n + 1 - \nu, m)$ transforms the system to state (n, m), since $n + 1 - \nu$ neutrons plus ν fission neutrons, minus one fission inducing neutron, yield a net of n neutrons. The last term in Eq. (10-7) is similar to the preceding term, but transforms $(n + 1 - \nu, m - 1)$ states into the (n, m) state by creating ν fission neutrons and one precursor, losing a fission inducing neutron in the process.

Taking the limit of zero Δt in Eq. (10-7) yields the corresponding differential-difference equation. It is

$$\frac{\partial P}{\partial t}(n, m, t) = -\frac{n}{l}\,P(n, m, t) + \frac{n + 1}{l}\left(1 - \frac{k_p}{\nu_p}\right)P(n + 1, m, t)$$

$$+\sum_{\nu=0}^{\infty}\left(\frac{n+1-\nu}{l}\right)\frac{k_p}{\nu_p}(1-\beta\nu_p)p(\nu)P(n+1-\nu,m,t)$$

$$+\sum_{\nu=0}^{\infty}\left(\frac{n+1-\nu}{l}\right)\frac{k_p}{\nu_p}\beta\nu_p p(\nu)P(n+1-\nu,m-1,t) \tag{10-8}$$

Now, define a two-dimensional generating function as

$$G(u,v,t)=\sum_{n=0}^{\infty}\sum_{m=0}^{\infty}P(n,m,t)u^n v^m \tag{10-9}$$

The initial condition $G(u,v,0)$ is obtained from the above prescribed $P(n,m,0)$ which picks solely the $n=1$, $m=0$ term of $G(u,v,t)$ in Eq. (10-9) to give $G(u,v,0)=v$. Equation (10-8) is then multiplied through by $u^n v^m$ and summed on n and m. The resulting first-order partial differential equation for $G(u,v,t)$ is

$$f(u,v)\frac{\partial G}{\partial u}(u,v,t)-l\frac{\partial G}{\partial t}(u,v,t)=0 \tag{10-10}$$

where

$$f(u,v)=-u+1-\frac{k_p}{\nu_p}+\left[\frac{k_p}{\nu_p}(1-\beta\nu_p)+\beta k_p v\right]g(u) \tag{10-11}$$

with the auxiliary generating function $g(u)=\sum_{\nu=0}^{\infty}p(\nu)u^{\nu}$.

The first-order partial differential Eq. (10-10) is solved in the usual way by writing down the three corresponding characteristic equations, namely in this case

$$\frac{du}{f(u,v)}=\frac{dt}{-l}=\frac{dG}{0} \tag{10-12}$$

The two particular integrals are simply the first from the two leftmost equations above, and the second from the two rightmost equations above, and yield

$$\frac{t}{l}+\int_0^u\frac{dx}{f(x,v)}=a,\qquad G=b \tag{10-13}$$

where a and b are constants. The general solution can be given by $a=F(b)$ where the function F is determined by the initial condition. That is, $a=F(b)$ gives from Eq. (10-13)

$$\frac{t}{l}+\int_0^u\frac{dx}{f(x,v)}=F(G) \tag{10-14}$$

Using the initial condition $G(u,v,0)=u$ then yields the particular solution of interest, which is

$$\int_u^{G(u,v,t)}dx/f(x,v)=t/l \tag{10-15}$$

Before rendering Eq. (10-15) for $G(u,v,t)$ less ungainly, a slight digression concerning $g(u)$ is in order. The first few coefficients of $g(u)$ were experimentally measured by Diven et al.[5] That is, $p(0)$, $p(1)$, $p(2), \ldots$ were ingeniously deduced from statistics obtained by counting fissions and fission neutrons from fissionable nuclides of reactor interest. For ^{235}U Diven et al. *op. cit.*, obtained a polynomial fit to $g(u)$ as

$$g(u) = 0.027 + 0.158u + 0.339u^2 + 0.305u^3 + 0.133u^4 + 0.038u^5 \quad \text{(10-16)}$$

where $g(1) = 1$ and $\partial g(1)/\partial u = \nu_p$.

As discussed at the outset, the limit of zero prompt neutron lifetime is of importance. So that for times large compared to prompt neutron lifetimes, but small compared to delayed neutron precursor decay times and reactivity changes as discussed, Eq. (10-15) can be rewritten in the limit of zero prompt lifetime, as

$$\int_0^{G(0,v,\infty)} dx/f(x, v) \simeq \infty \quad \text{(10-17)}$$

where $G(0, v, \infty)$ is the proper asymptotic form of $G(u, v, t)$ for the approximation of zero prompt lifetimes. Now $G(0, v, \infty) = \sum_{m=0}^{\infty} P(0, m, \infty)v^m$ is the sought for quantity. Therefore, considering that the roots of $f(x, v) = 0$ provide the singularities for the above integral to diverge, $G(0, v, \infty)$ must be one of the roots of $f\{G(0, v, \infty), v\} = 0$, so that the integral will be very large. So $f(G, v) = 0$ implies from Eq. (10-11) that $G(0, v, \infty)$ must satisfy

$$G(0, v, \infty) = 1 - \frac{k_p}{\nu_p} + \left[\frac{k_p}{\nu_p}(1 - \beta\nu_p) + \beta k_p v\right] g(G) \quad \text{(10-18)}$$

As mentioned the statistical results are implicit in the moments of the generating function G, and evaluating various derivatives of G about $v = 1$ in Eq. (10-18) will yield them. For example, from the definition of G given by Eq. (10-9), and using Eq. (10-18) it is easily verified that the time asymptotic mean number of delayed neutrons, $\bar{m}$, is given by

$$\bar{m} = \frac{\partial G}{\partial v}(0, 1, \infty) = \beta k_p/(1 - k_p) \quad \text{(10-19)}$$

for the subcritical reactor ($K < 0$). It is also easily verified that $\bar{m}$ checks with the solutions of the proper deterministic kinetics equations.

10-4 DELAYED NEUTRON PROBABILITY DISTRIBUTION

Since another goal is the statistical behavior of the delayed neutron precursors, a differential-difference equation will now be derived for the corresponding delayed neutron probability distribution. Let $P(m, t)$ be the probability distribution for the delayed neutron precursors that are produced from a single instantaneously occurring prompt neutron chain at time t as discussed earlier.

Assuming, as already mentioned, that there is only one group of delayed neutron precursors with decay constant λ, then with a neutron source rate,

r, $P(m, t)$ satisfies

$$P(m, t + \Delta t) = (1 - \lambda m \Delta t)(1 - r\Delta t)P(m, t) + \sum_{n=0}^{\infty} (m - n + 1)\lambda \Delta t p(n, t)P(m - n + 1, t) + \sum_{n=0}^{\infty} p(n, t)r\Delta t P m - n, t) \tag{10-20}$$

The first term on the right-hand side is, given the probability of the system being in state m, that it stays in state m, since $1 - \lambda m\Delta t$ is the non-decay probability in Δt of the m delayed neutron precursors, and $1 - r\Delta t$ is the probability that no source neutrons are produced in Δt to create delayed neutron precursors.

The second term is, given the probability of the system being in state $m - n + 1$, that it decays in time Δt, with probability $(m - n + 1) \cdot \lambda\Delta t$, contributing one neutron to cause a fission to produce (with probability $p(n, t)$) an instantaneously occurring prompt fission chain yielding n delayed neutron precursors. This then transforms the system to state m. This is summed over all possible prompt neutron chain populations n. The third term is similar, given the probability of the system being in the state $m - n$, that $p(n, t)r\Delta t$, the source neutron production probability produces an instantaneously occurring prompt neutron chain of n neutrons, to transform the system from the state $m - n$ to the state m. This is also summed over all possible neutron chain lengths n.

Now, taking the limit of zero Δt in Eq. (10-20) yields

$$\frac{\partial P}{\partial t}(m, t) = -(\lambda m + r)P(m, t) + \lambda \sum_{n=0}^{\infty} (m - n + 1)p(n, t)P(m - n + 1, t) + \sum_{n=0}^{\infty} p(n, t)P(m - n, t) \tag{10-21}$$

The corresponding generating functions and their respective initial conditions are given as

$$G(v, t) = \sum_{m=0}^{\infty} P(m, t)v^m; \quad g(v, t) = \sum_{m=0}^{\infty} p(m, t)v^m; \quad G(1, t) = g(1, t) = 1 \tag{10-22}$$

Multiplying Eq. (10-21) through by v^m and summing on m yields the partial differential equation for the generating function of the delayed neutron precursor population

$$\frac{\partial G}{\partial t}(v, t) = \lambda\{g(v, t) - v\} \frac{\partial G}{\partial v}(v, t) + r\{g(v, t) - 1\} G(v, t) \tag{10-23}$$

This equation is pivotal for obtaining the desired statistical result to follow, as will be seen. First, a steady state solution is obtained, that is, where $\partial G/\partial t = 0$, for $G(1, t) = 1$, namely

$$G_{ss} = \exp -\frac{r}{\lambda} \int_v^1 \{1 - g(x, t)\}/\{g(x, t) - \lambda\} \cdot dx \tag{10-24}$$

where the subscript *ss* implies steady state. G_{ss} provides an initial condition for the case in which the subcritical reactivity is constant for a long time, followed by a reactivity excursion. For example, if the reactor is initially subcritical, then the $\lim_{t\to 0} p(m,t) = \delta_{m0}$. That is, with probability unity there are no precursors in the reactor in that state. Hence, from its definition $g(v,0) = 1$ as well. Then from Eq. (10-24) it follows that $G_{ss} = 1$ and that $P(m,0) = \delta_{m0}$ also.

Second, differentiating Eq. (10-23) partially with respect to v and evaluating the result at $v = 1$ yields, with $(\partial g/\partial v)|_1 = \beta k_p/1 - k_p$ from Eq. (10-19), and $(\partial G/\partial v)|_1 = \bar{m}$ from the definition of G above yields

$$\dot{\bar{m}} = -\lambda\bar{m} + (\bar{m}\lambda + r)\beta k_p/1 - k_p \tag{10-25}$$

This is the subcritical kinetics equation for the single group delayed neutron precursor concentration $\bar{m}$, with decay constant λ, and with subcritical multiplication of precursors from the source r, as well as from the existing precursors. There is no explicit neutron production term in Eq. (10-25), as there is no neutron population in the reactor except the source neutrons because, once a neutron chain produces a precursor, it disappears since its lifetime is assumed to be zero.

Third, what is the probability that a single delayed precursor will ultimately produce a divergent fission chain when the reactivity is between delayed and prompt critical, that is, $0 < K < 1$? Here, let $g(v,t)$ be constant and with no source ($r = 0$). Then Eq. (10-23) becomes, with the initial condition that $P(m,0) = \delta_{m0}$, that is, no precursors with which to start.

$$\frac{\partial G}{\partial t} = \lambda(g_0 - v)\frac{\partial G}{\partial v}; \qquad G(v,0) = v \tag{10-26}$$

This has a solution

$$G(v,t) = g_0 - (g_0 - v)\,e^{-\lambda t} \tag{10-27}$$

After an infinite time $G(v,\infty) = \sum_{m=0}^{\infty} P(m,\infty)v^m = g_0$, a constant. Since g_0 is constant, all powers of v except the zeroth vanish, thence $P(0,\infty) = g_0$ which is the probability that no precursors exist after an infinite time.

Fourth, and the *raison d'être* of this development. What is the probability, P_N, that the precursor population is less than N? It is given by the probability that the precursor population has any value from zero up to $N - 1$. That is

$$P_N = \sum_{m=0}^{N-1} P(m,t) \tag{10-28}$$

Since $G(v,t) = \sum_{m=0}^{\infty} P(m,t)v^m$, where

$$P(m,t) = \frac{1}{m!}\frac{\partial^m}{\partial v^m} G(0,t),$$

which are the coefficients in this power series for $G(v,t)$. Then

$$P_N = \sum_{m=0}^{N-1} \frac{1}{m!}\frac{\partial^m G}{\partial v^m}(0,t) \tag{10-29}$$

In the complex z plane, using the Cauchy integral theorem, $G(z, t)$ can be expressed in terms of its derivatives at $z = 0$, by

$$\frac{1}{m!}\frac{\partial^m G}{\partial v^m}(0, t) = \frac{1}{2\pi i}\int_c G(z, t)\, dz/z^{m+1} \tag{10-30}$$

where the contour c is the unit circle, and that G is analytic in and on the unit circle. This contour is chosen in order that $1 - z^n$ is analytic for reasons to become apparent. Now Eq. (10-30) is substituted into Eq. (10-29) to obtain

$$P_N = \frac{1}{2\pi i}\int_c G(z, t)\sum_{m=0}^{N-1} dz/z^{m+1} \tag{10-31}$$

Now, the sum $\sum_{m=0}^{N-1} z^{-(m+1)} = (1 - z^N)/(1 - z)z^N$. After substituting this back into Eq. (10-31), and noting that the factor $1 - z^n$ will contribute nothing, since it is analytic within and on the contour above, the result is,

$$P_N = \frac{1}{2\pi i}\int_c \frac{G(z, t)\, dz}{(1 - z)z^N} = \frac{1}{2\pi i}\sum_{m=0}^{\infty} P(m, t)\int_c \frac{(1 - z^N)z^{m-N}\, dz}{1 - z} \tag{10-32}$$

Beside poles at $z = 0$ and 1, which contribute nothing of substance to the integral, the integrand has a saddle point on the real axis. The saddle point method, sometimes referred to as the method of steepest descents, is a well-known approximation for the evaluation of the integral at a saddle point. It is to rewrite the integral as

$$\int_c e^{f(z)}\, dz, \quad \text{where } f(z) = \log(1 - z^N) + (m - N)\log z - \log(1 - z) \tag{10-33}$$

The saddle point occurs at z_0, where z_0 is the root of $f'(z_0) = 0$. This yields $0 < z_0 < 1$ on the real axis. Now

$$f''(z_0) = -\frac{N(N-1)z_0^{N-2}}{1 - z_0^N} + \left(\frac{Nz_0^{N-1}}{1 - z_0^N}\right)^2 + \frac{N - m}{z_0^2} + \frac{1}{(1 - z_0)^2} \tag{10-34}$$

which is seen to be very large and negative for large N and $0 < z_0 < 1$. Therefore $f(z)$ has a sharp maximum at z_0, i.e. a "razor" sharp saddle point maximum. This means that most of the contribution to the integral will then come from values of the integrand in the neighborhood of the saddle point. So that, with $f'(z_0) = 0$, the first two terms of $f(z)$ become,

$$f(z) \simeq f(z_0) - \frac{1}{2!}|f''(z_0)|(z - z_0)^2 + \cdots, \tag{10-35}$$

and will yield a good approximation to the integral when evaluated. This allows the limits on the integral to be extended to infinity, in both directions, when this $f(z)$ is substituted back into the integrand in Eq. (10-32). This yields the real integral

$$P_N = \frac{1}{2\pi i}\sum_{m=0}^{\infty} P(m, t)\left(\frac{1 - z_0^N}{1 - z_0}\right)z_0^{m-N}\int_{-\infty}^{+\infty} e^{-\frac{1}{2}|f''(z_0)|(z - z_0)^2}\, dz \tag{10-36}$$

For very large $N - m$, the result gives to good approximation

$$P_N \simeq \frac{1}{\sqrt{2\pi\,|f''(z_0)|}} \cdot \frac{G(z_0, t)}{(1 - z_0)z_0^N} \tag{10-37}$$

With $z_0 = 1 - \varepsilon$ it is seen that for z_0 very near unity $z_0^m = (1 - \varepsilon)^m = 1 - m(1 - z_0)$ for $\varepsilon \ll 1$, then

$$G(z_0, t) = \sum_{m=0}^{\infty} P(m, t)z_0^m \simeq \sum_{m=0}^{\infty} P(m, t)[1 - m(1 - z_0)]$$

$$= 1 - (1 - z_0)\bar{m} \simeq e^{-(1 - z_0)\bar{m}} \tag{10-38}$$

Now, this G also happens to be the generating function of the Poisson distribution, that is, of

$$P(m) = e^{-\bar{m}} \cdot (\bar{m})^m/m! \tag{10-39}$$

So that for z_0 near unity and large $N - m$, $P(m, t)$ tends toward a Poisson distribution, and that from Eq. (10-28)

$$P_N \simeq \sum_{m=0}^{N} e^{-\bar{m}}\,(\bar{m})^m/m! = 1 - \frac{\gamma(N, \bar{m})}{\Gamma(N)} \tag{10-40}$$

where Γ and γ are the complete and incomplete gamma function, respectively. For $\bar{m}/N \gg 1$, either the saddle point method, Eq. (10-37) or Eq. (10-40) gives for the probability that the precursor population is less than N, where $\bar{m}$ is the mean population, the expression

$$P_N \simeq \frac{1}{\sqrt{2\pi N}} \left(\frac{\bar{m}}{N}\right)^N e^{-\bar{m}} \tag{10-41}$$

P_N becomes constant for large times. This is easily seen by considering that after a relatively long time following startup, $\bar{m}$ becomes large as the reactor power level increases into the deterministic kinetics regime. For large N and m, z_0 approaches unity, as already discussed, and from Eq. (10-38) $G(z_0, t) \cong \exp - (1 - z_0)\bar{m}(t)$. From Eq. (10-37) then

$$P_N \sim \frac{1}{\sqrt{2\pi}} \frac{e^{-(1 - z_0)\bar{m}(t)}}{(1 - z_0)z_0^N} \tag{10-42}$$

which approaches a constant, since the influence of the time dependence of $\bar{m}(t)$ diminishes as z_0 approaches unity in the above exponential.

10-5 RESULTS

The partial differential equation (10-23) in the generating function of the delayed neutron precursor population, can be numerically integrated about the saddle point, which lies near unity, using the method of characteristics. In that neighbor-

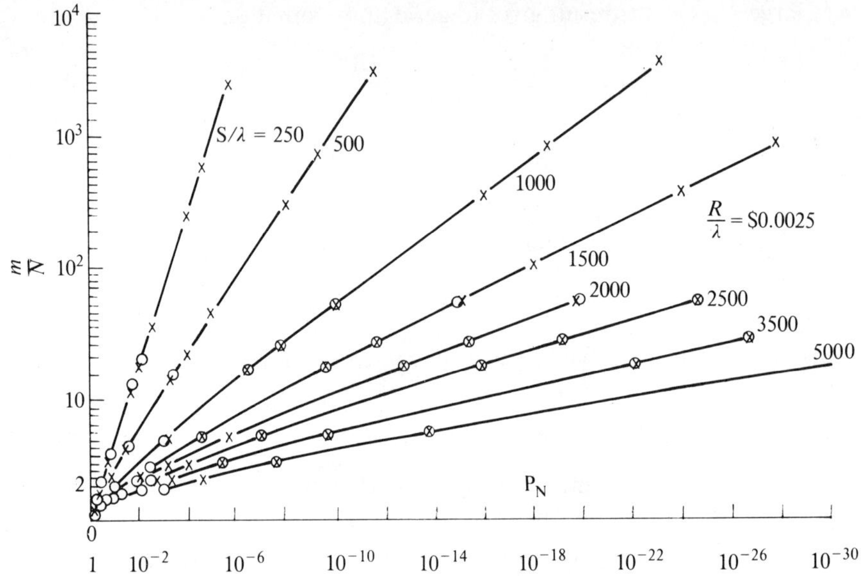

Figure 10-1 Mean precursor population, $\bar{m}$, versus P_N-probability of the reactor containing at least N precursors- for various ramp reactivity rates R and source rates.[4]

hood, $G(v, t) \sim \exp - (1 - v)\bar{m}(t, k_p)$. Using ramp reactivity increases from startup of R dollars per second, and source rates of S neutrons per second, P_N can be computed. Hurwitz et al. (*loc. sit.*) have made the computations, with a typical result presented in Fig. 10-1.

To use Fig. 10-1, suppose that deterministic kinetics studies have already shown that a shim rod rate of $R = 0.00025$ dollars per second (1/40th cent per second), and a minimum source strength rate $S_m = 10$ neutrons per second, are bordering on the safety threshold for adequately monitoring startup power. By how much would the startup phase be raised above such a threshold if the source rate were increased to $S = 100$ neutrons per second?

Answer: Now $S/\lambda = 100/0.1 = 1000$. $S/S_m = 100/10 = \bar{m}/N$, since the mean precursor population $\bar{m}$ is proportional to the source strength S, and let the a priori chosen precursor population N be proportional to the minimum source strength S_m. Also $R/\lambda = 0.0025$ dollars. Figure 10-1 is applicable, where for $\bar{m}/N = 10$ and on the $S/\lambda = 1000$ curve, it is seen that $P_N \simeq 10^{-5}$. That is, there is a likelihood of one chance in 100,000 that the mean precursor population $\bar{m}/N$ is less than 10 precursors. To attain these odds therefore, it might be prudent to increase the source strength rate by tenfold to $S = 100$ neutrons per second, to provide a sufficiently large neutron population for startup safety.

For ramp reactivity insertion rates an empiricism can be given for P_N, as (Hurwitz et al. (*loc. sit.*)).

$$P_N \simeq \exp - \left[10^{-3}\frac{S}{\lambda}\log_{10}(\bar{m}/N)\right]\left[5.4 - 30\left(\frac{k}{\lambda}\right) + 50\left(\frac{R}{\lambda}\right)^2\right] \tag{10-43}$$

where: S is the source strength rate (neutrons per second), R is the reactivity (shim) rate injected (dollars per second), λ is the delayed neutron precursor decay constant ($\simeq 0.1\ \text{sec}^{-1}$), $\bar{m}$ is the mean precursor population, N is an a priori numerical choice parameter, and P_N the probability likelihood that $\bar{m} < N$.

The accuracy of Eq. (10-43) suffers appreciably only for very small P_N, where it is ± 25 per cent. In such cases of small P_N, for example, $0.75^{-6} \leq P_N \leq 1.25^{-6}$, no wrong safety decision would be forced from such small numbers.†

10-6 EXPERIMENTAL VERIFICATION OF WAITING TIMES

Using the Godiva II reactor, Hansen and Wimett[6] performed a series of 89 startup experiments. Each experiment consisted of starting up the reactor from a very subcritical state to $K \simeq \$0.70$, as fast as possible. For an a priori fixed fission rate or power level, the waiting time to achieve that level, and the period, was recorded in each case. From the results, a cumulative probability distribution of the corresponding waiting times to attain that fiducial was obtained. That is, the abscissa of this distribution is the least amount of time that must elapse, that is, waiting time, before the reactor reaches that fiducial.

As mentioned, the computations of Hurwitz et al. (*loc. sit.*) consisted of trajectories in the characteristic curve solution space of the partial differential equation, for the precursor generating function $G(v, t)$. A given value of the precursor threshold N on the trajectory characterizes an asymptotically constant P_N—the likelihood that the mean precursor population is $m < N$. The time elapsed along such trajectories, to the asymptotic regime where P_N becomes constant, and the deterministic kinetics equations begin to hold, can be construed as the waiting time to achieve a given power level with probability P_N. Therefore the computed P_N can be compared with the experimentally determined waiting time distribution. The agreement is quite good as seen in Fig. 10-2.

It should be noted that part of the reason for the good agreement is that Godiva is a small closely coupled (fast) reactor. This implies that the point model yields a good description of its kinetic behavior in the above context.

10-7 WEAK SOURCE SYSTEMATICS

As alluded to in the previous sections, the early growth behavior of the neutron density, even in a supercritical system can depart significantly from the average neutron density time dependence, as described by the deterministic kinetics equations. For example, the startup neutron density due to a spontaneous fission neutron source in the Godiva reactor, only slightly supercritical, suffers from the lack of a reproducible peak fission rate time after assembly. That is, given the

† The exponents in the limits on P_N correspond to a shorthand for the appropriate power of 10, e.g. $1.25^{-6} = 1.25 \cdot 10^{-6}$.

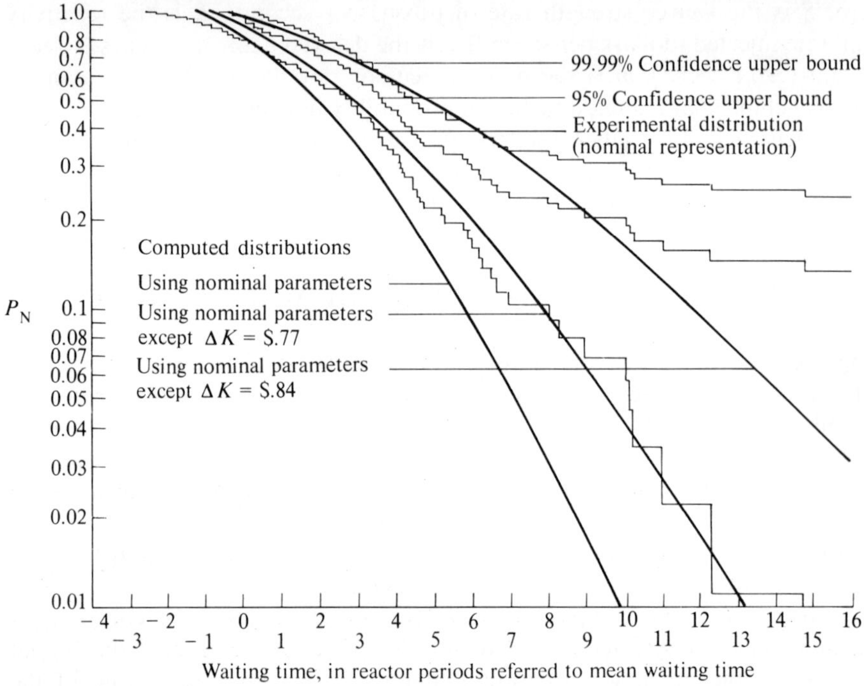

Figure 10-2 The waiting time cumulative probability construed as P_N versus waiting time to achieve a given power level.[4]

level of a burst of fission neutrons into the reactor, the subsequent number of fissions, and their time width are predictable and verified by experiment, but not the corresponding time of occurrence of their peak, which latter fluctuates. For other examples, cf. Hansen,[7] whose treatment is paraphrased here. This "weak source" situation will be examined, in that simple approximate expressions for the probability distribution of fission burst occurrences and energy release will be obtained.

As discussed in the earlier sections, the statistics of weak sources often play an important role in the subsequent temporal behavior of reactors of interest. This is especially so for certain accident analyses and much of what follows reflects such considerations. Consider a simple reactor, after Hansen (*op. cit.*), with a neutron source due to, for example, spontaneous fissions. It is assumed that for the times and reactivity magnitudes involved, that delayed neutrons are foreign to the discussion, and their effects are therefore negligible, as opposed to the delayed neutron precursor considerations in the previous sections. Let W_f be the conditional probability that given a source fission, it results in the initiation of a persistent chain reaction. A persistent chain is one that does not become finite due to the termination of the chain in non-fission absorption, leakage, or other non-neutron producing interactions. Also, let p be the probability of a neutron producing a fission, and $P(\nu)$ the probability of ν neutrons produced per fission.

Since $1 - W_f$ is the probability that a source fission does not result in a persistent chain, but in a finite chain, then none of the associated fission neutrons must be the progenitors of a persistent fission chain. This is embodied in the expression

$$1 - W_f = \sum_{\nu=0}^{\infty} P(\nu)(1 - pW_f)^{\nu} \tag{10-44}$$

If the reactor is but slightly supercritical, in the sense that the progenitor neutrons behave so that the compound probability of the initiation of a persistent chain, $W = pW_f$, is very small, then only the first three terms in the above binomial expansion need be retained. This gives

$$1 - W_f = 1 - p\bar{\nu}W_f + \frac{p^2 W_f^2}{2}\overline{\nu(\nu - 1)} \tag{10-45}$$

where the overbars stand for averages, as in $\bar{\nu} = \sum_{\nu=0}^{\infty} \nu P(\nu)$, and $\bar{\nu}$ is the mean number of neutrons produced per fission. Also, $p\bar{\nu}$ is recognized as the mean number of daughter fissions per fission, which is tantamount to the reproduction factor k, in an all metal reactor such as Godiva.

Using the Diven factor statistic $\Gamma_2 = \overline{\nu(\nu - 1)}/(\bar{\nu})^2$ then Eq. (10-45) yields

$$W = pW_f = \begin{cases} \dfrac{2\beta K}{\bar{\nu}\Gamma_2}; & K > 0 \text{ supercritical} \\ 0; & K \leq 0 \text{ subcritical} \end{cases} \tag{10-46}$$

where K, the associated reactivity in dollars, as defined in Chap. 2, is very small and positive, corresponding to a slightly supercritical reactor. $\Gamma_2 \simeq 0.8$ for fissionable materials of reactor interest, Diven et al.[5] W is taken to vanish in the above expression for a subcritical reactor, since persistent chains would not be supported thereby. W also has a more general validity in that it holds for space- and energy-dependent systems as well. The corresponding $W(\bar{r}, \bar{\nu}, K)$ is given in Hansen (*op. cit.*).

If the neutron source rate is S neutrons per second, the expected rate of initiations of persistent chain reactions is simply $pW_f S = WS$, so that the average time interval between successive initiations is $(WS)^{-1}$. For a first initiation occurring at a time t_i, the subsequent expected neutron density is

$$n_i(t) = n_{i0} \exp \frac{K}{\Lambda}(t - t_i); \qquad t \geq t_i \tag{10-47}$$

From the definition of $(WS)^{-1}$, it easily follows that

$$n_{i+1}(t) = n_i(t) \exp - (K/\Lambda WS) \tag{10-48}$$

Now suppose that $K/\Lambda WS \gg 1$, i.e. is very large. This then implies a rapid decay of persistent chains following the first at time t_i. The resulting total neutron density is then very nearly comprised of neutrons from only the first persistent chain. From Eq. (10-46), the above inequality can be rewritten as

$$2S\Lambda/\bar{\nu}\Gamma_2 \ll 1 \tag{10-49}$$

or equivalently $S\Lambda \ll 1$, since $2/\bar{\nu}\Gamma_2$ is of the order unity for fissionable nuclides of interest. These inequalities are taken to be the definition of a weak source in this context.

On the other hand, if $S\Lambda \gg 1$, then the neutron density at times during an accidental fuel assembly, which are large compared to $\alpha^{-1} = \Lambda/K$, is primarily comprised of the progeny from source neutrons emitted in the first e-folding time α^{-1}.

During the accident assembly time interval α^{-1}, the expected number of persistent fission chains, $\bar{N}$, is, using Eq. (10-46)

$$\bar{N} = WS\alpha^{-1} = 2\beta S\Lambda/\bar{\nu}\Gamma_2 \tag{10-50}$$

Since it is assumed that $\bar{N}$ is quite small for the slightly supercritical reactor under discussion, i.e. persistent chains are sufficiently rare events to be assumed uncorrelated, then the probability distribution of N is taken to be the Poisson distribution. Noting that this distribution has the property that the corresponding mean and variance are equal, the fractional variance of N, $\sigma_N^2/(\bar{N})^2$, is simply

$$\frac{\sigma_N^2}{(\bar{N})^2} = (\overline{N^2} - (\bar{N})^2)/(\bar{N})^2 = 1/\bar{N} = \bar{\nu}\Gamma_2/2\beta S\Lambda \tag{10-51}$$

The fractional variance of the corresponding neutron density is also assumed to be calculated from a Poisson distribution, so that

$$\sigma_n^2/(\bar{n})^2 = 1/\bar{n} = \bar{\nu}\Gamma_2/2\beta S\Lambda \tag{10-52}$$

as well, and further that the standard deviation (square root of the variance) of the time at which the neutron density attains a substantively large reference value is, where $\bar{t} = \alpha^{-1}$

$$\sigma_t/\bar{t} = \frac{K}{\Lambda}(\overline{t^2} - (\bar{t})^2)^{1/2} = K/\Lambda WS \simeq \bar{\nu}\Gamma_2/2\beta S\Lambda \tag{10-53}$$

10-8 FISSION BURST TIME PROBABILITIES

The following discusses the probability distribution of the time of occurrence of a fission chain burst, i.e. the associated subsequent neutron density "burst" from the initiation of a persistent chain. This time is defined by equating it to the sum of two time terms. The first, where zero time corresponds to the onset of reactor criticality, is the time t_p, of initiation of the first persistent chain. The second is the time $t_m > t_p$, at which the chain has grown to the substantively large neutron density level referred to in the preceding.

Two types of reactivity initiation of the above chain will be considered. They are the step reactivity increase, $K(t) = K_0 H(t)$ where $H(t)$ is the Heaviside step function ($H(t) = 1$ for $t \geqq 0$; $H(t) = 0$ for $t < 0$) and the ramp reactivity increase,

$K = mt$. Let $P(t_p)\,dt_p$ be the probability that the first persistent chain is initiated in dt_p about time t_p. This is readily seen to be the compound probability

$$P(t_p)\,dt_p = WS\,dt_p \exp - (WS\,t_p) \tag{10-54}$$

Since $\exp - (WSt_p)$ is the probability that no persistent chains are initiated prior to time t_p, and as before, WS is the expected rate of such initiations. $WS\,dt_p$ is the probability of an initiation in the interval dt_p, thence

$$\bar{t}_p = (WS)^{-1}, \qquad \sigma_p^2 = \bar{t}_p^2 - (\bar{t}_p)^2 = (WS)^{-2} \tag{10-55}$$

As mentioned, delayed neutron considerations are foreign to this discussion. The immediate implication is that the variance of t_m, the time at which the neutron density has grown to a substantive level, that is, $\sigma_m^2 = \bar{t}_m^2 - (\bar{t}_m)^2$ is small compared to the variance of t_p, $\sigma_p^2 = \bar{t}_p^2 - (\bar{t}_p)^2$. The fact that $\sigma_m^2 \ll \sigma_p^2$ can be appreciated intuitively, since the growth of a chain, given its time of inception, is well described by strongly deterministic laws as opposed to the gross statistics of when persistent chains might occur. Then, it can be asserted that the probability distribution of fission neutron burst times essentially apes the characteristics of $P(t_p)\,dt_p$ but with a time translation of amount $\bar{t}_m$. It can be shown that for one initial neutron at zero time that $\sigma_m^2 \simeq 1.64(\Lambda/K)^2$, Hansen (Ref. 7, App. II). Then the inequality $\sigma_m^2 \ll \sigma_p^2$ can be rewritten as

$$(\Lambda/K)^2 \ll (WS)^{-2} \tag{10-56}$$

or $S\Lambda \ll 1$, thus showing the correspondence to the weak source case.

10-9 TIME-DEPENDENT REACTIVITY CONSIDERATIONS

Continuing, if the reproduction factor is time-dependent, then so is $W = W(t)$. Equation (10-44) should be generalized so that it now becomes

$$1 - W_f = \sum_{\nu=0}^{\infty} \int_t^{\infty} P(\nu)[1 - pW_f(\tau)]^{\nu}\, e^{-(\tau - t)/\Lambda}\, d\tau/\Lambda \tag{10-57}$$

This simply implies that neutrons emitted from fission at time t are absorbed at later times τ about $d\tau$, where $\tau > t$, with probability $e^{-(\tau - t)/\Lambda}\, d\tau/\Lambda$. Again, with only the first three terms of the binomial expansion retained, Eq. (10-57) gives, with the reproduction factor $k(t) = p\bar{\nu}$

$$W_f = \int_t^{\infty} \left[k(\tau)W_f(\tau) - \frac{\Gamma_2}{2} k^2(\tau) W_f^2(\tau)\right] e^{-(\tau - t)/\Lambda}\, d\tau/\Lambda \tag{10-58}$$

With the fact that the reactor is assumed only slightly supercritical, $k \gtrsim 1$, differentiation of Eq. (10-58) yields

$$\dot{W}_f = \frac{W_f}{\Lambda}\left[\frac{\Gamma_2}{2} W_f - K(t)\right] \tag{10-59}$$

or the linear differential equation in W_f^{-1} obtained by dividing Eq. (10-59) by W_f^2, viz.,

$$\dot{W}_f^{-1} - \frac{K}{\Lambda} W_f^{-1} = \frac{\Gamma_2}{2\Lambda} \tag{10-60}$$

which upon integration results in

$$W = pW_f = \frac{2\Lambda}{\bar{v}\Gamma_2}\left[\int_t^{\infty} d\tau \exp - \int_t^{\tau} (K/\Lambda)\, d\tau'\right]^{-1} \tag{10-61}$$

For a step reactivity, $K = K_0 H(t)$.

$$W = 2K_0/\bar{v}\Gamma_2 \tag{10-62}$$

which coincides with Eq. (10-44), within the reactivity unit β, as it should. While for the ramp reactivity $K = mt$, Eq. (10-61) yields

$$W_f(t) = (\sqrt{8m\Lambda/\pi\bar{v}^2\Gamma_2^3} \exp - mt^2/2\Lambda)/\text{erfc}(x)\left(\sqrt{\frac{m}{2\Lambda}}\, t\right) \tag{10-63}$$

where $\text{erf}\, c(x) = 2/\sqrt{\pi} \int_x^{\infty} e^{-u^2}\, du$ is the complementary error function.

In a similar vein, Eq. (10-54) for time-dependent $W(t)$ is rewritten as

$$P(t_p)\, dt_p = W(t_p)\, dt_p \exp - S \int_{-\infty}^{t_p} W(\tau)\, d\tau \tag{10-64}$$

If W from Eq. (10-63) is inserted above, the result is

$$P(t_p)\, dt_p = \sqrt{\frac{2m\Lambda}{\pi}}\left(\frac{S}{\bar{v}\Gamma_2}\right)\left(\tfrac{1}{2}\, \text{erf}\, c\sqrt{\frac{m}{2\Lambda}}\, t_p\right)$$

$$\cdot \exp - [(2S\Lambda/\bar{v}\Gamma_2) - 1](mt_p^2/2\Lambda) \cdot dt_p \tag{10-65}$$

For the source strength parameter $2S\Lambda/\bar{v}\Gamma_2$, a family of curves of $P(t_p)$ versus t_p is plotted in Fig. 10-3.

From the graph, it is seen that for small values of the source strength parameter for the weak source, $P(t_p)$ is very small for $t_p < 0$, that is, prior to the onset of criticality. This means that persistent chains initiated prior to criticality are relatively unimportant in determining t_p, the time of persistent chain initiation, and similarly for $\bar{t}_p + \bar{t}_m$ the occurrence of a fission chain burst.

10-10 CRITICALITY ACCIDENTS

The foregoing is an important consideration in criticality accidents (alluded to in Sec. 10-7), Stratton,[8] whose severity depends strongly on the maximum reactivity inserted prior to the occurrence of an overriding negative reactivity quenching mechanism, which drops the core reactivity back to a subcritical level. Here, the maximum reactivity depends directly on the delay in fission power buildup time,

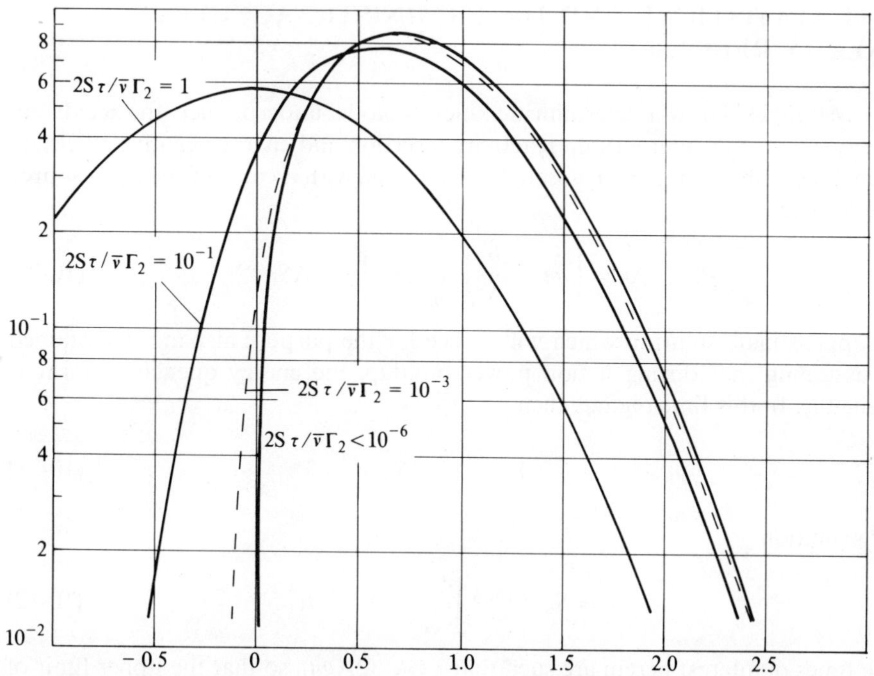

Figure 10-3 Probability per unit time $P(t)$ of the first persistent chain reaction initiated at time t for a ramp reactivity insertion $K = mt$, for various values of the source strength parameters.[7]

so that associated time moments, e.g. $\bar{t}$, or $\bar{t}^2$, as encountered in the preceding, are very important. To calculate such moments, $W(t)$ in Eq. (10-64) can be estimated for the weak source case, and ramp reactivity, by using a constant W in Eq. (10-46). That is

$$W(t_p) \simeq (2\beta K/\bar{\nu}\Gamma_2)H(t_p) \tag{10-66}$$

Then Eq. (10-65) under a ramp reactivity, becomes

$$P(t_p) \simeq (2mSt_pH(t_p)/\bar{\nu}\Gamma_2)\exp - (mSt_p^2/\bar{\nu}\Gamma_2) \tag{10-67}$$

The standard deviation of t_p is then

$$\sigma_p = \sqrt{\overline{t_p^2} - (\bar{t}_p)^2} = \sqrt{\frac{\bar{\nu}\Gamma_2}{ms} - \frac{\pi\bar{\nu}\Gamma_2}{4mS}} = \sqrt{\left(1 - \frac{\pi}{4}\right)\frac{\bar{\nu}\Gamma_2}{mS}} \tag{10-68}$$

or the fractional deviation

$$\sigma_p/\bar{t}_p = \sqrt{\frac{4}{\pi} - 1} \simeq 0.52 \tag{10-69}$$

This, of course, means that to this approximation, the "dispersion" or standard deviation of t_p, the time of initiation of the first persistent chain is about 52 per cent of the average t_p.

10-11 STATISTICAL AND DETERMINISTIC ACCIDENT ENERGY RELEASE

As an example of how a deterministic kinetics model underpredicts the accidental energy release, consider a ramp reactivity insertion and energy negative reactivity quenching embodied in the Hansen–Fuchs model, with a constant neutron source rate S_0, namely

$$\Lambda \dot{n} = \left[mt - \frac{b}{\Lambda} \int_{-\infty}^{t} n \, d\tau \right] n + \Lambda S_0 \tag{10-70}$$

An approximate solution which will suffice for the purpose at hand, is obtained by assuming that during fission power buildup, the energy quenching term is negligible. In this time regime, then

$$\dot{n} = \frac{mtn}{\Lambda} + S_0 \tag{10-71}$$

with solution

$$n = S_0 \, e^{mt^2/2\Lambda} \int_{-\infty}^{t} e^{-m\tau^2/2\Lambda} \, d\tau \tag{10-72}$$

The times of interest herein are such that $t \gg \sqrt{(2\Lambda/m)}$, so that the upper limit of the above integral can be extended to infinity. This yields the asymptotic form

$$n \sim S\sqrt{2\pi\Lambda/m} \exp mt^2/2\Lambda \tag{10-73}$$

This rapid increase of neutron density is assumed to hold until time t_m, at which the energy quench reactivity just cancels the input ramp reactivity, that is, the time at which the bracket in Eq. (10-70) vanishes. For times greater than t_m, the neutron density drops off to a negligible amount. Hence the energy released up to time t_m is from Eq. (10-73)

$$E_m = \int_{-\infty}^{t_m} n(\tau) \, d\tau = S\sqrt{2\pi\Lambda/m} \int_{-\infty}^{t_m} \exp(m\tau^2/2\Lambda) \cdot d\tau \tag{10-74}$$

Further, it can be assumed that t_m is sufficiently large so that the first term in the asymptotic expansion of the above integral will be sufficient. Then slightly re-writing the above, and retaining only the integrated term after once integrating by parts, gives

$$E_m = S\sqrt{2\pi\Lambda/m} \int_{-\infty}^{t_m} \frac{1}{\tau} e^{m\tau^2/2\Lambda} \, \tau \, d\tau \sim S\sqrt{2\pi\Lambda/m} \cdot e^{mt_m^2/2\Lambda} \cdot (\Lambda/mt_m) \tag{10-75}$$

Since the bracket in Eq. (10-65) vanishes at t_m, and with Eq. (10-75) one can obtain the three relationships, written as

$$\frac{\Lambda m t_m}{b} = E_m = S\sqrt{2\pi\Lambda/m} \cdot e^{mt^2/2\Lambda} \cdot (\Lambda/mt_m) \tag{10-76}$$

The two outer terms are seen to give a transcendental equation in t_m; the logarithmic form of the equation is

$$\frac{(mt_m)^2}{2\Lambda m} = -\ln bS\sqrt{\frac{2\pi\Lambda}{m}} + \ln (mt_m)^2 \tag{10-77}$$

This is solved for t_m, by successive approximations, where the first one is

$$(mt_m)^2/2\Lambda m \simeq -\ln\left(bS\sqrt{\frac{2\pi\Lambda}{m}}\right) \tag{10-78}$$

The next approximation is obtained by inserting the above in the rightmost term of Eq. (10-77) to yield

$$(mt_m)^2/2\Lambda m = -\ln\left(bS\sqrt{\frac{2\pi\Lambda}{m}}\right) + \ln\left\{2\Lambda m \ln\left(bS\sqrt{\frac{2\pi\Lambda}{m}}\right)^{-1}\right\} \tag{10-79}$$

or

$$(mt_m)^2/2\Lambda m \simeq \ln\left(2\Lambda m/bS\sqrt{\frac{2\pi\Lambda}{m}}\right) \tag{10-80}$$

since the second term of Eq. (10-79) amounts to the logarithm of a logarithm, and is assumed to therefore make a negligible contribution. Then, as $E_m = \Lambda mt_m/b$, the energy released according to the deterministic calculation is, from Eq. (10-80),

$$bE_m/\Lambda \sim \sqrt{2\Lambda m \ln (2\Lambda m/bS\sqrt{2\pi\Lambda/m})} \tag{10-81}$$

Statistically, the probability distribution of the energy release for a ramp reactivity insertion must be taken into account. This is contained in the probability distribution of the initial persistent chain. To reiterate, $P(t_p)\,dt_p$ is the probability that a persistent chain is initiated in dt_p about t_p. In keeping with the definition of W, as given earlier, the expected neutron density increase for a ramp reactivity insertion is

$$W(t_p)n(t) = \exp\frac{m}{2\Lambda}(t^2 - t_p^2) \tag{10-82}$$

with a corresponding expected energy release, in an analogously asymptotic form, as

$$W(t_p)E(t) \sim \left\{\exp\frac{m}{2\Lambda}(t^2 - t_p^2)\right\}\Big/ mt; \qquad E(t_p) = \frac{mt_p}{b} \tag{10-83}$$

Similar to the preceding development, the statistically determined energy release has the asymptotic form

$$\frac{b}{\Lambda}E(t_p) \sim \sqrt{m^2t_p^2 + 2m\Lambda \ln \{2m\Lambda W(t_p)/b\}} \tag{10-84}$$

Using the above mathematical apparatus, consider a ramp reactivity insertion rate into the Godiva reactor (cf. Chap. 9 for Godiva description) with a spontaneous fission source $S = 100$ neutrons per second. The prompt neutron lifetime is 6 nanoseconds, and the energy reactivity quenching constant $b = 5.10^{-20}$ per neutron absorbed.

Now the statistically expected energy release $\langle E \rangle$ is, from Eqs. (10-67) and (10-84)

$$\langle E \rangle = \int E(t_p)P(t_p)\,dt_p \sim \sqrt{\pi m \bar{\nu} \Gamma_2 / 4S} \tag{10-85}$$

and using Eq. (10-81), the ratio of statistical to deterministic energy release is

$$\langle E \rangle / E_m \simeq \sqrt{(\pi \bar{\nu} \Gamma_2 / 8S\Lambda) / \ln\,(2m\Lambda / bS\sqrt{2\pi\Lambda/m})} \tag{10-86}$$

For Godiva, with the above parameters

$$\langle E \rangle / E_m > 200 \tag{10-87}$$

for ramp reactivity insertion rates of one to ten percent per second. These ramp reactivity rates are sufficiently high to also insure that delayed neutron effects can be ignored.

In conclusion, for the type of unmoderated cores discussed above, criticality accidents would result in energy released of better than two orders of magnitude over that predicted by the deterministic reactor kinetics. Hence, due caution must be exercised in the interpretation of accident analyses results from deterministic models of the type discussed in the foregoing.

10-12 LANGEVIN METHODS IN REACTOR NOISE STATISTICS

As a prelude to the Langevin method for dealing with spacial reactor noise dependence, it is well initially to consider spacially independent reactor noise in a somewhat generalized manner. Paraphrasing the elegant treatment due to Cohn[9] the assumption is made that reactor noise obeys the Poisson distribution. This is only approximately true, as long as fission chain fluctuations which are one source of neutron noise, are not uncorrelated. The Poisson distribution implies uncorrelated events among other conditions. Nevertheless, under this approximation, a useful analogy can be constructed from other Poisson distributed fluctuation phenomena. One such is in electronics, and is the Schottky current produced in a temperature limited diode. This well-known Schottky noise formula can be written as

$$\langle |I_s|^2 \rangle = 2\,e^2\,\bar{m} \tag{10-88}$$

$\langle |I_s|^2 \rangle$ is the spectral density of the diode noise current, where spectral density is defined as the mean squared noise current amplitude divided by the sensible current band width (in MHz). e is the electronic charge, and $\bar{m}$ is the average diode current.

Using this Schottky noise expression as an analog for the noise producing processes in a reactor, the spectral density of the neutron noise equivalent source S_0 is given by

$$\langle | S_0 |^2 \rangle = 2 \sum_i \eta_i^2 \bar{R}_i \tag{10-89}$$

η_i is the net number of neutrons produced in the ith type of nuclear process which is within the ken of reactor noise kinetics, and $\bar{R}_i$ is the average number of the ith type of interactions per second. Delayed neutrons are neglected initially, so that the spectral density is frequency independent, that is, consists of white noise. This assumption will be discarded later. Here, the spectral density frequency upper limit is fixed from quantum considerations at about the order of 10^{20} Hz, so that white noise is an excellent approximation.

The prompt neutron interactions considered are absorption and leakage (lumped together), and fission. Then Eq. (10-89) is rewritten as

$$\langle | S_0 |^2 \rangle = 2 \left\{ (-1)^2 \cdot \frac{n}{\beta\Lambda} \cdot \frac{\Sigma_a}{\Sigma_a + \Sigma_f} + \frac{n}{\beta\Lambda} \cdot \frac{\Sigma_f}{\Sigma_a + \Sigma_f} \sum_{\nu=0}^{\infty} (\nu - 1)^2 P(\nu) \right\} \tag{10-90}$$

As is readily recognized, the first term on the right-hand side is an absorption term, and $\beta\Lambda$ is the neutron lifetime with the reactivity unit β, as defined in Chap. 2. Σ_a is the total absorption cross section for all types of non-neutron producing interactions including leakage. $(n/\beta\Lambda) \cdot \Sigma_a / (\Sigma_a + \Sigma_f)$ is the mean number of all such absorptions per second, where n is the neutron density. $(n/\beta\Lambda) \cdot \Sigma_f / (\Sigma_a + \Sigma_f)$ is the mean number of fissions produced per second, while $\sum_{\nu=0}^{\infty} (\nu - 1)^2 P(\nu)$ is the mean squared net number of neutrons produced per fission. Using the criticality relation $k = \bar{\nu}\Sigma_f/(\Sigma_a + \Sigma_f) = 1$ in Eq. (10-90) yields

$$\langle | S_0 |^2 \rangle = \frac{2n}{\beta\Lambda} \cdot \frac{\overline{\nu(\nu - 1)}}{\bar{\nu}} \tag{10-91}$$

For small neutron density fluctuations, the reactivity K represents the additional neutron source in amount $S_0 = Kn/\Lambda$ neutrons per second, if K is in dollars, or $K = \beta\Lambda S_0/n$ if K is expressed in dimensionless units (e.g. per cent). So that $\langle | K |^2 \rangle = (\beta\Lambda/n)^2 \langle | S_0 |^2 \rangle$, which when inserted into Eq. (10-91) results in

$$\langle | K |^2 \rangle = \frac{2\beta\Lambda}{n} \cdot \frac{\overline{\nu(\nu - 1)}}{\bar{\nu}} \tag{10-92}$$

To provide some simple examples of how the foregoing development is used, consider three examples. The first is the problem of the possible effect of reactor noise on criticality determination. In this connection, it is desirable to monitor a criticality indication fiducial, such as a control rod position in a series of experiments, where it is necessary to return to the critical reactivity fiducial following each experimental run. To determine the noise error so introduced, it is desired to calculate a particular statistic, namely the variance of the mean reactivity over a measurement time T, that is

$$\sigma_K^2(\tau) = \left[\left\{\frac{1}{T}\int_0^T K(\tau)\,d\tau\right\}^2\right]_{\text{av}} \tag{10-93}$$

where the subscript "av" indicates a time average.

The reactor noise is assumed stationary in that time averaged noise statistics yield the same values as ensemble averages, as discussed in Chap. 2. σ_K^2 will be calculated by taking the ensemble average. Let K be given by the Fourier series expansion

$$K = \sum_{n=1}^{\infty} (a_n \cos n\omega t + b_n \sin n\omega t) \tag{10-94}$$

where $a_0 = b_0 = 0$, since the reactivity is assumed to possess a zero mean. Inserting Eq. (10-94) into Eq. (10-93) yields the result that

$$\sigma_K^2 = \left[\left\{\frac{1}{T}\sum_{k=1}^{N}\int_{kt_0}^{(k+1)t_0} d\tau \left(\sum_{n=1}^{\infty} a_n \cos n\omega t + b_n \sin n\omega t\right)\right\}^2\right]_{\text{av}} \tag{10-95}$$

It is seen that the time integration above is divided into discrete intervals of t_0 seconds, to facilitate forming the ensemble average. So that

$$\sigma_K^2 = \left[\frac{1}{T}\sum_{k=1}^{N}\sum_{n=1}^{\infty}\frac{a_n}{n\omega}\{\sin(k+1)n\omega t_0 - \sin kn\omega t_0\} - \frac{b_n}{n\omega}\{\cos(k+1)n\omega t_0 - \cos kn\omega t_0\}\right]_{\text{av}}^2 \tag{10-96}$$

If the trigonometric terms are squared and ensemble averages taken for each harmonic, that is each n, then the ensemble averages are given by

$$\langle a_n^2\rangle = \langle b_n^2\rangle = \langle |K|^2\rangle / T; \qquad \langle a_n b_m\rangle = 0 \tag{10-97}$$

That is, the ensemble average of the cross terms vanish, while the ensemble average of the "like" terms yields the spectrum, which it should by the ordinary definition of spectrum. Also, $\langle a_n\rangle = \langle b_n\rangle = 0$ since the distribution of the ensemble $\{a_n, b_n\}$ will be gaussian with a zero mean. Finally, from Eq. (10-96)

$$\sigma_K^2 = 2\sum_{n=1}^{\infty}\frac{1-\cos n\omega t_0}{(n\omega)^2}\cdot\frac{\langle |K|^2\rangle}{T} \tag{10-98}$$

In the limit of this summation, in that as $n \to \infty$, $t_0 \to 0$, and $nt_0 \to T$

$$\sigma_K^2 = \frac{1}{2\pi^2 T^2}\int_0^{\infty}(1-\cos n\omega T)\frac{d\omega}{\omega^2}\langle |K|^2\rangle \tag{10-99}$$

Since the reactivity noise term is assumed white, it can be factored so that finally

$$\sigma_K^2 = \frac{1}{2T}\langle |K|^2\rangle \tag{10-100}$$

Then using Eq. (10-46)

$$\sigma_K^2 = \frac{\beta\Lambda}{nT}\,\frac{\overline{\nu(\nu-1)}}{\bar{\nu}} \tag{10-101}$$

Now, for ^{235}U, $\Gamma_2 = \overline{\nu(\nu-1)}/\bar{\nu}^2 = 0.745$, and from Chap. 1 $\beta\Lambda = (\nu\bar{\nu}\Sigma_f)^{-1}$, the above becomes

$$\sigma_K = (\Gamma_2/nv\Sigma_f T)^{1/2} \simeq 0.89(nv\Sigma_f T)^{-1/2} = 5.1\cdot 10^{-6}/\sqrt{WT} \tag{10-102}$$

where $nv\Sigma_f T$ is the number of fissions in the measuring time of T seconds, and W is the reactor equilibrium power level in watts, so that WT is the reactor energy released in time T.

The second example is the determination of the zero power (point model) source transfer function. Using the Langevin technique, which replaces the mean source term S by its fluctuating noise component S_0 to yield the corresponding fluctuating neutron density component in Eq. (1-47), results in

$$n = n_0 + G * (Kn + S_0) \tag{10-103}$$

where the asterisk denotes the convolution integral. For a critical reactor $K = 0$, and $n = n_0 + n_r$ where n_r is the fluctuating neutron density noise component, so that Eq. (10-103) becomes

$$n_r = G * S_0 \tag{10-104}$$

Its Fourier transform gives

$$\bar{G}(\omega) = \bar{n}_r(\omega)/\bar{S}_0(\omega) \tag{10-105}$$

which is the usual zero power point model transfer function, but resulting from neutron density noise due, in turn, from the source fluctuation noise in a critical reactor. From Chap. 3, and using herein electronic symbols to make the connection to the electronic series admittance analog of the transfer function $\bar{G}$,

$$\bar{G}(\omega) = [i\omega(\Lambda + \bar{f}(\omega)]^{-1} = (R + i\omega L)^{-1}; \qquad |\bar{G}(\omega)|^2 = (R^2 + \omega^2 L^2)^{-1} \tag{10-106}$$

where

$$R = \omega^2 \sum_{i=1}^{6} a_i/(\omega^2 + \lambda_i^2), \qquad L = \Lambda + \sum_{i=1}^{6} a_i\lambda_i/(\omega^2 + \lambda_i^2) \tag{10-107}$$

Then

$$\langle|\bar{n}_r(\omega)|^2\rangle = |\bar{G}(\omega)|^2\langle|S_0|^2\rangle = \frac{2n}{\beta\Lambda}\cdot\frac{\overline{\nu(\nu-1)}}{\bar{\nu}}\bigg/(R^2 + \omega^2 L^2) \tag{10-108}$$

The third example is that of the effect of the detector in the reactor. The mean current from an ionization chamber detector in a reactor is

$$\bar{i} = \varepsilon qn/\beta\Lambda \tag{10-109}$$

where q is the amount of charge resulting in chamber current per neutron absorbed in the chamber. ε is the chamber efficiency, which is the fraction of neutrons absorbed in the chamber that give rise to chamber current.

Superimposed on the mean chamber current $\bar{i}$ are two types of noise currents. One type, I_f, is that due directly to the fluctuating neutron population in the reactor, that is, with an ensemble average square

$$\langle|\bar{I}_f(\omega)|^2\rangle = \langle|\bar{i}|^2\rangle \tag{10-110}$$

Now, using Eqs (10-109) and (10-108)

$$\langle|\bar{I}_f(\omega)|^2\rangle = \frac{\varepsilon^2 q^2}{(\beta\Lambda)^2}\langle|\bar{n}_r(\omega)|^2\rangle = \frac{\varepsilon^2 q^2}{(\beta\Lambda)^3}\cdot\frac{2n}{R^2+\omega^2 L^2}\cdot\frac{\overline{\nu(\nu-1)}}{\bar{\nu}} \tag{10-111}$$

The second type of noise current is that from random pulses due to neutrons in the chamber. It is white noise current given by, I_d, where from Eq. (10-88)

$$\langle|\bar{I}_d(\omega)|^2\rangle = 2q^2\varepsilon n/\beta\Lambda \tag{10-112}$$

If the chamber efficiency is poor, that is $\varepsilon \ll 1$, this means that exceedingly few neutrons cause any ionic charge in the chamber to be available as chamber current Hence, these two types of current noise are essentially uncorrelated and their respective spectral densities can add in quadrature, that is like the squared lengths of two independent random uncorrelated vectors, so that

$$\langle|\bar{I}_{\text{tot}}(\omega)|^2\rangle = \langle|\bar{I}_f(\omega)|^2\rangle + \langle|\bar{I}_d(\omega)|^2\rangle = \frac{2q^2\varepsilon n}{\beta\Lambda}\left[1+\varepsilon\left|\frac{\bar{G}(\omega)}{\beta\Lambda}\right|^2\cdot\frac{\overline{\nu(\nu-1)}}{\bar{\nu}}\right] \tag{10-113}$$

From this, it is seen that the ratio of reactor noise to chamber noise is proportional to $\varepsilon|\bar{G}(\omega)/\beta\Lambda|^2$. It is then desirable that, at least $\varepsilon|\bar{G}(\omega)/\beta\Lambda|^2 \simeq 1$ hold in the frequency range of optimal detector sensitivity.

10-13 LANGEVIN TECHNIQUE FOR SPACIAL NOISE

In the preceding section, the reactor noise considerations have used two main assumptions. They are (1) that the noise fluctuations are characterized by a Poisson distribution, and (2) that the reactor kinetics point model is adequate. Here, the discussion will bear mainly on spacial reactor effects on the noise parameters, and will use the Langevin technique. Following Sheff et al.,[10,11] for a fluctuating noise stimulus whether due to source noise fluctuations, or reactivity noise, etc., the reactor neutron density will have a noise component. Then the spacially dependent neutron density can be written as

$$n(\bar{r},t) = \int_0^\infty d\tau \int_{\bar{r}'} d^3\bar{r}'\, G(\bar{r}'_1,\bar{r},\tau)S(\bar{r}',t-\tau) \tag{10-114}$$

where S is the fluctuating noise stimulus due to the fluctuating source. G is the Green's function which satisfies

$$LG(\bar{r}',\bar{r},\tau) = \delta(\bar{r}-\bar{r}')\delta(\tau) \tag{10-115}$$

L is generally an integro-differential operator, which represents the transport integro-differential equation corresponding to the transport integral equation (10-114). The former is then

$$Ln(\bar{r}, t) = S(\bar{r}, t) \tag{10-116}$$

For the usual deterministic reactor parameters, such as the time averaged neutron density $\langle n \rangle$, and a time averaged source $\langle S \rangle$, from Eq. (10-116) one obtains

$$L\langle n(\bar{r}, t)\rangle = \langle S(\bar{r}, t)\rangle \tag{10-117}$$

When these averaged quantities are replaced by their fluctuating counterparts, as in Eq. (10-116), then the latter is called the corresponding Langevin equation. The simple replacement of the average values by the fluctuations about their respective averages due to the noise, is called the Langevin technique or method, as mentioned in previous sections.

In Chap. 2, both autocorrelation and cross correlation was discussed and shown to play an important part in the calculation of reactor noise kinetics. Here, in keeping with the aim of investigating reactor spacial influences on the noise parameters, the space dependent neutron density correlation functions will be calculated. From Eq. (10-114) form

$$n(\bar{r}_1, t)n(\bar{r}_2, t) = \int_0^\infty d\tau_1 \int_{\bar{r}'_1} d^3\bar{r}'_1\, G(\bar{r}'_1, \bar{r}_1, \tau) \int_0^\infty d\tau_3 \int_{\bar{r}'_2} d^3\bar{r}'_2$$

$$\cdot G(\bar{r}'_2, \bar{r}_2, \tau_2) S(\bar{r}'_1, t_1 - \tau_1) S(\bar{r}'_2, t_2 - \tau_2) \tag{10-118}$$

The cross-correlation function of the neutron density fluctuations, and that of the source are defined, respectively, as

$$\phi_{12}(\bar{r}_1, \bar{r}_2, \tau) = \lim_{T\to\infty} \frac{1}{2T} \int_{-T}^{T} n(\bar{r}_1, t) n(\bar{r}_2, t + \tau)\, dt \tag{10-119a}$$

$$S_{12}(\bar{r}_1, \bar{r}_2, \tau) = \lim_{T\to\infty} \frac{1}{2T} \int_{-T}^{T} S(\bar{r}_1, t) S(\bar{r}_2, t + \tau)\, dt \tag{10-119b}$$

Now the required time average of the left-hand side of Eq. (10-118) will yield the cross-correlation function of the neutron density fluctuation. Since reactor fluctuation phenomena are assumed stationary in the ergodic sense, the ensemble average of the right-hand side of Eq. (10-118) will be taken in lieu of the time average. This will yield from Eq. (10-118)

$$\phi_{12}(\bar{r}_1, \bar{r}_2, \tau) = \int_0^\infty d\tau_1 \int_{\bar{r}'_1} d^3\bar{r}'_1\, G(\bar{r}'_1, \bar{r}_1, \tau_1) \int_0^\infty d\tau_2 \int_{\bar{r}'_2} d^3\bar{r}'_2$$

$$\cdot G(\bar{r}'_2, \bar{r}_2, \tau_2) S_{12}(\bar{r}_1, \bar{r}_2, \tau + \tau_1 - \tau_2) \tag{10-120}$$

where the order of integration was inverted in that the required cross-correlation integrations were performed first, as they commute with the other integrations.

Now, the source (i.e. input) cross-correlation function must be given. Even though it is not known precisely, it frequently has the characteristics of white noise, at least with respect to the relatively low frequencies of reactor kinetics interest, as opposed to those much higher frequencies of communication/radar signal processing fluctuation phenomena. This property is equivalent to the source cross-correlation function, being peaked about the zero of the correlation interval τ. Approximately then it can be written as

$$S_{12}(\bar{r}_1, \bar{r}_2, \tau + \tau_1 - \tau_2) \simeq A(\bar{r}_1)\delta(\bar{r}_2 - \bar{r}_1)\delta(\tau + \tau_1 - \tau_2) \tag{10-121}$$

which actually asserts that the source fluctuations are uncorrelated except for $\bar{r}_1 = \bar{r}_2$ and $\bar{\tau} = \tau_2 - \tau_1$. This implies that the neutron density (output) cross-correlation ϕ_{12} is limited to those correlation times that are large with respect to the actual width of the cross-correlation peak, for example, the full width at half maximum (FWHM).

10-14 CROSS SPECTRAL DENSITIES

The computation and interpretation of $A(\bar{r})$ will now be discussed. Inserting Eq. (10-121) into Eq. (10-120) results in

$$\phi_{12}(\bar{r}_1, \bar{r}_2, \tau) = \int_{t_0}^{\infty} dt \int_{\bar{r}'_1} d^3\bar{r}'_1 \, G(\bar{r}'_1, \bar{r}_1, t) G(\bar{r}'_1, \bar{r}_2, t + \tau) A(\bar{r}'_1) \tag{10-122}$$

with $t_0 = |\tau|\{1 - H(\tau)\}$ where $H(t)$ is the heaviside step function encountered earlier. It should be again noted that all of the above assumes that the noise fluctuations are superimposed on the mean values of the pertinent quantities. So that, say, the neutron density cross-correlation function implies that the product of the mean density at the two comparison points is subtracted. Similarly, the mean squared values would be considered subtracted from their respective auto-correlation functions.

As discussed in Chap. 2, the Fourier transforms of the autocorrelation functions are their respective spectral densities. If the Fourier transform with respect to the time variable, is taken in Eq. (10-120), the result is the neutron cross spectral density, where the overbar indicates the transformed variable, and the asterisk indicates the complex conjugate, that is

$$\bar{\phi}_{12}(\bar{r}_1, \bar{r}_2, \omega) = \int_{\bar{r}'_0} d^3\bar{r}'_1 \, \bar{G}^*(\bar{r}'_1, \bar{r}_1, \omega) \int_{\bar{r}'_2} d^3\bar{r}'_2 \, \bar{G}(\bar{r}'_2, \bar{r}_2, \omega) S_{12}(\bar{r}'_1, \bar{r}_2, \omega) \tag{10-123}$$

Also similarly, the Fourier transform of Eq. (10-121) yields the source cross spectral density as

$$\bar{S}_{12}(\bar{r}'_1, \bar{r}'_2, \omega) = A(\bar{r}'_1)\delta(\bar{r}'_2 - \bar{r}'_1) \tag{10-124}$$

Inserting $\bar{S}_{12}$ into Eq. (10-123) gives the neutron cross spectral density

$$\bar{\phi}_{12}(\bar{r}_1, \bar{r}_2, \omega) = \int_{\bar{r}'_1} d^3\bar{r}'_1 \, \bar{G}^*(\bar{r}'_1, \bar{r}_1, \omega)\bar{G}(\bar{r}'_1, \bar{r}_2, \omega)A(\bar{r}'_1) \tag{10-125}$$

Then the auto spectral density is immediate as

$$\bar{\phi}_{11}(\bar{r}_1, \omega) = \int_{\bar{r}'_1} d^3\bar{r}'_1 \, |\,\bar{G}(\bar{r}'_1, \bar{r}_1, \omega)\,|^2 \, A(\bar{r}'_1) \tag{10-126}$$

It should be noted that $\bar{G}(\bar{r}'_1, \bar{r}_1, \omega)$ is the spacial version of the source transfer function discussed earlier, Eq. (10-105), while simultaneously being a Green's function in the space variables. As can be seen from Eq. (10-124), $A(\bar{r}_1)$ is the source spectral density per unit volume about $\bar{r}'_1$, and is frequency independent to the approximation mentioned earlier concerning the peakedness of the source cross-correlation function. The determination of $A(\bar{r})$ will derive from a procedure used by Cohn (*loc. cit.*) discussed in the foregoing. Since, now delayed neutrons as well as spacial dependences are to be included, it should be noted that the Schottky white noise approximation does not apply. This is because a high frequency cutoff in the noise is introduced by virtue of the delayed neutrons. This cutoff is characterized by the delayed neutron decay constant corresponding to the shortest period of the delayed neutron precursors. For a ^{235}U fueled water reactor, this corresponds to about 4 Hz. Nevertheless, the Schottky noise approach will be used, but modified to accommodate delayed neutrons. Following the previously discussed Schottky noise model, but now introducing separate noise contributions from non-fission absorption, fission, delayed neutron precursor decay, source neutrons, and net leakage, yields

$$A(\bar{r}) = \frac{n(\bar{r})}{\Lambda}\left[\frac{\Sigma_a}{\Sigma_f + \Sigma_a} + \frac{\Sigma_f}{\Sigma_f + \Sigma_a}\sum_{v=0}^{\infty}(v-1)^2P(v)\right] + \sum_{i=1}^{6}\lambda_i c_i + S(\bar{r}) - \frac{L^2}{\Lambda}\nabla^2 n(\bar{r}) \tag{10-127}$$

The net leakage term is that from diffusion theory, $(L^2/\Lambda)\nabla^2 n(\bar{r})$, while the terms in the bracket above are the same as in the earlier treatment, and the inclusion of the last three terms is self-evident. Since the reactor is assumed critical in the mean, Eq. (1-37) asserts that $0 = \dot{c}_i = -\lambda_i c_i + a_i n/\Lambda$ implies that the term $\Sigma\lambda_i c_i = n/\Lambda$. Also, $\Sigma v^n p(r) = (1-\beta)\overline{v^n}$ and $\bar{v}\Sigma_f/(\Sigma_f + \Sigma_a) = (1-\beta)k$ to distinguish prompt neutrons from delayed neutrons. Also for a near critical system $\nabla^2 n(\bar{r}) + B^2\bar{n}(r) \cong 0$. All of this, when substituted into Eq. (10-127) yields

$$A(\bar{r}) = \frac{n(\bar{r})}{\Lambda}\left[\frac{(1-\beta)(\overline{v(v-1)}/\bar{v}) + \beta(1-K)}{1-\beta K} + 1 + L^2B^2\right] + S(\bar{r}) \tag{10-128}$$

To correct for the fact that the spectral density is not frequency independent, as discussed, all terms in $A(\bar{r})$ remain intact except the fission term. The fission noise contribution has a non-white frequency spectrum, and involves a further calculation (Sheff[10,11]) which is not presented here. However, the following modified fission term (Sheff[10,11]) should be used instead of the first term on the right-hand side of Eq. (10-128), namely

$$\frac{n(\bar{r})}{\Lambda}\cdot\frac{1-\beta}{1-\beta K}(1+(\beta\omega)^2|\bar{f}(\omega)|^2)\cdot\left(\frac{\overline{\nu(\nu-1)}}{\bar{\nu}}\right) \tag{10-129}$$

This is seen to be the fission source term, but multiplied by the frequency dependent factor $1+(\beta\omega)^2|\bar{f}(\omega)|^2$ where $\bar{f}(\omega)$ is the Fourier transform of the delayed neutron decay function discussed in Chap. 2.

10-15 APPLICATIONS

For pedagogical ease of presentation, the infinite reactor will be considered. For other geometries, cf. Sheff.[10] The Green's function for an infinite reactor is

$$G(|\bar{r}-\bar{r}'|,\tau)=H(\tau)\exp-\left\{[(\bar{r}-\bar{r}')^2\beta\Lambda/4L^2\tau]\Big/\left[(4\pi L^2\tau/\beta\Lambda)^{3/2}+\frac{K-1}{\Lambda}\tau\right]\right\} \tag{10-130}$$

Inserting this into Eq. (10-122) for the cross-correlation function and realizing that $A(\bar{r})$ is constant for an infinite reactor, gives

$$\phi_{12}(\bar{r}_1,\bar{r}_2,\tau)=\frac{A}{2}\left(\frac{M}{\pi}\right)^{3/2}\int_{|\tau|}^{\infty}dt/t^{3/2}\cdot\exp-\{\alpha t+M(\bar{r}_1-\bar{r}_2)^2/t\} \tag{10-131}$$

where $\alpha=(K-1)/\Lambda$, and $M=\beta\Lambda/4L^2$.

The point reactor autocorrelation function must be the sum of all correlations in an infinite system containing a uniformly distributed detector. If one integrates Eq. (10-131) over all space, that is, over all $\bar{r}_1$ and $\bar{r}_2$, the result is

$$\phi(\tau)=\int_{\bar{r}_1}d^3\bar{r}_1\int_{\bar{r}_2}d^3\bar{r}_2\,\phi_{12}(\bar{r}_1,\bar{r}_2,t)=\frac{A}{2\alpha}\exp-\alpha|\tau| \tag{10-132}$$

the well-known point reactor autocorrelation function. The coefficient A from Eq. (10-128) is inserted above to give

$$\phi(\tau)=\frac{n}{K-1}\left[\left\{(1-\beta)\frac{\overline{\nu(\nu-1)}}{\bar{\nu}}+\beta(1-K)\right\}\Big/(1-\beta K)+1\right]\exp-\alpha|\tau| \tag{10-133}$$

$\phi(0)$, the variance of the distribution is simply

$$\phi(0)=\sigma_n^2=\frac{n}{K-1}\left[\frac{(1-\beta)(\overline{\nu(\nu-1)}/\bar{\nu})+\beta(1-K)}{1-\beta K}+1\right] \tag{10-134}$$

From the above, the cross-correlation or cross spectral density between a detector at $\bar{r}_1$ and one at $\bar{r}_2$ in a reactor, can be computed if the Green's function of the reactor is known. Other applications, such as to counting statistics is discussed next.

10-16 DETECTOR COUNT RATE CORRELATION FUNCTION

This part of the development is a paraphrasal of Sheff.[10] An important physical aspect of reactor neutron detectors, whether they are ionization chambers, solid state detectors, or whatever within the present state of art, is that they consume neutrons from the population in order to operate effectively. Therefore, the contribution to the detector correlation function that would come from a neutron born at an initial zero time reference, that contributes at the correlation time interval τ must be deleted. Because, if it contributes at time τ, it must not have been absorbed, and so has gone undetected. The exception to this is the $\tau = 0$ contribution, that is, the neutron fluctuation variance.

It is important for purposes of count rate correlation function computation to distinguish between single neutron contributions and multiple, for example, pairwise neutron contributions. To illustrate the idea of a single neutron, and the pairwise contribution to the correlation function, consider the latter first. For example, within a fission chamber, let the probability of a fission occurring in time dt be given by $nv\Sigma_f\, dt$. As a result of this fission, the probability of a count in the detector at a later time T_1 is $\varepsilon v \Sigma_f v\, dT_1 \exp - \alpha(T_1 - t)$, where α is K/Λ as it appears in $\dot{n} = -\alpha n$, the prompt neutron kinetics equation for negative reactivities. ε is the detector efficiency. For the neutron that has been removed because of the count at T_1, the probability of a subsequent count at T_2 due to the detector fission at T_1, is $\varepsilon(v-1)\Sigma_f v \exp - \alpha(T_2 - T_1)\, dT_2$. Then the total probability due to past fissions is given by the integral on dt plus the average over $P(v)$, that is

$$\int dt(nv\Sigma_f)(\varepsilon v \Sigma_f v \exp - \alpha(T_1 - t) \cdot \{\varepsilon(v-1)\Sigma_f v \exp - \alpha(T_2 - t)\, dT_1\, dT_2\} \tag{10-135}$$

or

$$\frac{n(\varepsilon\Sigma_f v)^2}{2\beta K} \cdot \frac{\overline{v(v-1)}}{\bar{v}} \cdot \exp - \alpha(T_2 - T_1)\, dT_1\, dT_2 \tag{10-136}$$

which, exclusive of the factor $dT_1\, dT_2$ is the pair neutron contribution to the count rate autocorrelation function. Without the factor $(\varepsilon\Sigma_f v)^2$, the above is the pair neutron contribution to the neutron density correlation function.

The single neutron contribution to the neutron density correlation function is obtained by considering a hypothetical chamber that does not remove a neutron upon detection. This chamber is also assumed to have a detector efficiency such that $\varepsilon v \Sigma_f = 1$. Then the probability of a count in dt is $nv\varepsilon\Sigma_f\, dt = n\, dt$. The probability that the same neutron produces another count at dT is $\exp - \alpha(T - t)\, dT$. As before, integrating over all possible interval counts that produced chamber fissions gives

$$n \exp - \alpha(T - t)\, dT\, dt \tag{10-137}$$

Similarly, except for $dT\,dt$, this gives the single neutron-contribution to the neutron density correlation function. Adding Eqs (10-136) and (10-137) gives the desired function, which is

$$\phi_{11}(\tau) = n\left[1 + \frac{1}{2\beta K}\cdot\frac{\overline{\nu(\nu-1)}}{\bar{\nu}}\right]\exp - \alpha|\tau| \tag{10-138}$$

Notice for a pure moderator with distributed source $S_0 = n/\beta\Lambda$, $K = -\infty$, and

$$\phi_{11}(\tau) = \beta\Lambda S_0 \exp - |\tau|/\beta\Lambda \tag{10-139}$$

which is the contribution to the correlation from the hypothetical chamber. Since there is no such chamber, then for a real chamber in this moderator, $\phi_{11} = \beta\Lambda S_0\,\delta(\tau)$, that is, no real correlation would exist.

The spectral density of the hypothetical counter, Eq. (10-139), is

$$|\bar{\Phi}_{11}(\omega)|^2 = \beta\Lambda S_0/[(\beta\Lambda)^2 + \omega^2] \tag{10-140}$$

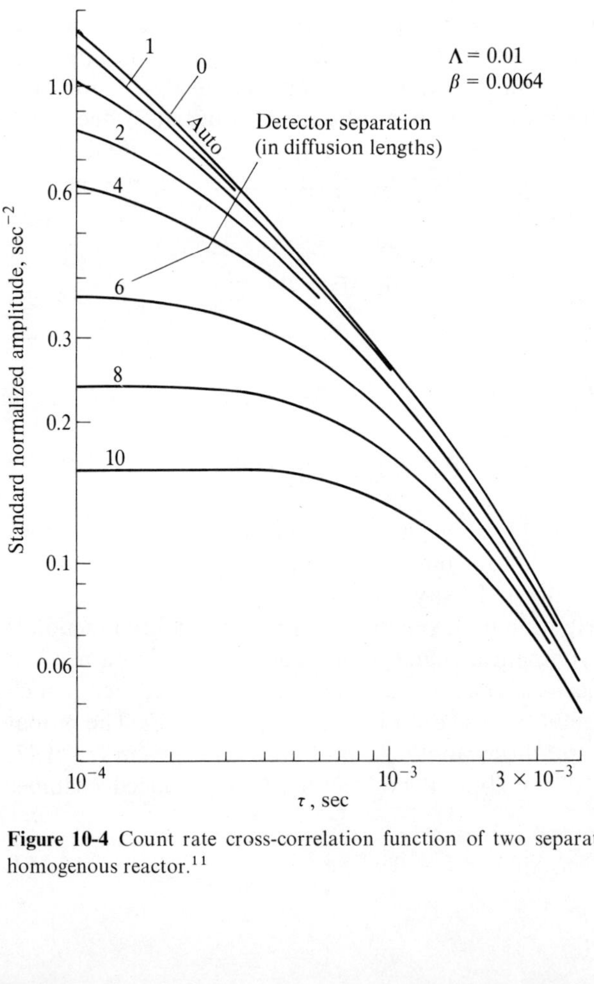

Figure 10-4 Count rate cross-correlation function of two separated point detectors in an infinite homogenous reactor.[11]

which behaves like the point model transfer function modulus of a homogeneous source impregnated moderator. This is measured as a matter of course in neutron wave experiments, cf. Sec. (7-2).

The structure of the correlation function in terms of single or pair neutron components, is contained wholly in the source equivalent input correlation coefficients A and A_s, or their corresponding Fourier transforms, $\bar{A}$ and $\bar{A}_s$, called input spectral densities, all of which is defined in the following. The fission terms in A are the only ones that contain factors corresponding to neutron pairs, e.g., each fission gives $\nu(\nu - 1)/2$ pairs. $(n, 2n)$ neutron production is not considered. The A_p and A_s in Eq. (10-127) can immediately be written out in terms of single and pair neutron contributions. So that from Eq. (10-127) for the total A, where subscripts refer to single and pair neutron contributions to the correlation function

$$A_s(\bar{r}) = \left(\frac{n(\bar{r})}{\beta\Lambda}\right)\left(\frac{K+2}{1-\beta K}\right) + S(\bar{r}) - \frac{L^2}{\beta\Lambda}\nabla^2 n(\bar{r}) \tag{10-141}$$

$$A_p(\bar{r}) = \left(\frac{n(\bar{r})}{\beta\Lambda}\right)\left(\frac{1-\beta}{1-\beta K}\right)\frac{\overline{\nu(\nu-1)}}{\bar{\nu}} \tag{10-142}$$

The non-white noise mode which corresponds to $A_s(\bar{r}, \omega)$ is given by Sheff[10,11] as

$$A_s(\bar{r}, \omega) = \frac{n(\bar{r})}{\beta\Lambda} \cdot \frac{\overline{\nu(\nu-1)}}{\bar{\nu}} \cdot \frac{1 + \omega^2 |\bar{f}(\omega)|^2}{1 - \beta K} \tag{10-143}$$

Using Eqs (10-141) and (10-142), and relationships given in Sheff (*loc. cit.*) for count rate correlation functions, themselves separated into single and pair neutron contributions, it is possible to obtain the corresponding count rate input coefficients $A_{cr}(\bar{r})$ which sums the single neutron and pair neutron contributions. For a homogeneous reactor $A_{cr}(\bar{r})$ is inserted into Eq. (10-120) with the appropriate Green's functions as already discussed, to yield a very complicated expression (Sheff, *loc. cit.*) for the count rate cross-correlation of two detectors, one at $\bar{r}_1$ and one at $\bar{r}_2$ in the reactor. Figure 10-4 is a plot of that count rate cross-correlation.

PROBLEMS

10-1 In the gasoline station queue example show, (*a*) that the variance of m, the number of cars in the system, is given by $\sigma_m^2 = G_{vv}(1, t) + G_v(1, t)\{1 - G_v(1, t)\}$, where $G(v, t)$ is the corresponding generating function, and the subscripts denote the respective partial derivatives. Also show, (*b*) that the variance evaluates to $\sigma_m^2 = (\lambda - \mu)[(2\lambda - 1)(\lambda - \mu) - 1]$ and is zero for $\lambda = \mu$. What is another relationship between λ and μ that yields zero variance?

10-2 For the two-dimensional generating function $G(u, v, t)$, of the probability that there are n neutrons and m delayed precursors, following the injection of a neutron into the critical reactor, show by differentiating the partial differential equation (10-10) and evaluating the result at $u = v = 1$, that the kinetics equation for the mean neutron population $\bar{n}$ results, that is, $\dot{\bar{n}} = -\{(1 - k_p)/l\}n$ with solution $\bar{n} = \exp - \{(1 - k_p)/l\}t$.

10-3 Considering $G(0, v, \infty)$ as a power series in v, show that the corresponding coefficients $P(0, m, \infty) = 1/m!(\partial^m G/\partial v^m)_{v=0}$, and in particular that

$$P(0, 1, \infty) = \beta k_p P(0, 0, \infty)/\{1 - (k_p/\nu_p)(1 - \beta \nu_p)p(1)\}$$

Interpret this expression.

10-4 Show that the steady state generating function $G_{ss}(v)$, given in Eq. (10-24) satisfies the partial differential equation (10-23).

10-5 In the expression for the variance of the mean reactivity over the measurement time T, as given in Eq. (10-93), why is there no second term as in the definitional sense of the variance, namely, $\sigma_K^2 = \bar{x}^2 - (\bar{x})^2$?

10-6 For a one megawatt reactor and a measurement time of 10 sec, what is the magnitude of the standard deviation σ_K, and what is the implication with respect to criticality errors due to noise?

10-7 If $\omega = 300\ \text{sec}^{-1}$ in a ^{235}U reactor, with a neutron lifetime of 500 μsec, what must be the chamber efficiency ε, so that $\varepsilon |\{\bar{G}(\omega)\}/\beta\Lambda|^2 = 1$ will hold? Also, how does ε vary with frequency for high frequencies, that is, $\omega \geq 300$. *Hint:* use the expression for $\bar{G}(\omega)$ for large ω.

REFERENCES

1. N. Pacilio: "Reactor Noise Analysis in the Time Domain," AEC Critical Review Series, 1970.
2. R. E. Uhrig: "Random Noise Techniques in Nuclear Reactor Systems," Ronold Press, N.Y., 1970.
3. M. M. R. Williams: "Random Process in Nuclear Reactors," Pergamon Press, 1974, chap. 3.
4. H. Hurwitz, JR, D. B. Macmillan, J. H. Smith, and M. L. Storm: Kinetics of Low Source Reactor Startups, Part I *Nucl. Sci. Eng.*, **15**: 166–186 (1963); Part II *Nucl. Sci. Eng.*, **15**: 187–196 (1963); Part III *Nucl. Sci. Eng.*, **16**: 369–380 (1963).
5. B. C. Diven, H. C. Martin, R. F. Taschek, and J. Terrell: Experimental Determination of the Probability Distribution of Neutrons from Fission, *Phys. Rev.*, **101**: 1012–1015 (1956); also J. Terrell: Experimental Determination of the Probability Distribution of Neutrons from Fission, *Phys. Rev.*, **108**: 783–789 (1960).
6. G. E. Hansen and T. F. Wimett: *private communication* to H. Hurwitz, JR.
7. G. E. Hansen: Assembly of Fissionable Material in the Presence of a Weak Neutron Source, *Nucl. Sci. Eng.*, **8**: 709–719 (1960).
8. W. R. Stratton: "A Review of Criticality Accidents" in "Progress in Nuclear Energy" (Series IV: Technology, Engineering and Safety), vol. 3, Pergamon, N.Y., 1960.
9. C. E. Cohn: A Simplified Theory of Pile Noise, *Nucl. Sci. Eng.*, **7**: 472–475 (1960).
10. J. R. Sheff and R. W. Albrecht: The Space Dependence of Reactor Noise, Part I, Theory, *Nucl. Sci. Eng.*, **24**: 246–259 (1966).
11. J. R. Sheff and R. W. Albrecht: The Space Dependence of Reactor Noise, Part II, Calculations, *Nucl. Sci. Eng.*, **26**: 207–221 (1966).

INDEX